TEAM AMERICA

TEAM
AMERICA

★ ★ ★ ★

Patton, MacArthur, Marshall, Eisenhower,
and the World They Forged

ROBERT L. O'CONNELL

HARPER

An Imprint of HarperCollins*Publishers*

HarperCollins books may be purchased for educational, business, or sales promotional use. For information, please email the Special Markets Department at SPsales@harpercollins.com.

FIRST EDITION

Library of Congress Cataloging-in-Publication Data has been applied for.

ISBN 978-0-06-288329-2

22 23 24 25 26   LSC   10 9 8 7 6 5 4 3 2 1

Joe Kett, Jim Sterling, Peter Kracht, Gordon McCormick, and Jonathan Jao—five gentlemen who changed my life. Thanks.

I'll go through [those Germans] like crap through a goose.

—George Patton

Young man, that bullet was not meant for me.

—Douglas MacArthur

I save my emotions for Mrs. Marshall.

—George Marshall

Make that 75 percent cold-blooded.

—Dwight Eisenhower's son John, when asked to balance his father's open and sunny side and his calculating nature

*"America f**k yeah!"*

—The cheer of the plush puppet U.S. special operators in the film *Team America*, after clearing Paris of terrorists, and in the process knocking down the Eiffel Tower

CONTENTS

INTRODUCTION
xi

ACKNOWLEDGMENTS
485

NOTES
487

INDEX
515

INTRODUCTION

There was once a real Team America, a military-industrial colossus engaged successively in what amounted to two World Series of Warfare (a third was suspended indefinitely due to the prospect of mutual annihilation).

Our country had many assets in these struggles, but among the most important was extraordinary military leadership, which got better and better until Team America forged a true murderers' row—George Patton, Douglas MacArthur, George Marshall, and Dwight Eisenhower—a string of warrior superstars unprecedented in military history. Probably, if they were in a position to do so, Caesar and Napoléon would point to themselves and their stellar henchmen and vociferously disagree. But in terms of the sheer magnitude of what they attempted and accomplished at such a critical point in human history—the transition from total warfare to strategic permafrost—it's pretty clear that the shadows cast by former lineups can't match the achievements of Team America, particularly these four superstars and the savvy managers who engaged them around the world.

Plainly, all of this requires further explanation. The idea of consistently comparing the struggles of the twentieth century to a series of gigantic sporting events might strike some as absurd, even obscene.

But as with any such artifice, it depends on how you look at it. For example, the matter of scale: On the one hand, the armies we will be considering numbered in the millions, which is inherently confusing, especially combined with military terminology unfamiliar to most. On the other, sports teams numbering in double digits more accurately approximate the scale in which key decisions are actually made in time of war, as does the push and pull of willful superstar athletes held in check and directed by managers with a larger view.

Yet sports are not about killing, or at least haven't been since the Romans. This book is all about excessive killing, how that came to pass, and what our four superstars did about it. So the analogy works only if you are willing to embrace another analogy, that of the big leagues in hell. Here in the infernal majors the score is kept mainly—though inaccurately, since everybody cheats—through body counts. Combat is often one-on-one, but ganging up is always encouraged. As we shall see, Alliance intersquad scrimmages produced some of Team America's most impressive wins courtesy of George Marshall and Dwight Eisenhower. In literal terms at least, these were bloodless victories, but they were made in the context of the constantly gorier, corpse-littered environment of World War II. The game our four protagonists had devoted their lives to getting good at was getting worse and worse. Fortunately, as befit the hellish locale, the game itself was doomed, destined to be replaced by something even worse—much worse, actually—but paradoxically much better. If hell could have a silver lining, this was it.

The fact was that Western societies and ultimately the rest of the world had blundered into a logical trap so simple yet so self-reinforcing that it amounted to a feedback loop for warfare intensification, a demonic treadmill of destructiveness always accelerating and with no apparent way to get off.

It began during the early stages of the French Revolution, when it was discovered that ordinary citizens, hastily assembled, armed, and trained, could make up for their military shortcomings with sheer enthusiasm and, in doing so, overcome the dispirited professionals of the ancient regime. This began with volunteers, but soon it was learned that the proposition basically held for those compelled to serve, so long as revolutionary fervor held and, later, Napoléon was around to lead them. Prior to this, armies had been very expensive and time-consuming to build and maintain. Now very suddenly the way was open for legions in the millions.

The second jaw of the trap was formed and sprung during the American Civil War. In this case it came in three forms—the telegraph, trains to transport troops, and rifled small arms with triple the accurate range of their predecessors. All were important, though not necessarily decisive, but it was their larger lesson that mattered: technology must be relentlessly applied to the means of warfare. The results would prove appropriate and effective.

When put in the hands of armies of ordinary citizens, those new means of warfare led to a series of ever-intensified bloodbaths, wars of such magnitude that they shook otherwise successful societies to their very foundations. Total war, periodic societal spasms during which everything was devoted to destruction, was the outcome of this simple but lethal proposition. Around one hundred million died prematurely during the first half of the twentieth century, caught in this perfect trap. Then without warning the grand mechanism self-destructed by creating something much more terrible but, in the end, far safer. Yet, at the time, who knew?

BY ANY MEASURE, FOR CIVILIAN OR SOLDIER, THIS WAS TOUGH GROUND TO have crossed. Yet our four protagonists not only successfully traversed the first, truly uncharted part of the Cold War, they also directed the rest of us toward safety. They were far from perfect human beings, but they did basically represent the values of their countrymen in some very tough circumstances. All four exhibited remarkable adaptability, which was exactly what tumultuous times demanded, but they never lost their basic respect for human life in an environment devoted to slaughter. They played in a very rough league, yet they never truly disgraced themselves, or us—not even George Patton.

No superstar is remembered primarily for being a nice guy. It's achievements that matter. Since this is Team America, we'll group them together: the most audacious general along with the best staff officer in the American Expeditionary Forces that finished off the Germans in 1918; the chief organizer of the US Army in World War II and that war's best field general, its victors in both the European and Pacific theaters; the organizer and first commander of NATO, the father of the Marshall Plan to rebuild Western Europe, along with the creator of modern Japan; two of the key architects of the US nuclear arsenal, and one of the most successful presidents in American history. Heavy on the military side, perhaps, but these achievements also get at the point that these men, all soldiers, cast a long shadow over the events of their times. It's a shadow that lasted long after World War II, one so long that Ike left office profoundly worried about its military-industrial implications.

That chimera of total war's imminence has passed; at the highest level of conflict, the environment has stabilized. And as this has transpired, it becomes possible to see our four protagonists and their ultimate significance in a clearer light. What stands out is how much they stood out. Among English speakers, World War II was largely an amphibious endeavor, yet the US Navy's key leaders were and remain completely overshadowed by the Army's four superstars. In large part, this was because they had the capacity to transcend their roles as purely military leaders and move into the realm of statecraft. Even Patton showed some diplomatic flare in North Africa, while Mac and Ike had no trouble more or less directing the fates of whole nations while very much in uniform. Also, it's important to note that doing this, and getting away with it without arousing the wrath of superiors, are two different things.

For most of their careers all four had to operate within the context of a rigid and potentially unforgiving military organization, one which in some respects was devoted to honing down outstanding personalities. Patton, MacArthur, Marshall, and Eisenhower were all deeply conditioned by their education and experiences as military officers; without question, it was the essence of their personal identities. So it makes sense to study them and their evolution as part of one team. However ambitious each might have been, they thought of themselves as part of a team.

But why four? All were born within the span of a decade, so the time frame would not change significantly by narrowing the focus to a more intensive study of a single individual. Yet this book is intended to cover a lot of history across a broad front, and it needs an equivalently varied cast of characters to propel the story beyond what could become a didactic swamp. But even if you did try to zero in on just one, the others would keep turning up, demanding our attention.

Their lives and careers were in fact intertwined, like a giant braid in time. Except for Patton and Ike (a bromance that ended badly), none were truly friends. Yet they used each other brilliantly in a series of symbiotic relationships of increasingly greater consequence, the net effect of which was to give all a thorough schooling in the tricks of the trade. MacArthur was the first to emerge: the right parents, good luck, and his own bravery cast him far out in front of the group. But the other three, left way behind in rank and importance for decades, nonetheless took on education and at times responsibilities far beyond their actual professional grade, thereby preparing themselves and each other to become what they all were destined to be: generals.

Being a general is important to consider at the outset, because to most of us they are a breed apart. In fact, they have to be. For ultimately a general must be prepared to face the prospect of convincing men to march to their deaths on his word alone. Rules can be invoked, institutional levers can be applied, but it still remains a rare individual who can make this proposition stick not just on a consistent basis but always.

Something beyond gravitas is demanded—the appearance of supreme authority, a facade that ordinary people would never dream of disobeying. Generals are by necessity insular, exaggerated figures, deliberately magnified images for soldiers and henchmen, particularly during the quintessential moments of combat, but also pretty much always, enforcing the drudgery and inconvenience of military life with a smothering projection of personal authority.

To do so, generals must necessarily live behind an artificial version of themselves, what the military historian John Keegan has aptly termed the mask of command. It is far from the only tool of a great military leader (strategic wisdom, tactical skill, and bravery help a lot), but it is perhaps his best trick—an illusionary presence meant to inspire crazy self-sacrifice. Some are able to construct it around a single act of heroism or some personal tick or even a nickname; for others it takes an entire career of careful calculation. In either case, the odds are the final product won't be that impressive. After all, the mask must empower the most difficult of all magic tricks, the levitation of entire armies into the cauldron of combat.

But this did not prove a problem for our four principals; the masks they fabricated and inhabited are among the most beguiling in the history of organized mayhem. They literally moved millions toward life-threatening experiences. It would definitely stretch credulity to say that what separated our four from their peers was that they wore better, scarier masks, but there is a grain of truth there. Those who dismiss the masks' import should consider why both George Patton and Douglas MacArthur habitually spent hours looking at themselves in the mirror, or why Dwight Eisenhower changed uniforms with the regularity of a fashion model, or why even the low-key George Marshall took the part of a Virginian even though he was from Pennsylvania. It was all part of the getup.

Still, nothing comes without a cost, and living behind a mask can easily lead to serious personality distortions. Specifically, one's head can swell sufficiently to fill the mask until the two become indistinguishable, the chief symptom being megalomaniacal self-assurance. This was the personal trajectory of both George Patton and Douglas MacArthur, and would ultimately lead to their respective downfalls. But along the way, as the swelling proceeded, it fueled absolute certitude and allowed them to project it infectiously across their legions even in the worst of circumstances—just what generals are supposed to do.

George Marshall and Dwight Eisenhower can better be seen as having concealed themselves behind their masks. Because they seldom confused their inner selves with their images, their vision outward was clearer and less subject to self-induced distortions. This fit also gave each man better control of his mask, since he took it for what it was. Yet observing others reacting to something you are not can also lead to cynicism and isolation.

For most of his life and career, Ike deliberately—almost compulsively—surrounded himself with people. But those closest to him, Mamie and his son John, were left with the strong suspicion that there was another Ike buried deep inside, and that's the way he wanted it. Ike placed great store in his ability to bluff, which by definition demands that others not know what you are thinking. And for this there was no better disguise than the smiling, trustworthy visage Dwight Eisenhower presented to the world.

George Marshall was even more mysterious. His mask, fashioned to be at once ferocious and imperturbable, almost totally hid the human spirit beneath. It's truly hard to get beyond what George Marshall wanted you to think he was. Partly, this was because he was not exactly emotive to begin with, but he also seems to have mastered anything potentially tumultuous that might have been left—except, of course, the fierce desire to be obeyed. It's telling that of Team America's four superstars, only Marshall refused to write a memoir, when even a hack job by a ghostwriter would have brought him millions. Far more valuable, it's hard not to conclude, was his privacy.

Still, masks can be revealing. Though exaggerations of the characters below, they basically represent them in a recognizable form. They have to.

It's the masks that are obeyed, that energize the troops and inspire suicidal urgency; it's the masks that formulate the strategies and fight over their application; it's primarily the masks that we remember. They are essential to understanding the people within.

MEANWHILE, OUR QUARTET'S SUCCESS AND SIGNIFICANCE ENSURED THAT their individual lives have been picked over by a succession of competent historians, ferreting out a variety of revelations our four heroes doubtless wanted forgotten, but enabling us to compose a group portrait based on enough information to be credible in both scope and detail. We also know enough to understand that their lives were archetypical and very much of the time and place they came from. How could it be otherwise?

But they came from an America of another era, the formative years of our four having unfolded during the last two decades of the nineteenth century. Whatever its internal societal and economic contradictions, America presented a face to the world based on democracy and opportunity, of mechanical ingenuity and industrial dynamism, a place bubbling with energy and ambition. This was a potent elixir to be raised on, and the strapping careers of our four certainly reflected its nutritious qualities, though some got more than others and the flavors varied.

George Patton, George Marshall, and Douglas MacArthur all sprung from what we would call "privileged circumstances"—Patton was rich; Mac's father was a high-ranking general; Marshall's father was an executive in the steel industry, until he lost his money. Still, these three were basically raised in an atmosphere of prosperity and considerable social standing.

Not Ike. By today's standards, Ike's family was not just poor, it was desperately so. But it was never squalid, far from it. The Eisenhowers were disciplined, well organized, and industrious in their poverty. Meanwhile, they were far from poor intellectually, Ike's mother and father being college educated, a rarity even among the upper classes of the time.

For Ike, schooling was a bridge to the other three, since the road ahead would be intellectually challenging, but also a barrier, since his parents' education was deeply pacifistic. Eisenhower went to West Point thinking of it as free college, and after four years hadn't really changed his mind. True enough, he would shortly embrace his military career with the same mighty and enduring grasp as the others; yet the seeds of doubt, planted in childhood, allowed Ike to see earlier and more clearly than the rest that what they were doing was truly crazy.

Still, to the end Ike remained a soldier, and of all his experiences Team America was the most formative. That certainly was true of George Marshall, the first general to become secretary of state and then secretary of defense. Ditto Mac, whose main problem after the shooting finally stopped was that he could never be anything other than a soldier.

Patton was so much the soldier that the US Army could barely keep him in its uniform and ready to fight. He was, in his own mind, the universal warrior, just the latest version of a long line of reincarnated combatants reaching back to the legions of Rome. Crazy George. He was destined to cast no shadow over subsequent events, since by that time he was dead; but while he lasted he was Team America's chief paladin, the cleanup man in the batting order, the best US field general of the twentieth century, maybe ever.

These were four gigantic personalities, all engaged in an epic race, but in the end when it mattered all managed to accommodate each other's eccentricities and operate effectively together.

Given their line of work, they were all basically cut from a similar mold, one shaped to succeed in a military environment. All were big, around six feet tall in an era when the average American male was below five seven. Three of the four were exceptional athletes—Patton went to the Olympics; Mac played varsity baseball at West Point; while Ike did the equivalent on the gridiron, in the process wrecking his knee and subsequently getting so good at coaching that he came to feel it was wrecking his career. Even George Marshall, who played no sport with any ardor as a young man, took up riding and then tennis in his own utilitarian way as a means of cooling down and staying fit.

Partly as a result of all this activity, but also thanks to Mother Nature, they were, all four, good-looking, and in Mac's and Ike's cases truly handsome. This not only helped them fashion engaging professional masks, it also had the same mesmerizing effect on the opposite sex.

All four were destined to have lifelong marriages to attractive, capable women, ones who were also willing to subordinate themselves to the careers of their husbands. Not much of a role model today, but in their own ways Beatrice Patton, Jean MacArthur, Katherine Marshall, and Mamie Eisenhower were strong and resourceful women. They had to be—they lived hard lives, sucked along in the wakes of their hyper husbands.

None of these women seems to have deeply affected the course their mates chose to plot—and also, you don't get the feeling there was a lot of steam propulsion involved. Really only Patton tried to play the warrior in heat; with the others you get the feeling that the enormous workloads they assumed as part of Team America tended to drain them of sexual energy. They were certainly all heterosexual, but in a period of not a lot of birth control and consequently big families, theirs weren't—Marshall had no children in two marriages (his first wife, Lily, died); Ike had only one son survive to adulthood, as did Mac; Patton did have three children but managed to be both a negligent and tyrannical parent. He also had multiple affairs. And though the extramarital record of the others is not spotless, their intent seems different. Mac presided over several embarrassing relationships before he found Jean but then remained completely faithful. Ike had one notable but sexually tepid affair during World War II, but when one ended, so did the other, and he returned to Mamie for life.

In the main our four looked to their marriages as sanctuaries, islands of emotional calm and encouragement, where they could finally relax and recharge—but always just as a respite from the savage ambition that would drive them across the globe to accomplish great things, deeds that changed history. As a personality trait, this lust for distinction more than anything united them. But its impact, as each of their careers wove in, around, and through those of the others, was the opposite.

Except for Marshall and Eisenhower, all of them shared a genial contempt for the others. In part, this was based on a surprising fact: neither George Marshall nor Dwight Eisenhower, two of the greatest soldiers in our history, ever saw actual combat. That was not something to be missed by "Blood and Guts" George Patton and Douglas MacArthur, whose contempt for bullets only grew with seemingly endless exposure. "Best clerk I ever had," Mac said of Ike, whose retort was equally acid: "I studied dramatics under him for five years in Washington and four years in the Philippines." Mac held a similar view of Marshall, nurturing combat officer–staff officer grudges that stretched back to World War I. While his relations with Marshall and Ike were closer, Patton's views on their fighting records, at least in private, were similar to Mac's. Meanwhile, both Marshall and Ike, both superior in rank, seem to have taken an almost sadistic pleasure in jerking Patton around, leaving him hanging after his various misdeeds, and leaving him to wonder if his chances to make war were over.

Actually, this was balderdash. Each of them was plainly a brave man. George Marshall looked for combat during World War I, he just never found it, and he was too good at staff work. Meanwhile, Ike desperately wanted to fight in that war, but he never made it to Europe, instead fighting the equivalently lethal Spanish flu stateside at his training camp in Gettysburg. He certainly took plenty of risks during the next great conflict, most of them in airplanes, ones lacking radar and the cabin pressurization to fly above the weather, not to mention the Luftwaffe. So did Marshall. All of them engaged in lots of death defying, so it's pointless to draw distinctions.

Meanwhile, the things they actually did that mattered—plot strategy, give orders, inspire others—were not in themselves violent acts; they were dependent instead on mental agility, mask-inspired gravitas, and an almost intuitive understanding of how armies of millions actually functioned and fought. This, not the ability to wield the proverbial sword, was what set them apart.

In an age of relentlessly improved swords, though, it's interesting to note that none of our four was intimately and consistently associated with the development of the weapons they were destined to use with such success. George Patton was involved with tanks early but then returned to the horse cavalry for nearly two decades, his sole arms innovation being an improved sword. Ike joined Patton's experiments with tanks but was chased away by the conservative Army bureaucracy, and then was turned toward strategy and command by a key senior officer who recognized his talent. Douglas MacArthur, when he was chief of staff of the Army in the thirties, did foster the development of three key weapons (the M1 Garand rifle, the 105 mm howitzer, and the B-17 bomber), but he was much more concerned with keeping the Army and especially its officer corps from shrinking still further under Depression-strangled budgets. George Marshall's career was consumed by strategy, operations, and tactics; when he was in a position to direct weapon developments, he delegated instead—even the atomic bomb.

This was characteristic behavior: even in an age of war increasingly dominated by weapons, great commanders didn't devise arms, they used them.

And our four did that brilliantly, each grasping almost instinctively the
revolutionary implications of the military tools put in their hands, weapons
without precedent. So, while the Navy's leaders continued to view technology through a lens obscured by their obsolescent battleships, Marshall, Eisenhower, Patton, and MacArthur all eagerly embraced mechanization and
airpower to create a new kind of army and a new way of fighting. In essence,
they all understood that mobility was the key to victory, and they provided it
any way they could. They threw at the enemy a profusion of everything that
moved, in the air and on the ground. Provided you didn't run out of fuel, it
was a formula that almost guaranteed success.

And our four were granted or otherwise carved out extraordinary latitude
to apply that formula. Because America remained separated by oceans from
all the major battles in both world wars, commanders were by necessity
given more freedom of action than would be the case today, an era of instant
and inescapable communications. Meanwhile, the militarization of American society for total war further enhanced their power and esteem—and,
being who they were, they took full advantage. Their masks became ubiquitous, and so did their influence. But it never dominated.

Except for Eisenhower's eight years as president, these men spent the entirety of their careers taking orders. True, they became masters at bending
them, but that went only so far. Functionally, even when they reached the
top, they remained the instruments of their democratically elected masters.
But it was also here that things got the most interesting personally.

Team America took the field under a string of dynamic managers—
Woodrow Wilson, Franklin Roosevelt, Harry Truman, and finally Ike. This
book is about generals, not presidents. But, just as it would be hard to write
a convincing religious text without considering God, our story would ring
hollow without dealing with the political equivalent.

By temperament the four generals' ultimate superiors ranged from doc-
trinaire (Wilson), to enigmatic (FDR), to forthright (Truman), but all were
politicians by training and inclination. Therefore, their perspective was different from that of our four—broader, since it had to be. Ultimately, it was
their responsibility to lead Team America in a league that included a string

of stalwarts: Lloyd George, Clemenceau, and the kaiser first; then Winston Churchill, Joseph Stalin, Hideki Tōjō, and Adolf Hitler; and finishing off with Nikita Khrushchev—all of them masters of the great game, or so it seemed at the time.

There would be setbacks, but ultimately Team America's managers guided us safely through one of the most dangerous stretches of human history, and did so in large part because they managed the interface between politics and the military deftly and, when needed, decisively. Technology separated Wilson from his army, probably a good thing; but airplanes made it possible for his successors to be, episodically at least, up close and personal. It was during these periodic, almost rhythmic interactions that the politicians took the measure of the generals and, being masters of the personal encounter, probed and pushed them in ways few others could.

This was a subtle and symbiotic process. Both Truman and FDR had some familiarity with things military, but in reality they were all but utterly dependent on the skills and knowledge of their military professionals—especially when the fighting started. The generals were the ones who could conjure up the strategies and generate the infinitely detailed plans that would send Uncle Sam's legions crashing against their adversaries. It only made sense to cultivate them, particularly the true stars, in part by giving them stars, but also, being politicians, by casting a charismatic spell over them. It worked practically not at all with Marshall, only occasionally with MacArthur and Eisenhower, and frequently with Patton, but only because, as his own worst enemy, he was so often in trouble.

JUST AS A COACH MIGHT NEVER EXACTLY TAME THE NARCISSISTIC ARROGANCE of a superstar athlete, he can nonetheless strive to maximize his performance. Besides the braided interaction of our four protagonists, managerial efforts in this direction provide a major thematic element of our tale. Scheming by masters of the art, reciprocated by resourceful minions, is a consistent push and pull that will drive our story through the decades. But all in a good cause . . . relatively at least, since they played in what amounted to an infernal league.

That said, being on Team America during the first half of the twentieth century was the place to be. Nationalistically, we were at the top of our game. There were certainly contradictions, but it was still plausible to entertain the national myth that we were a breed apart—not only geographically separated from the world's battles, but also perched on a higher plane politically. Until the end of World War II, freestanding democracies were rare and under siege, except on the vast and fertile ground of the United States, where the first modern model thrived fantastically. From our own less innocent perch, it's hard to grasp how confidently smug we once could afford to be.

But at the time, it certainly helped to feel you were on the right side. And not just because the opposition ranged from authoritarians to the truly evil. The experience of war had grown dramatically more horrible and lethal, killing nearly 700,000 Americans between 1917 and 1952. Under these circumstances, motivation really mattered.

So did bullets. And here we tilted the playing field still further in our direction. It was not just a matter of being far enough away not to get bombed; it was the American industrial base and the way it was run that made the United States such an excellent place to build lots of weapons. For one thing, it was huge, and not just in scale—it was spread across an entire continent, whereas European equivalents matched a fragmented map and were by definition smaller and easier to target. But success was not guaranteed.

Mostly because the United States entered the Great War in its fourth year, the country's hugely ambitious first attempt at total military-industrial mobilization proved too late to make a difference. Fortunately for Team America, the British and French had been churning out all manner of arms and had plenty for us to use. Meanwhile, at home the conversion had been picking up speed, and in the process lessons were learned about beating plowshares into swords on a gargantuan basis.

The next time it was tried these lessons were not only applied, but applied sooner to an industrial economy that was even better suited to absorb them. Given the sheer scale of the manufacturing base, it made sense that management had been studied earlier and with more rigor in America, resulting in multiple advances in productivity based on things like time-motion studies, precise cost accounting and inventory techniques, and supplier networks that were both reliable and flexible. True enough, the country had suffered through a decade of depression, but the latter also ensured that only the strong survived, and doing so while producing more steel, aluminum, oil, and cars than the world's other great powers combined.

All of this was epitomized by the US automobile industry. It had no choice: the combination of a rapidly changing technology and fickle public tastes had left the auto manufacturers capable of bending metal in new directions in what amounted to a moment's notice. Meanwhile all that bending—and also casting and forging—took place at a range of thicknesses nearly perfect for most land arms of the time and also for airplanes if you stretched things (quite literally, considering Ford's mile-long B-24 bomber production line at Willow Run). So, in 1940, if you wanted to arm and mechanize an army fast, you had only to look to Detroit, which is exactly what Team America's savvy manager Franklin Roosevelt did.

This time results came quickly—a veritable cornucopia of lethal devices—and despite MacArthur's carping about the Pacific theater being short-changed, our four protagonists never again suffered from a lack of firepower. Then there were the scientists, who promised something that would make all of this look puny.

But we are getting far ahead of ourselves. First, we must get to know our four much better, since it is through their braided and fated lives that our story of total war's demise will be told. Individually, they likely hold no secrets that a chain of biographers haven't already uncovered, but collectively they offer a tableau broad enough to capture history switching gears.

For the sum of these four lives remains decidedly greater than their individual existences: it was the multiplier effect of their braided interaction that drove them all to greatness and why corporately they are the subject of our story. Thus, early in their careers, the rich and relatively influential George Patton raised Ike from obscurity and introduced him to the right people, just as Ike later saved Patton's career from his own misdeeds and kept him fighting during World War II. Douglas MacArthur also cultivated the young Eisenhower, taught him the ins and outs of service politics, and how to operate at the highest levels in Washington. George Marshall then took over, put the finishing touches on Ike's operational skills, and then formed a partnership critical to winning the war in Europe. Even the epic, generation-long quarrel between Marshall and MacArthur had positive results, since it brought out the best in the former, who under constant provocation still managed to use MacArthur to best military advantage without regard for personal feelings. All told, the process of interaction was one of intensification, leaving all four better at what each did best. So, whenever possible, we'll look at them together, and also with the strong suspicion that, had they ever heard it, they would have joined the lusty cheer of the puppet special operators as they felled the Eiffel Tower . . . well, in spirit at least.

TEAM AMERICA

CHAPTER 1

DAWN

Before we can start weaving the lives of our four protagonists together, we should wait until meaningful interaction actually begins, a process precipitated by America's entry into World War I in 1917. Until then, we'll focus on spinning the early days of the individual strands, making occasional cross-stitches when experiences converge—for example, attending a military academy—but otherwise keeping them separate on the sound ground that they were all basically strangers. And since historian are suckers for chronological order, we'll begin with the first born.

ALWAYS EAGER TO GET THE JUMP ON THE OPPOSITION, DOUGLAS MACARTHUR entered this world prematurely at Arsenal Barracks in Little Rock. His mother, the redoubtable Pinky, had wanted to return home to Norfolk to give birth to what proved to be the last of three sons, but he got the drop on her, leading to the local headline: "Douglas Macarthur was born on January 26, [1880] while his parents were away."

The birthplace was certainly appropriate. If ever an American was preordained to be a soldier, little Doug already had a good shot at the title. Pinky's ancestors had marched with George Washington and Andrew Jackson, and her brothers had attended the Virginia Military Institute and fought with Robert E. Lee. But compared with the exploits of Doug's dad, Arthur MacArthur, they were amateur warriors.

Arthur's military career began during the Civil War when, as a teenager, he displayed virtually nonstop heroism (except when he was wounded, which was often), culminating in leading his regiment up Missionary Ridge against a storm of fire and clearing it of Confederates, all the while shouting "On Wisconsin," words immortalized, on the gridiron at least. At war's end, he was a full colonel, the youngest in the entire Union Army.

But that was a rank earned in combat, and in the peacetime Army he sank to captain, where he remained for twenty-three years. It was hardly time lost, though, as the elder MacArthur studiously, almost compulsively prepared himself for greater responsibility and in the process provided a remarkable template for his son.

We pick up little Doug's life at age three, at Fort Selden, New Mexico, a base of around fifty soldiers, about as far as you could get from anywhere else in America. It was his father's first independent command, and a period when the son built up a set of happy memories. "My first recollection," he liked to say, "is that of a bugle call."

That's not likely. The previous years had brought the death of Pinky's beloved mother (and a substantial inheritance of around $1 million in 2022 dollars); a father increasingly preoccupied with his study of all things Asian; and, above all, the end of Doug's middle brother and constant playmate, Malcolm, swept away by the measles. These were all things not to be forgotten by any of those involved.

But ever the romantic, young Mac morphed his youth into a tale of the Wild West. "It was here I learned to ride and shoot even before I could read or write," he wrote, conveniently forgetting the indoor part under a grieving and increasingly didactic mother. While Doug's remaining brother, Arthur, drifted temporarily into his father's orbit, Pinky's attention focused on him and would remain locked there until her boy was a general. So what if she occasionally dolled him up in skirts and left his hair in long curls until age eight? Achilles's mom also disguised her boy as a girl, and in both cases it made absolutely no difference.

Doug's idyll in the desert ended after three short years in 1886, when the scene shifted from the frontier to the confines of Fort Leavenworth, Kansas, by tradition the Army's professional schoolhouse. It also heralded the beginning of his formal education, entering second grade with mediocre results. Unlike his older brother, who thrived under regimentation, Doug seemed disoriented, missing apparently the freedom of the frontier. Pinky and Arthur wondered aloud why he couldn't be more like his elder sibling. But there was no need to worry. Things would change.

Arthur MacArthur's interminably stalled career suddenly snapped to life in 1889, when his commanding officer recommended him for promotion to major and a job in Washington, calling him "beyond doubt the most distinguished Captain in the Army." Better still was the proximity of Arthur's own father and Doug's grandfather, a politically well-connected superior court judge in the District of Columbia. As garrulous and engaging as his son was taciturn and remote, it was the judge who filled his grandson's head with the details of Arthur's heroics, and also gave him a glimpse of the psychological environment of the nation's capital—"that whirlpool of glitter and pomp, of politics and diplomacy, of statesmanship and intrigue," Doug remembered.

But that didn't help much with the schoolwork. Mac completed his elementary education at Force Public School on Massachusetts Avenue with nothing more than average grades, still, it seems, in the shadow of his older brother. But that vanished abruptly in 1892 when the first son received an appointment to Annapolis, and was essentially gone from the family. Now only Doug was the center of attention, and this being exactly what he always wanted, he rose to the occasion.

It coincided with another fortuitous transfer for the family, this time to sprawling Fort Sam Houston in San Antonio. They all liked the place. As always Arthur dove into his duties, and in the process won the complete confidence of his commanding general. Meanwhile, the surroundings allowed Pinky to give full vent to her southern-belle proclivities, making similarly inclined friends and lolling away lazy afternoons wandering San Antonio's shops and restaurants.

Yet it was their son who found the place transformative—or, more specifically, the ground occupied by what came to be known as the West Texas Military Academy, where his parents had enrolled him in hopes of lighting some sort of academic fire. The move resulted in a veritable conflagration of achievement. Although he was initially resented as a day student, something about the learning environment set his mind ablaze. "Abstruse mathematics began to appear as a challenge, dull Latin and Greek seemed a gateway to the moving words of the leaders of the past." Meanwhile, his jealous classmates gaped as MacArthur "was doing Conic Sections when the rest of us were struggling with Elementary Algebra."

But he didn't truly win them over until his third year, and he did that largely through sport. He was no natural athlete, but good enough, so that, when combined with his newly discovered work ethic and determination to win, he produced a reasonable facsimile of one—credible enough to become almost inexorably the school's tennis champion, shortstop, and quarterback.

By his senior year he was on full burn. He led both the football and baseball teams to undefeated seasons, organized the school's prizewinning drill unit, and was one of only four cadets to finish the year without a demerit in comportment. Already showing signs of verbal spellbinding, young Mac also was awarded the Lockwood Silver Medal in elocution for his stirring rendition of "The Fight of the General Armstrong." More impressive still, he grabbed the Academy Gold Medal, making him class valedictorian for 1897. A clean sweep, except for one significant category: he doesn't seem to have made any friends.

But even at this stage, it's likely he would have dismissed that as beside the point. He was focused on one objective, the official gateway to leadership in the US Army: West Point. Entrance required a presidential or congressional appointment, and that proved to be the first real challenge in Douglas MacArthur's young life. This is particularly surprising, since his father's career was about to accelerate as if it had been shot from a cannon. Promoted to lieutenant colonel the year before Doug graduated, Arthur would shoot to lieutenant general within four more years, and in the process put down a major insurgency in the Philippines. He also crossed swords with its civilian administrator, the future president William Howard Taft, and was sent packing back to the mainland. This was characteristic. Both father and son generally proved adroit at service politics, but when confronting the civilian version they were consistently tone-deaf and almost comically uncomfortable.

That was key to the West Point problem. Judge MacArthur, with just weeks to live, did what he could, gathering a stack of recommendations for his grandson, as did Arthur. But they didn't impress the outgoing Grover Cleveland enough to send one of his four presidential appointments in a MacArthur direction, and, even more depressing, the next year his replacement, William McKinley, remained similarly oblivious to Doug's virtues. Further complicating his way ahead was a preliminary academy physical, which detected a slight curvature of the spine, causing him to flunk. Finally, Arthur had to leave San Antonio, having been named adjutant general of the Department of Dakota. If Doug was going to get into West Point, Pinky would have to pick up the slack. It may have been coincidence, but with Arthur gone, she managed brilliantly.

The plan was elaborate, yet feasible. Wisconsin congressman Theobald Otjen was a friend of the late judge. Once Pinky established residence in his district, Doug would become eligible to take a competitive exam for an appointment. She did so by moving into Milwaukee's posh Plankinton House in October 1897, from where she ran an operation worthy of a professional strategist.

Except that it was all focused on an army of one. The first objective was to straighten Doug out, literally, by placing him in the hands of Dr. Franz Pfister, a renowned local spinal specialist, who later recalled that they "worked together for a year. He was one of the quickest fellows to obey orders I ever treated." He had to be, as Pinky left him time for little else. If her son had a slightly suspect body, it must have been obvious that he had a first-rate mind. But she was leaving nothing to chance. In addition to her own substantial reading list, she had Doug attend classes at West Side High School and receive special tutoring from its principal: "Every day I trudged there and back the two miles from the hotel to the school. I never worked harder in my life."

Arthur visited some weekends and the MacArthurs occasionally socialized with the family of Senator John Mitchell, who had also served in the Twenty-Fourth Wisconsin—sojourns that left Doug with a crush on one daughter and brief acquaintance with son Billy. The latter was never a friendship, but it would have some significance for the future of airpower. Still, for Doug the only future that mattered loomed in May, when the exam would at last be given.

The great day dawned inauspiciously. After a sleepless night, he promptly threw up breakfast, notable only because it would reoccur, but only at what he considered his most desperate moments. This time Pinky provided a pep talk: "Doug you will win if you don't lose your nerve. You must believe in yourself, my son, or no one else will believe in you. Be self-confident, self-reliant, and even if you don't make it, you will know you have done your best. Now, go to it."

That he did, crushing twelve other applicants, the best of whom did not come within twenty points of his 99.3 score. Dogged Doug, hard work and determination will triumph—that was the point of the story as he later told it. But in reality it was an exercise in overkill. When he finally did join the long gray line in June 1899, he would find himself both overprepared and overconnected. Not a good thing if you were a plebe.

THE SECOND OF OUR FOUR PROTAGONISTS, GEORGE CATLETT MARSHALL JR., also arrived in 1880, but characteristically sneaked in on the year's last day, December 31, far removed from the world of war and his future in it. His birthplace, Uniontown, Pennsylvania, was instead on its way to becoming the coke capital of America, which in the nineteenth century had industrial, not pharmaceutical, implications—specifically, the burning of bituminous coal to produce high-carbon ash for the blast furnaces of nearby Pittsburg. George's father, George Catlett Marshall Sr., was one of the key local "coke barons," which meant a prosperous if grimy existence in a place that was gradually becoming encrusted with its own byproducts. But if there was a perpetual cloud over Uniontown, part of it was of George's own making, at least from a historical perspective. His most scrupulous biographer, Forrest C. Pogue, who managed to produce four thick volumes on his later life, notes that accounts by contemporaries are nearly totally missing, and "almost all that can be reported of the boy is what the man at the age of seventy recalled and chose to reveal." This was not much.

In addition to Dad, George had a kindly mother, Laura, who showered him with attention since he was the youngest of her three offspring, and already somewhat isolated. Marie, four years older, treated her younger brother basically as a pest, an annoyance to be ignored when possible. Relations with his brother Stuart, six years his senior, were worse—so bad they would eventually lead to permanent estrangement. Meanwhile, as with MacArthur, the age gap in effect left little George to be raised as an only child. Still, sibling estrangement aside, the picture that Marshall painted was that of an upper-middle-class family "so normal," in Pogue's words, "that it could be remembered by Marie as well as by George in terms of something like folklore."

Childhood according to Marshall proceeded in a more or less Tom Sawyer fashion. Most of the pleasant memories he chose to relate were bucolic: playing in a stream with his friend Andy, or growing vegetables. Later there were hunting and fishing trips with his father into the countryside, including at least one visit to George Washington's ill-fated Fort Necessity.

But the great majority of his time must have been spent in Uniontown, and here the recollections grow scarce and negative. Even a child could see the place was being gobbled up by coal and greed. The mines had torn up the surroundings, flooded the place with newcomers, and generated true inequality—a growing plutocratic element (Uniontown would soon lead the nation in millionaires per capita) and reciprocal labor troubles, ones destined to explode in 1894 with the deadly bituminous coal miners' strike. All of this was hard to miss and must have been depressing, but George chose to relate his problems only in more personal terms.

First came school. Like MacArthur's, young George's elementary years were largely a wash, spent at a private academy and leaving him with only the most basic reading and computational skills. This became apparent when his father enrolled him in public school, and the admissions officer discovered he couldn't answer even simple factual questions. Inevitably, this led to ridicule by his fellow students, who "made fun of me a great deal." He grew to hate being laughed at. And as this happened, he grew more shy and remote, while his sense of humor retreated deep inside to a very dry and sardonic place. Recitation proved particularly excruciating, but in general Marshall remained a mediocre student throughout his education, blooming intellectually only decades into his professional career.

Then there were complications at home. In 1890, when George was nearly ten, the senior Marshall and his partners sold their coke operation to Pittsburg steel interests, his father's share amounting to around $4.5 million in today's money. Despite the strong objections of Mrs. Marshall, he almost immediately invested, then lost, much of it in a Virginia land speculation. After that blow, the son remembered, "we had to economize very bitterly," which, he also notes, meant no more servants. Very likely, this reflected his mother's perspective, given that she was now saddled with the housework after a lifetime of domestic help, but it seems to have left George perpetually tightfisted and inclined toward the security a steady income can bring.

Actually, the family doesn't seem to have been ruined, just taken down a notch, maybe two. The elder George Marshall appears far from defeated, using his extra time to expand his roles in Freemasonry and local politics, while still seeming to have money enough to occasionally buy his wife expensive gifts, albeit ones she invariably claimed they could not afford. Nevertheless, the years continued to roll past 130 West Main Street, the Marshall's capacious but ugly residence, much as they had before—socioeconomically, at least.

You get the same feeling about George, drifting along through his youth. He did make one true friend, Andy Thompson, but in adulthood they would reconnect only after he became a general; he described the experience as that of two middle-aged strangers who inexplicably shared common memories. He also refers vaguely to a sort of club of youngsters, names a dozen or so acquaintances, even brings up a heartthrob who made him want to spell better, but nothing more substantial in the way of social interaction. You get the impression of George mostly on the outside looking in, a kind of vaporous, slightly melancholy presence. "Flicker" was what one girl called him, a nickname soon forgotten, like so much of George in Uniontown.

At home, old tensions remained and new ones arose. As George moved into adolescence, hints of rebellion began to appear. Family tradition and possibly his economic takedown had left the senior Marshall more than a bit of a genetic snob. "My father was so keen in family interests that I was rather sensitive about it," George recalled. "I thought his continual harping on the name of John Marshall was kind of poor business. It was about time for somebody else to swim for the family." If this wasn't insulting enough, young George later proposed a feminine though seaworthy alternative to the former Supreme Court chief justice and secretary of state, a Marshall woman who had married the pirate Blackbeard, whom he proclaimed his only interesting relative.

Meanwhile, his older brother, Stuart, who left for the Virginia Military Institute when George was twelve, had graduated with a degree in chemistry, and was now back, gainfully employed and with a dim view of his brother's prospects, as his secondary education was nearing its unimpressive end.

The father, though, was far from losing hope. Given his son's academic record and apparent love of the outdoors—and perhaps any place that wasn't Uniontown—he had concluded his youngest son had a future as an army officer. But that, along with the associated free college education, required an appointment to West Point, and the senior Marshall was a lifelong Democrat swimming in a sea of Republicans. Quite predictably he failed. But there was an obvious alternative. VMI was cheap, had already provided his eldest son with a profession, and, if it couldn't give West Point's guarantee of a commission upon graduation, it was at least a step in that direction.

Young George's enthusiasm for a military future in general and this one in particular went unrecorded until he overheard Stuart discussing the possibility of his kid brother attending his alma mater, and the inevitable disgrace this would bring upon the family name. Nearly sixty years later he was still fuming, claiming his brother's prognostication "made more impression on me than all the instructors, parental pressures or anything else, and I decided right then I was going to 'wipe his face' as we say or 'wipe his eye' [get the better of him] and I ended up at the VMI." If nothing else, this slap across the genetic heritage cut through George's adolescent haze and revealed, in a way nothing else had, that he was already no one to cross. This was no small thing. It fueled, he claimed, his desire to excel for the rest of his career. At any rate, VMI's superintendent was impressed enough to accept him on sight; of course, there were few better names to have among Virginians than Marshall. He was certainly welcome there.

IF FATE MEANT TO PRODUCE AN AMERICAN ACHILLES, WITH ALL HIS ARISTO-cratic arrogance, bloodlust, and communication with the spirit world, then she couldn't have done better than sending George S. Patton his parents' way on November 11, 1885. They were as wacky as he would ever need to be.

Actually, there were three of them. When Ruth Wilson, the mother, married George Patton II, the father, Annie, the sister, was not only crushed, but took the concept of unrequited love to the point of moving in with them on a permanent basis. Economically, at least, this was not a problem, since the Pattons were among the richest families in Southern California. They were also uniquely religious, being in essence Reincarnation Episcopalians, proudly looking back on and, in desperate circumstances, convening with a long line of battling Pattons, beginning with Confederate war dead and stretching back to the dim reaches of history.

It was in this unique nurturing environment that our nascent warrior landed, subsequently known to the family as "the boy"—delineating him as the center of attention and from his sister, who was always Nita. Meanwhile, Annie became Nannie to the family, and dove deep into indulgent child care. Later, Patton would "remember thinking that I must be the happiest boy in the world. I was probably right."

He lived in a boy's paradise. While the family did spend considerable time in Los Angeles, where George's father pursued business and political interests, home base was Lake Vineyard, the ancestral home of Ruth and Annie (now Nannie) Wilson, a gigantic ranch that encompassed much of what is presently Pasadena. It was here, practically from infancy, that the budding Patton learned to ride a horse (actually, many horses from the family stables), shoot a gun (the place was a veritable arsenal), and swing a blade (Papa favored swordplay).

All of this had a purpose, of course. The elder Patton had attended VMI, had imbibed the heady ancestral brew of Confederate palaver and reincarnated heroics, then ended up as a man of affairs, a great disappointment in his eyes. The next generation would be different—it was a matter of Patton family honor.

Nor did it take too much convincing. From the beginning both children liked playing soldier, Nita labeling herself "major," while Georgie took "private" because it sounded superior. Papa would salute them every morning, inquiring into the status of both major and private. Soon enough the private caught on, and promoted himself, at least under his breath, to "Georgie S. Patton, Jr., Lieutenant General"—a title he never forgot, especially as he whiled away the hours in Papa's lap being read tale after tale of endless martial valor.

Years passed, and while he was no longer in Papa's lap, he was still being read to—that is, when he could sit still that long, or stop talking. Ex post facto medical diagnosis by historians is almost always a shaky proposition, but in this case it does seem clear that Georgie's symptoms were classic signs of a broad condition known today by the ominous acronym ADHD (attention deficit hyperactivity disorder).

His parents couldn't know that, but their wealth and singular agenda made it possible to bring him up in an environment that was least damaging to his self-esteem. Almost no matter what he did or didn't do, they showered "the boy" with love and approval, in the process setting in place a formidable bulwark of self-esteem. Critically, they kept him out of primary school, which would have been a shark tank for a child with Georgie's reading, writing, and sitting-still problems, relying instead on a string of tutors and "coaches" (we would call them therapists) to gradually work with his learning disabilities to the point where he was at least marginally literate.

Just short of his twelfth birthday his parents finally placed him in a private academy, where he joined around twenty-five other students from Southern California's elite. At this point Georgie was barely, if at all, shark bait. His academic difficulties certainly persisted, but he still thrived during his six years there, and through sheer grit gained a decent high school education. He was always bright and energetic, and he was growing into a big, handsome, athletic young man—plainly no one to be picked on. Yet significantly, there is no mention of even one close friend gathered during these years. Still, he was surrounded by his own class; he knew their ways and always would. So his initial clumsiness and later eccentricities would be tolerated in part because he knew how to get away with them along the upper crust.

And whatever Georgie's camaraderie deprivation, there was always his encapsulating family, and suddenly something much more attractive. Among their many properties, the Pattons had a place on Catalina Island, where they customarily summered for several weeks. It was there, in mid-1902, that the family came to know their distant (literally) Massachusetts relatives, the Ayer clan, and Georgie, age seventeen, first met their daughter Beatrice.

If this wasn't quite love at first sight, it was certainly a fated encounter. She was plainly more worldly, having studied in Europe and become fluent in French and on the grand piano, but she was also utterly fearless in the same gonzo way that came to epitomize George S. Patton. She rode horses and sailed boats with the same disdain for danger; she attacked life with his brand of volcanic energy. Like Georgie she could afford to: her father, Frederick Ayer, was an octogenarian patent medicine king and polymath capitalist who had accumulated a fortune far larger than the Pattons'. And she was almost as comely as she was wealthy, with rich auburn hair and large, piercing blue eyes—a "Pocket Venus," her daughter would call her, perhaps with a not-unintended double entendre. It's not hard to imagine that sparks flew immediately between George and Beatrice. It took years before the inevitable took hold. But plunk in the middle of adolescence, the nascent Patton, most concupiscent of our four protagonists, had already met his mate, and in doing so took a major step toward becoming who he would become . . . much to her detriment, incidentally.

Also, it was likely not the step he was most concerned with taking at the time. West Point stood like a citadel blocking his martial future, its rigorous entrance exam seemingly a bridge too far. Well aware of his son's academic deficiencies, Papa resorted to the oblique approach in the form of his alma mater, VMI, a less demanding institution with a number of relatives embedded on the faculty and staff: entrance was assured. At this point the strategy turned ingenious. George would stay in Lexington just a single year, after which he could apply to transfer to West Point and, as a regularly enrolled student in a recognized college, enter without passing the academy's dreaded exam. He would have to return to California and take a competitive test for an appointment, but this was presumably easier and would take place on ground carefully prepared by Papa.

If you were a war god or goddess, it was a scheme made in heaven. Yet on the way to Lexington Georgie was having unexpected second thoughts, wondering if he might in the end prove a coward. Papa was ready with genetic reassurance: "While ages of gentility might make a man of my breeding reluctant to engage in a fist fight, the same breeding made him perfectly willing to face death from weapons with a smile." That seemed sufficient at the time, but it was a question George S. Patton would ask himself over and over, no matter how many times he proved himself a hero.

ALWAYS THE OUTLIER AND DESTINED TO BE TEAM AMERICA'S GREATEST STAR, Dwight David Eisenhower was born on October 14, 1890, nearly a decade after MacArthur and Marshall, and almost exactly five years later than Patton. For most of his career he would be their junior, stuck in second gear; then very suddenly he wasn't, leaping over each until he had worked his way into the preferred slot in the batting order for a slugger and, finally, into the manager's Oval Office. In this story, the last shall be first.

Life for Ike began in Denison, Texas, which tried to claim him as one of its own. But that never stuck, since the future president came to know that his birth took place at an absolute low point in his family's life. Home base for the Eisenhowers was Abilene, Kansas; Denison was a place of exile for Ike's dad, who was playing the prodigal son.

David and Ida Eisenhower were an unusual couple to begin with, and they were certainly not the kind of people who might have been expected to raise a military man. The family, prosperous River Brethren Mennonites from Pennsylvania, had come to Abilene two generations earlier and flourished, working big plots of rich sod into substantial homesteads. Jacob, David's father, had done particularly well, but his son wanted nothing to do with farming, teaching himself Greek and taking a mind to study engineering at nearby Lane University. The father seems to have understood agronomy was a hopeless cause, and the college was operated by River Brethren, so in due course the son headed toward Lecompton to pursue an education. Instead, he found Ida.

She was a much more independent soul, particularly when she met him. Ida Stover was born in the middle of the Civil War in the battle-ravaged Shenandoah Valley, and spent her youth effectively orphaned, bouncing from relative to relative, until one set told her to memorize the Bible if she wanted intellectual stimulation rather than attend high school. She promptly ran away, somehow managed to complete her secondary education, then taught for two years, saving enough to move to Kansas and enroll at Lane.

Here, though opposites in many ways—she was sunny and plucky; he was solemn and stubborn—they fell together like lost souls found at last. Contextually, it's important to realize this took place in an intensely religious environment, and that Mennonites in general, and River Brethren in particular, were profoundly committed to pacifism—particularly Ida, who had grown up in war's aftermath. When they were married in the fall of 1885, the idea that a future offspring would become a famous general would have been not just an abomination, but the last thing on their minds.

Neither Ida nor David finished their studies at Lane. Likely this had to do with Jacob's custom of giving each of his children $2,000 (almost $53,000 in 2022) along with 160 acres of land upon marriage. David still disdained farming and instead convinced his father to mortgage the acreage and let him use the proceeds to open a general store in partnership with one Milton Good in the nearby village of Hope.

Roughly two years passed before there proved to be no hope in Hope, and according to Eisenhower family legend even less good in Good, who abruptly left the partnership with all the available cash. More probably, the two were simply incompatible, and when Good, who was an experienced salesman, broke the partnership, David proved incapable of running the store.

Its collapse left him very apparently mortified. Parking Ida, already pregnant with a third son, in Hope with one of his brothers, David bolted to Dennison, Texas, a place whose only excuse for existence was the railroad. Here he found a job as an "engine wiper," just about the lowest slot a trained engineer could occupy. No surprise, the pay was also miserable, but enough for him to rent the next thing up from a shack and bring Ida down to a very sooty and spare existence, where she promptly gave birth. It was at this point that Ike caught up with the family. So, eventually, did Jacob, who, when he finally visited Dennison, was sufficiently shocked by the young family's circumstances to promptly arrange a homecoming to Abilene. He found a place for David in the River Brethren creamery as an "engineer" (i.e., mechanic), and here he remained at low pay, six days a week, for decades in the land of milk, if not necessarily honey.

If this sounds pretty Old Testament, it wasn't necessarily all gloom.

Throughout their travails David and Ida appear to have remained a true couple, locked into their relationship and, in the process, fulfilling the "be fruitful and multiply" part of the covenant with a string of six sons (Arthur, Edgar, Dwight, Roy, Earl, and Milton). Ike stoutly maintained, and none of his brothers ever contradicted, that he never heard a cross word exchanged between the two parents. Energized by love, David and Ida formed the core of a resolute team, one that each of the boys would join in succession, and be assigned increasing levels of responsibility for the Eisenhower family existence, most of it lived in extremely tight circumstances, ones that Ike would later calculate at 818 square feet.

From the beginning (he had no memories of Dennison), Ike sprang from a dramatically different psychological environment than that of our other three. If none were actually only children, they were all basically brought up alone and without any responsibilities even remotely related to family survival. Ike was bred a teammate, part of a larger whole living close enough to the bone to make everybody's contribution seem vital. Real duties came early, and the Eisenhower boys were made to understand they were mandatory.

To drive this home, Ike learned—at around age ten—that the consequences of any lapse could be dreadful and terrifying. He had been told to watch his three-year-old brother, Earl, and carelessly left his knife in a place the toddler could reach, a lapse resulting in the loss of his eye. At the end of his days, Ike was still ruminating on the incident, insisting that Earl's life was nevertheless a success, which it was. Yet, at the time, it would have been nearly impossible for him not to conclude that even the slightest failure to fulfill a duty could lead to disaster, a lesson as indelible as a tattoo.

Still, the Eisenhower boys, at least the oldest four, lived a roughhouse existence, filled with fights and sports, and accompanying pratfalls—one of which, arising out of a simple abrasion and subsequent infection, had doctors talking about amputating Ike's leg. He would rather die, he told his brother Ed. Instead he recovered and joined him on the diamond and on the gridiron, especially the latter. Though smaller than his hulking brother, Ike loved football and got steadily better, until it broke his knee at West Point.

Meanwhile, being an Eisenhower was far from just a matter of hard knocks. Ida's prized possession (purchased when she married for around $10,000 in 2022 dollars) was an ebony piano, part of the household throughout, and one that provided Milton his road to the future. They remained always People of the Book, and not just in the literal sense—although regular Bible reading would have been part of it. These were bookish people, particularly Ida, who had put together what one Ike biographer calls "a remarkably solid little library of historical works." They acted like a lodestone to young Ike, so much so that his mother locked them in a closet, claiming he was neglecting his chores and homework.

Her disapproval was plainly more complicated. Ike's interest was focused on ancient history, which in turn is focused on one topic, war, which is exactly what Ida, a profound pacifist, would not have wanted her boy pondering. Although David appears to have fallen away from the River Brethren, Ida had gone the other direction, joining a pacific precursor to the Jehovah's Witnesses. It didn't matter: Ike found the key and soon "the battles of Marathon, Zama, Salamis, and Cannae became as familiar to me as the games (and battles) with my brothers and friends in the school yard." If there is such a thing as a hint of things to come, this was it.

In the meantime, life in the Eisenhower household, it is important to emphasize, provided a very stable base from which to spring. All six boys were destined for success—different forms, but still success. Even the family's economic deprivation can be exaggerated. Granted, the Eisenhowers' tiny house was on the wrong side of Abilene's railroad tracks, but when the aging Jacob moved in with them, his economic solvency facilitated the addition of a two-room wing and running water. Even David's glacial career started to grind forward. When the creamery moved into ice, ice cream, and cold storage, he took the opportunity to finish his degree through a correspondence course in refrigeration engineering. Eventually, he appears to have become the personnel officer of a local but substantial holding company; yet that would have been well after Ike left Abilene. In the upbringing he remembered, cash remained scarce, something he and his brothers never forgot. Yet David doggedly made sure it was regularly provided, and Jacob's fundamental prosperity also meant a considerable measure of security.

While his home base remained solid, Ike definitely had a life outside, one that helped shape his future while leaving Ida none the wiser. Right from the beginning he had a knack for working his way under the wing of an older man, who would then teach him a vital skill. In one instance, it was a neighbor and former deputy of Wild Bill Hickok, who filled Ike's head with tales of Abilene's bullet-infested days, along with teaching him how to shoot a pistol, occasionally in matches with the town marshal and a local Wells Fargo agent. Well out of Ida's earshot, Ike became a crack shot.

An even better one was Bob Davis, a local hunter and guide, whom Ike knew from ages eight to sixteen and who taught him the skills of woodland survival through a rough-hewn version of the Socratic method. But if Ike, long after, remembered the fifty-year-old illiterate as "my hero," it may have had less to do with nature than it did with poker, which Davis instructed in a way that introduced young Ike to the world of strategy. Though devoid of letters, Davis had a head full of poker percentages, and nightly around the campfire he beat them into Ike's head, forcing him to play a ruthlessly realistic brand of the game, one in which luck always plays a role, but in the long run the probabilities usually dictate—exactly the mental terrain a general has to face. At any rate, it left Ike a hell of a poker player at a young age, and without Ida ever knowing. Furtive Ike.

Still, the face he showed the rest of his world was far from that. He had grown handsome—contemporary photos show that, but still fail to distinguish his showstopping smile, a grin that would take him far. Ike always had a volcanic temper, and that resulted in a number of epic fistfights, but never grudges. Pretty much everybody he encountered in Abilene liked Ike, and that included his teachers. He was a good student from the beginning. He breezed through the primary grades, and picked up momentum in high school—literally, since he was joined by big brother Edgar, who had quit to work, then returned.

The Eisenhower boys, like the other students and the teachers, were proud of Abilene High: the building was brand-new, and that energized everybody. While most instruction was based on memorization, when Ike showed a flair for plane geometry, his math instructor had him derive proofs without benefit of the textbook. Yet his real academic strength was history, and that was obvious enough for his class prophesy to forecast a professorship at Yale for Ike, leaving the two-term US presidency to Edgar.

In either case that would require a college education, and they didn't need their commencement speaker to tell them that life without one would be a struggle "with one arm cut off." They already had a plan. Ed would go first, enrolling in the University of Michigan, using the money they both earned working furiously in the summer. Ike would continue grinding out the hours eventually in the creamery, building up funds against the day when their roles would be reversed and he could enroll.

Instead, he found Swede Hazlet and the path to what became Ike's mantra, "a free college education." Rediscovered him, actually. At Abilene High, Swede was a tall, gangly kid from the right side of the tracks, easy prey for a gang of bullies until Ike put an end to it. He also dubbed him "Swede." It was the basis of a friendship, until Swede was shipped to a military prep school in Wisconsin.

Now Swede was back, with an appointment to the US Naval Academy, but, having failed the math portion of the entrance exam, he was cramming to take it again. The two reconnected almost immediately. The creamery was not a bad place to hang out in the summer, and what began as penny-ante poker and ice cream raids almost immediately turned into serious study sessions when Ike learned that he too could compete for an appointment. And he didn't just study with Swede, he orchestrated a campaign. He enlisted the town's elite to provide letters of recommendation, and his former teachers at Abilene High to brush him up in selected subjects; he wrote Kansas senator Joseph Bristow, formally requesting an appointment, and the Navy Department for copies of past examinations—all on his own initiative, though probably with a little guidance from Swede.

Always a quick study, Ike emerged first among those seeking an appointment to Annapolis, only to learn that at twenty-one he would be too old for the next class. West Point had no such rule, but Ike stood second among applicants. Then number one failed the physical, and suddenly he had a slot along the Hudson. It's not hard to imagine his first thought: "Free college education."

It was not likely Ida's, though. Both parents plainly believed that, whatever their own travails, their offspring's road to success remained higher education. They also understood that it led out of town, a hegira that proved permanent for all six. And they seemed inclined to let the boys choose their respective routes as their talents dictated (Arthur became a banker, Roy a pharmacist, Earl an engineer, and Milton a college administrator).

Yet Ike's choice crossed an invisible barrier defined by his parents' religious pacifism. David may have fallen away, but old beliefs die hard, and nowhere is it recorded that he was enthusiastic about young Dwight's decision to follow the military path, even if it did include free college. But it was Ida who was truly appalled: her religious resurgence only reinforced the belief that war was a true anathema, an affront to God. That didn't say much for its practitioners, and by implication their prospects for salvation.

When Ike left for the railroad station, she went only as far as the front porch: "It is your choice," she told him, and waved goodbye. She then went to her room, Milton later told Ike, and could be heard weeping, the first time anyone had heard her cry.

Harping on the pacifistic background of one of history's most important generals can certainly be overdone, especially since Ike left for West Point with no particular reservations. But unlike MacArthur, Patton, and even Marshall, he also had no particular career aspirations. As with our other three superstar players, Ike would eventually and unreservedly take on the role of soldier. Still, when war and then its nuclear prospects reached what seemed to be the apocalyptic stage, Ike was decidedly more shaken than the others, eventually framing the problem as a vast trap into which humankind had blundered. He spent much of his presidency trying to figure a way out, never realizing the trap was the safest place to be. But until the end he never stopped trying, and that probably had something to do with his beginning.

COMPARING HIMSELF TO DOUGLAS MACARTHUR, IKE MAINTAINED: "I'M JUST folks. I come from the people, the ordinary people." While it should be clear that the Eisenhowers were far from ordinary, it's hard to imagine any of our other three protagonists even trying to get away with such a statement. They came from another world, born into the upper reaches and encouraged to think of themselves as special, all of them raised largely alone. Ike was born surrounded, and later, even as he retreated into himself, he did so beneath a penumbra of pals. Ike picked up friends like sweaters pick up lint, the others not so much. Mac preferred sycophants, Marshall had only "Black Jack" Pershing and "Vinegar Joe" Stilwell, and Patton was about as misanthropic as you could get and still stay in the Army. Yet Ike even befriended him, for a while.

Still, differences can also be exaggerated. An army is quintessentially a corporate endeavor—its components at every level must fit together. Our four, when circumstances demanded, always did. Meanwhile, at the brink of their respective academy experiences, each was already big, athletic, bright, and in his own way intensely competitive—an excellent profile for cadet survival. All would manage a good deal more.

CHAPTER 2

LONG GRAY LINES

When George Marshall entered the Virginia Military Institute in 1897, it was—besides West Point and Annapolis—the best-known and -respected martial academy in America. It was unique in a number of respects. First among them, it constituted a kind of Confederate theme park, its own gray line the last symbolic sinew of once mighty Southern legions. Here the memory of teen cadets sacrificed at the Battle of New Market to Yankee bullets mixed with the legacy of former professor Thomas J. Jackson (a.k.a. Stonewall) and, just a few feet away, on the grounds of adjoining Washington and Lee University, the last resting place of the Marble Man himself, Robert E. Lee—all of it combined to form a psycho-stereopticon of Confederate martial valor. But none of this necessarily implied renewed disloyalty to the union. VMI was instead a totem, a crown jewel of what has come to be known as the Lost Cause movement, a massive and concerted effort to romanticize the rebellion and, in the process, justify the subsequent Jim Crow South. Of course, this would not have been apparent to the students, who were literally immersed in it.

As with the two national academies, VMI was a degree-granting institution. While academics there were never stellar, neither were they abysmal. Using George Patton, who attended both, as our metric, it's clear West Point was harder.

But VMI may actually have been tougher. While the school couldn't promise a commission, and except in national emergencies few graduates actually became officers, that didn't mean the military part of the curriculum wasn't taken seriously. Mass drill and endless marching, obsessive attention to details of dress and comportment, all enforced by organized and relentless harassment, are murky and often dismissed parts of military education. But all these activities do appear to have a bonding effect, stimulate leadership behavior, and give participants an idea of what it's like to be truly miserable and still carry on. This was VMI's strong suit. Whereas West Point subjected its plebes to a summer of true abuse at the hands of upperclassmen, VMI's Rat Line lasted a full year.

Tall, diffident George began his year none too steady on his feet after a recent bout of typhoid. He seems to have stuck out like a sore blue thumb in a sea of southerners, who were almost all the student body. A natural target, or so it seemed, they mocked his Pittsburg twang and made sure to add a derisive note to the "Union" in Uniontown. But that was simply an introduction.

Within a few weeks of his arrival, a delegation of upperclassmen burst into room 88 of the barracks, shoved the butt end of a bayonet between the floorboards, and ordered George to squat over the blade until further notice. Still weak from typhoid, Marshall didn't last long, lurching to the side and in the process slashing his buttock. Even with Rats there were rules, and this bit of team sadism went far beyond the expulsion mark. Yet Marshall simply went to the infirmary, made up some excuse, and had himself sewn up. Gradually word rippled through the cadet corps: this was a Rat who didn't rat. Arguably, this transition from northern Rat to Brother Rat would have a major impact on George Marshall's life and also his identity.

Of course, he was still a Rat, and had to suffer through the requisite indignities for the remainder of the year. But, like our other three protagonists, he already was able to muster the self-possession to remain above it: "I think I was more philosophical about this sort of thing than a great many boys. It was part of the business, and the only thing to do was to accept it as best you could." The first real challenge he had to endure—duress, it seems—scattered the fog that had enveloped him in Uniontown, and gives us a glimpse of the real George Marshall and his personal specialty: being respected by everybody.

This gathering of gravitas was not necessarily accompanied by an

equivalent academic growth spurt. Even given VMI's only modestly demanding curriculum, Marshall remained a very mediocre student, floundering in French and German, and complaining later that the school left
him with an insufficient command of English, written and oral. That may
have been true at the time, but Marshall's true intellectual development
was driven by his subsequent career. He had a genius for one thing—giving
orders—and all else followed from that. "What I learned at VMI was self-
control, discipline, so that it was ground in. I learned also, the problem of
managing men."

It was on the broad fields fronting VMI that George Marshall first tried
on his personal mask of command, and realized he possessed that ineffable
quality known as leadership. To make it further apparent, he added ramrod-
straight posture to his lanky six-foot frame, and developed a bullhorn bari-
tone to bark out marching orders. His natural reserve grew into an imposing
solitude, all of which was not to be missed by faculty and fellow cadets, not
to mention brother Stuart, who was in the process of having his eye wiped.

He rose steadily in rank, from first sergeant to, eventually, first captain—
the school's highest student command—in his senior year, which must have
seemed like a miraculous dream to the rapidly maturing George Marshall.
He even made the football team and played first-string tackle for the entire
season. He had come to love the place, and in the future, when his career
in the real army flagged, he would repeatedly look into a strategic retreat to
Lexington and a VMI teaching post. By then, though, there was more to it
than that . . . a lot more.

While being a VMI cadet was decidedly not conducive to exploring the
surroundings, a four-year stay would have made it apparent to the meta-
morphosing George that Lexington was, as it still is, a charming town nes-
tled fortuitously in the soft green Virginia countryside. It's a place without
coal mines and coking ovens, nothing to gobble up the bucolic tableau. It's
also clear he was charmed by the studied gentility of the inhabitants, or at
least the upper-class Virginians he encountered. Uniontown was a boiling
pot of economic and ethnic turnover; these Virginians with a lot less money
but a lot more heritage struck exactly the right pose for Marshall's fortress-
like personality—a relaxed but assured Old Dominion sense of who was in
charge. Still, there was more to it than that.

There was romance, or George Marshall's version, at least. As told to Marshall's first biographer, Forrest Pogue, and largely repeated by those who followed, George's version had him meeting Lily Carter Coles "as a kind of romantic accident." She lived with her mother in a Victorian gingerbread house just outside VMI's Limit Gate, one frequently passed by George, with more freedom his senior year as first captain. He had been beguiled by the extraordinary piano solos emanating from within, and one evening stopped to listen. He was invited in, and one thing led to another with a kind of fated rhythm. The few photos that remain reveal Lily as a very attractive young woman, shapely and nearly as tall as George, with a full head of Titian red hair. She was said to be warm and engaging, with a sharp wit. Even better, she was firmly anchored to the upper crust of the Old Dominion with a chain of relatives that stretched across the state and far back in time to those near-mythic First Families of Virginia.

Soon, when the rules permitted, George was seen driving Lily around town in her horse-drawn phaeton. A bit later we catch him risking his zero-demerit record by sneaking off campus to see her. "I was very much in love and I was willing to take the chance," a trait he revealed repeatedly when he really wanted something. By the spring of 1901 they were engaged to be married, and at the school's final ball they were reported to be the most popular couple.

As idyllic as this moment must have seemed, there were a number of problems attached, the most pressing of which was that George's future and his newfound identity were not simply contingent, they were actually quite dubious. Before he could marry Lily, he needed to support her, which meant becoming a real officer in the real army. Here VMI's record was uniformly dismal. In 1890 there were but ten of its graduates actually serving in the regular army, and things hadn't gotten much better since.

On the other hand, George Marshall was a young man with a plan, determined to make his own luck. Instead, it seems, luck made him. In an early example of his knack for strategic simplicity, George headed straight to Washington and, using his father's political connections, wangled an appointment with the attorney general; crashed a reception to meet the chairman of the House Military Affairs Committee; and, when all of this didn't work, headed for the White House, where he surreptitiously joined a group on the way to the president's office and actually got to plead his case to William McKinley. It's hard to say if all of this helped—though he thought so—since luck was already on his side.

The Spanish-American conflict may have been a "splendid little war" to armchair strategists, but it revealed gaping deficiencies in the long-neglected US Army, a situation only made worse when the Philippines, our newly acquired colony halfway around the world, went into open insurrection. In order to cope, the peacetime army was more than doubled, and—critically for Marshall—the Act of 1901 authorized an additional twelve hundred officers, a fifth of whom would be selected by examination for immediate commissions. After sorting out a few complications, George's name appeared on the list of Pennsylvanians qualified to take the exam, which he did, and eventually received word that his score was among the highest, and that he would be commissioned on December 31, his twenty-first birthday. The way was clear to Lily and the future.

But before we get there, it's important to clear up a few things. For instance, on the long-remembered night of the final ball, George and Lily refrained from dancing due to the seriousness of her heart condition. She was also over five years older than George, which was significant in part because she had dated his brother Stuart back when he had attended VMI. Seen in this light, the "romantic accident" Marshall remembered doesn't look like any accident at all. Very likely he knew exactly who Lily Carter Coles was long before he was lured Ulysses-like by her irresistible piano. She was the old girlfriend of the key motivating factor in his life thus far, brother Stuart. And now he was opposed to the marriage. All the more reason to go forward, "wipe his eye" yet again.

The wedding was apparently remembered by Marshall as an informal gathering of family and friends, highlighted by Lily's sprightly "Come on, George, let's get married." Andy, George's boyhood friend from Uniontown, did attend, but the sole photograph of the wedding party seems far from mirthful, with nobody smiling, least of all Stuart.

They had a five-day honeymoon before George shipped off to the Philippines, and Lily returned to Lexington to live with her mother for the next three years. During their abrupt interlude as man and wife, George apparently learned that Lily's heart condition (basically, a leaky mitral valve) would not allow her to bear the strain of pregnancy. Whether sex was possible remains an open question, but the couple never had children. Until the end of their twenty-five-year marriage, when her heart finally stopped, Lily remained a kind of vaporous presence whose chronically poor health enabled her retiring husband to side-slip many a smoke-filled, alcohol-drenched social occasion his career would have ordinarily demanded he attend. Instead, he stayed home with Lily, and gave every indication of being content. He undoubtedly loved her, and was devastated by her loss. But it was a mild sort of love, perfect for a man who would be away a lot, focused on other things.

IN SEPTEMBER 1903, A YEAR AND A HALF AFTER GEORGE MARSHALL LEFT LEX-ington, George Patton arrived. Actually it was an invasion; the whole family came, and just to make sure "the boy" didn't get homesick, they left Aunt Nannie, who settled in with friends and prepared to hover. She needn't have bothered.

The place fit Georgie like his new uniform, exactly the same size in every dimension as those of his father and grandfather. Georgie was third-generation VMI; the Californian felt instantly at home, surrounded by southerners bred to his specifications of upper-class gentility. Being a Rat barely phased him. In fact, he took it as a challenge, becoming obsessive in every aspect of his appearance and comportment, until there was so little to criticize that he was the first in his class to be inducted into a secret fraternity that largely relieved the Rat onus. "I am treated almost as an equal," he wrote Papa. "Theoretically, I don't approve of this, but practically I do."

Meanwhile, acceptance also bore fruit in a surprisingly good academic performance. His learning disabilities hadn't disappeared, but neither did they prevent him from ranking at or near the top 10 percent of his class in most subjects and in the upper third in English and history. But as gratifying as his campaign in Lexington may have been, Georgie always had his eye on another objective.

Early in 1904 he received word from Papa that the competitive test for the appointment from California senator Bard to West Point would be given in Los Angeles in mid-February. Georgie spent the entire train ride to the West Coast and the short stay at Lake Vineyard hitting the books, working especially on his atrocious spelling, and memorizing a long list of state and foreign capitals—the sort of thing a senator might ask. He took the exam and immediately returned to VMI for an anxious two-week wait for the results. On March 4, Bard sent Papa a one-line telegram: "I HAVE NOMI-NATED YOUR SON AS PRINCIPLE TO WEST POINT."

Mission accomplished. Georgie had breached the citadel on the Hudson, was certified fit to join a new long gray line. In a burst of glee and megaloma-nia he wrote Papa: "Please thank the California Club for the literary effort on their part in my behalf, and tell them that when I become dictator I will send them my picture and autograph to be hung up along with the moose head, fish and the bear." He also made sure to inform his special friend from Catalina Island days, Beatrice Ayer, of his successful assault. On the other hand, when VMI's upperclassmen got the news, they considered it an act of secession and made sure Georgie spent his last three months in Lexington very much a Rat. But as it turned out, his military academy problems had just begun.

Perhaps sensing trouble, the family again deployed Aunt Nannie when Patton arrived at West Point in June 1904. She set herself up in Highland Falls, even finding it necessary to call in reinforcements in the form of Mama and sister Nita, who each visited several times during that critical plebe year. Quite probably Georgie needed some genetic Geritol at this point. He seems to have been suffering something like culture shock.

Established by a democracy to incubate officers for what had always been an aristocratic profession, the United States Military Academy occupied a kind of ambiguous middle ground. To add at least some egalitarian clarity, admissions had generally been open to all those deemed qualified (if you were white) through a relatively fair appointment process. Politics and free college educations being involved, there were bound to be those who gamed the system, but on the whole it produced a far more economically and so-cially diverse cadet corps than VMI's.

That was certainly Patton's impression. He wrote Mama that they were "nice fellows but very few indeed are born gentlemen . . . the only ones of that type are Southerners." Later that summer he informed Papa: "I be-long to a different class . . . as far removed from these lazy, patriotic, or peace soldiers as heaven is from hell. I know that my ambition is selfish and cold . . . but I have a firm conviction [and] I will do my best to attain what I consider—wrongly perhaps—my destiny."

The opposite of subtle, plebe Patton cultivated an image that combined this blatant class disdain with an obviously synthetic macho demeanor, to which he added a healthy dollop of naked ambition that had him loudly proclaiming he was fated to be the first general in his class. As a prototype mask of command, it did not go over well with his fellow plebes, who dubbed him "Georgie"—okay at home, but not here. Meanwhile the upperclassmen in charge of hazing during "Beast Barracks" and for the rest of the year zeroed in on Patton as soon as they found out he was from VMI, which they disdained as a "tin school" producing, by implication, only tin soldiers.

But they didn't break him—far from it. Georgie may have marched to the beat of a different drummer, but out on the parade ground drilling, or with respect to his military bearing, he already had it down, to the last polished button. Add to that a practically unassailably physical arrogance, and Georgie made it through the abuse part of the plebe experience practically without a scratch.

It was the academics that brought him down, pretty much in a heap. He knew this was his Achilles' heel, and the relentlessly detail-oriented West Point curriculum brought his learning disabilities sharply into focus. Despite obvious effort on his part, results in written grammar, French, and mathematics skidded toward disaster. Letters to Papa were filled with doubt almost to the point of self-loathing: "I am absolutely worthless and know that I should study and don't. . . . If I were only my self of a year ago I would ask nothing better for then I tried and took a vital interest but now, O! hell." It was the same with his missives to Beatrice, and later when she sent him a tiny silver warrior for his watch fob he worried she meant he was a tin soldier.

The time of reckoning arrived at the end of the school year when Patton, having barely survived French, failed his math final—a potential academic death sentence. But the authorities also took note of his exceptional military aptitude and bearing, along with a stubborn desire to improve his grades. This was not the sort of cadet you sent home. Instead they had him repeat first year.

The Rat-plebe-plebe progression, presumably dreaded by all cadets, was suffered, possibly uniquely, by George S. Patton. This amounted to an epic initiation into a military career, but as we shall see, he almost always ended up doing things the hard and painful way. Yet perseverance also usually brought success. He studied hard, and his grades headed higher. Significantly, he was given a cadet rank, second corporal, to break in the next year's plebes.

Just as significantly, he completely overdid it and was more severe and reported more infractions than any of the cadet corporals. In doing so, he also drew the attention of the tactical officers, generally captains sent from the real army to teach and supervise military training at West Point. They promptly demoted him from second to sixth corporal, a signal that left Georgie stunned.

Characteristically, he got the message, or at least enough of it to work his way back into their good graces, back up to second corporal by spring, and then promoted to cadet sergeant major his junior year, putting him in position to be named adjutant of the corps as a senior. It wasn't first captain, but it was an early recognition of military potential, and also likely a reflection of Georgie's obsequious side. He was already an authoritarian—tough and demanding of those below, but fawning to those above—yet so too was he a military romantic, determined to fashion himself into a reincarnation of the Patton family universal warrior.

It was an abrasive and eccentric profile, and it didn't win him friends among cadets. But it did win respect. He was big, probably at or near his eventual six feet two inches, and fleet afoot. He failed at football, but turned himself into an exceptional track athlete, setting the school record in the 220-yard hurdles, winning him the coveted varsity A.

But his real sporting self at West Point was more a matter of weapons and Patton's unique agenda. First, there was swordplay—fencing, at which he became expert—but especially the broadsword. "If I never amount to anything else, I can turn instructor with the broad sword, for I am the best of the best in the class," he wrote Beatrice.

He was already a crack shot when he arrived, so he excelled on West Point's firing range. But one day, he took it over the top, quite literally. He was down in the pit, replacing targets, when suddenly he stood up directly in the line of fire. Considering his penchant for serious injury, he was lucky not to be hit, but thought the experience necessary to test his courage under fire.

It's hard to say what Beatrice Ayer would have said about this stunt on the part of her future husband, which says something about her. The relationship had prospered, even through five years of largely enforced absence. One memorable dance at Theodore Roosevelt's inauguration, occasional weekends at West Point, and short reunions during the academy's truncated summer vacations were their only times alone. He was impressed when she climbed a cliff with him, and seemed interested in his endless stream of military lore. And she had the right attitude. "Don't argue with a man," she told him once. "If you can't convince him lick him. If you can't lick him keep still." Exactly the kind of advice Georgie might at least consider. Of course, he hardly ever kept still. But it was also clear these two remained compatible. If anybody could stand Georgie, it was Beatrice. They were practically joined at the hip, or soon would be.

Meanwhile, as his fifth year along the Hudson drew to a close, he had undoubtedly made his mark, and in doing so undoubtedly rubbed a number of cadets raw. But he was always funny, and his epic sense of his own destiny was in itself comical. So he was never ostracized, just categorized with a smile and a shake of the head—"Did you hear what Georgie . . . ?" Certainly a controversial cadet, but also a successful one. He even clawed his way into the top half of his class, graduating in June 1909, number 46 out of 103. Now a second lieutenant in the US Army, Patton chose the cavalry, not a branch that demanded the highest class standing. It didn't matter; had he been number one he still would have gone with the equestrians. As far as he was concerned he had always ridden a horse into battle swinging a sword . . . with Lee . . . with Napoléon . . . with Caesar. It had all happened before.

IF GEORGE PATTON EMERGED FROM WEST POINT AS A MEMORABLE CADET, A decade earlier Douglas MacArthur had arrived already in the spotlight. He was the son of Arthur, one of the Civil War's most heroic warriors, and now a general in the Philippines, fighting a guerrilla conflict, one he would eventually win with a brilliant strategy of toughness and accommodation. If this wasn't enough to mark Doug for special attention during Beast Barracks, he had arrived with Pinky, who joined the mother of cadet Ulysses S. Grant III at Craney's Hotel, just outside the grounds, a stay that, for her, would last four years. Both should have stayed home. For on the inside, word spread that both cadets Grant and MacArthur weren't just the scions of military moguls, they were mama's boys and therefore entitled to much more than their fair share of abuse. In Doug's case, it could have killed him.

His primary tormentors were southerners, who initially made him stand at attention and repeat his father's Civil War record over and over. Then things got increasingly physical, beginning with an all-night "sweat bath" and culminating in a truly brutal episode that entailed bringing him blindfolded to a darkened tent, where he was stripped naked and forced to do a series of exercises over a bed of broken glass until he went into severe convulsions. He was still jerking uncontrollably when they carried him back to his room, and he had to ask his roommate to wrap him in a blanket so the banging would not be heard.

The next morning, rather than going to the infirmary, MacArthur showed up for inspection and drill as if nothing had happened. On his way back, he was stopped by the cadet who had led the night's hazing: "By your plucky work last night you have a bootlick from the entire Corps," cadet-speak for "Keep your mouth shut: hazing is over."

But there was more harassment to come of an official variety, almost three years of it. Sadistic hazing went critical at West Point when members of Doug's own class managed to kill a cadet. Suddenly it was a national scandal, and in response the McKinley administration set up a special court of inquiry, while the House of Representatives launched investigatory hearings. By this time Doug's hazing heroics were well known, and he was on everybody's witness list. He knew what they wanted—the names of his tormentors.

When the superintendent demanded that he give them up, Doug refused. When the time came for his appearance before the president's court of inquiry, he prefaced it with one of his vomiting episodes, then stonewalled the panel, wrapping it with a histrionic plea for mercy: "I would do anything in the way of punishment, but do not strip me of my uniform." By this time Arthur MacArthur had just about triumphed in the Philippines; they were not about to rip the shirt off his son's back. Neither were the congressmen who grilled him to name names. He bobbed and weaved like a professional, refusing ever to be pinned down. In the end, he gave a very impressive list of inquisitors nothing they wanted.

This was an important episode in Douglas MacArthur's life. Being the son of a famous person is never easy, but in his case it put him in a position of exaggerated vulnerability. By sheer force of will he weathered the ordeal. But he had done something more important: he had won the trust of his fellow cadets, verified when a first-classman asked him to become his roommate—a practically unheard-of gesture. No matter how brave or talented, being a soldier demands becoming part of something larger, giving yourself up to it. For an individual as egocentric as MacArthur, this was no easy thing, but he had passed this crucial test with flying colors. A military career was now possible.

Meanwhile, Doug was burnishing his reputation as virtually the perfect cadet. A herculean appetite for work—for Doug the best part of rooming with a first-classman was the increased study time that lights-out at eleven p.m. offered—combined with what appears to have been an eidetic or total-recall memory led to mind-bending academic results. For a while Cadet Grant chased but never challenged his perpetual first in class. When all the numbers in the academy's excruciatingly detailed grading system were finally totaled, Cadet MacArthur had accumulated 2,424.2 points out of a possible 2,470, a record surpassed by only two cadets since West Point was founded in 1802, one of whom was Robert E. Lee.

His military demeanor was equally exalted. His uniform perfect, his posture the same, he was in the eyes of one real army tactical officer "the finest drill master I have ever seen." Each year he bounded to the highest possible cadet rank until he reached the ultimate, first captain, just like Lee and his future bête noire, Black Jack Pershing.

Yet rank and numbers provide simply an outline of the image that MacArthur managed to create while at West Point. A weak hitter and barely adequate fielder, he turned himself into "Dauntless Doug," captain of the baseball team and recipient of the varsity A—one he would wear on his bathrobe nearly fifty years later on the night before the Inchon landing.

This was telling: practically everything he did at West Point had significance for MacArthur later. It was there that he first tried on his military mask of invincibility—a force field of absolute self-confidence that would shield him from both bullets and bureaucrats while bending his image to make him appear larger than life. One upperclassman retrospectively wrote, "Without doubt, the handsomest cadet that ever came into the Academy, six-foot tall, and slender, with a fine body and flashing dark eyes." "To know MacArthur is to love him or to hate him—you can't just like him," another added. Still a third pointed to that ineffable quality that set him apart: "He had style. There was never a cadet quite like him." Contemporary photos also appear to capture what one biographer insightfully calls "a slightly far-away, visionary cast to his eyes that's familiar from the myriad photographs of him during World War II and Korea." No coincidence here. The mask made at West Point fit so well, he wore it for the rest of his life.

It also followed that it and he drew the interest of the opposite sex, or at least those who played a part in the highly supervised social life available to cadets. One rumor had him engaged to eight girls at once, which was ridiculous, but also symptomatic. Any young woman interested in a romantic relationship with Douglas MacArthur had to face several facts of life: one was Pinky, the other was that Doug loved Doug more than anybody. In the case of the former, he visited Craney's or thereabouts daily (with or without a pass, about the only rule he broke) to receive quotidian shots of praise and reminders of his personal destiny. For as long as she lived Pinky was a permanent fixture, cheerleader and fixer in chief, both serving to intensify the second condition: MacArthur's own self-absorption. Unless you were willing to love Doug more than Doug did—literally worship the man—he wanted you basically for sex. As attractive as he was, this was not an attractive profile, so it was a long time before he found the perfect mate.

Other than that, and also his looming experience in Asia, Douglas MacArthur left West Point fully formed, already a legend in his own mind, and never to be forgotten by those with whom he shared the Plain.

EVEN A BRIEF ACQUAINTANCE WITH THE ACADEMY EXPERIENCES OF THE FIRST three of our future heavy hitters leaves little doubt that they all left with a serious commitment to a career as an officer in the US Army. Perhaps more important, they had already assumed that identity and pulled on the mask. With Dwight Eisenhower, it really is hard to tell. Getting to know him is a bit like quantum physics—just when you think you have a good grip on the man, he turns into something else. Is he this? Or is he that? Or both?

On the face of it, Ike assumed the role of rogue cadet, one pioneered by William Tecumseh Sherman. The hairball of rules that enveloped West Point were not only made to be followed. War is cruel and mercilessly favors those most capable at deceit. Breaking the rules and getting away with it is good practice, and therefore was, and still is, enshrined as highly regarded, though necessarily sub-rosa, West Point behavior. Ike, already deeply schooled in the art of the bluff, excelled.

He arrived in June 1911, at nearly twenty-one more mature than most plebes, and, after two years in the creamery, a good deal more fit. Sharing tight spaces with other young males while being yelled at, though certainly more intense and sustained during Beast Barracks, was nothing new to Ike. "The calculated chaos" of his first day was certainly unpleasant, but it failed to knock him off course: "Where else could you get a college education without cost?" the mantra kept insisting. But when he took the Oath of Allegiance at the end of the day, Ike morphed into something else: "From here on it would be the nation I would be serving not myself. Suddenly the flag itself meant something." Student on the make, or officer in the making, the paradox of Ike at West Point would stretch through all four years. And at the end, like a quantum packet, he was still uncertain.

Yet he was never invisible. From the beginning, and as it always would be, Ike was well and widely liked—by his fellow classmates and, it seems, the rest of the academy, both students and faculty. Ike's cadre, the so-called class the stars fell on, due to over a third becoming generals, was plainly full of talent. Yet he managed to claim a respected place in the pecking order without stepping on toes or ostentatiously clawing for distinction like Patton and MacArthur. Instead, he deployed his smiling mask to camouflage his competitive instincts, and became a boon companion.

"I was, in matters of discipline, far from a good cadet. While each demerit had an effect on class standing, this to me was of small moment. I enjoyed life at the Academy, had a good time with my pals." His transgressions of choice were cigarettes and poker—both strictly forbidden—and his appetite for both proved monumental. Already a poker prodigy with a face to match, he nearly always fleeced his pals and kept careful record of their

IOUs against the day when they all graduated and had salaries. But he wasn't obnoxious about winning (not yet at least), plus he was great fun and always had a cigarette. These did come at the cost of some considerable quantity of demerits, but in the Sherman tradition, never enough to put Ike in serious jeopardy of expulsion—the essence of being a rogue cadet.

For Ike there was also a human side. When an overstimulated plebe accidentally rammed into him, Ike, now an upperclassman, blurted out angrily: "You look like a barber." To which the plebe replied softly, with some embarrassment: "I was a barber, sir." That night Ike told his roommate, "I'm never going to crawl another Plebe as long as I live . . . I managed to make a man ashamed of the work he did to earn a living." When West Point ritual reached the stage of contradicting older values learned in Abilene, Ike chose the latter, a basic decency he would carry with him throughout his life.

But he also bought into the system—that paradox again—mostly doing the things that characterized a good cadet, or at least trying. He already had an excellent if not eidetic memory, along with a firm grasp of written English and math, so if the minutiae at the core of the academy's curriculum sometimes infuriated him, he had little trouble digesting and spitting it back when required. His grades were good if not great.

Eisenhower did better in the professional areas. His days in the woods with Bob Davis had left him with a natural feel for terrain and a rare ability to translate the two dimensions of a map into the topography of an actual battlefield. This, along with Ike's already impressive though understated knowledge of military history, were indicators of true career potential, the qualities the real army officers who staffed West Point's tactical department were looking for. They also recognized the underlying glimmer of leadership beneath Ike's smiling mask, and cultivated it accordingly with a string of cadet ranks. Ike responded like Ike—busted twice, once for "improper dancing" with a Spanish instructor's daughter—yet none of this deterred them from naming him color sergeant for his final year. They obviously saw something in Ike.

So did the athletic department—though not enough of it. Cadet Eisenhower arrived at the academy weighing 152 pounds spread over a five-foot-eleven-inch frame, and determined to play football. Judged simply too light, he was relegated to the junior varsity in his plebe year, but returned the next, pumped up to a muscular 175 pounds, and was elevated to the main team. He started five games for the varsity that season, but then, against Tufts, he suffered the great crippler of football players, a wounded knee: "I couldn't get up, so they took me off the field, and I never got back on as a player again."

He should have stayed away entirely, but instead absorbed another career-threatening injury when he turned to coaching the junior varsity squad his last two years. He was a great success. He knew the game almost as well as he knew poker, and the cadets naturally did what he told them to do, so they won consistently and Ike picked up a reputation as a brilliant coach—one destined to hang over his career like a social disease. For a long time in the sports-crazed Army there would always seem to be a commander ready to divert Ike from a career-enhancing assignment and turn him into a football coach.

Meanwhile, Ike's knee almost dictated there would be no career at all. Shortly after he was hospitalized for the original injury, Ike fell off a horse in the riding hall and a temporary condition became permanent. Ike was still mobile and was soon marching without a limp, but running, especially at full speed, was a thing of the past. He kept in shape with gymnastics and calisthenics his last two years.

As his time at West Point wound down, it was probably the tactical department's assessment that best captured this stage of Ike: "We saw in Eisenhower . . . a man who would thoroughly enjoy his Army life, giving both to duty and recreation their fair values. We did not see in him a man who would throw himself into his job so completely that nothing else would matter." And as befit such an assessment, Ike was set to graduate sixty-first out of a class of 164 without a great deal of effort.

But then he was informed by West Point's chief medical officer that the bad knee and the possibility of a subsequent disability pension might prevent Ike from being commissioned, though he would still graduate. Ike's reaction was initially as enigmatic as that of a quantum packet: "I said that this was all right with me, I remarked that I had always had a curious ambition to go to the Argentine." The surprised officer decided to think things over for several days, then got back to Ike, offering him a commission in the less demanding coastal defense arm: "Colonel I do not want a commission in the Coast Artillery," Ike demurred. "He brought the interview to an end, and I thought, 'Well, that's that.'"

What was what? Was Ike really ready to become a gaucho, or was this simply a masterful bluff by an excellent poker player? Even in his memoirs over fifty years later, Ike never tipped his hand. Instead, he reports the negotiations over the colonel's next bid: "Mr. Eisenhower, if you will not submit any requests for mounted service . . . I will recommend to the Academic Board that you be commissioned."

"So I told Colonel Shaw that my ambition was to be in the Infantry. To which he said, 'All right I'll recommend you for commission, but with the stipulation that you will ask for no other service in the Army.'" Disregarding that infantry implied a lot of walking on a wounded knee, Ike plainly out-negotiated the colonel, twisting him like a pretzel. But the colonel had accomplished one thing nobody had in four years. "From the first day at West Point, and any number of times thereafter, I often asked myself: 'What am I doing here?'" Now he knew. The colonel had pinned down the elusive Ike, forcing him to define himself now and forevermore as an officer in the US Army, the last of our four superstars to join the team and begin carving a path forward.

CHAPTER 3

FIRST BASE

Attending a military academy, no matter how successfully, is only a simulation intended to encourage officer-like behavior. Learning to be an actual officer is different, more pervasive, filled with unanticipated lessons and pitfalls. It amounts to reinstruction in terms of military reality, and sometimes the learning curve can be very steep.

This was certainly the case with Douglas MacArthur, and as usual he met the challenge with his own brand of panache. But it changed him, made real what had become the family mantra: "The future is Asia, particularly the Philippines." That's where Doug was heading for his first assignment—one likely with the influence of Arthur MacArthur behind it. He would join the Third Engineer Battalion I Company, who were building a wharf and harbor on the island of Panay.

From the moment he arrived in Manila after thirty-eight days at sea, he recalled, "The Philippines charmed me. The delightful hospitality, the respect and affection expressed for my father, the amazingly attractive result of a mixture of Spanish culture and American industry, the languorous laze that seemed to glamorize even the most routine chores of life, the fun-loving men, the moonbeam delicacy of its lovely women, fastened me with a grip that has never relaxed."

He was even charmed by the heat-soaked island of Panay, which returned the favor with a strong indication that Douglas MacArthur led a charmed life. The Third Engineers needed lumber to build the piers and docks they were planning, and it became the duty of Second Lieutenant MacArthur to lead parties into the jungle to find and cut the necessary timber. Due to the danger posed by remnants of the former insurgency, he routinely went armed with a .38 pistol. This proved a lifesaver, literally, when one day, distracted by his search for suitable trees, he became separated from his work party and found himself confronted by two armed guerrillas, both front and rear along the narrow jungle path. The first managed to put a bullet through MacArthur's campaign hat before he killed them both. Very suddenly combat and the Curse of Cain found Douglas MacArthur. A sergeant, drawn by the shots, came upon the scene soon enough to see his commander's hat still smoking: "Begging thu Loo'tenant's paddon, but all the rest of the Loo'tenant's life is pure velvut." Not a bad prediction, particularly when it came to being missed by bullets.

But not by malaria mosquitos, which soon felled Mac and sent him reeling back to Manila to convalesce. While his health was slow to recover, being Douglas MacArthur, he still made good use of this time. He worked a series of desk jobs for the Army's chief engineer, and even managed to survey parts of the Bataan Peninsula in southern Luzon. In the spring of 1904, he was well enough to take the test for first lieutenant, during which he was asked how he would defend Manila Bay from a combined naval and amphibious attack, without any troops. "First, I'd round up all the sign painters . . . and put them to work making signs reading BEWARE—THIS HARBOR IS MINED. I'd float these signs out to the harbor mouth. After that I'd get down on my knees. Then I'd go out and fight like hell." Not far from the situation he would face in 1941, and had he not been removed to Australia, his solution probably would have been the same. No surprise, he was promoted.

Meanwhile, using his father's contacts, he began building his own network of connections in the city, including two young nationalist lawyers, Sergio Osmeña and Manuel Quezon, both destined to play important roles in Mac's Philippine future. Yet, for now, that future would be held in abeyance by his malaria, which sent him back to San Francisco in October for a yearlong recovery of light duties and Pinky's ministrations.

Health restored, on October 5, 1905, Doug received a telegram headed "Special Order 222" announcing his next step in the family project. He was ordered to proceed immediately to Tokyo, where he would assume the role of aide-de-camp to the acting military attaché, none other than Major General Arthur MacArthur, who had arrived seven months before to scrutinize the Russo-Japanese War. Effectively, it was a gridlocked preview of the technologies that would turn World War I into a ghastly stalemate—something almost all Western observers missed—but then, in late May 1905, Admiral Tōgō annihilated the Russian fleet in the Tsushima Strait, and the Japanese emerged as clear winners.

Expecting huge Russian reparations to cover their war expenses, the Japanese turned to American president Theodore Roosevelt to broker a settlement. What they got from the Treaty of Portsmouth was Manchuria, Korea, and an enormous debt. From top to bottom the Japanese were furious. Huge anti-American demonstrations swept the country; hundreds killed themselves over this perceived national humiliation, and Arthur MacArthur, once the toast of Tokyo, realized it was time to get out of town.

He thought about the Philippines, but then the secretary of war, William Howard Taft, who didn't want him there, proposed an irresistible alternative, an official tour of Asia, accompanied by First Lieutenant Douglas MacArthur.

Further reinforcements arrived in the form of Pinky, and the MacArthur family set off on an eight-month odyssey, which included visits to practically every locality of political or military significance in China, India, South Asia, and Japan. All the while Douglas was absorbing the experience filtered through the worldview of Arthur MacArthur, one he had heard him evolve and promulgate, probably as far back as his memory could stretch. Now Asia reeled out like a fantastically protracted cinematic experience, 19,949 miles of it, all the while accompanied by appropriate narration.

At trip's end in June 1906, Douglas MacArthur was not so much a changed man as he was a confirmed one. "The experience was without doubt the most important factor of preparation in my entire life," he explained later. "It was crystal clear to me that the future and, indeed, the very existence of America were irrevocably entwined with Asia and its island outposts." The family project had effectively passed hands; Douglas MacArthur's geopolitical compass was set.

So probably was his ego. In a letter to Pinky describing the firefight in the jungle, he wrote: "I heard the bullets whistle, and believe me, there is something charming in the sound," exactly young George Washington's words just after he set off the Seven Years' War, both a testimony to Doug's memory and his exaggerated self-image, even at this stage.

It didn't work well for him when, in August 1906, he arrived again in the city named after the aforementioned gentleman. He had been assigned to Washington Barracks and its elite engineering school, and better yet, in December was named as aide-de-camp to President Theodore Roosevelt. It would not have been hard for anyone who knew Douglas MacArthur to predict which of the two roles he would take more seriously. "He [TR] was greatly interested in my views on the Far East and talked with me long and often." How long and how often remains open to question, but it's likely MacArthur made himself available whenever possible. Or that's how the engineering school's commandant saw it in his efficiency report: "I am sorry to report that during this time Lieutenant MacArthur seemed to take but little interest in his course at the school." It was the first black mark on an otherwise stainless record.

It had been a bad year for the MacArthurs in general. In 1906 the top position in the Army, chief of staff, needed filling and Arthur, having acquired his third star and become the most senior general in the service, ordinarily might have expected the post. But his political nemesis—the man who had outfoxed him in the Philippines, and most recently turned him into a tourist, William Howard Taft—remained secretary of war and looked elsewhere. Enraged, Arthur retreated with Pinky to Milwaukee to await retirement in the comfort of a three-story mansion.

It was here, in the late summer of 1907, that Doug joined them, conveniently assigned "special duty" to work on plans to renovate several of Lake Michigan's harbors. But once again he found himself distracted. Arthur wanted to talk Philippine politics, Asian futures, and Taft treacheries, while Pinky wanted her handsome son at her side during Milwaukee's social swirl. He even acquired a girlfriend, a relationship intense enough to generate a twenty-six-page epic poem on his part. But as usual it didn't last.

Still, all of this took time, specifically, away from his duties. At one point, he told his commanding officer that he "wished to be undisturbed for about eight months." Predictably, his inattention was reflected in a scathing efficiency report that likely would have ended the career of a less well connected young officer: "Lieutenant MacArthur . . . did not conduct himself in a way to meet commendation . . . his duties were not performed in a satisfactory manner."

But rather than take this as a warning, Doug went on the warpath. Furious over his rating and also his rejection for a West Point teaching post, he lobbed a letter of complaint over his commander directly to the chief of engineers, a one-star general. No matter who did it, this was not done, and Doug got a stern rebuke in return. Pinky only made things worse when she contacted Arthur's friend E. H. Harriman, the rail tycoon, about a job for Doug, who responded by telling the recruiter he had no intention of leaving the Army, and if he did it certainly wouldn't be to work on the railroad. Even within the MacArthur family bubble, Doug seemed to be living in his own bubble of self-absorption.

Actually, the problem, at least at this point, was less him than his choices. Ironically, success at West Point naturally led him to pick, as did most high-ranking cadets, the elite Corps of Engineers. Now, six years later, it was perfectly clear that the nuts and bolts of the profession bored him. Engineers drew up plans and built things; there was no glory in that. He was a natural leader of men caught in the calipers of the Army's version of nerds, and his behavior suffered accordingly.

Fortunately, Arthur's network of connections provided a soft landing and also a new direction. His old comrade, now Army chief of staff J. Franklin Bell, posted the errant lieutenant to Fort Leavenworth to train a company of volunteers. In command of real troops for the first time, Douglas MacArthur found his true calling.

He took over K Company, rated worst of the twenty-one stationed on base. That didn't last. His enthusiasm and determination to improve the men's performance was contagious. Soon he was leading them on twenty-five-mile hikes, teaching them how to assemble a pontoon bridge fast and, if necessary, blow it up, turning them into real field engineers by the next general inspection, when they were ranked first. "I could not have been happier if they had made me a general," he remembered.

That would have to wait, but this posting, which stretched through 1912, stabilized and reoriented MacArthur's career. Once again, his efficiency reports used terms like "A most excellent and efficient officer," while Doug himself overheard a sergeant major tell his troops, "Boys, there goes a soldier." This was the affirmation MacArthur needed, that he was still part of the Army. When necessary his ego could be holstered and his career could move forward, which it did in February 1911, in the form of captain's bars.

Engineer training was a subsidiary function at Fort Leavenworth, which was known primarily as the "Army's Schoolhouse," the place where young officers learned the skills involved in ordering around large bodies of men—the command and staff function. Attendance was not mandatory for higher rank, but selection was generally reserved for hard chargers with perceived career potential. Though MacArthur was not directly involved with the school, his intellect and magnetic personality ensured that he got to know a number of the students.

Yet, in doing so, Doug would have been surprised to learn he was being introduced to his future. Two in particular, Robert Eichelberger and Walter Krueger, regarded him with something approaching awe, a spell that would weaken only with the experience of being his chief henchmen in the Southwest Pacific during World War II. Another, though, apparently remained unimpressed, which is of some significance to our story because he was George Catlett Marshall, and this marks the first time two of our protagonists crossed paths. Both were favorites of J. Franklin Bell, and this might have been the basis of a relationship. But, in reality, taciturn, retiring Marshall and voluble, emotive MacArthur were like oil and water, and all we really know about their interaction at Leavenworth is that they didn't bond. Nor would they ever.

On September 12, 1912, Arthur MacArthur got out of a sickbed to attend the reunion of the Twenty-Fourth Wisconsin, which he had led up Missionary Ridge almost fifty years before, then promptly dropped dead. Emotions aside, his demise posed practical problems for Doug. Pinky, otherwise prostrate, was refusing to move in with his brother's family, stay in Milwaukee, or settle anywhere near Leavenworth. She left Doug no room to maneuver—it was his job to find an appropriate perch.

Yet again connections to the top, this time in the form of General Leonard Wood, Arthur's comrade in the Philippines and now chief of staff, provided the necessary jolt to quickly propel Doug to an assignment with the Army's brain trust, the General Staff, and Pinky to an elegant apartment in

Northwest Washington, an address that remarkably improved her health, especially since she shared it with her son. Daily a chauffeur drove him to work, where, as the junior member of a staff of thirty-eight, he watched and learned how operational and strategic planning was formulated and promulgated. The process fascinated him, so much so that he turned down an offer to become an aide to the new president, Woodrow Wilson. Yet as good as things were, they were not heroic.

At least not until April 22, 1914, when Leonard Wood called MacArthur into his office for a chat. Relations with Mexico, already bad, had turned warlike in Veracruz, when, after the corrupt and assassination-inclined dictator Victoriano Huerta landed a shipment of embargoed arms, Woodrow Wilson ordered the Marines to take the city, which they did after heavy fighting. As they spoke, "Fighting Fred" Funston of Spanish-American War fame was on his way with a brigade of seven thousand to act as reinforcements. Other than that, intelligence was minimal. What Wood and the secretary of war wanted was someone on the ground to act as an independent source of information. It was a mission made for Doug. "How soon can you leave?" Wood wondered. "I can be off in an hour," said MacArthur, probably over his shoulder.

He arrived and was soon seen snooping around the battered streets of Veracruz in a prototype of his trademark series of campaign costumes—corncob pipe, battered hat, floppy tie between a set of captain's bars, topped with a cardigan sweater. How he was attired when he was received by General Funston goes unrecorded. But the general had bigger problems, and although he had reason to be suspicious of MacArthur, he leveled with him. His force was surrounded. Animal transport was lacking to deploy or obtain supplies, and beyond his own lines his intelligence remained next to zero. The city had a railway and sufficient cars but no engines to drive them.

Shortly thereafter Douglas MacArthur, on his own initiative, decided to go on a locomotive hunt, telling only a friendly captain, Constant Cordier. Recruiting three Mexican railway workers with knowledge of an engine cache thirty-five miles beyond American lines with the promise of $150 in gold, MacArthur set off at dusk on a harebrained hegira by handcar, canoe, and numerous stolen horses before reaching five locomotives, three of them "just what we needed." The trip home went downhill fast, featuring three separate attacks by armed Mexicans, during which MacArthur details having shot down seven while sustaining four bullet holes through various items of clothing. Cordier found him the next morning completely exhausted, bullet riddled, but otherwise unhurt. "It is a mystery to me that any of the party escaped," Cordier wrote Leonard Wood in a letter recommending Doug be nominated for the Medal of Honor. Leonard Wood did just that, but when the recommendation was forwarded to Funston, he executed what amounted to a bureaucratic shrug of the shoulders. "Captain MacArthur was not a member of my command at the time," and in fact the general knew nothing of the mission. So went Mac's opportunity for the medal.

He was furious and set off another letter bomb that did him no good— nor would they ever. But to neutral eyes the tale must have sounded more outlandish than real, the only available witness to the exploits of Douglas MacArthur being Douglas MacArthur. On these grounds, his chances were never more than slim. But still it's possible he had reason to be bitter. His Mexican interlude was exactly the sort of eccentric mission he would under- take repeatedly during World War I under much better observation, facing absurd odds with sheer audacity and emerging unscathed. It was his signa- ture combat move, one that left all others shaking their heads, but not in doubt.

For now, though, it was back to Leonard Wood and the General Staff, where he continued to shine, earning him a promotion to major in December 1914, and the reputation as one of the brightest young men in Washington. Mexico remained a concern, yet he and the rest of the city were already turn- ing their attention to a much graver threat, the horrible and futile war that was consuming Europe. But we are getting ahead of our story: we need to catch up with the nascent careers of our three other protagonists.

GEORGE PATTON RODE INTO HIS MILITARY CAREER ON THE BACK OF A HORSE, this despite the best efforts of internal combustion to signal his future lay elsewhere. Nevertheless, the canter began auspiciously with his assignment in late 1909 to Fort Sheridan, a small cavalry post just outside Chicago, his immediate superior, Captain Francis Marshall, having taught in the Depart- ment of Tactics at West Point during most of Patton's stay.

Marshall, who actually liked Patton, wasn't put off by his naked ambition.

Instead, he spent days integrating his new second lieutenant into his varied duties and made sure he understood the camp's routines. As usual Patton found his fellow officers lacking in breeding, referring to some of the militia members integrated into the regular army after the Spanish-American War as "the sin of 1898." Otherwise, he took to the place like a fish learning to swim. He liked working with the enlisted men of Troop K, whom he found quick and well trained. Under his commander's watchful eye, he took them on gradually longer and more realistic equestrian forays and patrols until each became comfortable with the other.

But he was also Patton. Consequently, after having cursed one trooper for not running at his command, he felt it necessary to apologize before the entire company—the first in a career-long string of apologetic episodes. Another time, as he was drilling the men, his horse threw him. When he remounted, the beast bucked, then fell, but failed to remove Patton, who emerged with a badly gashed eyebrow to resume leading the men as if nothing had happened, blood streaming down his face and sleeve. Love him or hate him—you never forgot him.

That was certainly true of Beatrice Ayer, who was ready to marry him. But her father, Frederick Ayer, hadn't built a huge fortune—part of it based on the heroin-laced elixir Cherry Pectoral—on sentiment; he did not relish the prospect of his daughter subjected to the harsh realities of military matrimony. Nor was he necessarily impressed by Patton's version of the grand gesture: riding his horse up the marble stairs at the entrance of the family's Massachusetts estate, Avalon, dismounting, then sinking into a deep bow before Beatrice. The old man made his position clear—either Beatrice or the Army. For once Patton said the right thing, explaining that his affection for the Army was anything but logical: "I only feel it inside. It is as natural for me to be a soldier as it is to breathe . . ." Georgie's true Army-bound bottom line did not immediately convince Ayer that matrimony was advisable. But Beatrice went on a hunger strike and locked herself in her room, Patton held firm, and soon enough the father agreed—subject to a statement of Patton's own financial position. Patton's reply left even Mr. Ayer impressed "at such a fine nest egg," counseling his now future son-in-law: "I would not sell the land." The deal was done, the match made.

The wedding, on May 26, 1910, was a high point in the Boston social

season, featuring a special train delivering guests out to Avalon. At the ceremony Beatrice wore an engagement ring that was a miniature of her fiancé's
West Point class ring, indicating that she had at least some idea what she
was getting into. But Georgie's extreme ardor that first night, and later their
wretched quarters at Fort Sheridan, definitely took some getting used to.
On the other hand, they were young and in love, and also really rich, which
can take the edge off even the most miserable environment. Frederick Ayer
may have been impressed by the Patton family finances, but his fortune was
much larger. By marrying Beatrice and her extremely generous monthly al-
lowance, George Patton established himself as among the wealthiest officers in the entire Army. That's exactly what he did, making no secret of
his money and social prominence, instead using them whenever possible to
advance his career.

Meanwhile, Beatrice, being tough and adaptable, made the best of Fort
Sheridan and began a lifelong campaign to move Georgie forward professionally, though sometimes unintentionally. For she was soon pregnant,
delivering a daughter in March 1911, whom they named Beatrice Jr. Patton
loved sex, but not reproduction. He vomited when asked to hold the baby,
and was subsequently more jealous than loving to the new arrival. To fill
the time he felt Bea and Bea Jr. were stealing from him, he dove deep into
Clausewitz, purchased a typewriter, and was soon writing professional articles for military journals. So it was that fatherhood gave birth to Patton
the military thinker, an offensive-minded one even at this nascent stage.
"Attack . . . push forward, attack again until the end," he advised in one early
piece.

He was similarly direct in career strategy. Fort Sheridan had taught him
all it could. Unlike our other three protagonists, Patton wanted to avoid
duty in the Philippines, where he saw little advantage. He turned to Captain Marshall, very bluntly asking him how strings were pulled in the Army.
The answer didn't really matter: the Pattons and Ayers were already pulling them, and in December 1911 orders arrived transferring Georgie to Fort
Myer, on the Virginia side of the nation's capital, the greenest of pastures
for a cavalryman.

Home of the chief of staff, Fort Myer was the Army's ceremonial center,
a showplace where its best gentleman horsemen put on impressive public

displays, managed military funerals, and played an excellent brand of polo. It was just the place for Patton, who considered Washington "nearer God than else where and the place where all people with aspirations should attempt to dwell."

Having arrived, Bea and Georgie immediately and emphatically began living large: a rented mansion, a ubiquitous social presence, the best in horseflesh. For official and otherwise upper-crust Washington, it was hard to miss this budding power couple. Secretary of War Henry Stimson first encountered Patton on Fort Myer's many riding trails. Impressed with his equitation and dash, Stimson was soon riding regularly with Georgie, the beginning of a relationship that persisted deep into World War II. Having upgraded his ponies, Patton also became a member of the Fort Myer's polo team, one of the best on the East Coast. Being rich, athletic, and very visible was a combination that would soon turn Georgie into an Olympian.

The modern pentathlon, a martial-oriented combination of swimming, running, fencing, pistol shooting, and steeplechase, was for the first time scheduled to be included in the 1912 Stockholm Olympics. Only military contestants could participate, so it fell to the US Army to pick one, who would then be entirely responsible for training and transportation. Though not a natural athlete, Patton was a strong but disinclined swimmer; he had been a record-setting hurdler at West Point, where he had also refined his swordsmanship; he remained a crack shot; and, basically, his job was to ride a horse—the perfect candidate, or at least the most visible one with the money to stage the campaign at his own expense. So Georgie was chosen, but not until the beginning of May, leaving him barely a month to train.

He immediately went on a diet of only raw steak and salad, quit smoking and drinking, and set up a punishing training regime, which emphasized swimming and running, his weakest events. The Patton clan (Bea, Papa, Mama, and Nita) joined the American team in the middle of June on the steamer *Finland*, where Georgie ran endless laps around its deck, practiced resistance swimming in a small canvas pool with a rope looped around his waist, shot at targets off the fantail, and practiced rigorously with members of the fencing team—a nonstop routine until the Atlantic was crossed.

The Patton delegation arrived in Sweden on June 29, where Georgie continued training, watched over by Papa, until the brutal five-day pentathlon began. Patton, the lone American competitor, initially faced forty-two other contestants. Possibly suffering from nerves, he did poorly in the first event, the pistol shoot, placing twenty-first, but then settled down, finishing sixth in swimming (they had to fish him out of the pool with a boat hook) and third in both fencing and steeplechase.

By the final day only fifteen contestants remained for the footrace, a 4K slog through a swampy course staged on a particularly hot day. Although performance-enhancing drugs were certainly not banned at this point, the American trainer's decision to give Patton a shot of opium for stamina was decidedly not among his best. Georgie came in third but collapsed and went unconscious at the finish. "Once I came to I could not move or open my eyes and felt them give me a shot of more hop [opium] . . . Then I heard Papa say in a calm voice, 'Will the boy live?' And Murphy [the trainer] reply, 'I think he will but can't tell.'" He did, but he was suffering from severe dehydration, which might well have been fatal. Instead, he took a very respectable fifth in the final standings, and soon embarked with the clan on a tour of Europe, one that included a two-week crash course in swordsmanship at the French cavalry school in Saumur, and the seed of an idea.

Back in Washington, the Olympian second lieutenant was soon recounting his adventures over dinner with Henry Stimson and Chief of Staff Leonard Wood, Doug MacArthur's patron, who was always on the lookout for promising officers. Soon, he too was riding in the morning with Patton, and by December Georgie was detailed to the chief of staff's office, about a month after MacArthur's assignment to the General Staff. Their paths may have crossed, but like Mac's meeting with George Marshall, it didn't amount to anything. MacArthur was fascinated with operational and strategic planning; Patton had another agenda.

He had become an evangelist for sword reform. At Saumur Patton had learned that the French planned to use their cavalry sabers not for slashing, as was the American practice, but for thrusting, which he believed was more effective. In an influential article in the *Army and Navy Journal* Patton argued that the curved US cavalry sword should therefore be replaced by a straight edge. He just happened to have a design of his own, twenty thousand copies of which the chief of staff soon ordered the chief of ordnance to manufacture. Wood even sent Georgie to Springfield Arsenal to make sure the new design, subsequently known as the Patton sword, met his specifications.

The next step on the career ladder was the Mounted Service School at Fort Riley, Kansas, and Patton accordingly received orders to report there on October 1, 1913. In the meantime, for months back in Washington, Georgie had been promoting a curriculum addition for the school, a rigorous course in swordsmanship taught by a true expert. That would be George Patton, who managed to convince both the school's commander and the War Department to let him return (at his own expense) to Saumur for a summer of fencing and saber rattling, and then proceed to Fort Riley as both a student and an instructor, the Army's first and only designated master of the sword.

It was pure Patton, in every way prophetic of his misguided nature, both astute and idiotic. During the twenty months he spent swinging a sword at Fort Riley, World War I had paralyzed Europe's best armies, and in the process made a mockery of swords and cavalry. In an atmosphere teeming with bullets and high explosives there was absolutely no room for equine-size targets armed with blades, either curved or straight. Actually, this had been true since the American Civil War, when swords and bayonets were used primarily to barbecue meat. But combatants on horseback had been a key part of human warfare since its very beginnings. It was a hard concept to give up, as witnessed by the tens of thousands of cavalry on both sides of the western front, waiting for the charge that never came. So the master of the sword was far from alone in his illusions; he was just more flamboyantly wrongheaded.

Yet it was this flamboyance that mattered for the future. Patton was in his own mind the universal soldier, or a reincarnated version; he would fight with anything you gave him. But by having himself designated master of the sword, this second lieutenant had turned himself into something larger, exactly the illusion that made the mask of command work. It was about this time that Georgie began spending long periods in front of a mirror working on his fighting face. His quotidian chatter became laced with obscenities, both creative and memorable, with only his high, squeaky voice betraying the image. Yet it was still formidable, even at this stage. And like Douglas MacArthur's, it would almost immediately be tested by real bullets.

In early 1915 rumor had it that Georgie's unit was about to be assigned to the Philippines, and once again he descended on Washington. Now apparently able to pull his own strings, he succeeded in getting himself sent to the Mexican border with the Eighth Cavalry, at sprawling Fort Bliss, headquartered near El Paso. If you were interested in fighting, it was the right place at the right time.

Since MacArthur's one-man mission, troubles with Mexico had persisted. After Veracruz, Huerta had resigned from the presidency and been replaced by Venustiano Carranza, causing the formidable guerrilla Francisco Villa, better known as Pancho, to go into open revolt, venting his anger and violence on the Americans, whom he felt had betrayed him. This resulted in a string of vicious raids that terrified border towns from Texas through New Mexico to Arizona. Given the circumstances, the US response was inevitable: substantial Army reinforcements for a potential punitive mission.

Leading it would be Brigadier General John J. Pershing, a pivotal figure in our story, one who would deeply affect the respective career paths of all four of our protagonists. The outstanding US soldier of his generation, he was also a hard man. Having served early in his career as a white officer with the African American Tenth cavalry, the Buffalo Soldiers, he had picked up the nickname "Black Jack," which exactly fit his nature. He was a sour individual, but a commanding presence. A middling student, he still left West Point first captain, and when he returned as a tactical instructor Pershing proved so unbending that the students twisted his nickname in an obvious direction. But they never came close to deflecting him—the man was a fortress of self-control and determination.

He had to be, considering the telegram that arrived at his headquarters at Fort Bliss in late August 1915 informing him that his wife and three of his daughters had been killed in a fire at the Presidio in San Francisco. He went to the funeral, brought his surviving son back to Bliss, and resumed command still unbowed, but an even blacker Jack.

Meanwhile, George Patton had established himself as a swaggering, harddrinking presence at the fort. He took and passed his exam for first lieutenant, led his cavalry unit on giant sweeps as far as ninety miles out to reassure ranchers and guard the Southern Pacific Railroad. He also helped

form a polo team, and brought Bea and the offspring—which now numbered two, had he cared to count—down to live in El Paso. But Georgie
wanted more, and this meant being noticed by Pershing. Fate may have intervened, but as he had shown with Stimson and Wood, Patton was a superb
sycophant, and he just happened to have a honeypot.

Sister Nita, tall, blond, still in her late twenties, and unmarried, on a
fortuitous visit to El Paso attended one of Fort Bliss's many social events,
where she was introduced to the general. It was attraction at first sight.
Though still grieving and ramrod stiff in demeanor, Pershing always had an
eye for the ladies, and in particular this one. But the relationship was just
getting started when things went critical.

In early January 1916, Villistas took eighteen American employees of an
Arizona mining company off a train near Santa Isabel in Chihuahua, and
murdered them. Then, on March 9, Pancho Villa hit Columbus, New Mexico, with a force numbering around five hundred. Although repelled by three
hundred American infantry and cavalry, he and his men left the town heavily damaged and another seventeen Americans dead. The punitive expedition into Mexico was now a certainty, and Patton soon learned his regiment
wouldn't be going. His only chance was to be detailed to Pershing's staff, a
possibility that set Georgie off like a string of firecrackers, and ended with
an unannounced evening visit to Pershing's quarters.

"Everyone wants to go. Why should I favor you?"

"Because I want to go more than anyone else."

"That will do," was the General's Delphic reply. Good enough, or enough
said? He left Georgie hanging until morning, when he telephoned telling him
to pack his bags, which they already were.

The whole thing turned into a wild goose chase, as Villa and his men took
to the Sierra Madres, familiar territory for them and a maze of precipitous
canyons to the Americans. His radios malfunctioning, his observation aircraft crashed, Pershing was left virtually blind, but still relentless in his pursuit of Villa. Daily he would take his open-top Dodge on forays, sometimes
fifty miles in front of his troops. There were a few contacts, and Villa himself
was seriously wounded, but by Carranza's forces, not the Americans. While
Patton remained the general's ever-upbeat aide, he privately concluded: "I
really think that Villa bad as he is, and he is unspeakable, was the man for us

to have backed; he was the French Revolution gone wrong . . . but old Villa is Dam[n]ed hard to find."

Some of Villa's followers, Patton began to realize after about six weeks in Mexico, were more stationary. On a number of foraging missions he gradually built up enough intelligence to locate the ranch of a noted Villista near a hamlet called Saltillo. On May 14, 1916, Patton, with three automobiles filled with riflemen, turned a maize-acquisition mission there into a sudden strike on the suspect rancho. They hit a hornet's nest, but a small one, and in the ensuing firefight the Americans killed three Villistas, two of whom Patton almost certainly had a hand in dispatching. He had his men tie the bodies to the cars and brought them back to Pershing, much like a cat brings home a dead mouse.

Pershing was delighted, and well he should have been. Georgie's foray marked one of the few tangible successes in the almost yearlong and ultimately fruitless search for the elusive Pancho. It also marked the US Army's first successful use of motorized warfare, and a strong hint to Patton that his future wasn't necessarily on the back of a horse. Still, he had won his spurs with Pershing: "We have a bandit in our ranks," he uncharacteristically gushed. "This Patton boy! He's a real fighter!"

Of course, by this time Pershing was subject to another Patton-based source of bias, Nita. In March 1917, Pershing paid a weeklong visit to Lake Vineyard, relaxing, probably reluctantly talking swords with Georgie, and then asking his sister to marry him. Because war with Germany was, at this point, almost a certainty, they agreed to postpone the nuptials. For by this time Pershing also knew, in the wake of Fighting Fred Funston's fatal heart attack, that he would in all likelihood lead the American Expeditionary Forces to France. And, if not exactly at his side, at least in his company would be brother-in-law designate George S. Patton Jr. Like Douglas MacArthur's, this was a career off to an impressive start.

Also, a sanguinary one. For in Mexico Patton had joined MacArthur in the fraternity of Cain, having slayed another human in combat. In the era of purely mechanical weapons, military leadership was almost synonymous with killing adversaries, but the introduction of the gun, by increasing the range of combat and the deadliness of personal confrontations, had led to a deemphasis of personal manslaughter in favor of supervision—giving orders, planning operations, plotting strategy. So it became possible to become a great general without having killed anyone, as Napoléon proved. This would be the fate of our next two protagonists: they would never actually join murderers' row; they'd just direct it.

THE LAUNCHING OF GEORGE MARSHALL'S CAREER WAS PLAINLY LESS DAZzling than the careers of MacArthur and Patton; instead, he laid his own foundation based on dedication, competence, and the capacity to lead. Not having gone to West Point probably hindered him somewhat, at least in his own mind, but the real Army was very much open to the kinds of virtues Marshall embodied. He got the job done, and did so repeatedly. And that was noticed, even though, at times, he didn't think so.

In his first stint in the Philippines beginning in 1902, on the island of Mindoro, in the temporary absence of a superior officer, he took over the disorderly American garrison, restoring not just discipline but injected enthusiasm into his men. He also held off a devastating cholera epidemic from his troops with a rigorous sanitation campaign that left none of them dead, while hundreds of Filipinos around them perished.

He departed the Philippines in late November 1903, and immediately headed for garrison duty at Fort Reno in Oklahoma Territory, the real Wild West and apparently no place for Lily—like the Philippines. Besides, he didn't stay put. By the spring he was out on his own with a platoon of soldiers, assigned to map a sprawling mass of southwest Texas's mountains, deserts, even dry riverbeds. It was "the hardest service I ever had in the Army"—exactly the kind of miserable mission designed to test Marshall's personal integrity and professionalism. He lost thirty pounds, but took no shortcuts, and according to the Southwestern Division's chief engineer the map he turned in was "the best one received and the only complete one." As he always would, he got the job done. But it left him sapped and the recipient of four months' leave to recuperate in 1905.

He visited Uniontown only to find his old house on Main Street torn down and nobody outside the family who remembered him—only the dog of his boyhood friend Andy, after some prodding. Things were better in Virginia, where he was at last reunited with Lily. They spent time together in verdant Albemarle County, in her aunt's clapboard manse south of Charlottesville.

He loved the place, and this visit likely marked the moment George Marshall became in mind and heart truly a Virginian.

He did return to Fort Reno, and this time brought Lily with him. But he had been nearly five years in the Army, and was still a second lieutenant stuck out in the military boondocks. In an effort to get back to the Old Dominion, he wrote the superintendent of VMI in March 1906, requesting a "detail" from the Army to Lexington as a professor of military science and tactics. The school wanted an older man, which left him one last and unlikely option.

He could be chosen to attend the "Army's Schoolhouse" at Leavenworth, then called the Infantry and Cavalry School. He had applied twice before, yet despite his being first on the qualifying tests, Fort Reno had sent more-senior officers. But they didn't like the school, so the third time Marshall applied he was alone; he was chosen in what would have been his last chance, since henceforth the school would accept only captains.

He arrived an overage second lieutenant with less-than-stellar academic credentials, swimming in a sea of officers more senior and better prepared—particularly the cavalrymen, who had been prepped for a year on the course material. Nevertheless, when Marshall overheard a gaggle of students discussing who might be chosen for a prestigious second year at Leavenworth, and did not hear his own name mentioned, he reacted much as he had to brother Stuart's disparaging remarks about his prospects at VMI. He'd show them.

Though he lived with Lily and her mother in the cramped married-officer quarters, he barely saw them between attending classes and hitting the books. "I taught myself to study very, very hard. If it was a simple statement I memorized the statement . . . It was the hardest work I ever did in my life." He avoided social contact when he could, using Lily's health as an excuse to stay home and study—also an obvious reason why he and MacArthur didn't connect at Leavenworth, not that they ever would.

His focus paid off. Not only was he chosen for a second year, he ranked first in his class, a number that caught the eye of J. Franklin Bell, the Army chief of staff and former Leavenworth instructor, who attended graduation. This in turn led to a prestigious summer detail instructing National Guard units in Pennsylvania, one that revealed Marshall's career-long talent and enthusiasm for turning civilians into soldiers.

His second year at Leavenworth was easier and less competitive; plus he took and passed his examination for first lieutenant—promotion at last! Even better, he was one of five students chosen to stay on as instructors, and when the question of a lieutenant teaching captains was raised, General Bell personally gave his permission.

That may have been the best news of all—George Marshall had found a patron, a guiding hand big enough to put him on the fast track. In 1908 Bell even tried to bring George to Washington as his aide for National Guard training, only to be overruled by the assistant secretary of war, who wanted a West Pointer. Instead, George rounded out his four years at Leavenworth by editing the *Infantry Journal* and even contributing an article himself on preparing for a mapping expedition. But Marshall was no scholar. Although he always encouraged professional writing among subordinates, his strength was teaching and telling people what to do.

George had been such a success training the National Guard that, after he left Leavenworth in 1910, he bounced to and from several such short assignments, at one point being fought over by the governors of Massachusetts and Pennsylvania. In early 1913, he found himself in Texas, part of the buildup over the troubles with Mexico. But in May he learned his future lay elsewhere, way elsewhere, back to the Philippines, where General Bell was now in command.

The threat to the archipelago at this point was hardly critical, but the Philippines were so far away and so vulnerable that the War Department thought it advisable to hold maneuvers in 1914 designed to test the Army's readiness to defend against a possible Japanese invasion. To simulate such an attack, a "White Force" was to land at Batangas Bay south of Manila and march on the capital, which the "Brown Force" would try to defend. But the senior colonel whom Bell had appointed to lead White was an alcoholic, and after he failed to procure sufficient landing craft, the umpires wanted him removed from command—a career crusher.

Into the breach stepped First Lieutenant George Marshall, offering himself as a face-saving alter ego to the tipsy colonel and running the operation while keeping him in charge. It was both a noble gesture and an awesome responsibility—sole command of almost five thousand men attempting the most difficult of military operations, an amphibious landing. Once the other colonels got used to accepting orders from a lieutenant, they came to marvel at their clarity and precision. Beyond that, Marshall seemed to have an uncanny grasp of the entire situation, how all elements were progressing. Keeping his units organized and together, he managed three successive simulated battles and several skirmishes until his forces successfully reached the capital. He had worked around the clock for two weeks, and in the process revealed himself as an operational genius. One superior wrote in his efficiency report that "there are not five officers in the Army as well qualified to command a division in the field," while another ventured that he was the best leader of large troop formations in the entire US ground force.

He also collapsed. After the maneuvers he went straight to a Manila hospital, where he stayed for ten days. In each of the prior two years Marshall had experienced what was called "acute dilation of the heart," but this was different, more on the order of a nervous breakdown. When he emerged, General Bell apparently saw he was in no shape to resume his duties, and gave him two months' leave, which he had to extend for another two.

It was a watershed moment for George. Fortunately, this tour he had been able to bring Lily out to the Philippines, so she was there to help him through his recuperation. Feeling somewhat better and being George Marshall, he turned it into a busman's holiday, with a tour of Japan, Manchuria, and Korea.

But it also seems clear he was thinking about himself. He was thirty-three years old and he had turned himself into a workaholic dervish, and in the process become a master of operations. Early in the trip this George couldn't resist wiping brother Stuart's eye with a rare letter recounting how he had carried "the entire burden" of commanding the White Force and "chewed the other side up," not forgetting to mention that Lily "looks very well" and "gained a number of lbs."

But this George was also killing him, and very soon thereafter he seems to have realized he was easily capable of working himself to death, and as a reward had acquired "the reputation of being merely a pick and shovel man." Faced with the deadest of dead ends, George Marshall "woke up" and resolved to "relax as completely as I could manage in a pleasurable fashion."

Plainly he was not the first compulsive worker to make such a vow, but George Marshall was a very determined fellow, and as his responsibilities grew he only redoubled his efforts to do less, until he became one of the great delegators in American history. This amounted to the profoundest of midcourse corrections and one that enabled him to be what he eventually became—always adaptive to changing circumstances, recognizing expertise, then using it. This was even truer of Ike, but not of Douglas MacArthur or George Patton, who had already cast their masks and were as adamant as steel.

Meanwhile, career frustrations persisted. In October 1915, Marshall was again in contact with VMI, this time talking about leaving the service entirely. The next year he expected to join Pershing's expeditionary force in Mexico for the Villa hunt, but instead paid the price of patronage and was sent to the Presidio in San Francisco under General Bell. He did get his captain's bars, but soon found himself running a summer officer training camp for twelve hundred SF swells on the grounds of the swank Del Monte Hotel in Monterey, complete with champagne breaks at lunch. He took it all with good humor, and retained enough authority to be a great hit. Next, he found himself near Salt Lake City setting up a similar program at Fort Douglas, after which the commander called him "a military genius" and recommended making him "a brigadier general . . . and every day this is postponed is a loss to the Army and the nation"—military-speak hyperbole for "this is a whole lot of talent being wasted."

But Bell knew what he had in Marshall. He also knew America was being inexorably drawn into the great conflict in Europe and desperately needed trained manpower, especially officers. For now, this was the best use for him; later, when war came, he would be included and excel, though not in combat.

IF YOU LIKE FOOTBALL AND LOVE AT FIRST SIGHT, YOU MIGHT SAY DWIGHT D. Eisenhower's career got off to a rousing start—but considering this man's abilities and his destiny, probably not.

Still looking to get away apparently, Ike initially applied for duty in the

Philippines, the only man in his West Point class to do so. Instead, the Mexico problem drew him to Texas like a big magnet, though well away from the border, to San Antonio's Fort Sam Houston, at this point the largest troop concentration in America and a city in itself.

Ike's first round of duties was prophetic, and not in a good way. The small size of the regular army had necessitated calling up a number of National Guard regiments to serve on the Mexican border. Not unexpectedly, they were shipped to Texas in largely a civilian state and pretty much devoid of training. It was the job of the fort's junior officers to turn them into soldiers at nearby Camp Wilson, where they were being assembled.

Second Lieutenant Eisenhower moved over as "inspector instructor" of the Seventh Illinois, an infantry regiment made up mostly of Irish Americans none too happy to be there, led by a colonel whose political skills plainly exceeded his military competence. With his poker-honed instinct for people, Ike sized up the situation instantly. "Colonel Moriarity . . . was happy to have me, as an instructor, take over in effect the running of his regiment. I wrote all his orders, prepared reports and other official papers for his signature, and became the power behind the Irishman's throne." It was an extraordinary admission for a man who generally kept his Machiavellian side behind his smiling mask, but it also revealed some singular capabilities for any officer, let alone one so junior. Like George Marshall he was a natural at running things, and when the situation demanded, which was often as a junior officer, he could be just as subtle.

Ike's success, as a trainer at least, probably did follow the paper trail back to Fort Sam Houston, but as far as the commander of this huge base, the aforementioned Fighting Fred Funston, was concerned, Ike's reputation was based on football. Cadet Eisenhower's success coaching West Point's JVs had apparently come to the attention of the headmaster of a local prep school, the Peacock Academy, who approached him soon after his arrival with a lucrative offer to take the helm of his team. Though cash strapped, Ike was still new to his duties and politely declined, not knowing the headmaster was an old friend of Fighting Fred's.

Soon thereafter the second lieutenant was at the officers' club nursing a beer when he heard a booming voice from the bar: "Is Mr. Eisenhower in the room?" It was Fighting Fred.

"Have a drink. . . . Mr. Peabody tells me he would like you to coach his team at the academy."

"Yes sir."

"It would please me and it would be good for the Army if you accepted this offer."

"Yes, Sir," I said.

He indicated the conversation was over.

It goes unrecorded whether Ike got his drink, but he definitely coached the Peacockers to a winning season, and the next year took over the program of a Catholic school that hadn't had a victory in five years, turning them around completely and nearly taking the city championship. If Ike was looking for a reputation in the Army, he now had it, but it was on the gridiron, not the battlefield, and for the next decade orders to coach assorted Army teams would follow him from posting to posting like a chronic infection.

Also, Ike had another very good reason for not wanting to spend his afternoons coaching Peacocks—he was in love. It had happened almost as soon as he arrived at Fort Sam Houston, when the wife of a major called him over to introduce him to her friends, the Douds from Denver but summering in San Antonio. One Doud in particular, nineteen-year-old Mary Geneva, known to history as Mamie, caught his eye: "A vivacious and attractive girl, smaller than average saucy in the look about her face and her whole attitude . . . I was intrigued by her appearance." Mamie was more succinct: Ike "was just about the handsomest male I had ever seen." So, when he asked her to join him on his rounds, Mamie, who hated to walk anywhere and was wearing fashion-sadistic boots, accepted instantly. Likely it was the walk that sealed it, when each discovered the extrovert in the other, their mutual bubbling optimism and ease with people. They were a couple, from the outside at least, meant to be together.

Then there were the Douds, somewhere near polar opposites of Ike's family. John Doud made a fortune in meatpacking by the time he was forty, then decided to retire and move to Denver, where he set up a very comfortable life for himself, his wife, Elivera, and his four daughters, enjoying just about every luxury life offered in early twentieth-century America. This consumptive lifestyle included the child-rearing. Mamie and her sisters were sheltered rather than stimulated; finishing-schooled more than educated; immersed in the debutante circuit, not in the rest of humanity; and otherwise pampered in a mansion sufficient to one day become the summer White House.

All of this was very impressive to Ike. Growing up poor in the Eisenhower family had been good for him and he probably knew it, but he didn't have to like it. By contrast the Douds arrived in San Antonio in a massive touring car, which wowed car-crazy Ike, then proceeded to treat him like a son. It was not simply hard to resist; it must have seemed like destiny. Throughout Ike's life people would appear at just the right time to suit his purposes—Swede Hazlet was the first, now it was Mamie and the Douds; the chain would lead him link by link to the presidency. By way of explanation, probably the most illuminating thing that can be said is: they all liked Ike.

Mamie also loved him, and on Valentine's Day, 1916, he gave her his West Point class ring—he didn't have money for a diamond—to signify their engagement. Initially, Mr. Doud was fine with Ike as a son-in-law, provided the couple pushed the wedding off into November, when Mamie would be twenty. Then the young officer announced to the assembled family that he had been accepted for aviation training. There followed what Ike described as "a large chunk of silence." Then Mr. Doud told him: "If [Eisenhower] were so irresponsible as to want to go into the flying business just when [he] was thinking of being married, [Mr.] and Mrs. Doud would have to withdraw their consent." Ike the poker player instinctively tried to bluff his way through, but after a few days conceded Mr. Doud held the winning hand and agreed to give up aviation.

It was a significant turning point for Ike, "a decision that brought me face to face with myself." Mr. Doud was right: military aviation at this point was somewhere between dangerous and suicidal. When he was in the Philippines during the late thirties working for MacArthur, Ike took flying lessons and got his pilot's license at age forty-nine, but he was far from a brilliant student and certainly not a born flyer. So if it was not to be aviation, Ike nonetheless vowed to double down on his own career, "to perform every duty given me in the Army to the best of my ability and to do the best I could to make a creditable record" and in doing so revealed some of his best qualities: flexibility, determination, and the capacity for growth. The days of Ike the Army skeptic were over.

And he also got Mamie in the bargain. She remained in most respects an adolescent with an excellent support structure, far less ready for the rigors of army life than Beatrice Patton. But, like Ike, Mamie could change. She would almost back out of the marriage on several occasions, but always returned to redouble her efforts until she became something like the archetypical army wife and ultimately the nation's grandma.

But that was far in the future, and as 1916 progressed Mamie worried that if they waited until November, Ike would be sent to Mexico or even the war in Europe. As things turned out he would not soon leave US soil, but it was a reasonable concern since the Army had already declared itself "on a wartime footing" and canceled all but emergency leave. The couple decided to pull the nuptials back to July 1, and Ike submitted his request for the time necessary to hold the ceremonies in Denver and visit his parents in Abilene. As it bounced upward Ike's entreaty failed to convince anybody that getting married warranted emergency leave, until finally he was ordered to explain himself to the general. That was still Fighting Fred Funston, who immediately recognized Eisenhower and smiled: "All right, you may have ten days. I am not sure that this is what the War Department had in mind, but I'll take responsibility." Whether Ike expected such a result or not, he would have had to admit this was one time football paid off.

He left for the wedding with the good news that he had been promoted to first lieutenant, but upon return this made little difference in the tatty two rooms and a bath provided for the young couple in the married-officers' quarters at Fort Sam Houston. In addition to the shabby surroundings, Mamie found herself immersed in a rigid and unforgiving social hierarchy that demanded that the wife of a young officer act as an appurtenance to his career—obsequious to seniors, helpful and outgoing to collaterals. Mamie had finishing-school manners, but she didn't take the first directive as seriously as Ike wanted; as for the second, she excelled from the beginning.

Although she couldn't cook (Ike could and gradually taught her) and had trouble making a bed, she was an instant hit with the post's junior officers and their wives. Before long she and Ike were staging regular Sunday-night buffets, the genesis of the roving but ever-present Club Eisenhower. From the beginning, they were the yin and yang of conviviality, she the master of party chatter and he the impresario of hospitality, bathing every gathering with his infectious smile. From the outside the facade could hardly have been more attractive, but beneath it there were troubles as Mamie discovered the rugged nature of army life and the long absences it demanded of Ike. Reciprocal retreats to the Douds' in Denver became something of a pattern. But Mamie was no quitter. There were rough times ahead, but the couple would survive.

Ike's career was another matter. Doubtless he would have liked to slip behind enemy lines at Veracruz like MacArthur, or chased Villa around northern Mexico like Patton. But the fact was that these exploits had been engendered through connections—MacArthur's inherited from his father and Patton's acquired. Ike was without a patron. General Funston really only liked him because of football, and he would soon be dead anyway.

Besides the gridiron, Ike's only Army reputation rested on his success training the National Guardsmen of the Seventh Illinois. The Mexican border was now quiet, but as the early months of 1917 passed it became completely apparent that the United States would join the war in Europe. Supplying the Allies what they needed most—fresh manpower—would demand a huge conscript-based Army, and also the equivalent task of turning these men into soldiers. Ike had more than a knack for doing so; he was a brilliant trainer and organizer of new units. This, along with his vow to perform whatever duty the Army assigned him to the best of his abilities, would swamp his efforts to join our three other heavy hitters in the catastrophe now known as World War I.

CHAPTER 4

TOTAL WAR

The war that America entered on April 6, 1917, had already been raging for thirty-two months, during which time it had become the greatest political and military disaster since the Thirty Years' War, nearly three hundred years before. But while the results of the earlier conflict were essentially stabilizing, the consequences of World War I would send shock waves of instability and revolution across the globe, setting the conditions for its even deadlier successor, World War II.

At the root of this, to a far greater degree than most military analysts and historians seem to realize, is the logical trap noted in the introduction: that vast conscript armies made up of ordinary citizens could compensate for their military shortcomings through better armaments. Yet what had happened thus far in World War I, in particular along a vast line of trenches stretching four hundred miles from Switzerland to the North Sea, was precisely the opposite. Here on the western front the arsenals of the era—primarily high-explosives-based artillery and the machine gun—enforced a stalemate so profound that from October 1914 until March 1918, neither side was able to move the front line more than ten miles in either direction, though millions would die in the attempt. Meanwhile, at home, entire economies were captured and dedicated to feeding the pursuit with all the arms and munitions they could produce—the military-industrial component of this self-amplifying feedback loop of lethality.

While there were extant technologies potentially capable of breaking the gridlock—aircraft and tanks, specifically—none were sufficiently developed at the time to do so. More to the point, perhaps, contemporary military leaders had a great deal of trouble recognizing what was happening, persistently blaming their operational failures on their men's lack of resolution or on faulty planning and execution. On the other hand, they remained stalwartly resistant to the proposition that it was the weapons themselves that had created an impossible environment for the men in the trenches, most of whom were not suited or trained sufficiently to be there in the first place. Although officers too were victimized by the trap, dying in huge numbers, they remained its enforcers, since almost universally they could not see through it or beyond it. There was a way out of trench warfare, but the war of movement that eventually replaced it was only a vastly more lethal version of the same trap, an episodic update of what promised to be a continuing cycle of ever more destructive total war.

This may sound a bit off the trajectory of our story, but it has great relevance to the futures of our quartet of protagonists. Patton, MacArthur, Marshall, and Eisenhower were as immersed in the logic and practice of total war as any equivalent four men in history. But great change was coming, something so powerful it was destined to blow the hinges off the feedback loop inevitably dictating larger armies and more lethal weapons, and unveil a future at once far safer and more dangerous than any would have imagined in 1917. Corporately, it would fall to them to lead us across this most difficult of all strategic topographies, the transition from one military era to another. But this was far in the future, for now each would embark on his own search for no-man's-land.

There were other causal factors, but on the face of events, America had entered the war against Germany as the result of a protracted maritime crisis based, appropriately enough, on a new and unprecedented weapon, the submarine. Basically, it had begun in May 1915, when a German U-boat torpedoed and sank the munitions-laden British liner *Lusitania*, killing over 1,100, including more than 120 Americans, and ended nearly two years later with the resumption of unrestricted submarine warfare, making America's entry inevitable. Whatever his idealistic motivations, Woodrow Wilson was well advised to steer clear of the European war for so long. At this belated point of entry, one when the combatants were nearing exhaustion, American leverage, both diplomatic and on the battlefield, would be magnified. Of course, that depended on having a viable force to fight with. In the spring of 1915 the US Army was utterly unprepared for war; two years later there were at least big plans. This is where we pick up the tale of our four fast-maturing warriors, each of them scrambling to find a role in the great enterprise.

IN THE SPRING OF 1917 THE PATTON FAMILY WAS ALREADY ON THE MOVE. Thinking he might snag a job in the nation's capital, Papa headed east with the family as soon as war was declared. But first they took a detour to San Antonio to see their prospective son-in-law, and arrived just as Pershing received orders to proceed immediately to Washington. They traveled east together. While he had been told he would take a single division to France, Pershing must have suspected he would be put in charge of the whole operation.

Fighting Fred was dead, and Black Jack was the Army's only officer with experience leading an expeditionary force. The chase across Mexico was not exactly triumphal, but Pershing had performed a miserable mission and gotten the force back safely, all without public complaint. MacArthur, in his *Reminiscences*, claims it was he who first suggested Pershing to Wilson and Secretary of War Newton Baker, but the choice was obvious.

Just how obvious became clear when Pershing met with the president (his only time during the war) and the secretary in what amounted to a pro forma affair. Not only was he named commander of the American Expeditionary Forces (AEF), but given a letter of instructions drafted by Baker granting him almost complete freedom to create and engage it, subject only to the stipulation that the "forces of the United States are a separate and distinct component the identity of which must be preserved." Beyond that, Baker told him: "I shall give you only two orders. One to go and one to return."

Still back at Fort Bliss, George Patton had been told unofficially that he would be included on Pershing's staff, and on May 18 a telegram came ordering him to report to the general immediately in Washington. He arrived to learn that not only had he been promoted to captain, but the whole party would be leaving from New York in four days. He also soon found out that his place on the staff would be somewhat less than exalted: overseeing sixty-five enlisted orderlies, chauffeurs, engineers, and medical and signal personnel designated to staff Pershing's advance headquarters. It was pencil pusher's work, bound to bore Georgie, but at least he was on his way "Over There."

George Marshall was worried he wouldn't be. Almost two weeks passed after war was declared before he received orders from his patron General Bell to head east and join him on Governors Island in New York Harbor, where he was in charge of officer training. This was an important job and Bell knew nobody better at it than Marshall, who soon found himself immersed in choosing candidates and setting up training camps. For all he knew, here he might stay. He tried but narrowly missed being attached to Pershing's staff. Then, as if to compound his humiliation, on the rainy morning of May 28 duty required that he accompany Pershing to a remote dock on Governors Island where a ferry would take him and his staff to the liner *Baltic* and Europe. Later Marshall and Lily watched the group, George Patton among them, from a window in his office: "Dressed in civilian clothes, coat collars turned up . . . not an imposing group," Lily remarked. "They were such a dreadful-looking lot of men. I cannot believe they will be able to do any good in France." But at that moment, they were going and he was not.

His concern was unwarranted. General William Sibert had been designated commander of the newly formed First Infantry Division, slated to go to France immediately despite its chaotic state. Desperately needing help with organization and training, Sibert remembered Marshall's success with the San Francisco swells at the camp in Monterey, and approached General Bell about releasing him to become the division's operations officer. Knowing this was exactly where Marshall belonged, Bell immediately agreed, and on June 11 Marshall joined Sibert and his staff on a converted fruit transport headed for Europe.

Both Patton and Marshall made their boats headed east because they were connected to superiors in positions to make it happen; Ike had nobody and as a result never left. To compound matters, his manifest talents as a trainer and organizer, combined with his reaffirmed commitment to his duties, worked to keep him in place, doggedly doing his part to put together an army of over two million, but always under the assumption that his next assignment would lead him to the battlefield.

It began in 1915 back in Fort Sam Houston, when he was one of a dozen officers chosen as cadre for the new Fifty-Seventh Infantry Regiment, Ike being designated supply officer. It was the start of Eisenhower the logistician, as he quickly put together a team that effectively fed and supplied the budding regiment. He was soon promoted to captain, but so was every member of his West Point class as the Army expanded exponentially. Mostly, Eisenhower looked forward to the day the regiment was fully trained and equipped. "We were sure that we were one of the best outfits in the whole Army and were confident that we were destined for overseas duty. Instead I got a special order detaching me from the 57th Infantry and assigning me to the training camp at Fort Oglethorpe, Georgia, to be an instructor of candidate officers. This was distressing."

To say the least, and it kept happening. Ike spent three wet months at Oglethorpe, relentlessly out in the field working with his tyros, when officer training was abruptly moved to Leavenworth. Ike had time for a short Christmas visit with Mamie and their ill-fated firstborn, little Icky, before heading to Kansas for more training. This time he oversaw phys. ed. and bayonet drills, and as usual excelled. One novice wrote home: "Our new Captain is one of the most efficient and best Army officers in the country. He is a corker and has put more fight in us in three days than we got in all the previous time we were here. He is a giant . . . and at West Point was a noted football player and physical culture fiend."

So it went for Eisenhower, but he was far from giving up. Ike had always been fascinated by anything with an internal combustion engine, so when he found out Leavenworth was the home of the Army's newly established tank school, he carved time out of the calisthenics and bayonetting to take a course aimed at a more mechanized future. Or so he thought, when his initiative was rewarded in February 1918 with orders assigning him to Camp Meade, Maryland, where he was to organize and prepare the 301st Heavy Tank Battalion, slated for European service in June. In mid-March Eisenhower was told that date had been moved up to immediately and that he would be in command. Their tanks, British Mark VI "Big Willies," would be supplied when they arrived. Given the short notice, Ike thought it wise to go up to New York himself and work things out with the port authorities. "Too much depended on our walking up that gangplank for me to take a chance on a slipup anywhere."

His diligence did not go unpunished. As soon as he returned to Camp Meade Ike learned he was not going with the 301st; instead, he was going to command Camp Colt, an abandoned base on the Gettysburg battlefield, where he would establish and run the tank corps's first formal stateside training facility. "My mood was black," Ike remembered as he first arrived at the ramshackle post, once again stuck in an assignment without the possibility of danger or heroism. He had no way of knowing that he would shortly be confronted by an invisible enemy even deadlier than the western front and that the lives of thousands of men would depend on his actions. But at this point influenza was likely the last thing on Ike's mind.

As befit his charmed existence, Douglas MacArthur had already found a spot deep in the military planning process on the General Staff, then adroitly earned the favor of the decidedly unmilitary but otherwise focused and efficient new secretary of war.

MacArthur had had a hand in crafting and publicizing the National Defense Act of 1916, which set the basis for a conscript army eventually numbering in the millions, but its fruition as a viable combat entity was a year or two away. In the meantime, to reassure its hard-pressed new allies Britain and France that American belligerence meant something, the United States had to send them troops in significant numbers and do it quickly. In actual fact, in April 1917 the regular army had just a single division remotely ready for combat, a mere appetizer on the man-devouring western front. The United States needed numbers, and there was one obvious solution: activate the National Guard, which operated in all forty-eight states and whose members had at least some military training; if they weren't up to European standards, they could be readied a lot faster than raw conscripts. Yet the regular army wanted to control the whole operation, and therefore every member of the General Staff voted against activation, save one, Douglas MacArthur.

Baker called him to his office: "I agree with you in this matter. Get your cap. We are going to the White House." For over an hour MacArthur and Baker took turns pressing their arguments for activating the Guard on Woodrow Wilson. Likely they were preaching to a choir of one, who readily gave his consent, not forgetting to thank the major for his frankness.

Still, there were problems. The National Guard was intensely political, and very uneven in terms of readiness. There wasn't time to wait until all state units were trained up to a uniform standard. Which state units should be sent first? Would the others feel slighted? With all these issues pending, Baker recalled: "I disclosed my puzzle to Major MacArthur. . . . He suggested the possibility of our being able to form a division out of the surplus units from many states." It wasn't Alexander cutting the Gordian knot, but it did defuse a number of issues. Intrigued, Baker sent Mac to prepare a list of available units, and was delighted when he returned with a potential roster that extended from coast to coast, proclaiming in a characteristic burst of grandiloquence, "Fine, that will stretch over the whole country like a rainbow." The name stuck and so did MacArthur.

Because the commander of the division would necessarily be the chief of the Army's Militia Bureau, General William Abram Mann—and he was nearing retirement age—Mac advised Baker to choose a senior colonel below him with superior organizational skills, preferably from the General Staff. "I have already made my selection for that post . . . It is you," Baker responded. When MacArthur (admittedly according to MacArthur) protested that he, being a major, was ineligible, Baker replied: "You are wrong. You are now a colonel. I will sign your commission immediately. I take it you will want to be in the Engineer Corps." "No, the Infantry," Mac shot back, claiming to be thinking of his father and the Twenty-Fourth Wisconsin.

How surprised MacArthur actually was remains open to question, but it definitely marked a turning point in his career, a moment when he burst out of the pack. The promotion and transfer were technically temporary, but for Mac alone of our four, they would be the springboard for permanently high rank at a very early age. That and his combat record.

While he must have been pleased by his sudden career jolt, very likely he thought of it as a means to an end, and that was getting into the war as fast as possible. The Rainbow Division—technically the Forty-Second Infantry—would be his chariot, but first he would have to build it and do so on the fly. It was already late August before National Guard units began arriving at Camp Mills on Long Island. While forty-two states were represented among the twenty-seven thousand men eventually assembled, the most prepared and the future combat core were from Ohio, Wisconsin, New York, Alabama, and Iowa. But this was a relative matter. Even the best units at Camp Mills were ill equipped and held little idea of what combat on the western front might mean.

From the beginning, though, the leadership was far better. During the Civil War the Army had segregated regular officers from volunteer units, and it had not worked. This time would be different. Regular commanders would send out some of their best young officers to train and lead what was now an embryonic mass army, and the Rainbow being among the first, it got a rare crop destined to include two Army chiefs of staff (Mac being one), six one- and two-star generals, and both a secretary of the army and of the air force. Add to that two mayors of major cities, two governors, and the founder of the CIA. And leading the mad charge at Camp Mills to whip the Rainbow into shape was MacArthur, who, grasping what he had in this group, began to delegate responsibility until he reached the point where he ceased even holding staff meetings.

But only so much could be done with the men in the short time allotted. When Secretary Baker and the Army chief of staff came to review Rainbow's progress on September 30, they liked what they saw, until one regiment almost ran down another on the parade ground. Things apparently hadn't gotten much better on a second visit on October 7, when an entire regiment got lost on the way to the review. But there was simply no more time. October–November 1917 marked a frightening low point in the Allies' fortunes: Russia had collapsed into the hands of the Bolsheviks and was out of the war, and the Italians would lose 300,000 killed, wounded, and captured at the disastrous Battle of Caporetto. "The Americans are coming" was the only bright note in a drumbeat of failure. There could be no waiting. On October 18 MacArthur and his staff headed toward Hoboken, where the Rainbow boarded the ships that would take them across the Atlantic and into the crucible.

PERSHING'S ARRIVAL IN PARIS WAS PURELY SYMBOLIC, SINCE HE BROUGHT NO troops beyond his staff. But his stern presence was reassuring to everyone he encountered, and the fortunate phrase of one of his aides—"Lafayette, we are here"—radiated hope that the Americans were more than a mirage. But at this point, that wasn't far from the truth. And Pershing's instincts told him it never would be, unless he escaped the politics of Paris, so he quickly shifted his headquarters to Chaumont, a village about 140 miles southeast of the capital on the river Marne. Here at least he could focus on the job at hand.

Faced as he was with the Sisyphean task of building an army in the middle of a war, it's no wonder Black Jack always seemed to be in a bad mood. For one thing, there remained tremendous pressure from the Allies not to bother. If the Americans basically had no weapons (in 1917 the Army possessed fewer than 300,000 Springfield rifles and around 550 field guns), not a problem, the other Allies had plenty and were glad to arm them, which they mainly did. In return, our allies simply wanted their bodies in the trenches, as stand-ins for the huge losses suffered since 1914.

From afar, Pershing had learned to hate trench warfare, believing that it had come about more through lack of resolve than high explosives and machine gun bullets. He came to France convinced of the fantasy that American troops could be trained by their officers to fight much more aggressively in open warfare with far fewer casualties. Besides Baker's orders to keep American forces "a separate and distinct component," this was why Pershing wanted them together—he thought they would fight better. But to do that they had to be organized, trained, and made ready absolutely as fast as possible. With Russia out of the war it was obvious a lot more Germans would shortly be headed west; the Americans were the Allies' best and indeed only hope. Pershing's plan was to initially bring four divisions (the First, the Second, the Twenty-Sixth, and the Rainbow) up to speed, then insert them into the quiet Lorraine front as placeholders for more to come. At this point we rejoin our three overseas protagonists.

While Pershing was in Paris, George Patton settled into an apartment on the Champs-Élysées, and when they moved to Chaumont he bought a twelve-cylinder Packard as his personal transportation. Pershing took him as aide-de-camp to a meeting with British field marshal Douglas Haig, and in the main treated him well. He in turn performed his duties scrupulously, rigorously supervising Pershing's personal staff as it grew steadily past two hundred, and in doing so put himself in position for promotion.

But he was bored. His war so far was long hours and the sensation he felt of "a rat chewing on an oak tree." But Pershing likely had not forgotten the dead Mexicans Patton had brought him, and reassured Papa: "George is eager to get to the front when the time comes, and I shall of course give him his chance."

But that time was not coming fast enough for Georgie, so he looked to alternatives. This began when, sick with jaundice, he found himself sharing a hospital room with Colonel Fox Conner, a wise and subsequently important figure in the careers of not only Patton, but also Marshall and especially Eisenhower—though decidedly not MacArthur. Georgie being an incessant talker and Conner a true scholar of war and a notably good listener—not to mention a captive audience—they probably covered a lot of ground along the western front, and more specifically its future and Patton's. Cavalry had done nothing in this war, and it must have been obvious that even the Patton sword wouldn't help. They weighed the possibilities of tanks versus the infantry, with Conner leaning toward infantry. At any rate, Patton left the hospital with a new determination to get out from under Pershing's thumb, and with the seed of an idea how to do it.

A nexus of American training schools was springing up in the towns and hamlets around Chaumont, aimed at preparing officers and men in the various military specialties they would need in battle. After considering establishing a bayonet academy, Patton fastened on the tank. Although the US Army had yet to field its armor corps, both the British and the French had enough heavy and light tanks respectively to supply the Americans, if they chose to establish such a corps in France. To Patton the choice was obvious: tanks could be the new cavalry. He went directly to Pershing requesting that he be named head of a contemplated tank school, and then be given command of the resultant corps of tyro tankers—not forgetting to mention that he was "the only American who has ever made an attack in a motor vehicle."

That was undeniably true, but even Patton may have been surprised how willing Pershing was to let him find his future elsewhere. For by this time Black Jack's roving eye had shifted away from Nita and fastened on a twenty-three-year-old Romanian artist he had met in Paris who would end up being the love of his life. Under these circumstances, Georgie's presence at headquarters would have been a continuing embarrassment. Instead, on November 10, 1917, when Pershing signed his orders, Patton the tanker was born.

He took his new duties extremely seriously. He tested the Renault light tank his units would eventually use, driving it and firing its gun. But mostly he was interested in talking to French and British officers about tactics and why armored assaults had usually run out of steam, not forgetting to consult J. F. C. Fuller, destined to emerge as one of the chief prophets of armored warfare. Within a month, and after a near-fatal automobile accident, he produced a fifty-eight-page paper, "Light Tanks," which distilled what he had learned into two keen insights: When faced with resistance, tanks failed when they got ahead of their infantry support; but "if resistance is broken and the line pierced the tank must and will assume the role of pursuit cavalry and 'ride the enemy to death.'" Those two conclusions would pave the way for George Patton's remarkable career.

Meanwhile, he ran his tank school at Langres, twenty miles south of Chaumont, with an iron hand, but also with an irrepressible enthusiasm that spread like a virus across his neophyte armored warriors. But whatever Patton might have thought initially, Pershing was not going to let him run the AEF's tank corps; that job went to a more senior officer, Samuel Rockenbach. He had no imagination, but he proved a steadying influence on Patton and ran bureaucratic interference for him through the remainder of the war.

During the first half of 1918, Georgie worked nonstop. By March the school at Langres had evolved into a vast training facility with a small fleet of demonstrator tanks and five thousand officers and men. Yet no detail in their maintenance and preparation seemed to miss Patton's steely gaze. By April he felt confident enough of his charges to hold their first field exercise, even inviting observers from the General Staff School, also located at Langres. Held in a driving rain, it proved surprisingly successful. "I ran this show. It was the first Tank Maneuver ever held in the US Army," he added to a memorandum, ever seeking to blow his own horn. He didn't need to. Within a week he was given a temporary wartime promotion to lieutenant colonel, and command of First Light Tank Battalion. George Patton and his fledgling tankers were almost ready to rumble, or so they thought.

While he was at Langres, Patton, always interested in professional credentials, found the hours to take a course at the General Staff School. Here he crossed paths with George Marshall for the first time. Like Marshall's meeting with MacArthur, nothing much came of it, but in this case, though polar opposites in personality, these two were not necessarily incompatible, just distracted.

George Marshall and the headquarters of the First Division had set themselves up in the town of Gondrecourt, just thirty-five miles from Chaumont, probably too close. Here General Sibert and his staff faced the utterly daunting task of turning a group of hastily organized civilians into a combat-ready division. So as to generate at least some unit cohesion, Marshall insisted on drilling the men, marching them around incessantly before beginning their training in trench warfare, which is what the impatient French wanted. Sibert went along, but this consumed a precious month. It was almost September before instructors from a crack French unit known as the "Blue Devils" arrived to begin actual combat instruction. A month later it was just beginning to take hold.

This was plainly not fast enough for the nearby commander in chief. As he saw it, blocking Allied schemes to bleed American manpower like vampires demanded getting fully trained combat divisions into the field immediately if not sooner. Or so it seemed on October 3, 1917, when a very bleak Jack arrived to witness a training exercise.

When it was over, Pershing asked Sibert for a critique and, after he fumbled it, let loose with a diatribe claiming the First Division hadn't followed directives, made poor use of its time, and didn't give much evidence of training. To George Marshall, as he watched the dressing-down, the worst part was "he was very severe with General Sibert in front of all the officers." This tripped the same reflex that had caused a younger Marshall to step in for the alcoholic colonel in the Philippines.

In this case, as Pershing turned to leave, Marshall stepped forward and started talking, and then, when the general tried to brush by him, put his hand on his arm. To say this was not done entirely understates the case: for a major to physically impede a four-star general, his commander in chief, was somewhere between felonious and simply career-ending. Although George Marshall would find his way to the front only briefly and without combat, this confrontation took courage truly of the over-the-top variety.

Fortunately for him, there was another John Pershing beneath the stern, unforgiving mask of command he showed the world—that one a shrewd judge of character. Instead of having Marshall arrested, Pershing stopped in his tracks: "What have you got to say?" Marshall spewed facts and figures. Pershing let him finish and left saying: "You must appreciate the troubles we have." But even then Marshall wouldn't let him get the last word: "Yes, General, but we have them every day and they have to be solved before night."

Henceforth, when Pershing visited the division, he would usually take Marshall aside to find out how things were really going. Before January 1918 was over the First Division was taking its place in the French line, Sibert was gone, and George Marshall was a lieutenant colonel with a patron at the very top.

Douglas MacArthur, meanwhile, arrived in France bent on heroics, and the Rainbow was to be his instrument. But just as the division was setting up in Vaucouleurs, he began hearing ominous rumors from Chaumont, fifty miles to the southwest—ones threatening to upset his entire plan. In fact, after considering the magnitude of his task, General Pershing had tentatively decided to scale back his original goal of four fully trained divisions to three—the First, Second, and Twenty-Sixth—leaving the Rainbow to be cannibalized and fed battalion by battalion as replacements when and where needed. This being the case, Doug might more likely find himself at a desk filling out rosters than at the front.

Outraged, MacArthur reflexively fell back on his Washington connections, sending cables to Secretary Baker and practically anyone he thought had a stake in the Rainbow. He next beat a path to Chaumont, where he burst in on General James Harbord, a friend from the Philippines who had introduced him to Manuel Quezon and was now Pershing's chief of staff. "I asked him to come and see the division and judge himself . . . whether such a splendid unit should be relegated to a replacement status. He came and saw, and revoked the order." Mission accomplished, but at a cost. "My action was probably not in strict accord with normal procedure and it created resentment against me among certain members of Pershing's staff"—not to mention the boss himself.

While the rickety militia chief General Mann was in charge, Colonel MacArthur had basically run the division. But no more. Pershing almost immediately replaced Mann with his West Point classmate Major General Charles Menoher. But if he thought this would stifle MacArthur, he was wrong. Menoher was an excellent but conventional commander who preferred to operate from a central headquarters, the better to keep track of orders from above. If MacArthur insisted on joining his men and sharing their danger, Menoher would enlist him to keep track of what was happening along the front. It proved an excellent combination, perhaps an inevitable one given the approaching storm.

Everybody knew the Germans would be coming in the spring, their legions packed with veterans now free of the Russian front. In preparation Pershing ordered the Rainbow into the Lunéville sector of the French front in Lorraine for a last month of training in what amounted to a combat zone. During the move and afterward MacArthur found himself buried in paperwork, but gradually he worked his way free, delegating his responsibilities for operations, administration, and even intelligence to the division's excellent young leadership cadre. For he had other plans, ones not dissimilar to his adventure in Mexico.

He learned that French troops in his sector were planning a raid on German lines, and he immediately asked the reluctant sector commander, General Georges de Bazelaire, if he could join them. "I cannot fight them if I cannot see them," he argued with an irrefutable logic that overcame the Frenchman's qualms.

On the night of February 26, 1918, Douglas greeted his Gallic companions, darkly dressed and their faces blackened, with a fashion statement: he arrived clad in his West Point letter sweater, topped with an officer's cap minus the steel liner so as to allow a rakish angle, and, rather than a gas mask and a pistol, carried a riding crop. Tally-ho!

The raid quickly turned serious when a German guard spotted them: "Flares soared and machine guns rattled. Enemy artillery lay down a barrage in front of the lines. . . . But the raid went on. . . . The fight was savage and merciless. . . . Finally, a grenade, tossed into a dugout . . . ended it." They returned with a pack of prisoners, including a German colonel Mac could be seen prodding forward with his riding crop. It was another storybook combat episode; while he no longer gunned down his adversaries, this time there were lots of witnesses to his courage. The French raiders crowded around, congratulating him; General Bazelaire further decorated his letter sweater with a Croix de Guerre, and back at headquarters Menoher—though it's uncertain if he knew about the raid in advance—still liked the results and awarded him a Silver Star for gallantry.

He also received the general's permission to join a brigade of Iowans in a much larger and more blatant raid set for the night of March 9. Too blatant, it seems, since the Germans anticipated it with an artillery barrage of their own. Amid the tumult, Iowans hunkered down in their trenches looked up to see a tall figure clad in a letter sweater and a squashed officer's cap calmly giving instructions to their commanders. Soon after, once the German barrage was replaced by French suppression fire, Mac mounted a scaling ladder and "went over the top as fast as I could and scrambled forward. . . . For a dozen terrible seconds I felt they were not following me. But then. . . . They were around me, ahead of me, a roaring avalanche of glittering steel and cursing men. We carried the enemy position." Menoher had little doubt who was primarily responsible, and this time Mac was awarded the Distinguished Service Cross, the Army's second-highest award for bravery.

Also, his reputation was spreading over the Rainbow and across the AEF; among doughboys he was alternatively the "d'Artagnan or the Beau Brummel of the AEF," or more simply "the fighting Dude," one credited with a sixth sense that granted him a charmed life in battle. Bulletproof he may have been (besides George Washington, it's hard to think of anyone more so), but as if to show there was no immortality along the western front, on March 11 Mac caught a face full of mustard gas, putting him in the hospital and threatening his eyesight.

But not enough to keep him from unpeeling his bandages eight days later and accompanying Secretary Baker on a tour of the front. A stray German shell allowed the secretary to revel in having come under enemy fire, while Mac presented him with a Bavarian officer's helmet and kept him entertained with tales of his raids. Later, Baker told a group of war correspondents that MacArthur was "the most brilliant young officer in the army," though by this time back in Chaumont they had tagged him "the show off." But all of this must have sounded very parochial two days after the secretary left for home and the Germans arrived.

This would be their last roll of the dice. The Grosser Generalstab and its chief, Erich Ludendorff, were characteristically both obtuse and thorough in their planning, convinced that new tactics based on elite "Stormtroops" and further intensified artillery barrages would create a breakthrough—in other words, more of what had not worked in the first place.

On March 21, 1918, more than four thousand pieces of artillery erupted from the German lines, soon to be followed by seventy-six divisions' worth of infantry. Within an hour they had opened up a twelve-mile gap in the British Fifth Army in the northern sector; within a week they had driven them back fifty miles, and almost half of Field Marshal Haig's sixty divisions were no longer combat effective. Meanwhile, the French were under equivalently heavy attack. Before the week passed they had been pushed over the river Somme and back forty miles west, where the front was now within striking distance of Paris—literally, since the city found itself under attack from what turned out to be a battery of giant Krupp cannons with ranges in excess of eighty miles, a chilling prophesy of how far ballistic projectiles could fly once free of the atmosphere.

But now the artillery threat to Paris was of little more than psychological import, compared with the strategic threat of the Allies being cut in half—the British Army rolled up to the North Sea and the French left to defend the capital. But the Allies reacted swiftly with a key choice. On March 26, they named French Marshal Foch supreme commander, enabling him to send the most-capable units where they were most needed regardless of nationality and leading the German attack to grind to a halt along a static but now fiercely contested front. George Marshall would never forget the ultimate impact of this move, but much more important for now was General Pershing's decision to get with the spirit of things by offering up his four divisions—provided, of course, they were not dismantled. Thus the Second and the Third Division (including two regiments of Marines) pivoted into the French sector, where they would eventually launch bloody counterattacks at Belleau Wood and Château-Thierry. The First Division, now considered fully combat ready, would join the British front at the point of greatest German westward penetration, near Cantigny.

As for the Rainbow, Pershing advised that they were still not ready to face the brunt of the Germans, so instead, on March 31, the Forty-Second was sent to the line around the town of Baccarat, relieving three French divisions being rushed to defend Paris. Here they would remain until June 21, the subject of no major offensives but still immersed in a continuous pattern of raids and counterraids involving anywhere from a few dozen to several hundred soldiers. "For eighty two days," Mac remembered, "the division was in almost constant combat," during which time its men suffered nearly two thousand casualties. Mac, as always, remained bulletproof, but he participated whenever he could, with orders or without, as did some of his subordinates, particularly the redoubtable future founder of the CIA, William Donovan.

This was fine with General Menoher, who looked the other way and assessed Mac to be "a most brilliant officer," maintaining that his "excellent staff work" had left the Rainbow "a complete, compact, cohesive, single unit which ran like a well-oiled machine." In other words, ready for the toughest combat. It seemed that Pershing now believed him when orders arrived on June 16 directing the division to board trains headed for the Champagne front, where they would join the French Fourth Army for the great combined Allied counteroffensive.

Five days later, just as they were leaving, Pershing showed up on the loading dock, with the Rainbow in the midst of packing pandemonium. The general walked directly up to MacArthur, who was supervising the effort: "This division is a disgrace. The men are poorly disciplined and they are not properly trained. The whole outfit is just about the worst I've ever seen. They're a filthy rabble. . . . MacArthur," he finally growled, "I'm going to hold you personally responsible for getting discipline and order into this division." He wasn't kidding. Five days later an envelope arrived from Chaumont containing Doug's promotion to brigadier general, signed John J. Pershing. As a leader, the commander in chief was a blunt instrument—first the stick and then the carrot. In his view MacArthur needed both, for his war was just beginning,

Meanwhile, George Marshall had been having his own and only brush with the front. By mid-May 1918, the twenty-eight thousand men of the US First Division had taken their place in the lines facing the village of Cantigny, and were being treated to a diet of sporadic German artillery fire. This being the point of greatest enemy penetration, it made sense to Marshall, as the division's operational planner, that orders would soon come to deliver a whack to this protrusion. Nightly, just before dawn, Marshall took to creeping around the no-man's-land in front of Cantigny, committing it to memory. So when orders did come to assault the village, he had a detailed understanding of the terrain and a Silver Star for his efforts. This was to be the first purely American mass offensive action, and Marshall's job was to plan it down to the last detail and then issue orders that exactly reflected all those details.

Even at this stage Pershing's vision of open warfare was recognized as a mirage. In Marshall's eyes, this operation had a "strictly limited objective": a single regiment of 3,700 men would advance along a 2,200-yard front, capture Cantigny, then push the line around a mile further eastward.

Apart from Marshall breaking his ankle when his horse fell on him, the assault went pretty much as he had planned. On May 28, the village fell to the Americans within an hour. "The success of this phase of the operation was so complete and the list of casualties so small that everyone was enthusiastic and delighted," he later wrote.

But then larger forces entered the picture to unhinge things. The day before the attack, word had arrived of a massive German assault spilling off the Chemin des Dames ridge to attack Château-Thierry and directly threaten Paris, less than sixty miles away. This being strategically far more important than Cantigny, it would henceforth bleed off resources from both sides, leaving the local issue in doubt. First to go was the French heavy artillery assigned to degrade counterattacks on Cantigny and above all to suppress German big guns. Instead, Marshall wrote, US doughboys were on the receiving end of a "continuous bombardment by 210-mm guns. . . . [A] 3-inch shell will temporarily scare or deter a man; a 6-inch will shock him, but an 8-inch shell, such as these 210-mm ones, rips up the nervous system of every one within a hundred yards of the explosion."

Nevertheless, the Americans managed to beat off a big German afternoon counterattack, and then, their lines reinforced by two more regiments, do the same in a second and third day of assaults. At this point the Germans realized their manpower was better spent on the more important push to the south, and withdrew.

A tiny village had been captured and then held, yet the cost was 300 Americans killed and 1,300 wounded. Marshall, who had sat through the action at headquarters in a cast, maintained that such losses "were not justified by the importance of the position itself, but were many times justified" by the positive jolt to "the morale of the English and French Armies." True enough, the Americans had shown they could fight, and within a month or so they would be arriving in France at the rate of 250,000 a month, a number that spelled doom for the Germans. But Marshall was wrong about the cost-benefit equation: such losses for equivalently insignificant places were the norm, not the exception, along the western front. Like the Rainbow, the First Division was headed for a still-bloodier future, but George Marshall, though he asked for a troop command, was going elsewhere. On July 12, he got orders to report to Fox Conner, chief of the Operations Section of General Pershing's headquarters at Chaumont.

SO IT WAS THAT OUR PROTAGONISTS WERE SET IN PLACE FOR WHAT WOULD BE World War I's final act: Douglas MacArthur facing the last of the German offensive outbursts; Patton with his tanks raring to go; Marshall at headquarters exercising his genius for giving orders on a gigantic scale; Ike back at Camp Colt about to face an enemy deadlier than even the most combat-tested brigade. All would rise to the occasion, each in his own very distinctive way.

MacArthur did it with what was becoming his patented blend of heroics and inspired leadership. On July 15, Ludendorff launched one last push, a dual convulsion: fifty divisions against the British in Flanders, preceded by forty-seven divisions along the Champagne front in a two-pronged attack, one headed south of Reims and the other toward Châlons-sur-Marne—exactly the point in the line the Rainbow now occupied.

But the French knew the entire plan from captured prisoners, and launched a devastating preliminary barrage as enemy infantry massed, followed by an even greater outpouring of both American and French shells as they groped their way across no-man's-land. "Their legs are broken," in the words of a French corps commander, which Mac began spreading to his own troops. Soon after, whistles sounded and dauntless Doug, as always in his distinctive battle garb, led the first wave of counterattacks that by nightfall left the Germans exactly where they had begun—except there were a lot less of them. But the Americans had also lost 750 men, more than MacArthur's father ever had in a single battle. For his part, the son was awarded a second Silver Star, but a few nights later in a bar in Châlons-sur-Marne to celebrate the victory, he found himself unable to get into the spirit of things. "It may have been the vision of those writhing bodies hanging from the barbed wire or the stench of dead flesh still in my nostrils. Perhaps I was just getting old;

somehow, I had forgotten how to play." Yet he was far from giving up his chosen profession, or its quintessential activity.

With the Germans definitively on the defensive, Foch moved quickly into attack mode, his immediate aim being to flatten the so-called Marne salient, the forty-five-mile bulge Ludendorff's spring offensive had yielded. The corps commander above the Rainbow, General Hunter Liggett, wasn't happy with his Twenty-Sixth Division, and replaced it with the Forty-Second for the coming operation. So, rather than being held in reserve, MacArthur and his comrades found themselves almost immediately thrown back into combat.

Specifically, they were ordered to cross the river Ourcq—a stream, really—down one long, sloping field and up another, easily done if there were no Germans on the opposing heights, a death trap if there were.

There were multiple enemy divisions and plenty of artillery waiting in the early-morning hours of July 28. The Rainbow's fire support was not responding; nevertheless, its units were ordered forward. MacArthur—as usual close enough to the action to feel stray bullets whizzing by—watched as three American battalions came under a storm of fire, causing two to drop back, leaving only William Donovan's, now trapped on the ridge with Germans shooting at them from three sides. Here they stayed, stubbornly resisting for three days until rescued, having sustained six hundred casualties out of one thousand men and won "Wild Bill" a Medal of Honor.

Meanwhile, MacArthur had been leading the up-and-down, forward-and-back ordeal to take the crest, which won him a third Silver Star and eventually devolved into savage hand-to-hand combat. On July 31 the situation seemed to stabilize, and General Menoher, dissatisfied with the support the troops had been getting, gave MacArthur direct command of the Eighty-Fourth Brigade (half of the Forty-Second's infantry). Being Doug, he marked the added responsibility with a nighttime adventure.

A deserter claimed the Germans were withdrawing, but days passed and there were no clear signs. Finally, during the early hours of August 2, Mac and an aide sneaked out into no-man's-land looking for clues. "The dead were so thick in spots we tumbled over them. There must have been at least 2,000 of those sprawled bodies. I identified the insignia of six of the best German divisions. The stench was suffocating." It was clear now, besides a few snipers and rearguard artillery, that this was all that was left of the Germans—a point driven home on the way back when a flare revealed a machine gun nest right in front of them, apparently ready to fire. Mac and the aide went flat on the ground, waiting interminable seconds for a burst that never came. Those Germans too were already dead.

It was almost dawn before MacArthur reached headquarters, where he found both Liggett and Menoher and reported the news. He hadn't slept in four days, so when they offered him a chair he promptly passed out. "Well I'll be damned! Menoher, you better cite him"—not for sleeping on duty, but for a fourth Silver Star.

The Germans were gone and the Battle of the River Ourcq won, but the Rainbow had suffered 6,500 casualties, with MacArthur's Eighty-Fourth having lost more than half of its effectives. Horrific bloodletting for insignificant objectives—that had been the norm on the western front for four years. But this actually was different. What MacArthur saw in microcosm—the insignia of six different divisions in such a small space being a sure sign of disorganization—was happening everywhere in the German ranks.

Combat experience is vital in war, but it's not everything and it can be easily overdone. These German troops were tired. Many had been in the trenches for four years. The spring offensive had drained them both in numbers and in spirit. They would continue to fight skillfully and obey orders almost to the last (unlike the High Seas Fleet, which almost never fought and went into open revolt), but at this point they must have known that facing a sea of feisty Americans they had no chance. Pershing's network of combat training centers was now fully operational and mass-producing doughboys ready to fight. Black Jack had turned the corner; he now had real soldiers he could move around and use to punch a hole in the Hindenburg Line or whatever defenses the enemy was preparing. From this point, as he saw it, Chaumont would call all the shots.

Mac saw things differently in early August, as the Rainbow finally had a chance to lick its wounds in reserve. To him both he and his men were being abused. Back home, an August 3 *New York Times* article reported that "it was officially learned today" that Brigadier General MacArthur would be ordered home to train and command a new brigade in Maryland. To him, the only source could have been Chaumont, and he sent a furious protest in that direction. But it was General Menoher's intercession that mattered; reminding headquarters that his men "are devoted to him" and MacArthur was "the source of the greatest possible inspiration," he succeeded in getting the decision overturned. Mac did nothing to smooth things over when he added a fuchsia scarf to his combat duds, or when he blew off a Chaumont-inspired ten-day training exercise for the now-veteran Rainbow, and instead began signing forty-eight-hour passes to Paris. But his men loved him for it, and apparently the Rainbow was too valuable to tamper with, even to Pershing's jaundiced eye. For he and his colleagues at Chaumont had big plans.

George Marshall spent most of August working on a scheme to eliminate the Saint-Mihiel salient, a bulge created in 1914, contested by the French in 1915, then pretty much ignored. The Germans built substantial defenses, but manned them with old and substandard troops, and now planned to withdraw if attacked. But Marshall didn't know that and painstakingly put together a crusher, sending all sixteen divisions of Pershing's newly organized First Army into the salient, and then on to the strategic city of Metz, the rail center that kept most of the German army supplied.

All went smoothly until September 8 or 9—Marshall lost track of time—when Pershing's new chief of staff, the appropriately named General Hugh Drum, called him into his office for modified instructions from Fox Conner. While the September 12 date of the Saint-Mihiel offensive hadn't been changed, he would now plan it for only seven divisions to simply straighten the salient, and instead of driving on Metz they would join the rest of the First Army for a secret march sixty miles northwest to mount a much larger attack along the Meuse River–Argonne Forest corridor on September 26 as part of Foch's united Allied offensive aimed at ending the war. Easier said than done—by an order of magnitude. He had to organize and come up with orders sufficient to move 600,000 men, 93,000 horses, 2,700 artillery pieces, along with 900,000 tons of ammunition and supplies to the new objective in less than two weeks—everything done at night to preserve secrecy.

His response was classic George Marshall. In less than an hour he turned in a coherent plan to move First Army into its assembly points in the Meuse-Argonne, while organizing the defense of the anticipated gains at Saint-Mihiel. Pershing called it "a fine piece of work"—high praise from Black Jack—and Drum labeled it "a dandy." No doubt Conner approved also. All three at this point must have realized that beneath Marshall's outline was a prodigy's understanding of what was required—a barrage of orders so clear and coherent that they could be replicated downward until the entire First Army knew what to do and when to do it. And if the devil is always in the details, it appears George Marshall was ready for him—ready enough to try to sleep through the artillery extravaganza that opened the Saint-Mihiel operation in the early hours of September 12. That's where we'll leave him, presumably tossing and turning, while we have a look at what sort of a day George Patton and Douglas MacArthur had at the front.

FOR A WHILE PATTON'S CENTRAL PROBLEM WAS GETTING HIS TANK BRIGADE into tanks. Since the American armored vehicle program had yet to manufacture its first tank, the British offered 150 heavy models to be shipped from England, while the French promised 144 of their own light tanks. Yet the English heavies were destined to arrive too late, and June had passed before all the Renaults had been delivered.

That left the problem of getting tanks to battle. It was almost September before all of them were at the main assembly point, at Bourg; then they had to be moved to the front by special rail, a logistical weight lift that ended only two hours before the operation began.

Since the Saint-Mihiel plan had been dramatically modified, George Patton learned only at the last minute that his tanks would operate in support of the First and Forty-Second Divisions. But the Rainbow had never trained with tanks, and showed it when they rejected Patton's reasonable request that a protective smoke screen be included as part of the initial artillery barrage, a decision that required the personal intervention of his commander General Rockenbach to get reversed.

Meanwhile, through all these travails Patton remained his usual hyperactive, hyperbolic self, finally accelerating into full bloodthirsty mode in his final prebattle orders: "If your gun is disabled use your pistols and squash the enemy with your tracks . . . remember you are the first American tanks. . . . AMERICAN TANKS DO NOT SURRENDER."

With these motivational thoughts he sent them off, and 5:00 a.m. found Patton sitting on a hill, watching his tanks move forward, he being under strict orders from Rockenbach to maintain telephonic contact with headquarters. As the light got better it became clear the Germans were falling back, which had allowed some of his tanks to advance far beyond the infantry they were supposed to be protecting. Even worse, he could see that one by one his Renaults were getting trapped in the rain-soaked German trenches, some fourteen feet wide. His telephone wire was fully unspooled; there was no way he could get closer unless he disobeyed orders. Leaving his adjutant to tend the phone, and taking a lieutenant and four runners, he headed toward the action, surrendering, as he always would, to the lure of battle.

In the meantime, after an extra-heavy barrage Mac had arranged for with the Rainbow's artillerists (along with the promise not to tell headquarters), the men of the Eighty-Fourth Brigade went over the top and cautiously moved forward to find only the sporadic resistance the Germans put up when they were retreating.

Close behind them, suffering from a raging fever, was Brigadier General Douglas MacArthur. Although the American opening barrage had caught much of the retreating Germans' artillery on the road and destroyed it, some remained in place to slow down the advancing Americans. Consequently, when a burst thundered close over their heads, the brigade took shelter in the ample number of shell holes the battlefield provided—everybody except Mac, who stood alone on a little hill gazing at the enemy lines.

It was at this point that George Patton arrived. "I joined him and the creeping barrage came along toward us," he wrote Papa. "Each one of us wanted to leave but each hated to say, so we let it come over us." MacArthur purportedly claimed Patton flinched, and told him, "Don't worry, Colonel, you never hear the one that gets you." True or not, no one disputes the incident happened, and this in itself reveals an interesting juxtaposition of personas.

Ever since the proliferation of firearms made it inadvisable for combat leaders to close within sword range, courage on the battlefield, at least among the officer class, has come to be defined essentially by ignoring the bullets. That's exactly what both did when an entire brigade didn't. Both were sneering at the possibility of sudden death when sensible souls took cover. If one flinched, it was likely because of a lifetime of getting hurt doing dangerous things. If the other didn't, it's equally attributable to a path of exploits without injury. Who was the braver: the one who took risks knowing full well what the consequences felt like; or the one who took them convinced there would be no consequences? It's a question without an answer, but still useful in helping us define each man and the choices he made.

As for the battle itself, Saint-Mihiel was a walkover, taking barely thirty-six hours and producing 15,000 German prisoners and 257 enemy artillery pieces, at a cost of 13,000 American casualties, now considered a pittance on the western front. As the fighting drew to a close, despite the loss of forty of their number, the US tank brigade was still advancing, a number of tanks personally led by Patton, a fact that infuriated Rockenbach and cost George any battlefield decorations. But nothing more in terms of consequences; nor did Patton have any intention of changing. Leaders led from the front: on this our two paladins, Patton and MacArthur, would always agree.

Although Mac did receive his fifth Silver Star for the action, he too was in the process of enraging the boss. The Rainbow had halted less than fifteen miles from Metz, its church spires within sight. That night Mac and his adjutant sneaked into the outskirts of the city and could find no sign of German troops. "Here was an unparalleled opportunity to break the Hindenburg Line at its pivotal point," he later wrote. "Victory at Metz would cut the great lines of communication and supply behind the German front, and might bring the war to a quick close."

Still feverish, he arrived at First Army headquarters with the mindset of an evangelist. Yet, he had a point, and George Marshall, who had included Metz in his original plan, was sympathetic, but quietly so. MacArthur's manner—statements like "the president will make you a field marshal"—served only to enrage Pershing, who had volunteered First Army to take up the Meuse-Argonne sector in Foch's combined offensive. What were the Allies supposed to do if he headed off in another direction? But he said nothing, only finally growling, "Get out! And stay out!" The Meuse-Argonne it would be.

The problem was he had picked the strongest point in the Hindenburg Line, a natural fortress protected by two rivers and deep forests filled with precipitous ridges, turned by Germans over four years into an attacker's nightmare with hundreds of machine gun nests and mortar emplacements, along with mile upon mile of layered barbed wire, all of it meticulously placed to do the most harm. The Germans could not afford to lose this place, as it shielded the Sedan-Mézierès rail link, which was the escape route for much of their army.

George Marshall's outpouring of orders effectively managed the massive but surreptitious movement of the giant First Army and all that went with it into their new positions stretched out over a twenty-four-mile front, but they said nothing about what came next. That was Black Jack's job, and as he was still captivated by the war-of-movement fantasy, in essence he roared, "Charge!"

On the misty morning of September 26, he sent nine fresh divisions, supported by 2,700 guns, 189 tanks, and 821 aircraft, storming forward. National Guard captain Harry Truman wrote home that it seemed "as though every gun in France was turned loose." The Germans were truly surprised and initially fell back, but then the strength of their position on the Montfaucon heights took hold, and things began to go wrong for the Americans.

It was going to be an eventful and nearly fatal experience for Lieutenant Colonel George Patton. He resisted the urge to move forward with his tanks until around 6:30 a.m., when he left with just a few staff looking to catch up. But there was a dense fog, and when it lifted he found he was 125 yards beyond his own tanks, which were now the targets of intense German fire from seemingly every direction. Back among them he realized the infantry support they needed to move forward was well behind, mostly taking cover wherever they could find it. "Some put on gas masks, some covered their face with their hands but none did a dam[n]ed thing to kill Bosch," he wrote Beatrice after it was over. "There were no officers there but me."

So he assumed the role of warrior shepherd, gathering a flock of doughboys willing to believe he knew what he was doing, or at least afraid of what he might do if crossed: "I think I killed one man here[;] he would not work so I hit him over the head with a shovel. . . . At last we got five tanks across and I started them forward and yelled and cussed and waved my stick and said come on. [A]bout 150 doughboys started but when we got to the crest of the hill fire got fierce right along the ground. We all lay down." It was at this point, Patton later told his daughter Ruth Ellen, that his ancestors arrived: "There were other faces, different uniforms, dimmer in the distance, but all with a family look. They were all looking at him, impersonally but as if they were waiting for him. He knew what he had to do and continued the tank action."

Patton leaped to his feet yelling, "Who will follow?" and when only six, including his orderly Private Joe Angelo, were willing, he still launched his charge against the tormenting machine gun nests. They didn't get far before Patton and Angelo were the only ones not gunned down. Then a slug smashed into Patton's leg, sending him cartwheeling to the ground. He told Beatrice the bullet "came out just at the crack of my bottom about two inches to the left of my rectum. It was fired at about 50 m so made a hole about the size of a dollar."

Angelo managed to drag him to a shell hole and bandage the wound. Here they lay for two hours, trapped. Fortunately, they had a means of keeping the enemy machine gunners' heads down, "one of my tanks guarding me like a watch dog." Eventually, more would come, along with better-led infantry sufficient to remove the nests and take the village ahead, so Patton could be evacuated.

On the way to the hospital he insisted on being taken to a division command post to provide his blood-soaked view of the situation at the front. Back in Mexico a gasoline lamp had exploded in his face, yet he had healed without scars. Now after surgery he was told that miraculously not only were no vital organs hit, but "the Dr. says that he can't see how the bullet went where it did without crippeling [sic] me for life." Patton healed fast, but this was still a serious wound, and by the time he returned to duty the armistice was only days away. He would be promoted to full colonel and awarded the Distinguished Service Cross "for extraordinary heroism" (so would Angelo), along with the Distinguished Service Medal for organizing the tank force, but George Patton's war was over.

Douglas MacArthur's would continue. Because of their exertions at Saint-Mihiel the men of the Forty-Second were not given an assault role initially at Meuse-Argonne, and were told instead to provide fire support and conduct raids, which, inevitably, MacArthur joined to win a sixth Silver Star. But that proved light duty compared with what was to come.

The doughboys finally drove the Germans off Montfaucon, but overall the initial phases of combat in the Argonne left Pershing's operational objectives blocked and his infantry hugging the ground. He had planned to move through the forest and then flow around and isolate the seven-hundred-foot Romagne Heights, which the Germans had fortified with the Kriemhilde Stellung, a defensive complex named after one of Brunhild's sisters and just as stout. But in the face of withering fire, his inexperienced combatants either ran to the rear or took cover, frequently unsuccessfully on both counts, since casualties were already horrific. Those few who moved forward were cut off and surrounded, such as the famous Lost Battalion, which had to hold out against constant pummeling without food or water for five days before being rescued, half the survivors carried out on stretchers.

Not a man to back away from a bad bet, Black Jack decided to double down on everything: this time by taking the Romagne Heights by direct assault, and doing it with seasoned troops. This is where the Rainbow came in, specifically on October 11, when it took its place in the line at the foot of the Romagne Heights, replacing First Division, which had already given its all attacking Kriemhilde's twin breasts, Hill 288 and the Côte de Châtillon, with no success. That night an unexpected visitor showed up at MacArthur's headquarters in an abandoned farmhouse; it was Major General Charles Summerall, his new corps commander. Mac offered him a cup of coffee but he was in no mood for socializing: "Give Me Chatillon, MacArthur. Give me Chatillon, or a list of five thousand casualties."

"All right, General, we'll take it or my name will head the list."

At dawn on October 14 both brigades of the Rainbow headed up the hill and were promptly greeted by a storm of projectiles. It was worse for the Eighty-Third, which was shot to pieces getting a tenuous hold on Côte de Châtillon, only to be pushed off the next day by a ferocious counterattack and its commander relieved by Summerall. Under equivalent fire the Eighty-Fourth, with MacArthur in the first wave, was able to drive the enemy off of most of Hill 288, but after five attacks the next day some still remained. Meanwhile, Summerall phoned saying that he wanted Châtillon by sundown the next day, and Mac could do nothing more than reassure him it would be done, raising the casualty ante to six thousand with his name still at the top. When the call was over he looked up at his staff: "Any ideas?"

In sheer desperation, he had been planning for the worst of ideas, a bayonet charge, but then an aerial photograph provided by Colonel Billy Mitchell's wing seemed to show a gap in the barbed wire to the northeast.

MacArthur reasoned that he might be able to get a battalion around the rear into that gap if he launched a diversionary attack in front, complemented by a massive machine gun fusillade to keep German heads down. Yet it could work only if the break in the wire really did exist.

That night he took a patrol to creep around in the dark until he could confirm the gap was real, but then the Germans let loose with artillery and shells began bursting all around them. Everybody dove for the nearest shell hole. When it was over, MacArthur crawled out and, moving from hole to hole, whispered to the prone figures: "Follow me, we're going back to the Rainbow lines." But none responded. They couldn't. They were all dead. Only Mac made it back. "It was God," he later told a friend. "He led me by the hand, the way he led Joshua."

Be that as it may, MacArthur's own Battle of Jericho was launched with a less-than-heavenly chorus of sixty machine guns sweeping German positions with more than a million rounds. Then began the dirty work. "The two battalions, like the arms of a relentless pincer, closed in from both sides," he wrote later. "Officers fell and sergeants leaped to the command. Companies dwindled to platoons. At the end Major Ross had only 300 men and 6 officers out of 1450 men and 25 officers. That is the way the Cote de Chatillon fell."

It was a high price to pay, but it did puncture the Hindenburg Line at a key point and opened the way to Sedan and apparent oblivion for the German army. Yet the larger strategic picture spoke otherwise. It was the total, near-simultaneous Allied bludgeoning across the entire front that finally seems to have pulverized the will of the German army and turned it into a crowd. For this the cost was much higher; the American share at Meuse-Argonne was 26,000 killed and a total of 120,000 casualties, some of the worst battle losses in US history. They also made the subsequent race to secure the Sedan rail link rather ironic, since there was no longer really a German army to evacuate.

On November 5, 1918, six days before the war's end, Fox Conner walked into the office of now Colonel George Marshall and dictated the following message: "General Pershing desires that the honor of entering Sedan should fall to the First American Army. He has every confidence that the troops of I Corps, assisted on their right by V Corps, will enable him to realize his desire." Marshall was skeptical—he knew Sedan was the site of a disastrous French defeat in the Franco-Prussian War of 1870, and strongly suspected French general Ferdinand Foch would want to get there first as a matter of national pride. But when questioned, Conner replied: "That is an order of the Commander in Chief. . . . Now get it out as quickly as possible." Still stalling, Marshall asked to have Hugh Drum look at the order, but he too approved it, adding the pregnant words for US units: "Boundaries will not be considered binding." And so began the antic race to Sedan.

The start found both Douglas MacArthur and the Rainbow close to running on empty; casualties had been brutal, but morale among the men remained high, and Menoher had recommended his wunderkind be given a major general's second star and also the Medal of Honor. So they lurched forward, only to find that in this race the lanes were far from clearly marked. The Rainbow and the First Division, the latter now known as the Big Red One, were angled to collide, and since boundaries were not to be considered binding, that's exactly what happened. Realizing the danger of unidentified friendly troops getting in each other's line of fire, Mac rushed forward with an aide, looking for someone from First Division.

It didn't take long before they met a patrol led by a young lieutenant who, after scrutinizing the letter sweater, the scarf, and the riding crop, decided they were Germans, pulled his pistol, and told them to march down the

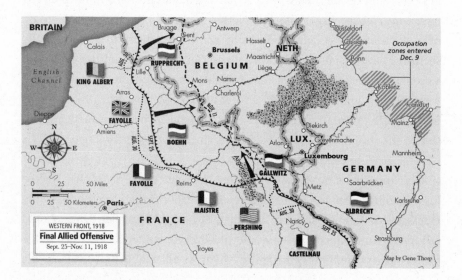

road. Ever cool under the prospect of gunfire, Mac adopted his most sooth-ing tone to explain why word of the colliding divisions had to be passed for-ward fast. It worked, of course, but just as they were getting the traffic jam unsnarled, word came that on the ominous eleventh hour of the eleventh day of the eleventh month the war would end. Death-defying was over for a while; they could all go home.

Except Ike, who was already there, but still went through an episode ev-ery bit as dangerous as any to be found Over There. The initial cases of what came to be known as the Spanish flu were actually observed in March 1918, at Camp Funston, Kansas, and were brought to Europe by the AEF; it then rebounded back through the port of Boston, and almost immediately, on September 8, 1918, the first case was reported at nearby Fort Devens. Days later 124 soldiers from Devens were transferred to Camp Colt. Some arrived sick, but they were initially diagnosed with the aftereffects from typhoid shots, until, once it was too late, Spanish flu was recognized.

Suddenly, without any warning, twenty-seven-year-old Dwight Eisen-hower, now in command of 10,600 men, found himself on the cusp of an epidemic destined to kill over 50,000 soldiers—just about as many as died in combat—along with 625,000 other Americans. Without a trace of medical training himself, Ike nonetheless already knew how to lean on subordinates, in this case his chief surgeon, and use them to maximum advantage.

Almost immediately a strict program of isolation and quarantine was instituted, every soldier was given a daily medical exam, and anyone with the slightest symptom was hospitalized. Still, the virulence of the disease shocked Ike. One day, he mentioned to one of his medical staff the vigor of a soldier unloading a truck. "The following evening," he later wrote, "the doctor came in and said: 'That man you thought looked so well yesterday morning is dead this evening.'" Another time a stampede of nervous soldiers erupted when a patient mistakenly placed in the morgue tent came to and wandered out naked, muttering: "Get me out of here."

He was right to be worried: within a week 175 in Ike's command were dead. Yet here the avalanche stopped; his soldiers continued to sicken un-til mid-October, but nobody died. When the War Department learned this, they ordered Eisenhower to send out a team of his doctors to explain exactly what measures had been taken to keep the death rate so low. He was also promoted to lieutenant colonel, one of only fifteen members of his West Point class to climb this high during World War I. By the Army's own metrics, he was a success. But not according to his own. "I suppose we'll spend the rest of our lives explaining why we didn't get into this war," he commiserated with a fellow officer on November 11, but being the new, dedicated Ike, he added: "By God, from now on I am cutting myself a swath that will make up for this." That he did, in the end one even wider than those of the other three, who had made it to France.

FOUGHT UNDER THE OXYMORONIC RUBRIC "THE WAR TO END ALL WARS," World War I proved anything but. For Europe, it was a nearly unmitigated disaster, one whose reverberations rumbled across the Middle East, Africa, and Asia, announcing the eventual doom of imperialism. The very weapons that had allowed Europeans to conquer these places had thwarted them utterly when used against each other—four years of operational futility, a sure sign their military omnipotence was finished.

Meanwhile, closer to home, the war brought the Bolshevik Revolution to Russia, and later Hitler to Germany and Mussolini to Italy, the latter two powered by private armies filled with angry, bitter war veterans. On democracy's side, equivalent bitterness undermined stability, while nominal victory allowed for the illusion of "never again" until it was almost too late. A political ecosystem emerged that made a sequel, a much worse one, practically inevitable and America's participation inescapable.

It's important to note that the war ended without a solution to the artillery and machine gun–based stalemate that killed millions on the western front (not to mention its equivalent at sea based on the torpedo and submarine). There had been progress with infantry tactical doctrine, particularly by the Germans, but not enough. The development of the tank pointed the way to the future and had achieved some notable tactical successes late in the war, but mechanically they remained in their infancy, and there was very little understanding, much less doctrine, of how to employ them effectively.

The same went for aircraft. They had evolved with remarkable rapidity during the war, and did provide the romance and swashbuckle of individual combat to what was otherwise a very grim business for the trench-bound masses. Future roles in intelligence, tactical air support, and strategic bombing (the Germans had staged multiple raids on London) were all there in outline, but none made the slightest difference to the outcome. This had begun and ended as a war of continuous slaughter; American participation in this regard made no difference. Doughboys do appear to have been more enthusiastic and aggressive than their European counterparts, but most fought for months, not years, and when faced with World War I's signature frontal assaults, they proved no more immune to random and copious death-dealing.

All these problems could and would be addressed to produce another kind of war, one largely of movement. But while this new variety of war might be more interesting and entertaining to read about, it proved much worse for the participants, still stuck in the same basic trap—caught by the notion that the military deficiencies of individuals within huge, conscription-based armies could be compensated for by better weapons. So the future would mean even more total war, one in which civilians would be enlisted not just to man endless battalions and produce the consumables, but also to suffer the consequences directly with raids from above.

Having emerged from the first of the twentieth century's lethal twin rapids, our four heroes remained fully on board. It was only when they hit the second and got their turn at the helm that they began to have second thoughts. At least Mac and Marshall and Ike did. Patton never did.

CHAPTER 5

HALFTIME

In late April 1919, George Marshall, as part of Pershing's command staff, went to Metz to receive the French Legion of Honor, which proved, actually, the least of the good news that day. As he marched to the ceremony one of Pershing's colonels slipped in beside him and theatrically whispered, in exact cadence to their step, "How would you like to be the General's aide?" An invitation, if not heaven sent, certainly from that direction. In the near term, it meant nonstop exposure to the European leadership structure he himself would one day have to deal with, while Pershing, the world's most victorious general, wandered from capital city to capital city collecting accolades in a daisy chain of parades and banquets that stretched into September, before coming home to equivalent blowouts in New York and Washington. It also meant being reunited with Lily back home, at last, almost a year after the shooting had stopped.

Now the future looked bright. Lean days were ahead for the Army, but it was hard to think of a better survival niche than a firm attachment to the service's only permanent four-star general, in effect his personal chief of staff, an assignment that would stretch out pleasantly and instructively until 1924 and Pershing's ultimate retirement. But Black Jack's influence would persist. However unpopular the military might become, he remained the living embodiment of America's victory in the Great War, and as such continued to have the ear of the powerful, and through that conduit, George Marshall's future was assured.

Pershing was a shrewd if brutal judge of men. With Marshall, uniquely for both, perhaps, close proximity resulted in close friendship. This gave Black Jack a perspective on George that the others who were destined to work closely with him lacked, left perpetually bouncing off the citadel of his oblique, cold-as-a-fish personality. From years of observation Pershing came to know exactly what his sidekick was capable of doing—in a word, organize, however huge the numbers involved—and in one notable instance, what he might not be up to: leading the largest amphibious force ever assembled. But until George Marshall was in a position to take control of his own career, it's important to remember that his vicissitudes and frustrations took place beneath the top cover of a very powerful wing.

That was decidedly not true of Douglas MacArthur, who from this point had to look out for Pershing's talons, not to mention his beak. His proposed second star was frozen by the armistice and then killed permanently by Pershing's staff, as was his Medal of Honor—at least, that's what Doug thought. And for the moment it seemed even Fortuna turned against her favorite bulletproof boy, slapping him down with a severe throat infection, compounded by the Spanish flu, a one-two combination followed by diphtheria in February that "the doctors pulled me through" but left him months in recovery.

It was in this chastened condition that Douglas MacArthur arrived in New York with his brigade of the Rainbow aboard the liner *Leviathan*, but nothing prepared him for the reception they got. It amounted to a single spectator. "One little urchin asked us who we were and when we said—we are the famous 42nd—he asked if we had been to France. Amid a silence that hurt . . . we marched off the dock, to be scattered to the four winds—a sad, gloomy end of the Rainbow."

This was what MacArthur thought, but more likely, the troops wanted it that way. Cumulatively, they had suffered 14,000 casualties and over 2,600 battle deaths learning the cruel lesson of what death machinery could now do. They had fought bravely, but on this battlefield death came randomly as often as not. They, like their countrymen, wanted to put this episode behind them—fuggedaboutit, as a New Yorker might say, a sentiment driven home to Doug the night of his arrival at the Waldorf Astoria when he and his date were told his boots and spurs "might injure the dance floor." He stormed out, perhaps vowing "I shall never return," or something to that effect, which wouldn't be a great prophesy, since he ended up living in the place.

But for the moment he had bigger problems. Being stateside meant that Pinky was back in his life, not that he necessarily objected at this point. Despite everything, she provided a bridge to Pershing, who had worked for Arthur in the Philippines and later Japan (when Pershing approved Doug's first star she wrote him, saying, "You will not find our Boy wanting!"). Yet her politicking in his behalf would become increasingly maladroit as she aged and her health deteriorated. He remained loyal and attentive to the end, but she was a sea anchor on his life and independence, and that would soon become apparent.

But regardless of Pinky, he still had a sword of Damocles perched just above his star. George Marshall's and George Patton's silver eagles and even Ike's oak leaf were similarly endangered species. As a mandatory part of the demobilization process, all wartime promotions were to be terminated on June 30, 1920, and those affected would revert to their prior grades. Military professionals take rank very seriously; almost by necessity it becomes central to their identity. Our four paladins had earned theirs by excelling in one way or another during wartime, and they gave every sign of pride in their elevation. Now it looked as if a general and three colonels would revert to a major and three captains—not a happy prospect for men as talented and dedicated to their careers as our four. More to the point, they were experienced enough to know the Army was entering a promotion wilderness, as it always had after big wars, and they likely would find themselves marooned for decades.

In fact, things turned out better for each of them. On July 1, 1920, Marshall, Patton, and Ike were all promoted to major. To further blunt the pain, demobilization stipulated that their pay remained that of the higher grade, so throughout the lean decades ahead all were guaranteed a comfortable, upper-middle-class lifestyle—except Patton, who was already rich.

Then there was Doug. Lady Luck was back in his corner. At point-blank range, the sword missed his star. Not surprisingly, it all had to do with West Point. The place was in turmoil. The events of the war had shown the curriculum was thoroughly outdated, and, on top of that, two classes of cadets who had prematurely graduated during the war emergency had been dragged back, many of them combat veterans, to be treated again like tyros. Finally, bringing the situation into sharp focus, a much-hazed plebe committed suicide on New Year's Day 1919. Congress was not happy with the situation, and there was open talk of closing the academy down. Something had to be done.

That responsibility ultimately fell on the intellectual, acerbic Peyton March, the chief of staff and—aside from Pershing—the Army's only other four-star. March was a reformer. Having had a primary role in putting the AEF together, he now wanted to pivot off its lessons to build a large, modern army. In his eyes, a key point of leverage was West Point, where the next generation of leaders would be shaped; reform there could mean a fundamental change in the psychology of command. He wanted the hazing curtailed and a modernized curriculum installed—one designed to impart an understanding of the latest military technologies along with a broad grasp of national and international affairs.

March realized that was a tall order at what amounted to an academic fortress, and that only a truly charismatic mover and shaker could accomplish it. He really didn't have far to look. Not only had he served under Arthur MacArthur in the Philippines, but Arthur's son had cut exactly the kind of swath across the western front he was looking for to install along the banks of the Hudson. But most of all, perhaps, Peyton March had come to despise John Pershing and knew nothing could infuriate him quite like appointing Douglas MacArthur superintendent of West Point.

MacArthur later claimed that when March made the offer, he demurred. "I'm not an educator sir. I am a field officer." It's hard to believe. Not only did he love West Point, but the job required a brigadier general, meaning that Doug's star would become permanent. While MacArthur himself was undoubtedly dazzled by the possibilities of coming to the rescue of the academy in distress, the salvation of his star had more long-term implications. In our field of four, he was still a general and they were all majors, an enormous gap that would take them decades to overcome. But it would forever condition their interactions. Doug was the star because he had a star, then two, then four, all while they had none. So, what was really nothing more than a stipulation in a job description would come to have a major impact on our story, and for that matter history.

Doug arrived in June 1919, installed Pinky in the superintendent's mansion, and set upon his West Point campaign, one destined for mixed results at best. He began well, gathering a few loyal staff, and ingratiating himself with the cadets by emphasizing sports and treating them like human beings, all the while injecting an infectious energy into practically everyone he contacted. During this time, he seems to have put the finishing touches on his signature office behavior—a brilliant, almost mesmerizing monologue, begun with a cigarette and soon replaced by nonstop pacing, until another cigarette was lit and abandoned to renew the cycle. But he was never a bore, never forgot a name, or much of anything, for that matter, and remained approachable.

Too approachable for the West Point faculty. They wanted to be left alone and he wanted to revolutionize them. In the proverbial contest between the irresistible force and the immovable object, the object usually has time on its side. The West Point faculty also had tenure, and determined to wait him out. MacArthur's already massive egotism also undermined him, as he proved unable to accept criticism with anything like equanimity, especially when it came from his presumed allies, the students.

But Doug's troubles at West Point were academic in more than the literal sense; from the beginning his tenure was doomed to be a short one. On June 30, 1921, March's term as chief of staff would expire, and John J. Pershing would replace him. This was no secret, and MacArthur must have understood the implications for his own survival at West Point. So, in the same spirit that caused him to play combat dress-up, he managed to truly insult the man.

Surprisingly, it was over a woman. To this point, MacArthur had kept his sexuality carefully in check, a project endorsed and, to the degree possible, enforced by the redoubtable Pinky, to whom no woman was good enough for her lifetime project of a son. Then, in late 1921, at a dinner party in nearby Tuxedo Park, he met Louise Brooks. Among her many facets, Louise was extremely wealthy, divorced, vivacious, sexually liberated, and mostly drunk. She had spent the war years in Paris, where she developed into a prototypical flapper and in the process met John Pershing. At the moment, she was acting as his official hostess in Washington, where rumors of matrimony were heard. But no more. By the time this party was over, Doug and Louise were engaged.

Besides Black Jack, there were other reasons for this lightning linkup. MacArthur was now past forty and had never had a serious relationship. Being married was standard operating procedure for high-level officers of the time. Then there was the undeniable physical attraction: "If he hadn't proposed the first time we met, I believe I would have done it myself," she later told reporters. Given Louise's nature, that might well have been true.

But it was a disastrous choice for both of them. Louise had no idea what she was marrying—a soldier of epic proportions—or what that required matrimonially. Instead, she would treat him as a kind of trophy husband, dragging him to parties he didn't want to go to, listening to talk of business deals that bored him, and where more alcohol flowed than he ever wanted. But not Louise.

After their marriage on Saint Valentine's Day 1922, he probably tried harder as a husband than she did as a wife. He was always kind and even loving to the children of her first marriage. He made an effort to adapt to her social circle. But in the end, what doomed their marriage was that Douglas MacArthur required a woman who utterly and unremittingly worshipped him. Louise Brooks was not that woman. Pinky boycotted the wedding, and an early marriage photo said it all—Louise with a feline grin, her eyebrow arched at a come-hither angle, and Doug waxing apprehensive, looking more worried than any shot of him during the war.

Two weeks after the engagement, MacArthur found out that Pershing was ending his term as superintendent; he also took the trouble of personally writing his fitness report, one damned with faint praise. Yet the wrath of Pershing proved decidedly milder than that of Achilles. For one thing, the love of Pershing's life was the Romanian painter Micheline Resco, whom he had met in Paris during the war, and was now his mistress but inappropriately young for a man over sixty. Women like Louise Brooks and Nita Patton were as much distractions as they were objects of Black Jack's affections. So Pershing's role, at least as jilted suitor, was probably as he claimed, "all poppycock, without the slightest foundation." It was simply time for General MacArthur to assume a foreign command, Pershing explained to reporters, and he had found the ideal slot for him: the Philippines. With one swift stroke he banished MacArthur to the other side of the world, and also sent him exactly where he wanted to be. For a very blunt man this was an elegant payback.

Meanwhile, for George Marshall, life under Pershing's wing proved chockablock with benefits. For a man with tepid emotional needs, it was perhaps the happiest time of his life. He and Lily could finally start living together on a permanent basis, and with the money to do it in a luxurious Northwest Washington apartment hotel. Here the couple met a precocious eight-year-old, Rose Page, the daughter of a widowed and distant economics professor. Before long, they had virtually adopted her, becoming her de facto parents. For George, it was the beginning of a forty-year relationship, the first in a series, one of them ending tragically. But for Rose, knowing the Marshalls intimately, especially George, was an entirely benign experience filled with hikes, horse rides, and wise advice on decorum.

Pershing also extended his growing friendship with Marshall to Lily, and they seamlessly became part of his tight social circle. Lily's weak heart still gave her husband an excuse for not socializing, but this was different; it might have been mandatory, but it was still based on affinity. And on propinquity—after Pershing became chief of staff, the Marshalls moved into the spacious Quarters No. 3 at Fort Myer, virtually in the shadow of the general's own house.

Professionally, George was also his shadow, accompanying him everywhere. In Washington, this meant sitting in meetings with the president and pretty much everybody on down. He made it his business to get to know Charles Dawes and Bernard Baruch, who had both been deeply involved in the underachieving national effort to arm the AEF. He also learned about Congress, what worked and didn't work with its members, concluding quietly that the general's style was a bit too partisan and grating. Marshall's mind was like a sponge, all of this apparently being absorbed and digested for future use—at least, that's what his later behavior suggests.

Pershing was a restless soul and this also translated into a lot of travel and a lot of speeches. Not only did all the logistics and scheduling fall on Marshall's shoulders, but the general was a notorious latecomer, so he often found himself in the role of tummler, speaking for the boss until he finally showed up. Combined with the constant movement, it added up to a lot of responsibility, continuous practice at improvisation, and, it seems, some good times. At one point they were on a train with a senator they knew—Marshall and Pershing in a drawing room, the legislator in a berth in the next car. After splitting a bottle of scotch, they decided the senator should have the last shot, and the two of them proceeded to his berth, only to be greeted by the head of a young woman, prompting a precipitous retreat. So hasty was it that Pershing knocked the scotch from Marshall's hand in his efforts to get back to the drawing room, where he collapsed in a fit of laughter until he cried. Black Jack could be fun.

But to Marshall he was essentially and fundamentally helpful. One day after he became chief of staff himself, he would have a giant portrait of Pershing mounted on the wall of his office, staring down at everybody who entered. Marshall was no sentimentalist, but this did not preclude gratitude—it seemed to be his way of saying, "Without you, I could not have gotten this job."

For the moment, though, like Douglas MacArthur, Marshall was headed west, even farther west, all the way to Tianjin, China, after five years as Pershing's aide, slated to become executive officer of the Fifteenth Infantry. It was his preferred assignment and Black Jack made it happen, along with a promotion en route to lieutenant colonel in August 1924—and this one was permanent.

IKE AND MAMIE WERE ALSO ON THE MOVE, ESPECIALLY IKE. AFTER HE CLOSED Camp Colt and proficiently took it off the books in late 1918, Eisenhower received orders to Camp Meade, Maryland, the home of the nascent Tank Corps and its uncertain future.

Almost immediately he learned the Army was planning to take the show on the road. A transcontinental mechanized column—mostly trucks but including a Renault light tank on a flatbed carrier—was set to roll from Washington to San Francisco. As we have seen, Ike loved anything with an internal combustion engine, so, when offered the opportunity, he and his pal Serano Brett were quick to join the caravan.

Calling it a road trip would be an oxymoron, since basically there were no roads. That was part of the point, to publicize the need for better main highways, but also to see if they could make it. The caravan did not pass unnoticed: an estimated three and a quarter million saw the column as it rolled at a stately pace through their villages, towns, and cities, or simply broken down and stationary—"Part of an audience for a troupe of traveling clowns," Ike later confessed. Mamie and the Dowd family, always drivers of big, fast cars, caught the convoy in South Platte, Colorado, and tagged along for several days, the first time Ike had seen his wife since Gettysburg. There were further adventures and even hijinks—Ike and Brett staged a fake Indian attack—but after sixty-one days the convoy finally rumbled into San Francisco on September 5, 1919, having averaged less than ten miles an hour on the way to great acclaim. "The trip had been difficult, tiring, and fun," Ike remembered, but it also left a profound impression. From this point, he understood the need in the coming automotive age to tie the continent together with something more than bands of steel; concrete and asphalt were called for—one day to be known, officially at least, as the Dwight D. Eisenhower National System of Interstate and Defense Highways. It had been two months well spent.

Plus, when he returned to Camp Meade, he found the place occupied by the returned tankers of the AEF, among them a conspicuous and fascinating new presence, "a fellow named Patton. . . . Tall, straight and soldierly looking. His most noticeable characteristic was a high squeaking voice. . . . He had two passions, the military service and polo." With his bum knee Ike was not about to join Georgie in a chukker, "but both of us were students of current military doctrine. Part of our passion was a belief in tanks. [So] from the beginning, he and I got along famously."

When they were in direct contact at least, Eisenhower had a way of bringing out the best in Patton. He was a highly decorated war veteran, still a scion of wealth and social position, and a notable snob; yet he truly befriended the unpretentious Ike, becoming his collaborator and even benefactor in the contacts he provided. Of course, Ike was always fun to be with, and Meade's officers' quarters were not yet woman-habitable, so in the absence of Mamie and Beatrice until the late spring of 1920, Ike and Georgie had a lot of quality time together.

Much of it they spent obsessing over tanks—the heart of their friendship would always beat to professional issues. Patton loved war; Ike had yet to see what it could do, and both believed the tank could rescue it from the trenches and return it aboveground to roam free.

At Meade, Patton commanded the light brigade of French Renaults, while Ike was executive officer of the heavies, American-made British Mark VIIIs, so they could enlist whatever machinery needed to populate their experiments. They took apart a Renault and, more impressive, put it back together. In search of better technology, they cultivated (and Patton may even have bankrolled) Walter Christie, the brilliant design engineer, whose concepts would be the basis of the Soviet T-34, World War II's best tank. In the field, they lashed Mark VIIIs to as many as three Renaults, towing them around until an inch-thick steel cable parted and "the flying end, at machine-gun bullet speed, snapped past our faces." On another occasion, an overcooked machine gun almost cut them down when they carelessly walked out in front to examine a target.

But all of this had a larger purpose. Patton and Eisenhower were trying to generate better tank doctrine, distilling what they learned into improved operating principles. Given Patton's prior success with publication, it's not hard to conclude that he was the instigator of what became dual manifestos. But this wasn't about swords. At a time of great military uncertainty, their approach was clever but also confrontational. They reanalyzed tactical problems used at Leavenworth's command and general staff course by adding tanks to one side or the other, inevitably resulting in victory for the side with the armor. Both then wrote parallel articles for separate journals—Ike's for the infantry, Patton's for the cavalry. To mix metaphors, they were shots across both bows.

Meanwhile, not all their time together at Meade was devoted to weighty matters. Both worked for General Sam Rockenbach, Patton's commander in Europe, who knew Georgie was best ridden on a loose rein, while Eisenhower interested him most obviously as a football coach. Still with time on their hands, the two took to nocturnal cruising in Patton's big Pierce-Arrow, the one he used to visit Joe Angelo, who'd saved him in the Argonne Forest. Now they were armed and claimed to be on the lookout for bandits, just the sort of scheme Georgie would instigate. Ike also played a lot of poker at Camp Meade, so much and with such skill that he finally hit a moral roadblock and vowed to quit and stop taking money from his brother officers.

Quarters improved and the ladies finally arrived, Beatrice with the two girls, and Mamie with little Icky. The wives came from two different worlds, one merely prosperous, the other truly elevated, one at home on the deck of a sailboat or the back of a horse, the other better perched on a couch. But

Mamie was attractive, gregarious, and truly likable, the distaff half of the forever reemergent Club Eisenhower. She couldn't cook but she could chat, and the Patton girls loved her, especially that she wanted them to call her Mamie. So was established the basis of a social connection, one that would lead almost immediately to a professional connection of fundamental importance to the career of Dwight D. Eisenhower.

The occasion was a Sunday luncheon and the guest of honor was Georgie's interlocutor from the infirmary and now Pershing's right-hand man, Fox Conner, whom he had invited specifically to meet Eisenhower. After the meal, Conner indicated he wanted to talk tanks, and the three wandered down to one of the workshops, an appropriate setting. Amid the spare parts Conner launched a series of questions, most hurled at Ike. It was nearly dark before he was finished with the inquisition. He thanked them, said it was interesting, and left. Ike had no way of knowing that he had just opened the path to his future, convincing a true Army bigfoot that he was something special.

But for the immediate future things turned miserable, and then, for Ike, tragic. The plucky duo's twin tank manifestos never got published, just forwarded to Washington, where they infuriated the respective commands. Ike was called before the chief of infantry, where he was threatened with a court-martial if he ever published "anything incompatible with solid infantry doctrine." He thought Georgie got the same treatment, but was unsure, since by summer he was gone.

The independence of the Tank Corps had been abolished by the National Defense Act, and Patton, still welcome among equine enthusiasts, bolted back to the cavalry. Had he been a true military reformer he would have rejected the horse arm as militarily useless. But Patton was a warrior, and being in the cavalry felt warlike, which for the moment sufficed. When the day came, though, he would have no trouble jumping out of the saddle and down the hatch of a tank. For in his eyes they did the same thing: chase the enemy and grind him into dust.

Ike stayed with what remained of the tankers at Camp Meade, coaching football and looking for his next move. Instead, he found himself suddenly entangled in a bureaucratic net that threatened his career. He had mistakenly claimed a $250 housing allowance for Icky, and was now in the clutches of the Army's inspector general, who was determined to bring him to trial on criminal charges. Ike claimed ignorance of the rules, but his signature on an official document was likely damning if the Army went forward.

That summer and fall the one bright patch in the Eisenhowers' path had been Icky. Ike doted on him, determined not to be the cold patriarch he had known. But then they hired a girl to help with chores, not knowing she was just recovered from scarlet fever. Shortly before Christmas Icky began running a temperature. When it would not come down, he was hospitalized and then quarantined at nearby Johns Hopkins. Scarlet fever transitioned into meningitis, and on January 2, 1921, the boy died. It drove a red-hot wedge of grief into the marriage, one that left scars those closest to them believed never faded. For Ike and Mamie, the innocent optimism of early matrimony was over.

It was at this grim turning point that the Eisenhowers discovered what having the right mentor could do. Pershing, now chief of staff, gave Fox Conner command of the Twentieth Infantry Brigade in the Panama Canal Zone, a plum assignment. Conner then informed a flabbergasted Eisenhower that he wanted him to be his own chief of staff. Just as suddenly, after a year of foreboding, the legal clouds cleared and the inspector general miraculously settled for a letter of reprimand added to Ike's files. Backstage, Conner engineered it all through Pershing, and literally before they knew what had happened Ike and Mamie found themselves aboard an Army troopship bound for Panama, landing on January 7, 1922.

They moved into large, screened-in, but ramshackle quarters, right next door to the Conners' duplicate but fully renovated version. The latter could afford to do this sort of thing. Like Patton, Fox had wed the heiress to a patent medicine fortune; but unlike the Pattons, he and his wife provided a model of what a sane, reasonably happy Army marriage might look like. Mamie arrived pregnant and had great trouble adjusting to the heat and insects, but Virginia Conner sensed more was wrong: "The marriage was clearly in danger . . . Ike was spending less and less time with Mamie, and there was no warmth between them. They seemed like two people moving in different directions." A shrewd prediction. When the Douds visited in June, they insisted that Mamie return with them and have the baby in Denver. That left Ike primed for Fox Conner's ministrations.

He was an unusual army officer, to say the least. A true military intellectual, Conner was intimately acquainted with all the classics in the field, having read many in the original French. But he was also a man who knew his own limitations—one day he would pass up the opportunity to become Army chief of staff himself. The son of a Mississippi planter, he worked beneath a mask of southern geniality so deftly that some have compared him to Cardinal Richelieu's legendary underling, François Leclerc du Tremblay, the gray eminence. Conner customarily wore olive drab, but was just as oblique and committed to the larger picture.

Above all he believed that the peace established by the Treaty of Versailles—particularly after America failed to join the League of Nations—would prove temporary, and that the United States would inevitably be dragged into the conflict that followed. Under the circumstances, the Army had a special responsibility to cultivate the careers of its most promising officers, and he took that as his personal crusade, one that led him to become Team America's chief talent scout. He had already mentored Patton and Marshall; now it was Ike's turn.

Like a genial spider he drew his object in slowly, inviting him to his quarters with increasing regularity, showing off his massive library, then, after learning West Point had smothered a nascent interest in military history, handed him a few war novels to read while Mamie was away. Once Ike gobbled them down in appropriately short order, he added more bait. "Wouldn't you like to know something of what the armies were actually doing during the period of the novels you've just read?" That set the hook, and when Eisenhower returned, having read the first of the borrowed military histories, he found himself bombarded with questions, particularly on the decision-making that might have been revealed. It was the beginning of a Socratic dialogue that stretched over the next three years, one that left Ike with an extraordinary knowledge of the martial past, particularly as it related to command. And as his future unfolded, Ike came to realize it was all part of Conner's plan.

The campaign to save the marriage went less smoothly. Back in Denver Mamie gave birth to a healthy boy in early August 1922, naming him John after her father. Ike was with her, thanks to the three weeks leave Conner gave him. When Mamie returned to Panama in the fall, she came reinforced by a Doud-funded nurse and an apparent new resolve to stick it out. "After Johnny was born and Mamie felt better, she began to change. I had the delight of seeing a rather callow young woman turn into a person to whom everyone turned," remembered Virginia Conner. Actually, she was getting ahead of herself.

Things were better between the two, but far from ideal. Ike was obviously delighted to have a new son, but he would never shower him with the affection Icky received. He was hardly a terrible parent like Patton, but at times he would turn strangely cold on John, particularly later, at points in his son's own military career when his life might be endangered. Meanwhile, in Panama, Ike was spending huge chunks of his time either on duty or reading Conner's never-ending stream of books. Mamie felt abandoned and looked it. "I was down to skin and bones and hollow eyed," she remembered. Ike begged her to stay, but once again she departed for Denver, bringing the baby with her.

Her granddaughter Susan thought it was the defining moment of her life. Dipped back into plush and familiar circumstances, Mamie quickly recovered her lost weight and spirits to the point of reconnecting with girlhood friends and classmates, all of whom had husbands with nine-to-five jobs and did familial things with the rest of their time. To her surprise she found them boring. Ike had ambition and purpose and, above all, gave her the sense that life was an adventure to be seized by the bold. So she gathered baby and nurse and returned to Panama ready to become the person Virginia Conner remembered. There would be further crises, but the marriage at this point reached a new level of maturity and bonding. That also meant becoming what her granddaughter titled her book, *Mrs. Ike*, an ego submerged. Excepting Louise Brooks, that would be the price the wives of our other three paladins also paid to live in their husbands' reflected glory. At any rate, this time Mamie stayed in Panama.

In the summer of 1924 Fox Conner was ordered back to the States, but before leaving arranged to have Eisenhower receive the Distinguished Service Medal for his work at Camp Colt, a significant addition to his service record. He also provided some advice, fragmentary as usual. He urged Ike to seek an assignment under George Marshall. "In the new war we will have to fight beside allies and George Marshall knows more about the techniques of arranging allied commands than any man I know. He is nothing short of a genius," Ike remembered him saying. Conner also recommended that he apply for the Command and General Staff School at Fort Leavenworth.

But when Ike did what he was told, he received instead orders back to Camp Meade as assistant football coach. When the season ended, Ike was next ordered to command a battalion of tanks—"the same old tanks I had commanded several years earlier." Having protested to higher channels, he then received what he called "a strange telegram": "NO MATTER WHAT ORDERS YOU RECEIVE FROM THE WAR DEPARTMENT, MAKE NO PROTEST ACCEPT THEM WITHOUT QUESTION. SIGNED CONNER." This was followed by an abrupt transfer to the adjutant general corps as a recruiting officer, traditionally one of the Army's least desirable assignments. Its only virtue as far as Ike could see was location, only seven miles from the Douds' home.

It was apparently all part of the plan. While Ike fretted in Denver, Conner was putting the finishing touches on a web that would prove his bridge to Leavenworth, the switch from infantry to adjutant being the silky keystone. After a suitable interval of recruiting, Ike received word the adjutant general had chosen him for the Leavenworth class entering in August 1925. "I was ready to fly—and needed no airplane."

Then he began to worry. He had never before excelled in a purely academic environment, and Leavenworth had the reputation of being hypercompetitive and a career killer for those who did poorly. He wrote Conner, wondering what he could do to prepare for the ordeal. "You may not know it," the general replied, "but because of your three years' work in Panama, you are far better trained and ready for Leavenworth than anybody I know." Just to make sure, Ike wrote to George Patton asking for his notes from the school.

Considering that a warrior without a war generally has trouble adjusting, George Patton managed quite well, largely because he got to live life on his own terms. Having rejoined the equestrians, his first assignment was commanding a squadron of the Third Cavalry, a ceremonial unit attached to Fort Myer, the leafiest and most prestigious post in the Army, a job that also came with a large brick house just a few doors down from the Marshalls and Pershing. Nita's engagement to Black Jack had ended mutually and with few recriminations, enabling relations between Pershing and her brother to remain cordial. Marshall and Patton, though close physically, remained distant. Marshall was involved with the wheels of governance; Patton was focused on the horses.

As with his first tour in Washington, George's job was to provide the equine component at parades and funerals, both of which were a frequent occurrence. But not frequent enough to preclude plenty of time for polo, his favorite sport and a pursuit that led to a steady accumulation of blows to the head and other shock-related injuries—so many that some believe they contributed significantly to Patton's mental and emotional instability. And there was also foxhunting and even some horse racing to be considered as Patton again cut an eccentric path across upper-class Washington—seldom at less than a gallop.

His party chatter, though animated and interesting, was frequently lewd, to include the occasional offer to show off his war wound. Still, the Pattons remained as earlier to the capital's elite: an attractive couple who could afford to entertain lavishly, and therefore be entertained. Yet the degree to which they were accommodated is still surprising—particularly in light of one rather flagrant incident. At a formal dinner party on Dupont Circle, to which Georgie wore his full-dress uniform, Bea heard a drunk make reference to fake heroes and jumped him, pummeling the man's face until Patton pulled her off. There were no apparent social consequences, but the incident says something of the marriage.

Beatrice Patton seems to have accommodated herself to living with George Patton by becoming, as best she could, a woman equivalent. Not only was she combative, but she was equally comfortable on the back of a horse and easily a better sailor. And if her husband retreated nightly to pore over military texts, she too had intellectual interests—she began translating military manuals from the original French, and would one day write novels. Meanwhile, she continued to develop into a very hard person, as adamantine in the end as he.

It was George's intellectual interests that had drawn the Pattons to Kansas, between 1922 and 1924, first to Fort Riley and the cavalry school and then to the Command and General Staff School at Leavenworth, but so long as he was in the cavalry he retained an inside track back to Fort Myer and the corridors of power, not to mention its polo fields. It was extremely expensive to take on such an assignment along with the social obligations it entailed; only a very few officers could afford to do it without crushing their solvency. George Patton was one of them.

Meanwhile, bad spelling was just about all that remained of Georgie's learning disabilities. At Fort Riley he excelled in his studies, lectured his classmates on Napoléon's marshals, and bubbled with extracurricular hands-on curiosity: "They say the machine gun is the killingest weapon on the battlefield. If that is so, I have got to know more about it," he told one cornered young instructor. "Will you give me some personal instruction on Saturday afternoons?" In return he taught him to wield a sword, a fair deal only in Patton's book.

His superior performance at Fort Riley springboarded Georgie to Leavenworth, while Bea, pregnant once more, returned to Avalon, the family estate, to await the birth. Patton's military erudition and professional dedication again paid academic dividends. Leavenworth may have been hypercompetitive and the student body the cream of the Army's young officers, but Patton with plenty of time to study managed a rank of twenty-fifth in a class of 248, just short of the top 10 percent but still an honor graduate. Meanwhile, on Christmas Eve Bea gave him a son, whom she named George, perhaps in the hope it might improve his parenting. He had that opportunity for the next eight months, courtesy of a well-connected temporary assignment to corps headquarters in Boston, an easy commute from Avalon and probably the closest he ever came to becoming a family man. But his subsequent assignment, the real one, was to Hawaii, a tropical honeypot that in Patton's case would overcook everything.

THEY WERE ALL HEADED WEST FOR MORE SEASONING. IN EACH CASE THE RESULTS would be inconclusive, but still very much a part of their evolution. For Patton, it proved corrosive. For Ike, an epiphany. For MacArthur, one more immersion in the pivot point of his life. For George Marshall, a glimpse at a world unintelligible from his perspective. All would learn something, but sometimes they were the wrong lessons.

For Doug it meant leaving Pinky in Washington with his brother's family and in October 1922 striking out for Manila with Louise, who took so much baggage that the other lower-ranking officers and wives aboard were limited to a single trunk apiece. After what must have been a frosty voyage to a warm place, he was greeted by his old mentor, General Leonard Wood, now American governor-general.

The news Wood conveyed later in his cavernous office to the beat of a tropical fan was not good. While the archipelago was now thoroughly pacified, the economy was in a deep three-year recession, and Manuel Quezon, Mac's acquaintance from two decades prior and now president of the Philippine Senate, was calling with increasing urgency for independence. While Mac (though quietly) had no problem with that, the strategic part of Wood's message troubled him. Specifically, the latest missive from Washington, War Plan Orange, dictated that the Philippines be defended against a hypothetical Japanese invasion for at least six months before help could arrive. America had 4,100 soldiers in the archipelago. Militarily it presented an impossible problem for defenders.

So began what one biographer called "the lowest point of his life" a view reinforced by his autobiography, *Reminiscences*, which devotes just one page to a tour that lasted two and a half years until 1925. Admittedly, as head of the Military District of Manila, a brigadier general's slot created specifically for him, he had command of just five hundred men and not a lot to do formally during much of his stay. He did continue to cultivate the local leadership and extend his range of contacts, which would one day pay dividends. But not yet.

Meanwhile, Louise tried to relieve the boredom by turning interior decorator, painting the walls of their sumptuous quarters on Calle 1 Victoria pitch-black and filling the rooms with cost-is-no-object furniture and servants dressed in sailor suits embroidered with MacA. It was an effect not calculated to wear well. Doug taught her son Walter to ride and bought a beach house. But they were all bored, especially Louise, who even tried volunteer police work—a true measure of her desperation.

Over dinner with Leonard Wood she frequently complained of her husband's exile, to which he just as frequently replied there was nothing he could do. Doug did get back to Washington in early 1923, but only because of an urgent cable from his brother's wife saying Pinky was in the hospital and might be dying. After receiving a pessimistic medical report from the attending physicians, the son—having dealt with this before—told Pinky the doctors said she was fine. In under a week she was out of the hospital, soon to resume her normal activities, among them importuning Pershing. "Won't you be real good and sweet—the 'Dear Old Jack' of long ago—and give me some assurance that you will give my Boy his well earned promotion before you leave the Army?" For Doug's part, there was little to do but get her a more savvy doc and retreat to the Philippines.

Back there a glimmer of hope appeared. In Manila, General Omar Bundy was seeking to inflate US military credibility with a new 7,000-man Philippine Division, built by integrating regular army units with two elite indigenous regiments, all under American command. Bundy knew MacArthur from AEF days and put him in charge of one of the division's two brigades.

It seemed a perfect match. MacArthur, a man remarkably free of racial prejudice, quickly took to these tough Filipino volunteers known collectively as Scouts. Always glad to be commanding troops, he was further buoyed by his Scouts' performance in field exercises, especially their marksmanship. Together they spent a good deal of time traipsing back and forth over the rugged Bataan Peninsula, with Mac mapping defensive perimeters and getting to know it intimately—a topographical preview of his date with destiny. Meanwhile, he was left to conclude his Scouts were "excellent troops, completely professional, loyal and devoted."

Then in late June 1924 they mutinied. The Scouts called it a strike over drawing less than half the pay of American troops and none of their other benefits. It was put down quickly and bloodlessly with the arrest of 222 mutineers. None of the pay issues were addressed, but Bundy was replaced as division commander by MacArthur, who remained widely respected by Filipinos and presumably the Scouts. Yet it was probably window dressing, since his command would last less than two months. On September 13, 1924, Pershing reached his mandatory retirement date, and his replacement, Malin Craig, almost immediately moved Douglas MacArthur to the top of the promotion list. Pinky's letter hadn't worked, but her son would soon be a two-star general and headed home.

At just about the same time, George Marshall landed at Tientsin (Tianjin) on the north coast of China. This was considered a cushy posting, witnessed by his bringing the delicate Lily and her aged mother along. In terms of amenities it didn't disappoint, with a ten-room house to occupy along with plenty of servants, and all manner of luxury products available cheap. Marshall's job also seemed comfortable: executive officer of the Fifteenth Infantry, here on a temporary basis to protect American lives, property, and the railway link to Peking (Beijing). These extraterritorial police occupied the former Imperial German concession surrendered after the war; the place looked familiarly Western and solid. It was anything but.

China was at the threshold of revolution, one of those interregnum stretches when the Mandate of Heaven is up for grabs and death and destruction prevails—in this case, not to end until 1949 when Mao and the Communists consolidated power. George Marshall had stumbled into a political maelstrom with virtually no idea of what was going on or who the major political players even were. And almost immediately he was faced with a crisis when the army of one of three rival warlords in the North collapsed, sending around 100,000 troops, both pursued and pursuers, spilling pell-mell toward Tientsin.

Since the regiment's colonel had yet to arrive from the States, as executive officer it was Marshall's problem, and he reacted as he always would. He analyzed the situation on the basis of limited information, came to a simple solution, then took care of the details. In this case he set up a cordon around the city and railroad based on five outposts, where fleeing soldiers were offered food for their arms or persuaded to go around Tientsin by Chinese interpreters recruited by Marshall. Hunger being a great incentivizer, the plan worked like a charm, and the mob army passed without harm. Soon after the new colonel appeared and rather abruptly cut back on Marshall's authority. But it didn't bother him. "I snaffled a nice letter of commendation out of the affair which is worth my three years in China," he wrote an acquaintance, betraying his frame of reference.

In his own way George Marshall did attempt to stretch his gaze beyond Tientsin and try to grasp the Chinese environment better. He immediately signed on for language lessons, doggedly reaching the point where he could carry on simple conversations in Mandarin. Yet probably the most useful thing he did in terms of decoding the place was renewing his acquaintance with Major Joseph Stilwell, who came to Tientsin late in Marshall's tour. Stilwell had helped him plan the Saint-Mihiel offensive, but now his main attraction, at least initially, was his fluency in Chinese and the opportunity to practice. Yet over the next eight months Marshall came to realize that beneath the blunt and profane exterior was an officer of rare ability to figure out what was going on in this strange place, and then report it with absolute truthfulness. An honest man, Stilwell got a large check in Marshall's mental black book of officers he could look to in the future.

But in most other respects Marshall spent his time in China focused within the wire; professionally that meant seeing to the soldiers. The Fifteenth Infantry had the highest rate of venereal disease in the Army, a badge they wore proudly until Marshall tried to replace it with intramural sports and other wholesome activities, a signature move, but one that earned him the reputation as something of a martinet among the troops in Tientsin.

On the softer side of life, George almost unconsciously settled into the colonialist lifestyle, one where at social gatherings or at the numerous clubs all the guests and members were white and all the servants and laborers Chinese. Part of his later attractiveness as holder of high rank and office was Marshall's commitment to democracy, but he was never one to question the social order, not back in Virginia, and especially not here. So it remained opaque to him. Before leaving he wrote Pershing: "How the Powers should deal with China is a question almost impossible to answer. There has been so much wrongdoing . . . so much of shady transaction . . . so much of bitter hatred in the hearts of these people . . . that a normal solution can never be found." Both cogent and bewildered, George Marshall's view of China when he left in May 1927 was one destined never to change.

Meanwhile, Lily had the time of her life, a "three year shopping trip" during which she assembled a house full of elegant furnishings she looked forward to using in the dwelling that came with George's choice new job as lecturer at the Army War College back in Washington. Instead, almost as soon as they arrived, Lily's heart condition worsened, complicated by a diseased thyroid, and she was soon hospitalized at Walter Reed, where a long and complicated operation ensued. Her recovery was slow but steady, and she was in the midst of writing her mother that she would be discharged the next day when she slumped over the desk dead.

The marriage had floated serenely for a quarter century; Lily had been a pale presence but a beloved one; and George Marshall was left devastated and once more alone. Pershing, who had lost his wife and daughters in the Presidio fire, wrote: "No one knows better than I what such a bereavement means," one more thing the two men now had in common. And like Black Jack he would bury himself in work and the Army.

Our other George's sojourn in Hawaii was not much better. Patton in paradise proved a corrosive experience, exposing and even exacerbating his worst tendencies.

He arrived in March 1925, followed by Bea and the children just after Christmas. Once they had settled into the local social structure, they both came to love the place. Beatrice in particular. "It was as if she had been waiting for Hawaii all her life," her daughter Ruth Ellen remembered. Immersing herself in local culture, she began translating Hawaiian legends into French (later to be published in Paris). Meanwhile, her skills as a mega hostess were complemented by Georgie's rescue and elevation of the Schofield Barracks' polo team, landing the couple a comfortable space at the center of Hawaii's social pecking order.

But while the hierarchy was a rigid one, it was also dedicated to what amounted to a relaxed, even louche lifestyle. How louche it got for the Pattons is hard to say, but when husband informed wife of several obvious attempts to seduce him, she seemed more amused than anything else. It may have been fun, but Hawaii was not good for the marriage, and one day it would prove much worse. Meanwhile, it doesn't seem to have done a thing for Patton's attitude toward parenting, complaining when one child misbehaved: "Beatrice Ayer Patton, how did a beautiful woman like yourself ever have two such ugly daughters?"

To set him further adrift, his own parental bedrock disappeared while he was in Hawaii. They visited with Nita once in the spring of 1926, and Papa told George he was dangerously ill. Still, he was shocked and inconsolable a year later when he received a telegram announcing his death. George arrived at Lake Vineyard too late for the funeral, which was probably fortunate, since Aunt Nannie lost it, crying out that it was she, not Mama, whom Papa loved. Immersed in grief, apparently to the point of hallucination, Georgie did find solace in several visions of Papa making silent but encouraging gestures. Mama died the next year, and although his immediate anguish went unrecorded, one profound act of generosity was that he signed over his half of his mother's massive estate to Nita, who had always been in his corner. Besides, he already had enough, barely a quarter of Bea's income being necessary to support their lavish life.

Professionally, Patton's stay proved no more felicitous. At Schofield Barracks he was initially put in charge of administration and also intelligence, decidedly not his preferred slots. Finally, in November 1926, he was given the job he wanted, G-3, head of operations and chief adviser to the division commander. It seemed only to bring out his worst qualities: as an observer at war games he was brutally and indiscreetly critical; he issued directives aimed at improving discipline but seemed to imply his collaterals weren't doing their job, and in general struck everyone as an arrogant know-it-all. He didn't last long. Calling him "too positive in his thinking and too outspoken," the general relieved him and sent him back to intelligence.

This was done in a routine manner to disguise its impact, but being relieved at any level was a serious, if not fatal, blot on any Army record. "Invaluable in war . . . but a disturbing element in time of peace," went the division commander's spot-on appraisal. Fortunately for Team America, Fox Conner swept in as his replacement, and Patton's final fitness report would read: "I have known him for fifteen years, in both peace and war. I know of no one whom I would prefer as a subordinate officer." So Georgie would soldier on.

But Ike thrived, particularly during his year at Leavenworth from August 1925 to June 1926. It's hard to imagine anyone in the class coming better prepared to the Command and General Staff School—armed with Fox Conner's curriculum, George Patton's notes, and even copies of past assignments, which he had worked out and compared against approved solutions. But the coursework at the school wasn't just designed to challenge the intellect; students were deliberately placed under extreme time and competitive stress, the aim being to determine who would crack and who would rise to the challenge. So Ike still had to perform under pressure. Instead, he excelled.

From the very first he charted his own course. Students were encouraged to form big study groups, and most joined one. Not Ike, who thought they were a waste of time. Instead, he teamed up with Len "Gee" Gerow, a pal from Fort Sam Houston days, and turned the third floor of the Eisenhowers' roomy quarters into what he called "a model command post." Mamie took care of all logistics, kept the house running smoothly, and made sure Ike was in bed by nine thirty. Now comfortably sunk into the rut of being Mrs. Ike, she more than made the best of it. Surrounded by the wives of similarly distracted husbands, she did what she always did: drew them to her; and before long the Club Eisenhower was resurrected, mostly the distaff branch, but when the men had a few hours on the weekends, Ike as ever played the smiling host.

At all other times, he wore his poker face and absolutely killed the competition. When the course was over, not only did Ike finish first in his class, but Gerow was a close second—both a testimony to Eisenhower's organizing instincts and also his lifelong knack of drawing out of his vast bag of friends exactly the right person at exactly the right time.

But, as usual, his virtuoso performance at Leavenworth initially seemed to be a springboard into something like the Army's version of an empty swimming pool. The chief of infantry had not liked Fox Conner's fancy maneuvers to get Ike to Leavenworth, and now that he was back under his command he would suffer accordingly. He was ordered to Fort Benning for six months in August 1926, where he would take command of an all-Black regiment with white officers, and in addition coach the football team. Ike and Mamie actually enjoyed Georgia and their quarters, but from a career perspective he felt himself locked on a gridiron.

No need to despair: yet again Fox Conner swept in from above to deliver Ike beneath the wing of a still-bigger eagle, John J. Pershing himself, now head of the Battle Monuments Commission, his job being to beautify the European cemeteries of American war dead, and having the parallel desire for "a battlefield guide, a sort of Baedecker to the actions of Americans in the war." Ike wrote well, almost by instinct, and Conner must have known this; along with his easygoing personality, not likely to ruffle the feathers of a touchy eagle, Ike was the ideal candidate. But after just seven months on the commission, then located in Washington, Eisenhower was accepted at the prestigious Army War College at Fort McNair just across town, for the class scheduled to begin in August 1927. It seemed for a time that he would have to choose. But then an invisible hand—likely Conner's—smoothed the way, and Ike was given the opportunity to attend and graduate from the yearlong course (his thesis was on mobilization) and rejoin Pershing and the commission, now headquartered in Paris.

Mamie loved Washington, luxuriated in its social life and shopping, but the transition to the City of Light in August 1928 sent her over the moon. Once she found them an elegant apartment on the Right Bank, it became a launching pad for even more epic shopping expeditions, and a magnet for American officers in Paris as the Club Eisenhower sprang yet again into existence.

And while Mamie handled the social side, Ike at last found his way to the battlefields of World War I. It was an opportunity to examine them and the overall terrain in detail over a period of many months, an invaluable prequel to the real thing in 1944. Ike was relentless, covering the entire western front from the Swiss border to the English Channel, and by the spring of 1929 turning some of his outings into rather grim family adventures. Then seven-year-old John later remembered a visit to the Trench of Bayonets, where a squad of French soldiers was buried alive by a German shell, leaving only the knife-topped ends of their rifles sticking up to commemorate them. That was just one of countless other impressions of France gathered and digested by Ike to supplement his detailed study of the places where Americans fought and died, a process that enabled him to produce a revised battlefield guide that met Pershing's exacting expectations.

The general was sufficiently impressed by Ike as scribe to draw him into his own writing project, a war memoir, specifically asking him to revise his descriptions of Saint-Mihiel and the Argonne. Ike read the chapters and realized they were based solely on the general's diaries and lacked a narrative thread, which he added in his redraft, one that Pershing seemed to like but wanted to show to his former aide, Colonel George Marshall.

Within days Marshall appeared and immediately huddled with Pershing for several hours. As Ike later wrote, Marshall then stepped into Ike's office and stood looming over his desk, announcing he had read the chapters. "I think they're interesting. Nevertheless, I've advised General Pershing to stick with his original idea. I think to break up the format right at the climax of the war would be a mistake." When Ike defended his narrative concept, Marshall conceded it was a good idea, then delivered the real message: "He thought General Pershing would be happier if he stayed with the original scheme"—meaning "That's what the general told me to tell you." Ike being Ike and no fool, understood, and, apparently with no further small talk, the meeting ended. Ike was probably left quietly fuming. Yet he had made a good impression on Marshall, a check in his mental black book, one soon enough manifested in a specific request for Ike's services at Fort Benning, where he was starting to gather his other checks. But Ike had already struck out on a separate path, one that would wind a long way before it finally connected with Marshall's.

Bored and convinced the Battle Monuments Commission was a career cul-de-sac, Ike got in touch with Fox Conner, and in August 1929 he received orders curtailing his overseas tour and assigning him to the Army General Staff at the War Department. Pershing's chief of staff, Xenophon Price, thought Ike was abandoning a brilliant future and gave him his worst performance rating in years. It mattered hardly at all, as demonstrated in 1946 when Ike, now a five-star general and Army chief of staff, had his minions look up Price, and when told he was then a lieutenant colonel, muttered, "Hell, he's not that bad," and instantly promoted Xenophon to full colonel. In the meantime Ike was on his way back to Washington, his career again on the right track, if not necessarily the fast track.

DOUGLAS MACARTHUR'S RETURN FROM EXILE WAS FAR FROM A SMOOTH CHARiot ride to the top, but with Pershing retired, and given his talent and combat record, his rise was pretty much inexorable. Personally, though, it was a miserable stretch.

His first assignment in May 1925, as major general, IV Corps Area in Atlanta, turned sour almost immediately, though it was hardly his fault. Actually, it was Arthur's. The locals hadn't forgotten the Battle of Kennesaw Mountain or the Civil War, and walked out en masse the first time he and his officers attended services at a local Episcopal church.

MacArthur responded with a furious telegram to Washington demanding a transfer, which he got three months later, landing atop III Corps Area, headquartered at Fort McHenry in Baltimore—just a twenty-minute drive from Louise's estate, Rainbow Hill, and also close to Washington and Pinky. But the experience proved more than sour; it was a bitter one.

Almost as soon as he walked through the door, "one of the most distasteful orders I ever received" was delivered to his desk. He was to serve on the court-martial of his acquaintance from Milwaukee and colleague on the western front Colonel Billy Mitchell. MacArthur and much of the public knew it was a put-up job.

Mitchell was America's leading proponent of airpower. To prove his point he and his bombers had sunk the surrendered German dreadnought *Ost-friesland* in 1921, thereby poking a huge hole in naval strategy and allowing Secretary of State Charles Evans Hughes to call the Washington Naval Conference, which would put an end to the nonsensical international race to build battleships after World War I primarily through freezing further construction until 1931. The Navy never forgave Mitchell, and just to make sure they didn't, he continued to launch widely publicized verbal darts their way, a process he extended to the Army's fledgling Air Corps. Finally, after the crash of the dirigible *Shenandoah* in 1925, he crossed the line by announcing to the press that the accident was "the direct result of the incompetency, criminal negligence, and almost treasonable administration of the national defense"—hanging words for an officer in uniform.

As the trial began Mitchell was overheard observing that "MacArthur looks like he's being drawn through a knothole." Probably it was more like being drawn and quartered—ripped asunder by countervailing allegiances. At heart, he was a technological progressive, though he wasn't drawn to flight like Ike, nor did he say much about airpower during the interwar period. But when the next war came he would pick it up instantly, in part because he was a brilliant strategist, but also, it seems likely, because he had been thinking about its implications for a long time—the great distances involved in the Pacific, always his primary focus, almost demanded it.

On the other hand, although MacArthur nearly always played the fool in civilian politics, his instincts for the service equivalent were better. Knowing the panel was stacked against Mitchell and that his words were damning, Doug did what nobody would have expected: he said absolutely nothing, "his features as cold as carved stone" to Mitchell. Conviction was inevitable, though it was a split vote, and the dissenter may have been Doug—a reporter claimed to have found a discarded not-guilty ballot that seemed to be in MacArthur's handwriting. But it was never clear.

What was apparent, or soon became so, was the demoralizing nature of his assignment at Fort McHenry. His first tour of the III Corps Area revealed abandoned harbor defenses, worn-out vehicles, and a largely land-bound Air Corps—all a consequence of nearly a decade of shrinking budgets and doomed to further deterioration during MacArthur's three-year tour. In no mood for bigger military budgets, the public was not only preoccupied with domestic affairs but increasingly inclined to listen to pacifists, along with those blaming the Great War on military industrialists and therefore bent on disarmament. Since MacArthur's duties included a considerable public relations component—a seemingly endless series of rubber-chicken luncheons and dinners—he had a ready-made platform to weigh in on these subjects. Not only did he denounce the movement to outlaw war (soon the Kellogg-Briand Pact of 1928), but it was during this period that MacArthur began equating disarmament with radical politics, and in general took a swing to the right that would incline him to question the loyalty and patriotism of those on the other side.

Perhaps not coincidentally, the emergence of his paranoid-tinged politics coincided with a period of personal misery. After the Philippines, Louise determined that Doug was stuck in a dead-end profession and wanted him out. He was equally adamant about staying. Their sex life deteriorated into acrimony. "The General is a buck private in bed," she later claimed, failing to concede that she was generally drunk when it was time to retire. Still, MacArthur tried desperately to keep the marriage going, in part because he had come to adore Louise's children, but also because divorce could be a dangerous blot on his record. It was no use. In the summer of 1927 Louise decamped with the children to New York, leaving Doug alone at Rainbow Hill. The paperwork would require until 1929, but the marriage was effectively over.

Then, as if to divert Doug's attention from domestic disaster, Fate threw something entirely different at him. In mid-September the president of the US Olympic Committee dropped dead. With the games set for the following spring, a replacement was needed pronto. Mac was already nationally known for his emphasis on sports while at West Point, so the choice may have been obvious, but it was still inspired.

Douglas MacArthur was nearly fifty when he followed George Patton to became the second of our big four to head for the Olympics. But leadership has little to do with age. "The outlook was not bright for our entrants, but I was determined that the United States should win at Amsterdam." So he turned it into a crusade and himself into head coach. "Athletes are among the most temperamental of all persons, but I stormed and pleaded and cajoled. I told them we represented the greatest nation in the world, that we had not come 3000 miles just to lose gracefully." "The fighting Dude" once more, his standard became their rallying point, and they won twenty-four gold medals, more than the second- and third-place countries combined, and set seven world records. Mac and the team returned to the open-armed welcome the Rainbow Division had missed, and were feted cross-country, while he provided a florid recounting for President Calvin Coolidge that began: "To portray adequately the vividness and brilliance of that great spectacle would be worthy even of the pen of Homer himself." Overcooked prose aside, the 1928 Olympics added no stars to his shoulder or decorations to his chest, but it was definitely a feather in his cap and seemed to have reenergized MacArthur after the breakup.

Further brightening his horizon was news, coming just after the games, that he would be returning to the Philippines, this time as commander of the entire department. "No assignment could have pleased me more," or put more miles between himself and Louise. But his third tour was destined to be a short one.

Things there were much the same, only more so. The friends he had made earlier had all moved forward, particularly Manuel Quezon, who remained president of the Senate and had gained political power steadily. By virtue of their respective positions their friendship deepened, and as they talked strategy the conversations always devolved to Japan and the Japanese. The US Joint Army-Navy Board (predecessor to the Joint Chiefs of Staff) had just estimated Japan could put 300,000 men ashore within a month of a war's beginning. Facing them would be 11,000 regulars and Scouts, a weak 6,000-man constabulary, along with an air force of nine outdated bombers and eleven fighters. Meanwhile, War Plan Orange hadn't changed its tune: hold Bataan and the island fortress of Corregidor and wait. Worse still in MacArthur's eyes were the provisions of the 1922 Washington Naval Treaty that prohibited more fortifications in the archipelago—not that he had the money to build them. And further complicating the situation, thousands of migrant Japanese sugar workers were pouring into Mindanao, something that pleased Quezon and the Philippine business community, but looked like a Trojan horse to MacArthur and Washington.

About the only thing MacArthur was able to accomplish in terms of shoring up the defenses was to finally secure the once mutinous Scouts their pay raise and benefits—a finger in the dike, perhaps, but one that did enhance his reputation among Filipinos.

He also managed to make two significant new friends. Henry Stimson has to be considered a serious candidate for the title "ultimate Washington survivor," holding cabinet posts intermittently for well beyond three decades. Besides a network of old-boy connections stitched at Andover, Yale, Harvard Law, and then Wall Street, he seems to have built his longevity on a kind of sturdy rectitude and basic good judgment. He already knew Patton, his riding companion back in Washington, and now it was Mac's turn, in this case because Leonard Wood died suddenly in August 1927, and Stimson was his replacement as governor-general. It was a brief acquaintance, cut short when newly elected president Herbert Hoover called Stimson back to be his secretary of state, but an important one for the future. Like Fox Conner, Stimson had an excellent eye for talent, and MacArthur's brilliance and dedication to the Philippines left him thoroughly and permanently impressed.

Doug wanted to take his place as governor-general, even managing to get the *New York Times* to announce his availability in a story. For there was part of him that longed to remain in Manila—in this case, the one below the belt. Louise Brooks had been a creature of both sexuality and wit; now Mac seemed bent on eliminating the latter. There may have been more to the Eurasian movie starlet Isabel Rosario Cooper than history remembers, but in the case of Douglas MacArthur she played her role as kept kitten with considerable authenticity—even calling him "Daddy," appropriate since he was almost fifty and she not yet eighteen. This was a relationship destined to end up in a bad place—Washington, DC—which his other mistress, Lady Luck, arranged as Doug's next destination.

Charles Summerall, Army chief of staff, scheduled to retire in the summer of 1930, recommended Douglas MacArthur to take his place. Not only had he gotten him Châtillon during the war, but the other candidates were too old to serve a full four-year term before mandatory retirement. Doug was the logical choice; but a controversial one that Secretary of War Patrick Hurley was initially against. He too had been to France and heard a lot about the Fighting Dude, both good and bad. But when he interviewed one of Mac's corps commanders in Europe, half expecting a negative assessment, he got instead a sustained burst of praise, which turned Hurley into a supporter.

Meanwhile, when aging eagle John Pershing heard rumors of the recommendation, he immediately took flight, directly to the president and desperately pleading "anybody but MacArthur." Hoover too had his doubts about MacArthur. An accomplished engineer himself, he was annoyed when Mac had earlier turned down an offer to become chief of engineers. But when the president went to Hurley with the idea of a compromise candidate, he now found his secretary of war firmly in MacArthur's camp and refusing to budge. So Hoover did: "I therefore searched the Army for younger blood and finally determined upon General Douglas MacArthur. His brilliant abilities and his sterling character need no exposition from me." Pershing's was even briefer: "He's one of my boys. I have nothing more to say."

However faint the praise, Douglas MacArthur had reached the Army's pinnacle, the youngest ever to make the climb. He was the first of our four to take his place in the starting lineup at the top, and the first to learn just how tough the majors could be.

LILY'S DEATH LEFT GEORGE MARSHALL REELING, AND THE WAR COLLEGE AND Washington suddenly felt stifling. He needed space and had enough pull to get orders for the Infantry School at Fort Benning, Georgia, as head of the Academic Department, with full responsibility for curriculum. He would look back on the four and a half years he spent at Benning as among the happiest and most fruitful of his career. They certainly witnessed the further evolution of the commanding figure history came to know as George Marshall.

He came to Benning in late 1927 intent on applying the Great War's lessons, preaching simplified orders, tactical flexibility, and field exercises over classroom time. His aim was to overcome the conditions that led to the western front and replace it with what he called "fire and flank"–based maneuver. But the Germans had already tried much the same in their last, failed offensive. Tactics alone wouldn't do it. Tanks needed to be integrated with tactics, and to be fair to Marshall that was hard to do with the single tank company he had to work with. But if Marshall's tactical oeuvre at Benning had little lasting impact, he certainly did.

Almost unconsciously but with an intuitive certitude George Marshall quietly used his time at Fort Benning to recruit and gather much of the command staff of what would become a decade later the largest army in US history. One hundred and fifty future generals would pass through Benning during his tenure, and another fifty were instructors, including Omar Bradley, Joseph Stilwell, J. Lawton Collins, and Walter Bedell Smith—key henchmen in that great but still imaginary army, figments in George Marshall's black book. All, it seems, instructors and students alike, were impressed with the commanding figure preaching a new style of war with the certitude of Ahab. The seed was planted in the minds of these future leaders that George Marshall had grown into something extraordinary—that the mask he now wore demanded deference and obedience to the voice that came from behind it. He was now truly a commanding presence.

But also a lonely, hyperactive one. During this period Marshall developed a thyroid condition that triggered an irregular pulse, coronary arrhythmia, and excitable behavior. At social occasions the characteristically taciturn Marshall would talk compulsively to what amounted to a captive audience of junior officers. He also had a rather telling penchant for staging mass pageants, using them to convey to visiting dignitaries his own exaggerated vision of the wholesomeness of Army life, a vision he would never cease to impress on the American public, even after a successful thyroid operation in 1937 calmed him down. Meanwhile, it went unrecorded what the marching soldiers carrying tennis rackets, baseball bats, and polo mallets thought.

Another steadying influence came into Marshall's life in 1929, as spring turned to summer and he met Katherine Tupper Brown, brought together by a local couple on the otherwise dubious grounds that George and Katherine had both recently lost a spouse suddenly. Her story was the more horrific— her husband, a prominent Baltimore lawyer, had been gunned down by a deranged client, leaving her with three teenage children. Like Lily she was bright and accomplished, having had a promising career on the legitimate stage cut short by a nervous disorder and subsequent marriage. Also like her predecessor, she was well bred, but had a lot more money, and was plainly healthier and more assertive—a sort of intensified Lily for a more prominent Marshall.

The relationship deepened apace, especially after Katherine gave him a five-week tryout with her two sons at her summer place on Fire Island, and Marshall passed with flying colors. By mid-October they were married, with Pershing standing as George's best man, which he certainly was.

It would prove an excellent pairing. Katherine was at once a homebody and more than capable of socializing when the occasion demanded, exactly in George's social sweet spot, always a small one. Yet the marriage would also be marred by tragedy. Like Lily, Katherine had a hidden flaw—her problem was with violent endings. Otherwise, the couple proceeded happily, and also away from Washington, at least for a while.

His next assignment, a yearlong sojourn ending in mid-1933, was suggestive: command of Fort Screven, Georgia, a truly tiny base manned by an understrength battalion of less than four hundred men. It certainly gave George and Katherine time to get to know each other better. But George was plainly too big a fish for this pond, and in the words of one biographer "took over the place like a paterfamilias," or as a young subordinate wrote, "as would a Southern planter his domain." These were increasingly hard times in terms of the economy, and so Marshall encouraged his men to plant gardens and raise pigs and chickens, along with serving them extra-large lunches to bring home to their families. Helpful as that may have been, Marshall's capabilities far exceeded Fort Screven, yet the assignment did not prevent him from being promoted to full colonel at its end, a time when such promotions were scarce. You have the feeling he was being taken care of, and also lying low and out of range of Douglas MacArthur. For him, that was wise, but for our other two rising stars, Ike and Patton, it was a posture they had no intention of assuming.

George Patton galloped in from Hawaii to Washington in 1928, attached to the chief of cavalry's staff. Patton could write and had ideas, so he was naturally enlisted into what was the central issue among equestrians: Was the horse to be replaced by smoke-belching machines? Georgie's job was to defend the steeds, to speak straight from the horse's mouth. Yet he was torn. Although he loved practically everything having to do with horses, Patton was too bright and experienced not to suspect that tanks were the future.

By 1930 he was advising the cavalry, without much success, to at least be open to mechanization if it became inevitable. That year, Chief of Staff Summerall ordered that a combined-arms force be established at Fort Eustis, Virginia, one in which tanks would operate together with infantry, cavalry, and artillery. It looked like the future until late in 1931, when Douglas MacArthur was forced to abolish it as the deepening Depression put tighter constraints on the Army's budget. So, further work on mechanization went back to the individual branches of the service, the opposite of combined arms, while Patton retreated to the War College at nearby Fort McNair.

Here, where the key element of the yearlong course was a research paper, Patton's thesis, "The Probable Characteristics of the Next War and the Organization, Tactics, and Equipment Necessary to Meet Them," argued that a small, professional force opposed to a large conscript army was the key to rapid and decisive victory in the future. However shaky the premise, Patton's erudition enabled him to turn the piece into an impressive essay on the history of warfare, one that was forwarded to the War Department as "a work of exceptional merit."

That jump cleared by a wide margin, Georgie cantered over to the greenest of Army pastures, Fort Myer, to rejoin some real horse soldiers as executive officer of the Third Cavalry. That perpetuated his stay in the nation's capital for three more years and put himself in a position, on a day not too far off, to show America his true colors.

Meanwhile, the Pattons resumed the congeniality campaign where they had left off: a rented house big enough to require nine servants, a string of polo ponies, foxhunting, and nonstop dinners and parties—so many and with such exalted guest lists that a number of local socialites took umbrage. But the Pattons were the social equivalent of an armored column, and barely noticed the frowns as they plowed ahead, scattering their largesse on the likes of General Pershing, Vice President Dawes, and Secretary of State Stimson as Georgie pursued his own distinctive version of networking. Thus, one evening when the phone rang and a cultured voice asked for Major Patton, Georgie, without hesitating to ask who it was, replied: "Why Francis, you damned old nigger lover! What in hell are you doing in Washington, and who the hell let you out of jail?" only to be corrected: "This is the Secretary of State. I called to see if Major Patton would care to come over to

Woodley for a game of squash rackets and a drink." Presumably, a game did ensue, and the long horse rides together persisted: Stimson knew what he was getting with Patton, but apparently, like a good judge of horseflesh, he recognized something special about this middle-aged major, something that distinguished him from the other men who wore the same uniform.

Dwight and Mamie Eisenhower were frequent guests at the Patton table; Ike's career was on the move too; and the friendship between the two blossomed accordingly. Washington had been good for, and to, Eisenhower.

He had joined the Army General Staff in November 1929, just after the stock market crashed, to work for Fox Conner's old friend Major General George Moseley, professionally a brilliant officer but politically much darker, so far to the right as to almost define paranoia. Characteristically, Ike chose to see the brighter side and dove into the project Moseley had picked for him.

Apparently aware that Ike's research paper at the War College had been on mobilization, he wanted him to look into the industrial component of that process, how America could gear up to produce weapons on a massive scale. This was an important question, especially since the previous effort to arm the Army for the western front had proved too little and too late. But at a time of constantly diminishing Army budgets and a climate of growing pacifism, one that tended to blame the previous war on "merchants of death," conducting a survey of industry's capacity to generate arms would seem to have been an almost fanciful task.

Ike took it very seriously, but the only person who really took *him* seriously was Bernard Baruch, the canny financier who had led the War Industries Board and had been waiting for someone who was interested. Through their talks Ike broadened his perspective and became an admirer of Baruch, who was widely influential, and in particular with New York governor Franklin Roosevelt.

Ike worked on the project for nearly a year, condensing his findings into a lucid 180-page document, "The M-Day Plan," which would soon be gathering dust at the War Department. But Moseley read it closely, and was impressed enough to recommend it to his boss the Chief of Staff. And so it was that Ike met Mac.

Eisenhower's report, though fiscally absurd at a time of shrinking budgets, was exactly the sort of "big thinking" Douglas MacArthur craved but found so scarce among his subordinates. That's why he liked Moseley—he had imagination, and now he had brought him another gem. Forthwith Mac personally put Eisenhower to work, expanding his report into a comprehensive plan for mobilizing US industry. It was the genesis, some argue, of what would become the "arsenal of democracy"—but more useful to MacArthur was Ike himself.

Douglas MacArthur may have been self-absorbed, but he was nothing if not alert. He can't have missed Ike's almost magical ability to get along with people, the magnetic smile, his instinctive ease swimming through the corridors of power; but probably best, the man could write. In Ike Mac must have seen someone who could lift the burden of prose off his shoulders, both bright and responsible enough to assume his voice and broadcast his message. That, in essence, was MacArthur's plan for Ike—to be simply an extension of MacArthur. That's how it was by 1932, and that's how it always would be, though it took a long time for Ike to figure that out, or at least to admit it to himself.

He was dazzled. Being around MacArthur made him feel as if he was finally at the center of things. Many years past the initial effect, Eisenhower still had to admit, "On any subject he chose to discuss, his knowledge, always amazingly comprehensive, and largely accurate, poured out in a torrent of words." Ike also remembered that, in an Army of micromanagers, MacArthur's sole office requirement was that the work be done. His staff was trusted to keep their own hours, and if anybody needed a week's leave, he had only to ask.

Still, Ike was not entirely blinded. He wondered why everybody spoke of MacArthur reverently as "the general," as if there were no other generals, or why he referred to himself in the third person like some kind of deity. These were just quirks to Ike when he first noticed them, not realizing they represented something deeper, or that it was about to blow up in everybody's face.

AS A NOD TO THE POWERFUL VETERANS' LOBBY, CONGRESS IN 1924 PASSED THE Adjusted Compensation Act, giving all those who had served during the Great War a "bonus," set up like an insurance policy, which could be claimed at full value in 1945. Over three and a half million of these certificates were issued, and as the Great Depression deepened, public pressure grew, as did support in Congress, for allowing immediate redemption. The payout was potentially substantial—up to $500 ($9,400 in 2022) for domestic veterans and $625 ($11,800 in 2022) for those who served in Europe. The Hoover administration and congressional Republicans argued it would require a recovery-stifling tax increase, and were therefore dead set against it. This did not prove to be simply an academic or political debate; this one was to be played out in the streets.

In June 1932, the rowdy but generally good-natured offspring of the AEF, the BEF—Bonus Expeditionary Force—descended on Washington. Roughly twenty thousand strong counting family members, they set up camp in Anacostia Flats and in some abandoned government buildings along Pennsylvania Avenue to add some up-close-and-personal pressure on Congress as it considered redemption. On June 15, the House passed its version by a vote of 209–179, but two days later the Senate rejected it decisively, 62–18. Legislatively, this left the bonus on life support, but it also left a lot of vets camped out just across the river.

Among them was Joe Angelo. Now an unemployed riveter about to lose his home, he had walked 160 miles from Camden, New Jersey, to testify for redemption in full uniform before Congress, and in the process recounted how he had won a Distinguished Service Cross saving George Patton's life. Joe was a virtual archetype of the BEF. Later analysis demonstrated that they were almost all bona fide veterans, the majority having served in Europe. Among them there were certainly left-wing agitators, many of them Communists, but the rank and file of the BEF just wanted the money. Even after the congressional defeat and June stretched deep into July, the memory of once being a real army held the vets in good order and relatively good humor. DC police chief Pelham Glassford—a year behind MacArthur at West Point and once the youngest brigadier general in the AEF—did his part to keep them that way, providing tents and bedding, medicine, and help with sanitation to keep the campers disease-free. The situation was far from optimal, but neither was it anywhere near revolutionary.

This was not the BEF seen from the White House. Herbert Hoover was convinced the majority weren't even veterans and that the whole thing had been in his words "organized and promoted by Communists." Thinking to prove his point, he supported, and Congress passed, a bill funding rail fare for veterans to go home (to be deducted from their 1945 bonus redemption), an offer that six thousand took up, leaving Hoover further convinced those remaining were left-wing rabble.

The War Department—Secretary Patrick Hurley, Douglas MacArthur, and his principal adviser George Moseley—were all rock-solid behind the president. As far back as May 24, Mac had met with Moseley and been warned of incipient revolution. After taking a night to consider his options, MacArthur told his subordinate to begin activating "Plan White," the General Staff's scheme to defend Washington against domestic insurrection, on July 28.

No believer in half measures, Moseley transferred a detachment of tanks from Fort Meade to Fort Myer, where he also put the Third Cavalry, George Patton's unit, on full alert, canceling all leaves and intensifying crowd-control training. He also arranged "to put a small force at a moment's notice in the White House grounds," a move he considered all the more necessary after J. Edgar Hoover, from the Justice Department's Bureau of Investigation, reported that some in the BEF had stockpiled "dynamite in a plan to blow up the White House."

Given the paranoia prevalent at the top, the results were all but inevitable. Prodded by the White House, the commissioners of the District of Columbia ordered Glassford to clear the occupied buildings, and in the process two veterans were shot and killed. Thoroughly spooked, the commissioners told the White House they needed federal troops to maintain order.

That did it. For the first time, three-quarters of the greatest gathering of military commanders in American history would take the field, working together. Like many a rookie outing, the results were not pretty.

MacArthur, being who and what he was, took the lead. There was no reason for him, as chief of staff, to be directly involved in the forthcoming operation, but he announced to the designated commander, the astonished General Perry Miles, that he had decided to be on the scene in person. Mac had always been drawn to the front lines like a moth to a flame, but this time politics were involved. "The incipient revolution was in the air," he told his bright sidekick Dwight Eisenhower, before sending his Filipino valet to fetch his uniform, the one, it turned out, festooned with all his decorations.

Unlike MacArthur, Ike was innately politically astute, and probably knew from the beginning that this was all a horrible idea. More than three decades later he remembered telling MacArthur exactly that. "By this time our relationship was fairly close, close enough that I felt free to object," and for his troubles he was ordered to put on his own uniform. Except for the fashion recommendation, this is almost sure to be balderdash, as a recent biographer, Jean Edward Smith, astutely recognizes. Put simply, no major vocally second-guesses a four-star general, especially when that spun up. Ike was part of the team, and we can assume he took his place on the field without complaint.

At Fort Myer the Third Cavalry quickly assembled and galloped over Memorial Bridge to the Ellipse to await the infantry and tanks. At its head was Major George Patton, who as executive officer had no official reason for being there either. But like MacArthur he sniffed revolution and potential combat and was helpless to resist, not that it ever would have crossed his mind.

By 4:00 p.m. all was ready, and the anti-BEF juggernaut headed down Pennsylvania Avenue, Patton and the cavalry taking the lead with sabers drawn. Thousands of spectators lined the way, some of them shouting to the soldiers, "Shame! Shame!" and, "You goddamned bums." The BEF demonstrators made no effort to resist as they grudgingly gave ground, but were still tear-gassed and whacked by the troopers' sabers' flat sides when they didn't move fast enough. Finally, as the evening began, the last of the BEF were chased across the Anacostia Bridge and back to their camp, leaving the Army and Douglas MacArthur lined up on the other side of the river.

It was roughly this state of affairs that confronted Herbert Hoover back at the White House. He was far from the brightest star in the political firmament, but even a dim one shed enough light on this situation to see a political disaster in the making. Not only did he give explicit orders to MacArthur not to cross the bridge, but he sent them in duplicate, one copy to be delivered by Moseley himself. MacArthur and his defenders later claimed he got neither message, but Eisenhower, who was there at the time, told it like it probably was: "In neither instance did General MacArthur hear these instructions. He said he was too busy and did not want either himself or his staff bothered by people coming down and pretending to bring orders." So over the bridge they went.

Almost at once the camp burst into flames; whether they were set by vets as a gesture of defiance or by the Army has never been established. But it was hard to argue that the whole scene wasn't grotesque. "The veterans . . . were ragged, ill-fed, and felt themselves badly abused. To suddenly see the whole encampment going up in flames just added to the pity one had to feel for them," wrote Ike remorsefully. Before it was over two babies were suffocated from the tear gas, and one little boy was bayoneted in the leg trying to save his pet bunny from the flames.

His night's work done, Douglas MacArthur got into his staff car and headed back to the War Department, warned by Ike that reporters were likely awaiting him. Apparently that was the point of his destination, staging a midnight press conference with Hurley at his side: "That mob down there was a bad looking mob. It was animated by the essence of revolution. . . . If there was one man in ten in that group today who is a veteran, it would surprise me." Then Hurley piped in: "It was a great victory. Mac did a great job, he's the man of the hour." That he was, and always would be—the man of this hour.

Back at the campsite the operation continued through the night, with a cordon being set up around its charred remains. One account had Patton hit by a brick and knocked from his horse during a melee in the darkness. Though unconfirmed, it was the sort of thing that happened to him. At any rate, he was not in a good mood the next morning, remembered one of his future combat stalwarts, Lucian Truscott. An infantry sergeant approached the regimental staff with one of the BEF members under guard. It was Joe Angelo. "When Major Patton saw them, his face flushed with anger: 'Sergeant I do not know that man. Take him away, and under no circumstances permit him to return!'" He then explained who Angelo was. "'Can you imagine the headlines if the papers got wind of our meeting here this morning!' Then he added, 'Of course, we'll take care of him anyway!'" Such was Patton's version of public relations, not to mention loyalty.

Douglas MacArthur remained similarly delusional that morning until he reached the White House and was thoroughly upbraided for not following the commander in chief's explicit orders—so thoroughly that he offered his resignation. But Hoover was in no position to accept it; he was trapped. Publicly he had to back Mac to the hilt or admit the whole thing was a horrible mistake.

It was more than that: it was a disgrace. The US Army had just led a cavalry charge against the pitiful remains of its older brother, the force that had won World War I. Worse still, Team America's best, or at least three-quarters of it, were at the core of this ignominious operation. Patton was out front with his sword drawn, MacArthur gave the orders, and Ike, though grumbling, was at his side to do his bidding. Only Marshall escaped by lying low in Georgia. In an antimilitary era the Army was already unpopular, and as the Depression deepened, the repeated use of National Guard troops to break strikes only added fuel to the fire. Now this travesty completed the picture, making the Army look like a bunch of Cossacks. It's safe to say its reputation had never been lower.

Some maintain that the episode cost Hoover reelection that November. It certainly didn't do him any good. But since he was facing one of the greatest politicians in American history, it seems likely he would have lost anyway. Up in Hyde Park the man who would succeed him, Franklin D. Roosevelt, was very much aware of what had happened to the BEF. Shortly after the incident he took a phone call from Huey Long, the radical and soon-to-be-assassinated governor of Louisiana, and upon hanging up told an aide he had been talking to "one of the two most dangerous men in the country." When queried who the second might be, Roosevelt replied, "The other is Douglas MacArthur."

The future did not look bright for the soldier side of Team America. But then initial impressions can be deceiving.

CHAPTER 6

SPRING TRAINING

Team America had a new manager. To Franklin Delano Roosevelt, when he took office in March 1933, the players who might someday take the field were of far less importance than the societal equivalent of the fans in the stands—or rather on the street because they couldn't afford a ticket. The Depression had hit rock bottom. He was focused on saving the banking system and, in pre-Keynesian mode, balancing the budget and lowering taxes. Ever flexible, he would soon reverse himself on the parsimony, but for a time it meant tight federal budgets, particularly the Army's.

FDR was a Navy man, had been its assistant secretary during World War I in his pre-polio days as a rising young political star. He loved ships in general, from the little models he collected to the new superdreadnoughts the Navy wanted to build now that the Washington Naval Treaty's moratorium on their construction had passed. They were expensive and took a long time because of the huge castings required for their armor, and meanwhile other elements of the Navy continued developing aircraft carriers, and even submarines, all of which cost money. FDR was in no position to be munificent, but if it came to a choice between warships and armored vehicles, fleets and ground forces, he could be depended on to come down on the nautical side. Worse still, the Army had just compounded its own unpopularity by chasing the BEF out of Washington, and was still commanded by the man who had come to personify the shameful episode, Douglas MacArthur. It's hard to imagine a less propitious starting point for what would become within little more than a decade by far the largest army in US history, still under the same wily manager, with the same Douglas MacArthur as one of his key paladins.

Along the way the relationship undeniably had its rocky moments, but it was destined to survive. They had a history that stretched back to their days in World War I Washington, where they knew and liked each other. They were both fundamentally men of the upper classes—distant cousins, actually—and though they had profoundly different views of the social order, they spoke the same patrician patois and acted the part in a comfortably predictable manner. There was another thing about Douglas MacArthur that soon served to put FDR at ease when he dealt with him. He was after all a political wizard, and must soon have realized that when it came to the civilian brand of the dark art, MacArthur was, if not quite a buffoon, enough of one to be wound almost effortlessly in or out of any scheme FDR envisioned for him. No longer seeing him as a threat, the president commenced to play MacArthur like a harp.

Nevertheless, there was still a lot of dissonance when he initially sat down with the instrument. No surprise, it was a tune about money. Hoover's last projected defense allocations had already envisioned severe cuts for the Army, but now FDR's budget director mandated further incisions that together amounted to a 51 percent reduction in total funding for the regular force. Not content with the surgical results, Congress voted a 15 percent pay reduction for all federal and military salaries, and was threatening to give the president the right to furlough on half pay up to four thousand army officers. MacArthur managed to scotch the furloughs in the Senate, but FDR remained adamant on the remaining budget cuts. A showdown was looming.

While MacArthur's name was poison among most New Dealers, he did have an ally in the secretary of war, George Dern, who realized that a 51 percent cut would all but disembowel the service. They both agreed on the urgency of convincing the president of the error of his ways, and in late March 1934 headed for the White House.

At the meeting that ensued, Dern led the way with a somber series of arguments that bounced off FDR like a string of annoying rubber balls, until he lashed out with his patented combo of patronizing sarcasm. Dern grew white, then silent.

His turn at bat, Douglas MacArthur "felt it my duty to take up the cudgels. The country's safety was at stake, and I said so bluntly," he recalled decades later. Blunt hardly describes what followed and then ended with the memorable phrase: "When we lost the next war, and an American boy, lying in the mud with an enemy bayonet through his belly and an enemy foot on his dying throat, spat out his last curse, I wanted the name not to be MacArthur, but Roosevelt."

It was at this point that the general realized just who he was confronting: not only the President, this was Franklin D. Roosevelt. His taunt had stripped away all the polite camouflage and indirection and revealed the inner man. FDR was gigantic and he was furious. "You must not talk that way to the president!" The words were simple but the delivery was crushing; it was as if he reached up from his wheelchair and sent Douglas MacArthur's elaborately inscribed mask of command spinning across the room. "I told him he had my resignation as chief of staff. As I reached the door his voice came with that cool detachment, . . . Douglas; you and the budget must get together on this." As they departed Dern told him he had saved the Army, but MacArthur was left vomiting on the steps of the White House, a posture that made it pretty clear who had won this encounter. And also, who was in charge.

The next time MacArthur would be more circumspect. FDR had come into office believing that the government's contracts with the airlines to deliver the mail were more a matter of sweetheart deals than a means of fostering long-range aviation, both commercial and military. When the head of a Senate investigation claimed to have found evidence of "fraud and collusion" between the Hoover administration and the airlines, Roosevelt stepped in, canceling the airmail contracts effective at midnight February 19, 1934. He did so after Dern, without consulting MacArthur or the head of the Air Corps, had assured him that the Army could step into the role with a minimum of preparation.

The fact was that the airlines and the aircraft manufacturers had used the subsidies to invest in technology to become increasingly efficient carriers—driven home when the World War I fighter ace Eddie Rickenbacker delivered the last bag of commercial airmail in the prototype of the revolutionary DC-3, the first truly successful airliner, and in the process broke the transcontinental record by five hours. The Army's effort at replacing commercial service he labeled "legalized murder."

It was certainly lethal. Not a great deal better funded than the ground forces, the Army's Air Corps was basically a collection of fair-weather flyers and obsolete single-engine fighters. Now they were asked to fly in winter and frequently at night. Dern's minimum of preparation translated into eight fatal crashes in the first eight days of airmail delivery. As the pilots adapted, the casualties declined through the spring, but it remained a rolling disaster. Roosevelt personally tongue-lashed MacArthur and his air chief—"When are those airmail killings going to stop?"—as if they were to blame. This time Doug had every reason to be furious: yet again the press and the public were blaming him.

When called before the Senate he defended himself, but cautiously. "I knew nothing about carrying the mails until I was told by [the] Associated Press." Yet he was careful to add that the Army always obeyed orders, it was a matter of duty. Nor did he attempt to deflect any more blame on Roosevelt for what was undeniably an impulsive and bonehead directive. By mid-May the airmail was back in the fuselages of the airlines, and everybody could move on, including Franklin and Douglas, working together with something that smelled like team spirit.

There was no denying that MacArthur was a diligent and farsighted chief of staff. He streamlined the Army's mobilization process in ways that would prove vital sooner than anyone expected. He also strengthened the Reserves and the college-based ROTC program, both key to a rapid expansion of the force.

Given his fiscal handcuffs, MacArthur also did a remarkable job fostering the development of three critical weapons destined to become icons of what would soon enough grow into the arsenal of democracy: the M1 Garand rifle, the B-17 bomber, and the 105 mm howitzer. All were revolutionary at the time of their introduction—a four-engine aircraft with a two-thousand-mile range and a three-ton delivery capacity; a true semiautomatic infantry weapon, once cocked delivering eight high-velocity rounds by simply pulling the trigger; and an artillery piece portable, accurate, and simple enough to be produced in numbers sufficient to create a wall of fire in front of American infantry units. During the course of World War II none of these weapons would be matched by America's enemies, and the latter two frequently provided the edge in sheer firepower that made up for deficiencies in training, experience, and bringing replacements up to speed. Even conceding the blemishes, just these three very efficient implements of death make it hard to argue MacArthur's tenure wasn't positive, at least in terms of the nation's security.

Nobody had a better view than Dwight Eisenhower; he saw everything, warts and all. After the BEF fiasco the bedazzled phase was definitely over, but Ike still retained an awe of his boss's knowledge and insight into military affairs. He was an acute student of people, and his role as MacArthur's amanuensis drew him inside the man's head to capture his thoughts, even his diction. "Ike got so he could write more like MacArthur talked than the General himself," one general staff officer recalled. And as Mac wandered the corridors of power, Ike was there with him, like George Marshall under Pershing, reconnoitering everything and quietly learning the ins and outs. Like a good poker player he always downplayed his understanding of Washington's ways, but Eisenhower was a keen student, and he learned in the footsteps of MacArthur. His esteem was even on display in a special letter of commendation from the general, which Mamie had framed.

Yet it now came with skepticism. Ike was always careful about his personal appearance, but his boss was capable of staring at himself in a mirror for hours. For a personality as highly developed as Eisenhower's, the man's vanity must have been akin to fingernails scraped across a blackboard. And the BEF episode was impossible to forget, in large part because MacArthur kept bringing it up. Specifically, in the spring of 1934 he filed a $1,750,000 lawsuit against a liberal columnist, Drew Pearson, for slandering him over his treatment of the marchers. Ike remained troubled over the incident, and probably disapproved of the legal counterattack. But he had no way of knowing that his boss was working hand-and-glove with the president, who considered Pearson a "chronic liar" and told his cabinet he had "authorized" the lawsuit. Then again, perhaps Eisenhower had strategic insights into the soundness of MacArthur's own defensive position that were missing at the top.

Isabel Rosario Cooper had followed Doug to Washington, where he set her up in the lap of luxury—a suite at the posh Chastelton Hotel, a fur coat, even a limo. Yet her heart did not necessarily belong to Daddy. At least not after the initial opulence rush wore off and boredom set in. Like everything in the general's life, she had her time in the day, and the object of his visits never varied. The relationship cooled. She took to clubbing and other men; Doug looked to his bank account. He tried to end it with a steamship ticket back to Manila, but Isabel had no intention of leaving. Shortly after, Drew Pearson, a man not without sources, caught up with her.

The first MacArthur heard of it was when his puzzled attorney informed him that the defense in the Pearson case planned to call a mystery witness, one Isabel Rosario Cooper. No mystery to Doug, he immediately turned to Ike, telling him to run down Cooper and talk her out of appearing in court. Of course that had to entail explaining exactly who she was, presuming the very perceptive Ike didn't already know. What he authorized Eisenhower to offer in return goes unrecorded, but it doesn't matter, because he never found her, since Pearson had Isabel stashed in Baltimore. No surprise, Ike didn't elaborate on his unsuccessful girlhunt, but it was not the sort of mission that makes a subordinate look up to his commander rather than fasten on his clay feet.

The squalid affair ended with Doug's surrender, dropping the lawsuit and paying Isabel $5,000 for a packet of his letters while Daddy. Admiral William Leahy, the future chairman of the Joint Chiefs and a friend of FDR's, speculated that the reason he settled was "he didn't want his mother to learn about that Eurasian girl." Pinky was always a stick in the mud of her son's sex life, but it likely had more to do with his position and image of himself. People of his stature in that era simply could not have their names dragged across the tabloids without consequence. Douglas MacArthur was in danger of being reduced to a caricature; he had no choice in the matter.

FDR likely agreed. A man of multiple affairs himself and so good at manipulating the press that through his long presidency the public never realized he was largely confined to a wheelchair, he found his sympathies in this matter and increasingly others with MacArthur. They may have been polar opposites politically, both in skill and orientation, but FDR could plainly work with the man, and he understood that in the military sphere MacArthur's knowledge and strategic instincts were unique. FDR also gave every indication of enjoying the experience, matching wits and jousting with a truly adroit and skillful sparring partner.

So it was that, as Doug neared the end of his term, FDR stunned Washington in mid-December 1934: "I have sent a letter to the Secretary of War directing that General Douglas MacArthur be retained as Chief of Staff until his successor has been appointed. . . . I am doing this in order to obtain the benefit of General MacArthur's experience in handling War Department legislation," and also because he didn't like the available candidates, one of whom was Moseley.

But the point was still well taken; this was a budgetary turning point for the Army. It wanted to increase enlisted strength from 119,000 to 165,000, the biggest force in fifteen years and requiring a budget increase up to $332 million. New Dealers in the Bureau of the Budget were not happy and hacked away big chunks of the proposal, but MacArthur took his case to Congress and got the force increase written into law and all but $6 million. In the War Department's annual report for 1935 he (actually Ike) was able to state: "This year definitely marks the beginning of a long-deferred resumption of military preparation on a scale demanded by the most casual regard for the Nation's safety and security." He got that right. Though antimilitarism was far from finished domestically, it was not hard to see that peace was falling apart both in Europe and Asia.

And Mac would soon have an opportunity to observe a Far Eastern version of the landslide close at hand, though from a familiar perspective. His friend and indefatigable campaigner for Philippine independence, Manuel Quezon, had at least some recent congressional success with the Tydings-McDuffie Act, which promised full sovereignty in 1946, at once way too far in the future but paradoxically reassuring. Soon to become the Philippines' first elected president, Quezon was a strategically worried man. Japan had already invaded Manchuria and northern China; its navy was expanding explosively; and Luzon, the northernmost island, was within three hundred miles of Nipponese air bases on Formosa (Taiwan). Meanwhile, his own defenses were the same, a rickety trip wire made up of enough American troops to keep up the pretense that they had to be reinforced, but little more in terms of actual protection. Militarily it was a gossamer lifeline, and Quezon was looking to reinforce it with a strand of rebar named Douglas MacArthur.

It was an appealing scheme for all parties. There was some back-and-forth, but the agreement that emerged had MacArthur remaining in the US Army but also becoming "military adviser" to the Philippines for an $18,000 salary, plus $15,000 annual expenses (combined, around $620,000 in 2022), and, it was later discovered, a percentage of the country's defense spending as a performance bonus. For Quezon, although he never would say it, his friend constituted a sort of strategic hostage: From his perspective, at least, who in America would abandon its most famous and flamboyant warrior? It also worked for FDR. To him Douglas MacArthur was the cheapest means available to shore up Philippine defenses, albeit mostly on a cosmetic basis. FDR was a profound anti-imperialist, but he viewed Quezon personally as a pest. If this gambit served to placate him, then he was all for it. Meanwhile, he now had a suitable candidate for chief of staff, so MacArthur could be dispensed with, though not forgotten. Hardly forgotten.

As his final meeting with MacArthur wound down and he rose to leave, the president looked up from his desk grimly: "Douglas, if war should come, don't wait for orders to return home. Grab the first transportation you can find. I want you to command my armies." This was not the usual smoke derived from his famous cigarette holder—in Roosevelt's eyes MacArthur had passed a crucial test. He was being sent to the minors for a while, but when the time came he had a rock-solid spot in the starting lineup. As it happened, he would not even have to return home, nor would he again for sixteen long years.

But before leaving in the fall of 1935, he turned to his right-hand man, and, to hear Ike tell it, made him an offer he could not refuse: "Although I was not ecstatic about the prospect of going to the Islands . . . General MacArthur was still Chief of Staff and was very insistent that I go along with him for a year or so. He said that he and I had worked together for a long time and he didn't want to bring in somebody new." Meanwhile, Mamie, never fond of hot places after Panama, was sticking to Washington, at least until son John completed the eighth grade. Nor did Ike ever succeed in pinning MacArthur down as to the length of his assignment, his sole concession being to allow him to pick a fellow officer to accompany them. Ike obligingly reached into his vast bag of friends and produced exactly whom he needed, Jimmy Ord, fluent in the Filipinos' colonial Spanish, a pilot capable of island-hopping, and a fellow big thinker. If Ike was dragging his feet in the great man's footsteps, his actions don't seem to reflect it. But by now he never did anything halfheartedly, and at least there were no Philippine football teams to be coached.

THE MOST POPULAR OF ALL NEW DEAL INITIATIVES AND ONE BELOVED BY Franklin Roosevelt was the Civilian Conservation Corps (CCC), a vast program to enlist 300,000 unemployed young men (over three million would eventually participate) and set them to work on an array of projects dedicated to improving the environment. Army officers were instinctively attracted to a program that gathered the nation's youth and pointed them in a supervised direction somewhere between the Boy Scouts and the armed forces. MacArthur jumped on the CCC with enthusiasm and also ulterior motivation. Major Eisenhower predicted: "We will lose no officers or men (at least this time) and this concession was won because of the great number we are using on the Civilian Conservation Corps and of Gen. MacArthur's skill and determination." But he was too optimistic, the cuts kept coming and his boss accordingly lost interest, transferring CCC responsibility to the Reserves when the fiscal tide finally turned in the Army's direction.

One officer who never lost faith in CCC was George Marshall—you might even say it gave him an entrance ramp to the top. Right from the beginning of the program in 1933, when he was back at Fort Screven, he embraced it wholeheartedly and continued doing so during his successive assignments, spending a very substantial portion of his time establishing camps and keeping track of their progress. "The greatest social experiment outside of Russia," he told a local Charleston club, his next assignment being nearby Fort Moultrie. He might have added Germany and Italy, since his version of a social experiment had more to do with vast armies dedicated to industrial warfare than with conservation projects. "I think he regarded civilians and military as part of a whole," one staff member concluded. But this remained a carefully hidden agenda, obscured by Marshall's infectious enthusiasm for the CCC and his success in expanding the program wherever he went.

FDR first met George Marshall when he visited Fort Benning as New York's governor-elect in 1928. Chances are the activities included one of George's civil-military mass pageants; it was the sort of thing FDR liked. And this initial good impression can only have been compounded by Marshall's herculean and well-publicized efforts on the CCC's behalf. It would be a vital connection for the future, but before he got there a formidable obstacle still loomed: Chief of Staff MacArthur.

The blow came in November 1933, in the form of orders to become senior instructor with the Illinois National Guard. They arrived through proper channels, but Marshall knew from whence they came and complained to MacArthur directly, only to be told that his well-known and recently demonstrated talent for working with civilians had been critical in his choice for this important post. He may have been needed in Chicago, but what George needed as MacArthur knew was more time with regular troops if he was ever to make brigadier general before retirement. It may not have been the perfect trap, but it was sufficient to hold George securely in the Midwest for two dramatically uneventful years as the sand ran out of his hourglass.

It was at this point that John J. Pershing took wing on behalf of his favorite junior eagle. He spoke first to Secretary Dern and then, critically, to FDR. In late May 1935 came the magic words: "General Pershing asks very strongly that Colonel George C. Marshall (Infantry) be promoted to general. . . . Can we put him on the list of next promotions?" It would not happen until Mac left town in October, but as soon as a new chief of staff—Malin Craig, a Pershing man—was installed, George became the second of our four stargazers to capture one and pin it on his shoulder. At last the future was wide-open.

His next posting would be with Regulars at Vancouver Barracks in the Pacific Northwest, and was preceded by a leisurely transcontinental auto trip with Katherine in a new luxury Packard he bought for the occasion. The couple spent twenty happy months among the tall pines of Washington State while George continued to give plenty of quality time to the CCC. But it was simply a holding operation. His real destination and permanent base would be the other Washington.

His orders assigning him as the Army's chief of the War Plans Division arrived in May 1938, but before he departed Marshall participated in an exercise in which he was assigned a much inferior force and sought to compensate with a night operation. At the time, there was some grumbling among subordinates, but the officer in charge of the formal critique praised Marshall's approach as imaginative and reflecting World War I experience. That officer was Major Mark Clark and was promptly filed in George's mental black book. This was not insignificant, since very soon, if you were an Army officer and your name was up there, it likely meant a place in the lineup to die for. For George Marshall would no longer be just a notional talent scout.

Good news greeted him in Washington. Craig told him his job at War Plans was simply a stopgap. As soon as possible Marshall would be made deputy chief of staff, then in a year, when Craig retired, put in position to succeed him. There was competition—most notably the relentlessly self-promoting Hugh Drum—but like Marshall none were combat officers, nor did they have his special connection with Pershing, not to mention his talent—at least from the Army's perspective. Yet the chief of staff position and the dazzling four stars that went with it were in no way guaranteed. This was the New Deal and still a time of powerful antimilitarism, so the Army could not just dictate who it wanted in the position. George would have to earn his extra stars.

Admittedly, his approach at his first big meeting with the man making the decision was unorthodox. It came at a November 1938 gathering in which the president called together his key Army and civilian advisers to outline his plan to build at least ten thousand military aircraft—the notion being to not only dramatically expand the Air Corps, but also potentially make substantial numbers available to post-Munich Britain and France, both being intimidated by the apparent growth of the Luftwaffe. As Marshall listened from the back row, the president "did the major portion of the talking . . . most of them agreed with him entirely[,] had very little to say and were very soothing." Not George; his contemplated army didn't just need airplanes, it required tanks and artillery and small arms, not to mention the men necessary to operate all this equipment. "He finally came around to me . . . I remember he called me 'George' (I don't think he ever did it again . . . I wasn't very enthusiastic over such misrepresentation of our intimacy) . . . and said, 'Don't you think so, George?' I replied, 'I am sorry, Mr. President, but I don't agree with that at all.' I remember that ended the conference. The President gave me a . . . startled look and when I went out they all bade me good-bye and said that my tour in Washington was over." That proved far from the case. The president didn't like being crossed in front of others, but he also knew an honest man when he saw him, and also a dedicated one.

That was the key to his relationship with Harry Hopkins—technically head of a string of three-letter agencies devoted to economic recovery, but really Roosevelt's most intimate adviser. He was an unlikely figure in any age, and to our own almost incomprehensible. He combined the heart of a social worker, which he was by trade, with the political brain of a card shark. Money meant nothing to him, nor did his health, which was chronically awful, made worse by his party-animal proclivities. He loved women, but his only true passion was public service, a fact recognized by FDR and the basis of a friendship so close that Hopkins would actually move into the White House for long stretches. He was more than FDR's most trusted adviser; he was almost his alter ego.

And Marshall not only won him over but, just as he had with Pershing, turned him into an actual friend—one he let call him George. Dedication attracts dedication, so it was actually Hopkins in the closing days of 1938 who took the initiative. He knew the deputy chief of staff was deeply concerned about the Army's capacity to expand and rearm and wanted to find out more. They instantly fell into deep discussion, one that quickly expanded from building more airplanes to all the shortcomings of an Army neglected for nearly two decades. Marshall was a spellbinder at this level; when it came to what was needed to expand and rearm on a gigantic scale, nobody in Washington could touch him. Hopkins was stunned and also convinced. On the spot, he urged his new acquaintance to visit the president at home in Hyde Park or at his retreat in Warm Springs, Georgia, so he could present his case one-on-one. This was not the sort of invitation Hopkins extended lightly, and again he was stunned when Marshall turned him down flat. His approach to FDR was the opposite of Hopkins's embrace, more like approach-avoidance, keep the man at arm's length always. Each groove was destined to spin brilliantly, but for the moment the result of their talk seems to have morphed into a job interview, the one Marshall believed sealed the deal with FDR.

On a quiet Sunday in April 1939, without telling anybody, including the secretary of war, the president summoned George Marshall to his second-floor study in the White House. He apparently got to the point immediately, offering him the chief of staff diadem, whereupon the two fell into a spirited discussion of how to defend the country in a world full of war. "It was an interesting interview," Marshall remembered. "I told him I wanted the right to say what I think and it would often be unpleasing. 'Is that all right?' He said, 'Yes.' I said. 'You said *yes* pleasantly, but it may be unpleasant.'" Again Roosevelt acceded, and the general rose. "I feel deeply honored, sir, and I will give you the best I have." With that, being George Marshall, he left. The appointment would become official only on September 1, 1939, but the new chief of staff had unofficially settled into his post months before, one he was destined to occupy until the fast-approaching season of war was finally over. Unlike MacArthur, the job fit him exactly. Meanwhile, FDR had his team captain, though his role would be largely limited to home games. The traveling schedule would be reserved for his other three sluggers.

One of them was George Patton, though at this point it would have been hard to predict. After his cameo role versus the BEF, Georgie would slide further into tawdry times, not surprisingly in Hawaii, where he was reassigned in the spring of 1935. It was a measure of his mood that he, now fifty and embroiled in what amounted to a midlife crisis of heroic proportions, announced to Bea that he planned to solo sail their forty-foot schooner out to his new posting. She was the sailor in the family and told him she wasn't going to let him "go and drown without her." Actually they left port reinforced by a crew of six, with Patton, who knew not a lot of celestial navigation, asking his wife, "We can learn, can't we?" It was fifteen days and 2,240 miles before they finally spotted Mauna Kea's volcanic cone and knew they had made it. The dock was filled with Hawaii friends, hula girls, and a band. It was a preface of things to come. The Pattons were back in paradise— never a good thing.

As before, the couple plunged into socializing with the islands' social elite. For Bea it also provided the impetus to dive again into Hawaiian lore, enabling her to finish and publish the novel *Blood of the Shark*, suggestively titled since she had plenty to be bloody-minded about. Meanwhile, her husband was fiercely traversing the polo pitch, and, while practicing for the climactic interisland games, took another epic fall on his head, one that left him asking Bea hours later, "Where the hell am I? The last thing I remember is seeing the ground come up and hit my face!" It seems to have had residual effects; as daughter Ruth Ellen recalled: "Whatever had happened to his head, it cut down on his capacity to carry alcohol . . . just a couple of drinks would make him tight. This upset Ma no end . . . and she told him so. . . . Sometimes he would take an extra drink while she was looking at him; just to 'show her.'" Trouble in paradise, and soon there would be more.

It arrived in the form of Jean Gordon—Ruth Ellen's best friend and also the beautiful twenty-one-year-old daughter of Beatrice's half sister—who sailed in for a family visit, one that decidedly did not enhance domestic tranquility. Ostensibly there to see Ruth Ellen, from the beginning Jean only had eyes for Georgie. He was low-hanging fruit, over fifty, balding, and, especially, in his own mind, nowhere near the man he wanted to be. Not just flattered, he was primed for action. Neither made any attempt to hide the attraction, but Bea remained clueless and then got sick just before a family expedition to another island to buy horses—Patton still being a member of the cavalry and this one of his duties. So Bea remained home with Ruth Ellen caring for her, while Georgie and Jean went equine-shopping. When they returned, it was apparent to mother and daughter that the wild horses had not dragged them apart, but together, and in the biblical sense. "It's lucky for us that I don't have a mother because, if I did, I'd pack up and go home to her now," Bea tearfully told Ruth Ellen. But then a note of steel crept into her voice. "Your father needs me. He doesn't know it right now, but he needs me. . . . So, if your husband ever does this to you, you can remember that I didn't leave your father. I stuck with him because I am all that he really has, and I love him and he loves me." With true Patton grit she would stand by her man. This was not the last of Jean Gordon, but in the end Bea would have her revenge, and it wouldn't be pretty.

Professionally, Patton's second tour in Hawaii was similarly tumultuous. He had arrived looking forward to working for Major General Hugh Drum, whom he had known in France, and was assigned to as his G-2 intelligence director. It was not his specialty, and it showed. Yet, through his torpid interwar years, Patton's military instincts remained saber-sharp. Thus in 1937 he generated an internal document he called "Surprise," in which he predicted a Japanese attack that startlingly resembled the real one at Pearl Harbor; nor was he impressed with War Plan Orange, the Navy's fantasy scheme to reinforce the Philippines.

But, like Drum, Patton was more concerned with internal subversion of Hawaii by the local Japanese community, and at the general's direction drew up a plan to prevent it. But in classic Patton fashion it was over-the-top, a scheme to "arrest and intern certain persons of the Orange race who are . . . inimical to American interests . . . [and] . . . retain them as hostages (not prisoners)." He even included the names and addresses of those to be taken. It's not clear how Drum reacted, but what Patton was suggesting was clearly illegal and not likely to be missed by his lawyers.

Meanwhile, Drum was increasingly put off by Patton's louche and luxurious lifestyle. The priggish son of an officer who had been killed in the Spanish-American War, Drum plainly resented his subordinate's easy connection with Hawaii's social elite, and, in particular, Patton's patented foul mouth when doing so.

Late in Patton's errant G-2 tour, this ball of resentments went critical and exploded very publicly. Fittingly, it came at a polo match, the Inter-Island Championships, played before packed stands, which included Drum. Out on the field Georgie was fiercely leading the Army team, and in passing was heard to yell with his unmistakable high-pitched warble: "Goddammit, Walter, you old son of a bitch, I'll run you right down Front Street."

Drum was enraged. He called Patton off the field, reprimanded him in front of everybody for using foul language, then relieved him as captain of the Army team. Truly knocked back in his stirrups, all Georgie could do was stiffly salute and begin to ride off the field. But then the cocaptains of team Maui, among the richest men in the islands, rode over and informed the general they weren't continuing without their friend, blandly claiming they had never heard him utter even the mildest oath. Some staring ensued before Drum called Patton back, warned him again about cursing, then left with his family as the game resumed.

You didn't do this sort of thing to George Patton, embarrass him in front of a crowd. Major general versus lieutenant colonel, it mattered not a lick. Drum lacked one thing Patton was bulging with—military talent. And soon enough he would use it to destroy Drum's career. Like Beatrice, revenge was a meal Georgie relished, be it hot or cold.

Some claim that Drum savaged Patton's efficiency report, but this is unlikely since the ambitious general also saw him as a way to Pershing and the chief of staff job—he actually asked Georgie to put in a good word for him. Yet at this point even a horrible rating probably wouldn't have been a career killer: whatever his blemishes, Patton had already impressed several who would truly matter that nobody knew the trade of war better.

Far more threatening was what happened next. After sailing back to the mainland through the storm that disappeared Amelia Earhart in June 1937, Patton began a long leave at the couple's Massachusetts estate before reporting to his next assignment at Fort Riley. But it turned out to be a lot longer than he expected. Early in the vacation, disaster, or at least near disaster, struck—as usual on the back of a horse. He was riding with Bea and got too close just as her mount's rear hoof lashed out at a fly, breaking Patton's leg instead, which sounded "like a dry stick snapping," he remembered.

They rushed him to a nearby hospital, but the injury was serious enough to unleash a wave of blood clots into his system, one embolism coming within seconds of killing him. Drugged for the pain, Patton bounced back along his previous lives as a warrior, he told his brother-in-law, until he landed on a large shield as two armored Vikings hoisted him up for the fiery journey to Valhalla. But then one shook his head, and they dropped him. "I guess they're not ready to take me yet. I still have a job to do." But this meant getting better.

He was lucky they didn't amputate the leg, which would have finished him in the service. But to get back on the team, being still in the cavalry, he had to literally climb back on his horse and look like a soldier. That required three months in the hospital and a further three at home in therapy, during which time a board of medical officers from Boston came to investigate if he might have been drunk or on drugs when the accident occurred.

He was cleared, but this still marked a true low point in Patton's life, one in which Beatrice caught him in the barn beating the horse that injured him with one of his crutches. Otherwise, being around his family didn't do much else to raise his spirits. Finally, an exasperated Beatrice implored her daughters: "One or both of you girls go down and talk to your father! He's lying brooding in a hammock about not knowing his children and the fact that no one loves him, and he's very depressed and I can't do anything with him!" The girls drew straws; Ruth Ellen lost, then found herself standing in front of her father with nothing to say. Finally, she remembered their favorite poem and asked him to recite it. As the stanzas rolled off his tongue his mood brightened; he even hugged her and exclaimed: "My, I certainly enjoyed our talk. I don't see enough of you girls!" The effect was likely temporary, but for Patton progeny this was about as good as it got.

And things did get better for him personally, a career turning point reached. The medical examiners pronounced him fit for limited duty, though he was still limping and wearing an iron brace that made his leg swell. So in early 1938 he was finally able to report to the cavalry school at Fort Riley, where he taught and, through rigorous exercise, was quickly able to regain his full fitness and mobility. Several months later, he was promoted to full colonel and sent to Fort Clark, Texas, where he commanded a regiment in the First Cavalry Division along the Mexican border. This too would prove a short assignment, during which he preached with increasing urgency that horses were obsolete. It was made most memorable perhaps by a family meal in Piedras Negras. When one of the guerrillas he had ambushed in 1916 recognized him and picked up his tab, Patton marveled: "He was one of Pancho Villa's officers! That goddam Yaqui shot at me and nearly winged me!"

In December 1938 the chief of cavalry called Patton at home, telling him he had to come back to Fort Myer and take over the Third Cavalry, since Jonathan Wainwright, a talented officer and a service favorite, was ruining himself with the bills and socializing the posting entailed. Only Patton among officers available was capable of picking up this tab. But he honestly didn't want to move. War clouds were looming and he was intensely involved in a series of combat experiments that gave vent to his energy and imagination. This was no time to be leading a bunch of show ponies. Still, the rapid movements during this phase of George Patton's career are also suggestive of some premeditation, a sense that he was being maneuvered into position, just as Eisenhower would be. Douglas MacArthur likely would have pointed to what he called the Chaumont clique—Pershing's boys from the Great War and in particular its fastest-rising star, George Marshall.

In late July 1939, Patton wrote Beatrice from Fort Myer about having "just consummated a pretty snappy move. Gen. George C. Marshall is going to live at our house!!! He and I are batching it. I think that once I can get my natural charm working I wont need any letters from John J. P. or anyone else." In all likelihood Patton's name was already listed in Marshall's proverbial black book. The general was happy to accept his hospitality while his own quarters were being readied, and perhaps get to know him a little better.

But by this time he probably had Patton pegged: give him enough rope, then jerk him around like a string puppet—a trick he later seems to have passed on to Ike. Here at Fort Myer it took the form of repeatedly ignoring Patton, fully saddled up, and riding instead with Molly, Katherine's daughter, in the mornings. Direct when he needed to be, Marshall could also be subtle and not above using his wife to convey a message. After Patton played the vulgarian at one too many Fort Myer gatherings, Katherine confronted him: "George, you mustn't talk like that. You say these outrageous things and then you look at me to see if I'm going to smile. Now you could do that as a captain or a major, but you aspire to be a general, and a general cannot talk in any such wild way." Clearly a warning, but Patton just laughed, which is exactly why Marshall and later Ike would have to continue jerking him around to keep him in the lineup.

That was the whole point—the man had talent, and Marshall never forgot it, even to the point of putting up with a certain amount of Georgie's nagging. "General Patton was up in years and, incidentally, would always talk to me about the age question all the time for fear we would apply it to him," Marshall recalled. "Well, he was the epitome of vigor and leadership and that sort of thing. He was the exception, and there were not many like him." The Army's leadership was aged, laden with World War I veterans, many long past their fighting peak, if not their expiration date. George Marshall had every intention of cleaning house, and he would use Patton and a few others as his sweepers.

THE LONG STEAM OUT TO THE PHILIPPINES IN LATE 1935 PROVED AN EVENTFUL one for MacArthur. "Out with the old, in with the new" would be the theme aboard the SS *Hoover*, definitely for Mac and perhaps even Ike, who left Mamie in Washington with some mixed feelings.

As Jean Faircloth boarded ship in San Francisco she couldn't help noticing a tall, imposing figure carefully leading a tiny aged woman dressed entirely in black toward her cabin. That was the last she saw of her, but Jean soon learned the man was General Douglas MacArthur and, as she dined at the captain's table, couldn't help fastening on him across the room sur-rounded by his entourage, all of them looking very military despite their civilian garb. Then at a cocktail party she suddenly found herself chatting with the general and his aide, finding him surprisingly charming. But just as quickly he excused himself, leaving her to conclude he was just being polite, at least until she returned to her cabin the next day to find a large floral ar-rangement and a note reading "With compliments of Douglas MacArthur." It proved the best bunch of flowers he ever proffered; for the two of them, it was the bouquet of a lifetime.

Like Pinky she was a Southern girl, from Murfreesboro, Tennessee, and it's not hard to imagine people referring to her as "cute as a button" practi-cally from birth. But beneath that pleasing and perky exterior there was con-siderably more. Despite her small-town background, she had already been around the world and also on separate journeys to the Orient and South America, using an inheritance of $200,000 (around $3,500,000 in 2022) to finance her peripatetic ways. Also testifying to her independent streak was the fact that she was nearing forty and had never married. Plainly waiting for the right man, she now had found him.

Yet it hardly proved a whirlwind courtship; rather, it took place at the stately Victorian pace that suited them both. The general dropped his mask only gradually, in large part because this time Pinky at eighty-four really was fading fast. By the time they reached Manila she was in desperate condi-tion and a month later fell into a coma and died. Ike understood how much his mentor depended on her emotionally, and remembered: "Her departure from his side, and from his counsels, affected the General's spirit for many months."

Meanwhile, Jean very naturally fell in with the members of his inner cir-cle, checking in as they suggested at Manila's Bayview Park Hotel, lunching with Ike frequently, and joining him and the others on the golf course or at the bridge table. A ticket for Manuel Quezon's inauguration, the highlight of the social season, arrived, and so did she in the general's limo with Ike and Ord. Later in the evening, MacArthur, decked out in his formal whites, personally introduced her to his friend Quezon on his big night.

Finally, Mac felt free to start the dating process, first to the movies or the theater, then to dinner, and only slowly toward intimacy. It stands to reason that Doug was tentative, having had his fingers recently and badly burned by the blowtorch of love, but this time he had found a soul mate, or his version at least.

Jean was looking for a real hero, one she could worship, and that was exactly what he needed in a woman—one who would call him "Sir Boss" without a trace of sarcasm. Should this sound demeaning to a contemporary reader, it's important to note that at times she would do her worshipping not from afar like the mates of our three other heroes, but right there with him amid strafing runs, bomb blasts, and an occasional Japanese warship. Through it all she barely flinched and never complained. Douglas MacArthur had found a partner virtually as brave as he was, yet he couldn't have known it at the time.

Apparently for propriety's sake he asked her to return to the States, then intercepted her liner in Hawaii. She did make it back to Murfreesboro in April 1937, but only temporarily. The phone rang, and at lunch she announced to her sister: "I'm going to New York City to marry General MacArthur." To Jean at least the call came as no surprise; they had been secretly engaged for nearly five months. Now and for the next twenty-seven years they would remain inseparable, moving in lockstep through what would amount to a minefield of tumultuous events.

Less than ten months after the nuptials, Jean gave Doug a very special gift, a son, whom they promptly named Arthur IV. Doug was nearly sixty, and when a friend wired "I didn't know you had it in you," he replied, "You know I didn't know it myself." Arthur proved to be his one and only bull's-eye, but, combined with Jean, this new addition left Doug happier than he ever expected at this late date. He responded by shamelessly pampering the boy, showering him with just about everything a kid might want, at least until things got dangerous. At any rate, Pinky's passing was now just a memory.

But while MacArthur's personal life was getting a huge boost in the Philippines, his professional performance had lagged. In his own mind—or at least his dreams—he had come to re-create the Philippine military. But emotionally distracted and always a delegator, MacArthur reflexively shifted the bulk of the work to Ike and Ord, a load that amounted to generating the military budget and defense plan for an entire country. Both self-starters, they did exactly that, only to be faced with fiscal reality. They initially believed they could engineer a marginally credible Philippine defense for around $24 million; instead, authorities in Manila could barely scrape together $8 million. On this basis, Ike and Ord thought they could support a Philippine Army consisting of one regular division and ten in reserve. MacArthur from on high insisted instead on thirty divisions as his bottom line.

It proved an archetypical dispute. At this point Douglas MacArthur had entered his grandiose phase. Besides his personal vanity and an enormous ego, there were practical reasons why he began to strut like a peacock. As in the case of the flashy bird, he was sending a message, trying to swell up well beyond his real size to impress an ambitious competitor, specifically Japan. It was barely a facade—more like a balloon—and one he mistakenly convinced himself would fly, or at least might fly for a while before being punctured. But it was basically all he had in his hand, that and the smoke-and-mirrors promises of aid from the United States.

For a good poker player, Ike surprisingly missed the bluff part and chose to focus on the grandiose. Of course, it was hard to miss, much less live with on a day-to-day basis. The first irreparable fissure in Eisenhower's disaffection had to do with rank—not his own, but his boss's bizarre scheme to be appointed field marshal in the Philippine Army. Quezon later told Ike it was MacArthur's idea, but this is debatable since it was in his interest to tie an active US Army general as tightly to the Philippines as he could. In the discussion that followed Ike remembered asking, "'Why in hell do you want a banana country giving you a field-marshalship?' Oh, Jesus! He just gave me hell."

Then, at an impressive ceremony at Malacañan Palace on August 24, 1936, Mac assumed the rank. Decked out in a uniform worthy of Eisenhower's description, one festooned with decorations, stars, and gold cordage, and wearing a special richly braided cap, the very one he would make famous in World War II, MacArthur became the first and only such dual-hatted general in US history. Ike, in the audience, found the whole thing "rather fantastic," funny even, since the field marshal had no army to field.

In that regard, when MacArthur continued to insist on thirty divisions, Ike and Ord calculated that this would blow up the Philippine defense budget by roughly 200 percent. When Quezon ultimately learned of the cost overruns he personally quizzed MacArthur, who responded by claiming it was his two subordinates who had come up with the plan without his approval, effectively shooting his own messengers. But he did so apparently without realizing how close Quezon and the ever-ingratiating Ike had grown.

When he got word of the conversation Ike was left reeling. "Every scrap of evidence furnishes ample proof that he is again executing one of his amazing 'about faces,' . . . I've got to decide soon whether I can go much further with a person who, either consciously or unconsciously, deceives his boss, his subordinates, and himself so incessantly as he does." If Eisenhower was ready to jump ship, he would soon have the chance.

Back in the United States, news of the event, complete with photos of Mac in his Ruritanian clown suit, did not go over well with the public, or with key members of the administration who wanted to smother any attempt to build up Philippine defenses. But it was the Army that brought matters to a head. Chief of Staff Malin Craig, not uncoincidentally a Pershing man, wrote MacArthur that his two-year detail in the Philippines would end on the last day of 1937 and that he would be ordered home to assume another command.

Mac's days in two hats were numbered; he'd have to shed one or the other. Notably, he chose the one with the most braid, informing Craig he preferred to retire from the US Army at the end of the year. Whether this "I shall not return" declaration exactly surprised official Washington is hard to tell, but in early October Craig informed MacArthur that the president had accepted his request.

He would retire as a four-star general, which, combined with his Philippine field marshal's four, added up to eight. Doug did like rank and regalia, but this was still a rather startling end to thirty-four years of continuous service. We now know it was vastly in his financial interest, that he was by a good margin the best-paid military officer in the world. He also truly loved the Philippines and the Filipinos, and his desire to defend them is not to be questioned. But you also get the feeling that back in Washington, at least in the White House, Roosevelt took this resignation with a wink and a nod and that he never exactly erased Douglas MacArthur's name from his starting lineup. At least, that's the way it turned out.

Quezon had extended similar feelers to Ike and Ord, offering to make them Philippine generals with appropriate salaries, but neither bit. By this time Ike was first and foremost a US Army officer, so the attractions of two hats, or another hat, were negligible. Yet, when presented with the obvious opportunity to return to duties at home, he chose to stay with MacArthur. It's also interesting that the Army let him stay. By this time Eisenhower was known widely to be a valuable officer, one whose career needed to be enhanced by command assignments with troops. Instead he remained in the tropics, the minder of a man in another uniform, which perhaps was the point.

From Ike's end at least, part of the explanation for staying was that he liked the place. With Mamie missing, the ever-gregarious Eisenhower quickly made social connections ranging from Mac's staff and their wives to Quezon and his circle, frequently spending weekends on the presidential yacht. Bridge, golf, and daily flying lessons filled his recreational hours, which seem to have been considerable. As always people were irresistibly drawn to him, ones useful for the future like Captain Lucius Clay, the son of a senator, or ones pleasing to the eye while playing golf or bridge, like Marian Huff, the attractive wife of MacArthur's naval aide.

Meanwhile, back in Washington, the almost equally gregarious Mamie made it clear to Ike in her letters that she too had a social life, parties to attend, and even escorts to take her. This went on for upward of a year before she realized that absence was not making Ike's heart any fonder, and if she wanted to preserve the marriage, the Philippines had to be her destination. School was out for young Johnny, and Mamie was out of excuses. So she and the boy headed west for the Far East.

They arrived in October 1936, just a month after their twentieth anniversary and Ike's promotion to lieutenant colonel. But he was not in a good mood when he met them at the dock. In the taxi, his version of the conversation was: "I gather I have grounds for divorce, if I want one." Quite a hello. Although son John remembered the statement as "terse but jocular," it was certainly edged with jealousy, this despite Ike's own wandering eye, particularly for Marian Huff.

Still, Mamie's presence did seem to rekindle the relationship, and she settled comfortably into life in the tropics, particularly after they moved into the air-conditioned wing of the Manila Hotel. She could not compete with sportif women like Marian Huff or later Kay Summersby; Mamie's strengths were in the social domain. With her in the picture the Club Eisenhower roared to life, particularly the distaff wing, now featuring luncheon and shopping expeditions with a new set of companions, Jean MacArthur and Marjorie Clay, Lucius's wealthy wife. Invitations were endless, and so were the silks and brocades and Chinese seamstresses who turned them into high fashion for what seemed like pennies. "We've been financially on easy street for the first time in our married life . . . We're having pretty near anything we wish," Mamie told her parents in early 1938.

But as his domestic life stabilized, Ike's job satisfaction and his relationship with MacArthur continued to deteriorate, as did the whole mission. Efforts to get the United States to provide more military aid had fizzled, typified by the 1937 joint visit by MacArthur and Quezon to Washington, one that yielded nothing but a meeting with FDR and the Navy's promise to send a flotilla of PT boats—fated ones, as it turned out, but still a pittance. Shortly after, Quezon's disenchantment was intensified when his field marshal resigned from the US Army, thereby losing his most attractive attribute.

It was in this context that MacArthur's next attempt at deterrence through sheer illusion was received. In January 1938, he informed Ord and Ike he wanted to bring units of the Philippine Army from all over the archipelago to Manila, where they would camp around the city and then hold a massive parade. When the two balked at the cost—money far better used to actually train those units—Mac responded by ordering them to start the planning. But as Ike soon learned when the Philippine president asked him what was going on, MacArthur had neglected to clear the parade with Quezon, who promptly scotched the festivities. An enraged Mac blamed Ike, claiming he never ordered his aides to actually begin planning the parade, only to investigate it quietly. "General all you're saying is that I'm a liar, and I am not a liar," Ike roared. "So I'd like to go back to the United States right away." MacArthur did manage to calm him down, even got him to extend for another year. But this marked a decisive turning point. Ike took his job seriously, and this sort of chicanery could earn only his long-term contempt.

It also appears to have weighed on his health. During this period Eisenhower suffered from a number of digestive problems, culminating in a bowel obstruction that required his hospitalization in January 1938. Although it cleared shortly before the scheduled surgery, while he was recuperating he learned that veteran pilot Jimmy Ord had been killed in a freak plane accident. It didn't stop Ike from flying—he earned his pilot's license a year and a half later—but it carved the heart out of his work environment.

As part of his one-year extension, Ike got to choose Ord's successor as well as undertake a three-month trip back to the United States beginning in June 1938 to see if he could scare up some modern weapons, or at least leftovers. To fill the vacant slot Ike seems to have known exactly who he wanted, Major Richard Sutherland from the Fifteenth Infantry in Tientsin. Although Eisenhower was largely apolitical at this point, he was still attracted to officers like Lucius Clay with political roots. Sutherland too was the son of a senator, and Mamie and Ike had socialized with him and his wife back in Washington. Unlike Ike, he'd gone to Yale, served in the AEF, then earned a commission in the regular army after the war. Also, he was not simply political, he was coldly and ruthlessly political, as Eisenhower would soon learn.

The trip did Ike's health a world of good. He got to take Mamie—who needed a minor operation—this time aboard a luxury liner for the long trip home. MacArthur even saw them off with a bottle of scotch. And Ike's reception in Washington was way more cordial than almost any lieutenant colonel could expect. Key was a meeting with Malin Craig, the same chief of staff who had forced MacArthur to retire. But Ike had his own network of connections to make time for him with the Army's first soldier. Craig not only met with Ike, but also made a series of phone calls to subordinates ordering them to help Ike on his scrounging expedition. Save for mortars, which were in short supply, Ike got decent quantities of every surplus infantry weapon on his shopping list. Moreover, Craig arranged for him to purchase a few aircraft from Boeing in Wichita and more guns and ammunition from Winchester in Connecticut. Given the necessarily low expectations, as Ike steamed back to Manila he had every reason to believe his mission had been a rousing success, one deserving MacArthur's gratitude.

Instead, he returned in September to find he had lost his job: Sutherland had replaced him as MacArthur's top subordinate, while he was relegated to Ord's old position. According to Lucius Clay, the switch had been prompted by rumors that a group of Philippine legislators wanted to replace MacArthur with Ike, who they believed was doing all the work. Meanwhile,

Sutherland was proving a dedicated sycophant, loading MacArthur with adulation in place of Ike's growing criticism. A furious Ike spilled his rage into his diary: "It is almost incomprehensible that after 8 years of working for him, writing every word he publishes, keeping his secrets, preventing him from making too much of an ass of himself, trying to advance his interest while keeping myself in the background, he should suddenly turn on me. . . . He's a fool, but worse he is a puking baby."

The next day he wrote his friend Mark Clark, apparently aware of his connection to the fast-rising George Marshall, asking that he get him assigned to the Third Division, hopefully as a battalion commander. But with Conner and Moseley retired, it took a while to arrange things to put Ike in a position for battalion command, though instead of the Third it would be the Fifteenth Division at Fort Lewis, Washington. He also learned that it was George Marshall, now acting as chief of staff, who had personally decided to terminate Ike's tour in the Philippines at the earliest possible date. "I feel like a boy who has been promised an electric train for Christmas," he wrote Clark gratefully. He and Mamie would leave on December 13, 1939. Later he said, "I got out clean—and that's that."

It was a notable parting. But also an inevitable one. Like two very big dogs occupying the same kennel, the personas of Douglas MacArthur and Dwight Eisenhower were too large and intense to coexist on a permanent basis. Nor did they share a common strategic perspective: MacArthur would always remain focused on Asia, while Ike was increasingly preoccupied with the European crisis.

He clearly knew what was going on in Germany. It is notable that he reached out to Manila's Jewish community, a number of them refugees from Nazi oppression, and it is also apparent that they took his concern seriously, offering him a job wandering Asia looking for countries willing to act as havens for other escaping Jews at the fantastic salary of $60,000 ($1,106,000 in 2022) a year for five years. The money was certainly tempting, but it required that he resign his commission, and as the situation in Europe continued to deteriorate, Ike felt even more compelled to get back to the real Army. Nonetheless, the offer was a clear indication of confidence in Eisenhower's capacity to operate at the highest levels of statecraft.

It was hardly a secret in Manila that Ike had grown close to Manuel Quezon, which in itself was quite remarkable. In the world of power politics a major/lieutenant colonel simply does not operate anywhere near the top of the heap. Quezon may not have been exactly a head of state, but Eisenhower's ease and adroitness in dealing with him was suggestive. The higher Ike climbed, the more sure-footed he got. Churchill, de Gaulle, Stalin—he would never hesitate to deal directly with any of them.

With Ike gone, Mac was left holding a very unsavory strategic bag largely alone. As with his women, he wanted worship and obsequiousness from his subordinates, and Sutherland and the others he gathered around him were perfect. They got things done, and they bathed him in praise, but they were hardly idea men. From this point, for better or worse, all the big thinking would be done by Doug.

Though his relationship with Quezon never completely broke down, the two of them were clearly moving along divergent paths. In short, Quezon wanted to save his country, while MacArthur wanted to defend it. Neither one had much chance, but both felt compelled to make the effort.

In the summer of 1938, notably without his field marshal, Quezon sailed for Tokyo with neutrality on his mind. This was not a season for appeasement, and talking the Japanese out of the Philippines was a bit like talking a hungry tiger out of a fresh kill. Still, returning with the few platitudes he obtained in Japan, Quezon concluded that further increases in Philippine defenses could only antagonize the tiger, and so he continued slashing budgets until the field marshal's army had shrunk to under 500 officers and 3,700 men.

But MacArthur was no quitter, and in the face of evaporating funding, he continued conjuring fanciful if ingenious schemes to defy the inevitable in his strategic imagination. He objected to the various iterations of War Plan Orange, which dictated that the thirty thousand US Army Regulars he no longer commanded retreat to Luzon's Bataan Peninsula and the island fortress of Corregidor and defend them until help came from the other side of the world. Instead, he asked the visiting correspondent Clare Booth Luce, "Did you ever hear the baseball expression, 'Hit 'em where they ain't'? That's my formula." When she asked him his formula for defensive warfare, he answered with one word: "Defeat."

Prophetic words, but premature. MacArthur believed that a force of fifty

PT boats and 225 aircraft, combined with 400,000 irregulars, could effectively thwart a Japanese invasion of the key island of Luzon. He never got anything like those numbers, but it really didn't matter since his plan was never instituted. Instead, he would find himself caught on Corregidor, or so it seemed.

AS ARMY CHIEF OF STAFF IN 1940 GEORGE MARSHALL FACED A HERCULEAN task, and one that was bound to be thunderously underappreciated. Although most of the country didn't want it, and FDR was not above pretending he didn't either, Marshall was both duty bound and intent on building a gigantic army capable of meeting and defeating the rampaging forces of what was now a world-girding Axis, but in particular those of Nazi Germany. Remarkably, while this was always an uphill slog, it was never a Sisyphean one.

That was largely due to Marshall's unique combination of organizational wizardry and wholesome ruthlessness. In the politico-military environment, he operated at a Zen-like level of awareness, one based on an uncanny ability to prioritize any problem however large, and then to delegate its solution. It didn't always work out, but in an era nearly devoid of computers, it was the only way forward, and Marshall was the master navigator. There is a temptation to think of the coldly unemotional George as computer-like himself. But actually, he had moved in the opposite direction. Computers are blazing fast at dealing with near-infinite detail, but after a thorough immersion in the facts and figures, Marshall had realized that the patterns they formed were far more important—the waves, not the water molecules, pointed in the right direction. The key to everything was to simplify and know what's important.

First in order of priorities, fleshing out Team America required millions of players for the enlisted ranks, gathered from a country entirely reluctant to give them up. There was but one solution—that great enabler of the era of military megadeath—conscription, and Marshall took the unenviable role of Pied Piper. But quietly at first, and taking a leaf out of his manager's book, he operated through others.

It was an election year and FDR, though not sure he would run for a third term, didn't want to preclude his options by declaring for a peacetime draft, so he effectively passed the buck back down to Marshall. But when approached by an elite group of conscription supporters, including the general manager of the *New York Times* and the highly influential Wall Street attorney Grenville Clark, he politely told them he too could not publicly support the group's proposed legislation for Congress. That was plainly for the record, since he had already assigned a general officer along with a conscription specialist—one Major Lewis Hershey, soon to become the czar of Selective Service, a position he would maintain for over three decades—to assist Clark and his team of lawyers in crafting a mandatory service bill. "I very pointedly did not take the lead," he later reflected on his cat's paw approach with the draft supporters. "I wanted it to come from others."

There was another problem. The secretary of war, Harry Woodring, was an isolationist and an enemy of conscription. Clark had a simple solution: replace him with his friend Henry Stimson, an ardent supporter of the draft, whom he just happened to know was available for another bout of cabinetry. It appears that Woodring was then set up by means of a memo authored by Marshall objecting to the sale of urgently needed B-17 bombers to Britain, then forwarded with the secretary's signature to the White House, where FDR, suitably enraged, could demand his resignation. The president wanted a bipartisan cabinet as the United States ramped up its military, and he already planned on bringing in Republican Frank Knox as secretary of the navy. Stimson, a revered patriarch of the GOP, would be his bookend at the War Department. FDR appointed them both on the same day in July. From the president's perspective, it all fit together perfectly. But years later Woodring still maintained the coup had come from elsewhere: "Marshall would sell out his grandmother for personal advantage." Decidedly not the case, but if it was for the draft, Granny (had there been one) might have had reason to worry.

Very suddenly conscription's political possibilities had been transformed by the news from Europe, where the Wehrmacht had swept across Holland, Belgium, Luxembourg, and France in the six weeks from May to the middle of June 1940. Soon after, photos were taken of Hitler staring across the English Channel.

The very magnitude of this catastrophe strongly suggested to members of the House and Senate that the United States needed a vastly enlarged army—even to those dead set against using it. Marshall only helped his case by retaining his cautious, nonpartisan approach. In early July, he warned the Senate Military Affairs Committee what a huge undertaking would be tripped by passage of a conscription bill: "We do not have the trained officers and men—instructors to spare . . . We lack the special training set-up at the moment, and we cannot afford to create it." Although actually a subtle way of asking for more money, it made George sound judicious, nothing like a rabid militarist or partisan shill. Soon after, the president went public with his support of the draft and the bill passed.

That done, Marshall apparently felt free to provide a franker perspective in his radio address on the day of the Selective Training and Service Act's enactment in September 1940: "It is only through discomfort and fatigue that progress can be made toward the triumph of mind and muscles over the softness of the life to which we have become accustomed." George Marshall not only believed in the military necessity of conscription, he endorsed its redemptive potential—the idea that the rigors associated with learning to kill other people would promote serendipitous qualities of civic virtue. His mental bedrock also told him that with the proper training and conditioning any young man within the ages stipulated could be turned into an effective combatant. This was the same sort of thinking that pervaded Europe prior to World War I, and history would prove to be a very poor witness for the viability of either proposition—in the prior or the coming conflict.

But in the meantime George Marshall remained convinced, yet also very concerned. Although Selective Service, by drawing on all males between twenty-one and forty-five, opened up a pool of 16.5 million, it also stipulated that no more than 900,000 could be in training at one time. Far worse, the legislation limited the obligation to a single year. This was the same ridiculous situation George Washington faced during the first several seasons of the Revolutionary War—at the end of each year his army was free to evaporate unless he could get its members to reenlist. One hundred and sixty-five years later our George found himself caught in the same trap, and with the same potentially slim prospects.

Now that it was actually vacuuming young men out of towns and cities, the draft got only less popular. Nor were the conscripts finding the uplifting qualities of basic training as obvious as Marshall thought; in fact, many were intent on OHIO—not the state, but "Over the Hill In October."

His infant army in danger of mandatory infanticide, paterfamilias Marshall flung himself into the fray to extend its lifetime. Notably it was largely a solo effort. Marshall was learning that his arm's-length approach to Roosevelt had definite advantages with Congress. His ostentatious fair-mindedness stood in stark contrast to the ever-maneuvering White House, his passion so obviously devoted only to the nation's defense. It was exactly the posture to convince skeptics to keep the boys in. He was never more convincing than on the night he held a dinner at the Army and Navy Club for forty Republicans and got a dozen to risk their seats to extend the draft. Arguably, that proved to be the difference in the 203–202 House vote on August 12, 1941, to continue conscription and keep the conscripts in the Army. The public may not have been happy, but it would have been even unhappier to discover four months later, on the morning of Pearl Harbor, that most of its army was in OHIO.

Marshall's first objective was reached: he had lots of bodies to work with. He wrote FDR that "the morale of the hostile world must be broken not only by aggressive fighting but as in 1918 by the vision of an overwhelming force of fresh young Americans being rapidly developed in this country." By December 31, 1941, the Army would grow to nearly 1.7 million troops, and by the end of 1942 it would number 5.4 million, drawn from a draft pool now up to 32 million—a true tribute to the proposition that more is better.

And given the circumstances, hard to deny. That central conundrum of modern warfare—the logical trap dictating that the individual shortcomings of citizen soldiers in mass armies could be made up for by ever-better weaponry—didn't persist for no reason. When properly applied, it continued to serve military ends, just at a huge and ever-growing cost in manpower and destruction. And at the then-current levels of technology—all based on the explosive power of chemical energy—it was likely inescapable, modulated only by the cumulative damage it inflicted.

Something else was coming that would change everything. But until it became blindingly obvious, George Marshall—whom Winston Churchill tellingly labeled the "organizer of victory"—was hardly one to question the soundness of a faulty tower built on better guns and ever more bodies, this being profoundly ironic, since he would soon include in his deep bag of re-sponsibilities the enterprise destined to blow it to smithereens.

But in the meantime, the manner in which he managed his conventional behemoth tells us a good deal about how he thought and operated. When it came to the issue of race, it was strictly in Black-and-White terms. Ignoring numerous instances of heroism by African American troops reaching back to the Revolutionary War, along with the antidiscrimination provisions of the Selective Service Act, Marshall announced, "It is the policy of the War Department not to intermingle colored and white enlisted personnel in the same regimental organization . . . this is not the time for critical experi-ments which would inevitably have a highly destructive effect on morale—meaning military efficiency."

Spoken like a true son of the Old Dominion and VMI, but also of the US Army that won the Civil War, which remained, despite sporadic efforts at reform, a rigidly segregated institution. Nor was Marshall alone among our four-star players: it's safe to say that had Patton or Eisenhower been put in Marshall's position at this juncture, neither would have moved to desegre-gate the Army. Only Douglas MacArthur, whatever his other political fail-ings, seems to have avoided the pox of prejudice, an immunity that can be counted as a significant factor in his success operating in Asia, a place that pretty much dumbfounded Marshall.

At home, though, he was on surer ground when he took on the next and concomitant challenge of building a megaforce on the fly: leadership. With-out it any army, no matter how large, is just a crowd. Some contemporary military authorities do maintain that the vital element here is actually pro-vided by the noncommissioned senior enlisted, but that would not have been Marshall's view or that of any of our key players. They, like practically all military professionals everywhere, were firmly committed to the tradi-tional view that officers not only ran the army, but were its critical com-ponent. And like an ashlar arch, the possibilities for success or failure only grew with each step toward the top, until the keystone was reached.

It was from this perch that George Marshall rebuilt the officer corps to accommodate the burgeoning force structure, and also to clean house. His first step was to employ the resident cadre of professionals as a basic framework, but elevated to a higher level of responsibility. Thus, in June

1940, responding to his prodding, Congress agreed to promote all officers one rank, subject to time in grade. This accomplished, he could then fill in and dramatically expand the vacated space below, adding 400,000 new officers in two years. Meanwhile, it took Marshall until the summer of 1941 to persuade Congress to give him what he really wanted—the authority to promote junior officers of exceptional promise regardless of seniority, and, no less important, to retire those he considered superannuated. With these utensils in hand, George Marshall could begin the cleaning process.

And soon enough George Patton and Dwight Eisenhower would reveal a stunning capacity for sweeping out the old, and in doing so set themselves up for prime spots in Team America's lineup.

The possibility of combat always improved Patton's mood, and the jolt supplied by the Nazi blitzkrieg of Poland in 1939 seemed to knock him out of his professional and even his personal funk. Still restlessly stuck with the show ponies at Fort Myer, he at least had time to think and let his imagination give vent to what was possible for tanks operating as the prime offensive instrument.

In the spring of 1940 he served as an umpire of the Third Army's maneuvers in Louisiana, during which an entire cavalry division was thrashed by an ad hoc combination of tanks and mechanized elements. Once the exercise was over, a group of insurgent officers, including Patton and his rival and steadfast tank supporter Adna Chaffee, met secretly in the basement of a local high school and reached a consensus that an independent armored force, free of infantry or cavalry control, should be formed.

Given what was happening in Europe, this was an obvious conclusion, and just a few weeks later, when the proposal by the "basement conspirators" reached George Marshall's in-box, he approved it without hesitation. Chaffee was put in command and tasked with the immediate formation of two armored divisions, one at Fort Knox and the other at Fort Benning. Apparently prompted by Marshall, Chaffee specified he wanted Patton to command one of the Second Armored Division's two brigades, a one-star general's slot and his ticket out of Washington and to the tanks, or at least the promise of tanks gathering at Benning.

In August 1940 Patton hit the place like a tornado on methamphetamine, establishing a presence that could not be missed if for no other reason than he seemed to be everywhere at once, racing around by car, plane, or a light tank he had fitted with a steamboat siren. He was a fanatic for realistic training and an absolute stickler for discipline and proper military dress, but this was a martinet with a clear purpose—getting his men ready for war—and they got the message and also more than a bit of his manic energy.

For Patton, they learned, no detail was too small, including ordering at his own expense nuts and bolts from the Sears and Roebuck catalog for the broken-down assortment of six hundred or so tanks and armored vehicles he inherited. He even designed his tankers a special uniform, one that reminded one biographer of a football player masquerading as a bellhop, and when he posed for photos, newspapers took to calling him the Green Hornet. He dropped the duds quick, but they were still revealing. His commitment was total, and that, he convinced his men, was what war demanded, so in a remarkably short time they began behaving like an actual fighting force. Besides his battlefield skills, this was what separated George Patton from even good commanders—he wore a spellbinding mask. By sheer force of personality he could weld a crowd of thousands and eventually hundreds of thousands into a single, psychologically coherent whole, and he could do it fast.

That was important at Benning because Chaffee had cancer and was dying, allowing Patton a window of time to command the whole Second Armored Division and thereupon work his magic. In doing so he snagged his first star in October, the third of our four to go celestial. In the spring of 1941, during maneuvers in Tennessee, he and the Second Armored roared through a tactical problem scheduled for two days in nine hours. Not long after he would pin on his second star, and it's not hard to imagine George Marshall's fingerprints on them—or at least Patton did.

In any case, things were finally breaking for Georgie. Even his personal life clarified. Both daughters had married, to his extreme satisfaction, West Pointers. Meanwhile, back in Washington, Bea didn't just continue to stand by her man, she also took to riding for him. An ultra-accomplished horsewoman, she drew the attention and soon companionship of Eleanor Roosevelt. So Patton, the archconservative with the stench of the BEF still recognizable, might have something like a friend in the White House, or at least not an enemy.

Ike too, though still bringing up the rear, was starting to move. Virtually from the moment he and Mamie reached stateside in December 1939, they both had the sensation that he was being pushed from job to job like a chess piece.

It began portside in San Francisco, where General John L. DeWitt ambushed him with a special assignment to figure out a plan to comply with the chief of staff's order to concentrate all California National Guard and regular army troops at a single site for a training exercise. Ike quickly realized that this was impossible and recommended two sites instead, but the colonel positioned between Ike and the general insisted on sticking to the original order. Ike's response was to steer the colonel to a beach exercise he knew DeWitt would be observing, but then to his surprise he found George Marshall there also: "Have you learned to tie your own shoes again since coming back Eisenhower?" was all he said. But DeWitt would have lots to say to the colonel as a result of Ike's conversation with him on the beach. Two sites were established and Ike was free to proceed to his assigned posting at Fort Lewis in January 1940.

At some point in his truncated stay there, he had a long conversation with his son John, resulting from the boy's decision to try for an appointment to West Point. He told him that realistically the highest rank a graduate could expect was full colonel, using himself as an example and rating his chances of making general near zero. "Of course, in an emergency, anything can happen—but we're talking about a career, John, not miracles."

Maybe not, but his fate still seemed directed from above. When George Patton wrote him about commanding a regiment in his Second Armored Division, Ike was excited, since it would practically guarantee combat should the United States enter the war. But the post never materialized. Then his study partner at the War College, "Gee" Gerow, whose career had moved ahead of Ike's, wired him: "I NEED YOU IN WAR PLANS DIVISION DO YOU SERIOUSLY OBJECT TO BEING DETAILED ON THE WAR DEPT GENERAL STAFF AND ASSIGNED HERE. PLEASE REPLY IMMEDIATELY." Ike wanted to stay with the troops at Fort Lewis, but he didn't have to make the decision, since a few days later Gerow abruptly withdrew the request, indicating there were other plans for Ike.

He would stay at Fort Lewis but not commanding troops—he would be the division commander's chief of staff, beginning in November 1940. That lasted three months, whereupon Ike was kicked upstairs to occupy the same post at corps level, still at Fort Lewis but now looking over multiple divisions and doing so as a full colonel, the presumed terminal promotion coming in early March 1941.

The next move was a real move, down to San Antonio in August and becoming the chief of staff of three-star Walter Krueger, the commander of Third Army, whose domain stretched from New Mexico to Florida. Like a number of the Army's top generals, Krueger was old (sixty), had risen from the enlisted ranks, and had an obvious shortcoming—his being born in Prussia and speaking with a heavy German accent. Still, he showed every sign of being, and would prove to be, an outstanding commander. The others, George Marshall suspected, weren't, and he had every intention of finding out before it was too late.

As we have seen, Marshall loved pageantry; indeed, the idea of moving large numbers of people was almost reflexive with him. At this point, as far as he was concerned, the behemoth of an army he was organizing remained a static entity, a collection of well-organized bodies. To stand any chance against a force like Germany's—just then rampaging across the Soviet Union in the early stages of Operation Barbarossa—it had to be able to move, and move fast and decisively.

So in August and September Marshall staged a vast exercise in the bayous and pine barrens of Louisiana, what he called a "combat college for troop leading." It was staffed with nearly half of the entire army, almost 500,000 soldiers, along with a full cadre of closely monitored commanders—most of the usual suspects, plus Ike and Patton.

The Louisiana Maneuvers were divided into two stages. In the first, General Ben Lear—who had also risen through the ranks but was known without affection among the enlisted as "Yoo-hoo"—had around 160,000 men, including both armored divisions, gathered east of the Red River, with the mission of crossing it and then destroying the 270,000-man defending Blue force assembled in the vicinity of Lake Charles and commanded by Krueger with Ike at his side, the two working hand in glove. True to his nickname, Yoo-hoo made an easily detected and slow effort at getting his troops across the river. Then he persisted in using the tanks mainly to protect his infantry rather than forging ahead, allowing Krueger and Ike to bottle them up. George Patton made a desperate attempt at a breakout, but Blue reacted quickly and turned the standoff into a rout.

In the second phase roles were reversed; Krueger and Ike had Patton with his tanks and were on the offense, while Lear had the larger infantry element to defend Shreveport. Forced to move forward in the midst of a category 2 hurricane, it took Blue two and a half days and a lot of high water before finally making contact with Lear's defenders, who had moved back in an orderly fashion. But then Krueger and Ike wrong-footed them by sending their most mobile elements, spearheaded by Patton's tanks, on a vast flanking movement. Exactly one day later he had covered two hundred miles and was on the outskirts of Shreveport nearly unopposed when the chief umpire, Lesley McNair, declared the exercise finished.

This was exactly the sort of result George Marshall had been looking for. As expected, Patton had performed strongly in both phases, giving every evidence he was worth his two stars and maybe more. More surprising perhaps were Krueger and especially Ike, whom his commander had the good sense to let do all the debriefings and press contacts. During the final critique, deputy umpire Mark Clark, who had designed the exercise for Marshall, was handed a telegram listing the officers slated by FDR for promotion to general. He announced every name without Ike's, even adjourned the meeting, before adding as everybody filed out: "I forgot one name—Dwight D. Eisenhower." The room roared with laughter and Ike broke into his patented smile: "I'll get you for this, you sonofabitch." They would remain friends, but it was not the sort of thing you did to Ike, and it said something about Clark's judgment.

Of course, this was a mere bagatelle compared with what happened to most of the high-level participants. Of the forty-two army, corps, and division commanders who took part in the Louisiana Maneuvers, thirty-one, including Lear, were either relieved or shunted away from combat roles. But still Marshall was not finished maneuvering or housecleaning.

In November 1941, less than a month before Pearl Harbor, he staged another huge exercise, this time in the Carolinas with a slightly smaller cast of 300,000, the idea being to test the durability of infantry with strong antitank defenses against armor. This time Patton's two tank divisions were to be paired against the vastly larger Third Army and VI Corps, commanded by none other than Hugh Drum.

So eager was he to give Patton a drubbing that Drum was already flagrantly violating the rules before the exercise even began. But it didn't do him much good, since on the first day of the final stage he was captured by Patton's men at a roadblock: "Good morning General, will you join me?" he was asked politely. The umpire spared him the humiliation of being taken like a trophy to Patton, but Georgie got his revenge—it was effectively the end of Drum and pretty much all the other deadwood, or at least what George Marshall considered dry rot.

The starting lineup was looking more and more like his black book, and on the whole that was a very good thing. Meanwhile, for the junior officers, the senior enlisted, and the troops themselves, these megamaneuvers proved a useful intermediate step toward combat, revealing perhaps most of all the logistical challenges of mobile warfare. Though still not ready for prime time with the Wehrmacht, most everybody was performing better, getting used to being an army.

And a modern army at that. These maneuvers not only included a heavy motorized component, but also significant air operations, including experiments with very light planes. All the new land arms categories were represented—not just tanks, but also emerging hardware like mobile infantry carriers, self-propelled artillery, and antitank and antiaircraft guns, not to mention the host of new warplanes the Army Air Corps had waiting in the wings.

But as yet it was mostly all imaginary. The truth was that Marshall's maneuvers had too frequently included trucks posing as tanks, broomsticks as heavy machine guns, and perhaps other useful implements impersonating weapons, none of which could disguise the fact that this was still an army largely without arms. In this regard, George Marshall mostly looked the other way.

IN 1482 A CASH-STRAPPED LEONARDO DA VINCI WROTE THE DUKE OF MILAN offering his services, "unfolding you my secrets," including a chemical weapon, a machine gun, and a vehicle "safe and unassailable, which will penetrate the enemy and their artillery." This was a job application, but he wasn't simply spinning fantasies ad hoc. Scattered through his still-astonishing notebooks are precise descriptions or drawings of not only these weapons, but also cartridge-based small arms, light mortars, rocket launchers, submarines, and aircraft—strongly suggesting that, when combined with his protean imagination, Leonardo's freakish technical and scientific aptitude allowed him to grasp with one sweeping glance the course of arms until the mid-twentieth century.

Of course, none of it would come to fruition in Leonardo's lifetime or for a long time to come, in part due to human and political reservations. But another reason was that the application of chemical energy to instruments of destruction—the gun being the first example—still awaited two key advances: the exploitation of the explosive possibilities of nitrogen-heavy molecules, and the development of hydrocarbon-fueled power sources that were both light and durable enough to propel his contemplated weaponry through a variety of challenging environments, frequently at high speed.

By the opening of World War I, both had progressed to the point that they would significantly alter the course of events, particularly in terms of casualties inflicted and also the regimes in which the fighting took place. The battlefield was now dominated by designer molecules, from the ultra-slow-burning cordite that so increased the range and accuracy of artillery, to the high explosives that made their projectiles so lethal, with pretty much every other weapons type also made a better killer through chemistry. Meanwhile, the internal combustion engine was turning Leonardo's dreams into reality, its Diesel cycle powering the German submarines that drove England to the edge of starvation while the Otto cycle not only motorized land warfare, but also brought forth a host of aircraft designed to fight from above. Intuitively, one way to escape death by nitrogen-heavy molecules might be through the mobility provided by internal combustion, but pretty much the opposite happened along the western front—nobody moved, and a lot of men died.

Although military aircraft and armored vehicles were barely out of infancy when the war ended, both had enormous potential, not simply in terms of technical advances, but also in the ways they might be used to accomplish tactical and operational, even strategic objectives. Despite worldwide antimilitarism and Depression-starved military budgets, the combat capability and employment of both—not to mention submarines—leaped ahead and were plainly dominating the gigantically rekindled conflict in Europe. Together airpower and armor were delivering a war of movement, but also delivering much greater quantities of those explosive molecules and at such ranges that mass bombings of civilian populations became a regular endeavor—death-dealing now being extended practically everywhere.

Fortunately out of bombing range, Franklin Roosevelt watched what was happening and realized that playing in this league demanded the latest in death-dealing equipment, and lots of it. As assistant secretary of the navy during World War I, he had been part of the effort to turn America's industrial might into the tools of war—one that had largely failed, despite our Allies' willingness to share the plans and technologies associated with their own well-developed arsenals. The so-called Preparedness movement had done little in that direction prior to America's belligerency, and by the time industry was mobilized and starting to spew out weapons, the war was over.

This time FDR knew that if the full capabilities of American industry were to be enlisted, he would have to start early and go big, for he had to equip not only a giant new army of his own, but also those of Hitler's enemies, increasingly desperate for arms. As far back as the fall of 1938, after listening to Hitler's Nuremberg speech, he had sent Harry Hopkins to the West Coast to quietly assess aircraft production capacity. But given the isolationist tilt of the moment, mobilizing American industry had to proceed sub-rosa, as if his guiding hand were not behind it. Also, FDR's progressive politics had embittered many key industrialists, some sufficiently to preclude—at least at this point—working directly with him. So, he would operate through an intermediary.

Instinctively, he turned to his friend Bernard Baruch, who had run the earlier industrial mobilization effort and brought it back from near chaos. But Baruch, who was nearly seventy, turned him down. As a fallback, FDR asked him for the three top industrial production men in the country to call. "First Bill Knudsen. Second Bill Knudsen. Third Bill Knudsen," was his reply, and a very good one.

In retrospect, three key industries spearheaded US war production: shipbuilding, aircraft, and automobiles. Critical in the first were the efforts of Henry J. Kaiser to build Liberty ships by the thousands, and Andrew Higgins to generate landing craft by the tens of thousands. So too were the gaggle of aeronautical entities that sprang from a callow adolescence to produce nearly 300,000 warplanes. But it was the automotive industry that basically filled in all the rest, and William Knudsen, as president of General Motors, operated at its fulcrum.

It is important to note that at this point the network of 850 American automobile producers and suppliers centered on Detroit constituted not only the country's biggest employer and pool of engineering expertise, it was also the most progressive industrial conglomerate on earth. Henry Ford had introduced the assembly line, which made millions of cars per year possible. But that in turn demanded the introduction of advanced inventory and auditing techniques, along with the time-motion studies critical to optimizing production sequences—all of them products of a singularly American partnership between business and academia. Add to that the rapidly evolving nature of automotive technology, and the fact that sales were demonstratively driven by rapid changes in style, and you have one complex industrial problem. Bill Knudsen's answer was what came to be termed "flexible mass production": assembly lines specifically designed for constant product change. Since cars at this point averaged around 15,000 parts each, this was no mean accomplishment, and one that forced the hand of competitors to follow suit, leaving the industry as a whole with an unprecedented capacity to mass-produce complicated, rapidly evolving machines—exactly what was needed to arm an army facing the crescendo of industrial war.

How much of this FDR understood is open to question, but Baruch's recommendation was enough for him in late May 1940: "Knudsen? I want to see you in Washington. I want you to work on some production matters. When can you come down?" That in turn was enough for the patriotic Knudson to immediately quit his job at GM: "I don't know, exactly, what the President has in mind," he told Alfred Sloan, the chairman of the board. "And still you go?" was his incredulous reply.

That was a good point. With FDR things were almost never cut-and-dried, and Washington was a bewildering place for a newcomer. Even before Knudsen spoke to FDR, Harry Hopkins, now living in the White House, was waiting for him, sizing him up with the whispered information that he had to work gratis. Knudsen didn't bat an eye: "I don't expect any paycheck." To a man like Hopkins, for whom public service was everything, these were, coming from a businessman, magic words; right there an alliance was formed with the man who had the president's ear. The meeting itself took place in a smoke screen of FDR's unique mix of charm and flimflam, and Knudsen probably left thinking he had been crowned czar of war production.

That was far from the case. Actually, he had been named to an amorphous body FDR called the Council of Defense Advisory Commission, made up of seven dignitaries, including labor leader Sidney Hillman and the University of North Carolina's dean of women, who didn't have any idea what she was supposed to do. It didn't take Knudsen long to realize this body had absolutely no legal status, but also that that allowed it to become whatever it made itself into. If it gave a coherent picture of what it took to turn America's butter into guns, then it would gain an audience and leverage—as long as it had the right leadership, of course, which he had every intention of providing.

Actually, he was operating in something close to a vacuum. When he asked what was available in terms of prior planning, he was given not Ike's comprehensive "M-Day Plan," but an eighteen-page document originally put together in 1922 and periodically revised. Knudsen worked with tangibles; he didn't need overviews, he needed specifics as to what was required. Yet when he went to George Marshall that summer and asked what exactly he wanted, the Delphic reply he got was: "Our greatest need is time." He was left to conclude that the reason he was having such trouble finding out exactly what the military needed was that nobody was sure what the American economy could produce, and in particular how long it would take. "The first thing to do," he remembered thinking, "was to get started on the weapons that required a long cycle in manufacturing"—naval vessels, tanks, warplanes, propellants, and high explosives. That done, shorter-cycle items such as trucks, soldier support equipment, and small arms could be acquired. Lead time was critical in making the whole thing work, and therefore a key factor in establishing priorities.

It was around this time that Knudsen began navigating Washington according to his own compass. Following the money, he reconnected with Jesse Jones, the dual-hatted secretary of commerce and chairman of the Reconstruction Finance Corporation, a man in charge of spending $50 billion ($915,000,000,000 in 2022) of the taxpayers' money and one whom FDR took to calling "Jesus Jones." If anybody in Washington could work miracles coming up with the money needed for America's arsenal, it was him. The feeling was mutual. Jones remembered marveling at Knudsen's knowledge and connections: "He seemed to carry in his head a picture of the whole manufacturing business in the United States," off the top of which he could tell which plants "would have to be greatly enlarged, which with only a little retooling were ready to go to work," and if more details were needed, he had the phone numbers of those in charge.

But if those corporate heads were going to nod in the affirmative, they had to be presented with an attractive proposition to build weapons. Retooling and hiring new workers was expensive, and the only practical alternative in what were still hard times was to offer them cash advances, and also the promise of making a profit—7 percent of costs was the figure stipulated. Such cost-plus contracts had earlier enraged Congress against so-called merchants of death, but on July 2, 1940, at Knudson's urging, it reluctantly gave its approval to the practice, one that would ensure the solvency of those undertaking to build America's arsenal.

By late 1940 it was obvious to the newly installed but vastly experienced secretary of war, Henry Stimson, that Knudsen and the advisory commission needed some actual power, the capacity to make decisions and issue directives that were binding. FDR liked the idea but insisted that the new Office of Production Management (OPM) have two heads, Knudsen and labor leader Hillman. "I told Knudsen that he was to be the chief figure," Stimson noted in his diary; "[he] had won his position during the last six months by his outstanding work here." Knudsen wasn't exactly happy, but he was a self-starter, already aware that in Washington, especially, it was always easier to ask forgiveness than permission.

On December 29 FDR gave his "arsenal of democracy" speech—Knudsen's own phrase, and a clear signal that the United States was now open for a lot more business. Add to that the earlier commitment to rearm Great Britain—one that would syphon off half the tanks and a third of the warplanes produced in the States in 1941—and the magnitude and the sheer ambition of the program becomes not just apparent, but stunning.

Yet the manner in which it spun out, and also Knudsen's role, are easy to misunderstand. He was never some all-powerful administrator working from a gigantic master plan. In late 1941 he revealed the Big Book, a weighty compilation of everything warlike being produced, likely to be produced, and the cost, around $50 billion (one-seventh of the $350 billion the United States actually spent). But it was more a compilation of statistics than a blueprint, never more than a rough guide. So Knudsen, far less a dictator than a central information node, was networking, putting people and capabilities together, and then letting them work things out on their own. With good reason the historian Arthur Herman considers America's mushrooming military-industrial complex, which so quickly began turning out a cornucopia of arms, an emergent entity, one formed and made complex by its own simple rules and constant iteration through interaction.

How Knudsen and the emergent system worked was demonstrated in the fall of 1940 when the British offered the plans to their Rolls-Royce Merlin aircraft engine—a brilliant design whose two-stage supercharger offered superior high-altitude performance. Edsel Ford offered to produce the engines, but then his father, Henry, an isolationist and antimilitarist, overruled him. After a red-faced confrontation with the old man, Knudsen simply called the head of Packard Motor Car Company: "Alvan, I'm leaving for Washington in a few hours and we need to talk." It was a productive conversation, eventually adding up to 55,000 Packard-built Merlins, many of them powering P-51s, often considered the era's best fighter.

The desperate days a month after Pearl Harbor, and the need to get the auto industry to produce an additional $5 billion in arms quickly, only intensified Knudsen's impromptu approach. He flew to Detroit and called a meeting of every auto baron he could get hold of, which was most of them. He then held something resembling an auction. "We want more machineguns. Who wants to make machineguns?" Hands were raised, and he moved on. "Who wanted to make turbine engine blades. Someone ought to be able to forge these things." More hands followed until he had everything he wanted. And so the arsenal was formed not with a plan, but hand by hand.

But wheeling and dealing on the fly is always a high-risk proposition, and Knudsen could stay one step ahead of bureaucratic retribution only for so long. "Look here Judge, I've been fired!" he announced shortly before he was to give a national radio address on OPM's accomplishments by early 1942. When Jesse Jones found out, he went straight to the White House and Harry

Hopkins, whose solution was to fireproof Knudsen by making him a briga- dier general. At midnight Jones called Hopkins back and told him Knudsen would accept nothing less than three stars, which he got, the only civilian ever to be appointed lieutenant general. They gave him a plane and told him to go practically anywhere, the start of a three-year, quarter-million-mile in- spection tour of over 1,200 factories and military units as far as New Guinea, a hegira whose only schedule was prompted by emergency and possibility.

Meanwhile, emergence was not simply evident in procedures for getting things produced, it was also apparent in the products themselves and, most significantly, in the motorized land armada it engendered.

At the primary level—the weapons to be generated—Detroit's engineers and those of the aircraft industry were not passive entities obediently trans- lating into steel and aluminum the designs given to them in the form of mil- itary requirements. They were the ones who knew how things worked, and when asked to produce something that wouldn't, they reflexively changed it. Therefore, when Chrysler engineers took a look at the M3 Lee tank the Army wanted them to produce, they completely reengineered it, as they would its far better successor, the M4 Sherman, this time with the help of GM and Ford.

Most ambitious in this regard was Ford's program to adapt the B-24 heavy bomber and its 1.5 million parts so it could be produced along what amounted to a vastly enlarged, one-mile-long auto assembly line at Willow Run. While the first ones produced leaked so much gasoline they terrified test pilot Charles Lindbergh, eventually the fuel tanks were sealed, other bugs removed, and thousands of serviceable bombers manufactured. But these examples only mark big and notable steps in a process better defined by countless little ones that were generating something unique and unex- pected.

As the automotive and aeronautical industries shifted to war produc- tion—a process completed in February 1942 when all commercial work ceased—line managers and engineers reached out to lower levels of the Army, its Air Corps and naval aviation, responding to their suggestions and requirements with new designs for parts and modifications, up to even en- tire self-generated weapons systems. Before long a host of possibilities be- gan to materialize, and when they worked were added to the whole, which itself changed. It is true that the Air Corps, because of its high-level commitment to strategic bombing as a war winner, more closely controlled the development of its force structure, especially when it came to multiengine platforms, but even here successful fighters such as the P-47 and P-51 (by adding a Merlin) were largely the result of industry initiatives.

Meanwhile, the Army on the ground was learning that when you hired Detroit to build your weapons, you were likely to find a motor in it. George Patton could proclaim, "We must fight the war by machines on the ground, and in the air, to the maximum of our ability," but it was the engineers who gave him that capacity, and likely more than he ever expected. Just about everything was assigned internal combustion: a legion of trucks was joined by half-tracks for rough terrain and amphibious "Ducks" for water obstacles—enough to ensure most troops would ride to the battle. Meanwhile medium and light tanks were backed up by self-propelled artillery and antiaircraft guns, not to mention mobile command, kitchen, and hospital units, and all of it supplied by still more motorized transport.

Then there was the jeep. Almost as an afterthought, the Army decided it wanted a four-by-four reconnaissance vehicle, and in June 1940 put out a fairly specific requirement with a deadline to produce a prototype in forty-nine days. The proposal was sent to over one hundred concerns, but only two responded, and there followed a very rapid and multiparty development process, so much so that by the fall of 1941 one participant claimed that "credit for the original design of the Army's ¼ ton, 4×4 may not be claimed by any single individual or manufacturer." Nor would reconnaissance be its sole purpose; very soon jeeps were everywhere, over 700,000 being produced by Ford and Willys.

So it was, without anybody really planning it, that the US Army achieved almost total mechanization, well beyond any force in the world, in particular the Wehrmacht, which in spite of the renown of its panzer units was still primarily dependent on soldiers' feet and horses' hooves for transport.

Obviously, this gave US ground forces the potential for extraordinary strategic mobility. Though they still had to establish themselves and break through prepared defenses, theirs was a vehicular juggernaut that—so long as it was covered from above by the tactical support planes soon pouring out of America's aircraft factories—begged its commander to move, and move fast. In the hands of a warrior as bold as George Patton it was almost the perfect instrument of aggression, but also one that would stop dead in its tracks (and wheels) if it ever ran out of fuel.

Being Detroit built, this was a gas guzzler of a force structure, one completely dependent on its logistical support, sometimes called "the pipeline," though it wasn't one, just a multitude of ships and trucks rushing back and forth. For the most part it worked spectacularly well, though on several memorable occasions it brought US forces to a grinding halt. Nonetheless, once the kinks were worked out, world-leading American logistical practices were quickly adapted to the military context, and soon proved capable of accurately delivering and distributing the enormous quantities of war matériel the American economy was generating—all at a time almost utterly devoid of computers.

But then, the numbers were easy to grasp, and if plotted on a graph showed a precipitous ascent among representative weapons systems produced. The United States made just under 4,000 tanks in 1941, over double the production of the previous three years. In 1942, the number rocketed to just under 25,000. Similarly, 1941's 617,000 small arms and 97,000 machine guns became, respectively, 2.3 million and 660,000 the next year. As did 2,000 20 mm antiaircraft guns become 92,000. Among heavy bombers, 318 B-17s and B-24s in 1941 became over 2,600 in 1942, along with 136,000 aircraft engines and a total aircraft production of just short of 50,000—a number that made FDR's November 1938 proposal of 10,000 look puny. And the weapons would just keep coming; the United States would not simply outnumber the opposition's arms, it would overwhelm them with numbers, all produced in a landmass free of bombing and the threat of invasion. This was definitely Team America's edge in the coming season of all-out competition.

But it wasn't necessarily a manifestation of American exceptionalism, as some might have had it. At the time, George Patton could plausibly proclaim: "The Americans, as a race, are the foremost mechanics in the world. America, as a nation, has the greatest ability for mass production of machines." But seven decades later the words ring hollow, a mythic delusion dispelled by a European and Asian world caught up. The vertical integration of American industry's golden era turned to silicon-powered, just-in-time logistics and global supply chains, much of it invented elsewhere.

But this misses a larger point. This great outburst of productive energy—one we tend to look back on favorably as manifesting qualities now lost—was devoted to generating a profusion of death machinery serving the logical trap of industrial war into which humanity had fallen. Thanks to the benefits of internal combustion and high explosives, there was now no such thing as a noncombatant. Worse yet, the fantasies of Leonardo made real and turned against everybody gave vent to a terrifying human capacity for killing, one that grew only larger through ease of operation and labor-saving devices. True enough, Team America's opponents were far worse, but playing in this league demanded a complete devotion to winning no matter how high the death toll, leaving humankind apparently locked in an endless feedback loop of synchronized devastation. Fortunately, something much more powerful would soon put an end to the game, but not before many innings of terrible death-dealing.

CHAPTER 7

DEVIL SEASON

For America's Team, and particularly its manager, Pearl Harbor constituted the wrong game, at the wrong time, in the wrong theater. FDR was a keen student of naval history, and was very much aware of the Japanese proclivity for surprise; he had even predicted they would open a war with the United States in this manner. But in his eyes a far larger menace loomed: the Wehrmacht had conquered most of Europe and was presently running rampant over Russia, while the US Navy for all practical purposes had already joined the fight for the Atlantic, where German submarines were threatening to cut England's lifeline to North America. That larger menace probably acted as a cognitive filter, which left FDR and Team America awash in information, including intercepted and decoded Japanese diplomatic instructions, but still unable to quite perceive the shape of an attack on the way. Then surprise, like a hard right to the jaw, left them reeling, confused and slow to react appropriately.

Among our four all-stars, Ike—being the last to pin his on in September and still with the Third Army in San Antonio—went to the kitchen when he learned of the Japanese attack, and made vegetable soup. George Patton, who had predicted an attack on Pearl Harbor, was predisposed to be happy about the coming of any new war, but far from this one in Fort Benning, Georgia, and soon to be sent to the California desert, in the middle of nowhere, to train troops. George Marshall should have realized that Army commander general Walter Short in Hawaii had reacted incorrectly to the warning message he had been sent—concentrating his planes against sabotage rather than scattering them—and instead went on a leisurely Sunday-morning horse ride. Meanwhile, in Manila, Douglas MacArthur had plenty of warning and a lifetime of combat experience to guide him, but reacted as if he were sleepwalking—allowing the Japanese to shatter his air force on the ground and thereby making their conquest of the Philippines practically inevitable. Altogether, not a promising reaction.

But there is a difference between stunned and stupid, between knee jerks and planning, between deep planning and planning so deep it would be revealed only at the last moment. That was the hidden dimension of Team America, and it played into not just strategy, but grand strategy.

To illustrate, we might consider why Patton was almost immediately sent out there in the desert, and we would likely conclude that one way or another it was because George Marshall wanted him there. This is suggestive in light of Marshall's insistence on a Europe-first strategy, to the point of proposing Operation Sledgehammer, an entirely premature and probably suicidal cross-Channel invasion for the fall of 1942. But if he was truly serious about this, and not simply about levering a commitment from the British to eventually invade France, why was his ace paladin already headed for the desert and the terrain that resembled no other strategic objective save North Africa? Strategic feints are hard to document, but in this case prima facie evidence points to Marshall understanding from the beginning that North Africa would be our first invasion point, and also to his being, to say the least, a deep player.

And there was a much deeper game being played, one that truly marked the changing of history's gears, a game so fraught with significance that it is worthy of digression even in the face of the rush of events that followed December 7, 1941.

THE ORIGINS OF THE ATOMIC BOMB PROGRAM HAVE BEEN RECOUNTED MANY times, but for our purposes the process began in October 1939 when Albert Einstein, the formulator of $E = mc^2$, wrote FDR about what might happen should the Nazis manage to make it work for them in the form of an atomic bomb, also suggesting the Americans might want to build one of their own. One sign of political genius is the ability to move earlier than the rest in situations of great danger or great gain. In this case it was both, and FDR responded immediately, setting up within the month a committee with military and scientific members capable of exploring the issue.

On the basis of what they found and further research, two years later the president created the Top Policy Group to advise him on atomic energy and oversee the vast project to build a nuclear weapon. The Top Policy Group had five members: Vice President Henry Wallace, two scientists—Vannevar Bush and James Conant—along with Stimson and Marshall. The vice president had a formidable intellect and still had FDR's trust, but the Army team dominated—it brought organization and competence, along with the sheer size necessary to cloak the program in secrecy when budgets started to climb. The day before Pearl Harbor, the Army Corps of Engineers was named to manage what would become the Manhattan Project, and from that point George Marshall was its administrative focal point, but in an utterly Marshall-like way. If history's gears were shifting, he wanted an automatic transmission. With his Zen-like knack for delegation, he left everything to a key subordinate, then kept an eagle eye on him from a distance.

By the summer of 1942 none of the Top Policy members were satisfied with the progress made toward developing the bomb, so the Army changed management. Leslie Groves was chosen largely because he was successfully overseeing the construction of the Pentagon, easily the world's largest office building, in record time. That was good enough for Marshall. And Groves proved to be fully up to his mark; both ruthless and relentless, he had no trouble making big decisions and telling his boss later, exactly what Marshall wanted in subordinates.

As long as progress was being made, which it was, Marshall viewed his own role largely as a facilitator, primarily making sure the program was adequately but otherwise invisibly funded. Critically, he approved channeling 20 percent of all Army weapons development money to the nascent project. He also played a vital role with Congress, essentially telling its members: "Give me the money and don't ask why," and actually getting away with it—even with Senator Harry Truman, the congressional point man for waste, fraud, and abuse.

Otherwise, Marshall ruled the program with a loose but occasionally abusive hand—much in the way he dealt with Patton. This sort of passive-aggressive management was illustrated when Groves arrived at Marshall's desk to ask for another $100 million with the promise of quadrupling the output of fissionable material. Marshall told him to wait until he had finished a correspondence. When he was done, he looked up and Groves quickly got Marshall's signature approving the money. "Then I said to General Groves, 'You may have wondered why I kept you waiting. . . . Well I was writing out an order to Burpee for flower seeds to the total of $3.94.' Without batting an eye Groves said, 'You don't need to apologize . . . perfectly all right' got up and walked almost to the door before he turned around and gave me some sign that he got the point of my remark"—in other words, "I'm completely behind you; but don't ever forget who is in charge."

Vannevar Bush later complained that neither Ike nor Marshall were "interested in the evolution of modern weapons and their impact on warfare," and at least in the Chief of Staff's case he was right. How much of the science Marshall actually understood remains unclear, but he himself admitted, "I would spend so much time with the Encyclopedia Britannica and the dictionary trying to interpret . . . that I finally just gave it up, deciding that I would never quite understand."

On the other hand, it's not hard to get the point of $E = mc^2$: the potential bang equaled the mass of fissionable material times the speed of light multiplied by itself, a monster number by practically anybody's reckoning. Since the program was shrouded in secrecy it's hard to exactly pin down Marshall's motivation, but this explosive potential he seems to have understood, and it serves to explain his consistent support for the bomb project's huge and secret budget. It was a high-risk bet, but with a potentially huge payoff, though Groves consistently lowballed the possible yield in his progress reporting. And the expense, while huge at around $2.2 billion ($33 billion in 2022), was markedly less than was thrown at the B-29 long-range bomber, which ended up costing $3 billion ($44 billion in 2022). Finally, there was the threat of a German bomb to keep Marshall and the other Top Policy members motivated.

Meanwhile, under Groves's iron hand the program accelerated from theory toward reality when, on December 28, 1942, FDR approved the transition from research and development to actual production of fissionable material—plutonium at Hanford, Washington, and highly enriched uranium at Oak Ridge, Tennessee, both huge facilities. Ultimately, the Manhattan Project would employ around 500,000 people and consume nearly 1 percent of the nation's electricity, all of it under a deep but far-from-impenetrable cover.

Yet one thing the Manhattan Project did not include was measured consideration of its implications. With the possible exception of a few scientists, no one was able to wrap his or her mind around the magnitude of what was being attempted, what it might mean strategically, much less for the very future of war. Granted, no matter what the science indicated, until the thing actually went off and revealed its true nature, it was easy, even reasonable, to suspend thinking about if and how it might be used. So it was that America's military brain trust marched in lockstep and obliviously toward a mushroom cloud that would change everything.

MEANWHILE, WORLD WAR II WAS NOT GOING WELL FOR TEAM AMERICA. HITler did obligingly declare war on the United States four days after Pearl Harbor, removing at least that necessity. Otherwise, the situation was grim and nowhere worse than in the Philippines, where Douglas MacArthur—now back in the US Army again, wearing four stars—and 31,000 of his American troops were trapped and half a world away from relief.

Basically, Mac had made a mess of things. Not only had he lost his air force, but his whole strategic approach had been a mistake. He correctly concluded that Lingayen Gulf—a large bay located halfway down the east coast of Luzon and promising easy access to Manila—would be the Japanese objective. But the decision to stop them there, on the beaches, was entirely misguided. It wasn't his fault that the three US submarines that had the Japanese invasion fleet in their periscope crosshairs were equipped with Mark 14 torpedoes that refused to detonate. But other than that, MacArthur's strategic undoing was his own. While his beloved Scouts fought well, the rest of the Philippine troops he had counted on to hold the beaches simply melted away, and the invasion that had begun on December 22 quickly metastasized and headed toward Manila.

Less than two days later, MacArthur switched back to War Plan Orange's original scenario: turn Bataan's defensible terrain into a killing zone for Japanese; use Corregidor, the fortress islet just off the coast, to block Manila

Bay; and then hang on and wait for help. It was a much better plan, but MacArthur had fatally neglected one key element: food. The time he wasted on the ill-conceived Lingayen Gulf defense could have been far better spent bringing supplies down to Bataan. A single depot on the central Luzon plain held fifty million bushels of rice, enough to feed all Mac's troops for four years, but now it and just about everything else was out of reach.

American soldiers under General Jonathan Wainwright struggled desperately and skillfully, conducting an extremely difficult fighting retreat onto and then down the Bataan Peninsula, until they finally succeeded in stabilizing their line about two-thirds of the way to the bottom. "With its occupation," Mac cabled George Marshall on January 24, 1942, from his 1,400-foot tunnel headquarters beneath Corregidor, "all maneuvering possibilities cease. I intend to fight it out to complete destruction." But his troops would do it on half rations, and then half of that, and then still less until slow starvation robbed them of their energy and will to resist. Their only hope was help from home.

On December 12, 1941, the Third Army's direct line to the War Department rang: "Is that you Ike?" asked "Beetle" Smith, George Marshall's flinty make-it-happen guy: "The Chief says for you to hop a plane and get up here right away." Early Sunday morning, barely two days later, Ike walked into the chief of staff's office, fairly certain he was there only because of his MacArthur- and Philippines-tinged past.

But what he got was intimidatingly broader. No greeting, no small talk; Marshall immediately got down to business, outlining the dismal situation across the entire Pacific. He finished with a simple but potentially paralyzing question: "What should be our general line of action?" Ike was stunned, realizing suddenly this wasn't an information session, this was a test. And he might have passed it right off when he had the good sense to stall: "Give me a few hours," which elicited an immediate "all right."

Ike borrowed a typewriter in Gee Gerow's War Plans Division, tried to think what Fox Conner would do, and in three hours generated a three-page memo, which he kept folded in his pocket while he briefed Marshall in exactly the way he liked, briefly. "General, it will be a long time before major reinforcements can get to the Philippines, longer than the garrison can hold out with any driblets of assistance. . . . We must do what we can. Our base must be Australia, and we must start at once to expand it and to secure our communications to it. We must take great risks and spend any amount of money required." Marshall said simply: "I agree with you," but he might have added: "You passed." He then got instructive: "Eisenhower . . . I must have assistants who will solve their own problems and tell me later what they have done. The Philippines are your responsibility. Do your best to save them"—meaning, "It's okay if you don't, but you're right about Australia."

And Ike operated on exactly this basis, meeting with the chief regularly but telling him only what he had already accomplished. His first move was both decisive and telling. A seven-ship convoy guarded by the cruiser *Pensacola* and packed with five thousand troops and eighteen P-40 fighters was already on the way to relieve the Philippines. But with much of the Pacific Fleet now sitting on the bottom of Pearl Harbor, the Navy wanted to call it back to Hawaii to bolster the islands' defenses. Instead, Ike moved swiftly to have the convoy diverted to Australia, where it arrived safely in Brisbane on December 22—the first tangible step in turning the continent Down Under into a base of operations. Marshall was further impressed when he learned in January that Ike had managed to enlist from the British the 80,000-ton luxury liner *Queen Mary*, filled her with most of an infantry division, and then sent her to Australia without an escort, banking successfully on a nearly thirty-knot cruising speed to avoid U-boats.

But what to do about the Philippines? The public at home, already depressed by the grim march of events, anxiously followed the fate of Mac and his men—narrated by 142 official dispatches from Corregidor, 109 of which mentioned only General Douglas MacArthur, now America's most famous soldier. Some effort had to be made.

As a substitute for the *Pensacola* mission, former secretary of war Patrick Hurley was sent to Australia in early February to try to put together a rent-a-convoy. But there weren't many hulls available, and few captains liked their odds of getting through. In the end Hurley got just six ships started toward the Philippines, and only three made it to the southern islands, where they unloaded ten thousand tons of cargo, only one thousand of which ever reached the American garrison on Luzon.

To compound matters Manuel Quezon—now a sick and bitter man on a cot in the Corregidor tunnel—was looking for an exit strategy for his tortured country. If no aid from America could be sent, he proposed to FDR that the islands should be neutralized and the Japanese invited to withdraw their forces: an absurd proposition considering Nippon was on the attack all over Asia, but also a problem when MacArthur wrote Marshall that Quezon's proposal "might offer the best possible solution of what is about to be a disastrous debacle." With some trepidation he and Stimson took Mac's words to FDR, whose quicksilver loyalties were hard to predict. But the president barely blinked. "We can't do this at all." That moment got Marshall's attention like nothing before in his dealings with FDR. "I immediately discarded everything in my mind I had held to his discredit . . . Roosevelt said we won't neutralize. I decided he was a great man."

Forthwith, the president ordered MacArthur to "keep our flag flying in the Philippines so long as there remains any possibility of resistance." But then Mac countered by informing Washington that he was arranging for Quezon to be evacuated by submarine (before leaving, he awarded MacArthur $500,000, plus $75,000 to Sutherland) while he would remain to fight and die with his troops. Both Marshall and Roosevelt knew this was no bluff; they had to consider what the demise of America's first soldier would mean, not to mention that of his wife, Jean, and young son, who were right there with him. Both agreed something had to be done.

But not Ike. As far as he was concerned Quezon and Mac "are both babies." When Patrick Hurley indicated that the general would have to be ordered off the Philippines, Marshall and FDR began to believe that was their only option. Ike disagreed. "He's doing a good job where [he] is," he wrote in his diary on February 23. "Bataan is made to order for him. It's in the public eye; it has made him a national hero; it has all the essentials of drama; and he is the acknowledged king on the spot," a diatribe that says as much about Ike as it did about Mac. But when it became clear that nobody was listening, Eisenhower argued that the general's evacuation didn't necessitate ordaining him with another command, especially not after his performance in the Philippines appeared to be that of a spent entity. But this too had little resonance, in part because Ike's anger blinded him to the larger picture, one he had just months before played a key role in outlining.

Citadel Australia was not simply a matter of soldiers and supplies; it demanded leadership and inspiration. Seven million Australians were on the verge of invasion by a nation ten times its population, and having earlier sent its only three divisions to the British to fight Germans in the Middle East, the country was now virtually defenseless. "Without any inhibitions of any kind," Prime Minister John Curtin wrote in a Melbourne editorial, "I make quite clear that Australia looks to America, free of any pangs as to our traditional links with the United Kingdom." Not only did he want his three divisions back, but Curtin wanted them and all forces in the Southwest Pacific commanded by an American.

Roosevelt and Marshall suddenly found themselves with the perfect candidate, one who could inject instant charisma into any situation, however grim, and was about to be available and in need of a new future. The problem was getting him there.

MacArthur was plainly willing to die with his troops, perhaps even looking forward to one last firefight as a means of restoring his reputation with them. In the months he was on Corregidor he only once took the short boat trip across to Bataan to visit them, and they bitterly took to calling him "Dugout Doug." Actually, he insisted on eating and sleeping in a cottage atop the fortress, oblivious to Japanese air raids, but the fact that they could question his courage was a measure of his alienation from them, and he knew it. Yet MacArthur remained a soldier, and he would obey orders even if they told him to leave. Management back in Washington knew that, and on February 23 FDR directly ordered Mac to depart for Australia, where he would take command of the Southwest Pacific theater. His only option now was a new lease on life.

How he chose to exercise it is revealing. Probably the safest and surest means of getting him there was by submarine, but he looked elsewhere. There is no record of Douglas MacArthur going anywhere in a submarine, so conceivably there was something warlike that actually frightened him. But more likely he wanted a more dramatic exit, one that was even symbolic. So he looked to the small squadron of PT boats still available to him, the survivors of his original dream to defend the archipelago with a swarm of such craft.

The initial leg of his escape, which included Jean, little Arthur, a Chinese nanny, and his entire personal staff—all of them packed into four of the plywood craft, each powered by three Packard aircraft engines able to plane and bounce them over the waves at speeds in excess of 40 mph—that bumpy segment is remembered today as adding a dash of splash to the MacArthur myth. But in actual fact, it may well have marked the lowest point in his life. For most of the trip he lay silent below deck in a business suit, looking like a soggy mannequin. Then deep into the night he beckoned his naval aide Sid Huff, who, as they departed from Corregidor, heard him mutter, "I shall return," but now endured an hours-long soliloquy that left him convinced that his boss had reached a true nadir, having lost all self-respect.

Yet there were still heroes on the journey. Jean remained absolutely imperturbable and not seasick as she convinced her son that this was all a great adventure—a momentary role reversal her husband likely never forgot. Then there was the mission commander, Lieutenant John Bulkeley, who piloted the little flotilla through mechanical breakdowns, periods of separation, and patrolling Japanese warships, to arrive after thirty-five hours and 560 miles continuously at the helm exactly on time at Cagayan de Oro, still in American hands on northern Mindanao. His dignity at least apparently restored, Mac emerged to announce: "Bulkeley, I'm giving every officer and man here the Silver Star for gallantry. You've taken me out of the jaws of death and I'll never forget it."

There ensued an on-, then off-, then on-again flight schedule to take the party to Australia and safety in B-17s worn-out from constant use and lack of maintenance, followed by a long train ride from Alice Springs to Melbourne, across outback terrain boring enough to render MacArthur at last unconscious. "I knew this train trip would be best," Jean told Sid Huff. "This is the first time he's really slept since Pearl Harbor."

HE ARRIVED ON MARCH 21, 1942, TO BE GREETED BY THRONGS OF ADORING Aussies, a new man, or rather much like the old one; his attitude and situation epitomized those of Archimedes two millennia earlier: "Give me a place to stand and a lever long enough, and I will move the world." That and the advice of baseball's "Wee Willie" Keeler—"Hit 'em where they ain't"— would constitute his grand strategy, the aim of which was and always would be to get back to the Philippines.

Likely he would have been annoyed had he realized that the foundations for citadel Australia had already been laid by Dwight D. Eisenhower, his errant staff officer, in the form of men and supplies originally intended for his doomed troops on Bataan and Corregidor. Back in the tunnel he thought he recognized Ike's voice in the excuse-laden cables he kept receiving, but actually this would be nearly the last time these two heavy hitters would take swings at each other. George Marshall had brought Ike to Washington on Walter Krueger's advice that he could make things happen in a big way, as he had in the Louisiana Maneuvers. Ike's Philippine involvement was part test, part emergency; the chief of staff really wanted him facing in the other direction, working on the major effort, the war against Germany.

The conflict in the Pacific would remain a sideshow, garnering around 15 percent of total resources, and MacArthur would have to share that with the Navy, which laid claim to the central zone and a revenge-driven strategy ultimately aimed at meeting and defeating the Japanese fleet.

This left Mac in the far southwest corner of the vast Pacific, about as far from Washington as possible and still remaining on this planet, with the difficulties of obtaining supplies also bordering on the extraterrestrial—in other words, an ideal environment for an ultracreative military mind maximally suited to making something out of nothing, and doing it fast.

With word that no more supplies were presently on the way from the United States, MacArthur concluded that he would not only have to rely on the Aussies, he would have to dominate them.

The Japanese were proving horrific conquerors, and the terrified Australians clung to MacArthur as their best hope of survival, but as far as he was concerned they were in need of a fundamental attitude adjustment. When he learned that in the face of an actual invasion they planned to fall back to the population centers on the southeast coast, he rejected their logic completely. The best defense was a good offense, he stoutly maintained, a dubious proposition at this stage, but one he charged with enough charisma and optimism to sound convincing to desperate ears.

Back in Washington, with nothing much else to give him and Goebbels, Hitler's chief propagandist, and Mussolini calling MacArthur respectively a "fleeing general" and a "coward," George Marshall conjured up a propaganda version of Mac's advice to the Aussies and decided to award him the Medal of Honor. Ike disagreed and told him so, but the recommendation went forward and FDR approved it. So it became part of a very pivotal moment.

MacArthur had yet to be formally introduced to the Australian political establishment, and on March 26 he flew up to Canberra, the capital, for the reception. It was at this gala that he first met PM Curtin, whom he promptly hypnotized, then seduced with the suggestion that the two of them henceforth personally coordinate all Australia-US military planning. Despite having no military background, Curtin jumped at the opportunity, made all the more enticing when the US ambassador unexpectedly rose to announce that Douglas MacArthur was being awarded the Medal of Honor. Blindsided, Mac at that moment was more abashed than honored, believing still that he had deserted his command, albeit under orders. But it put exactly the right cast of heroic authenticity on MacArthur, one that crystallized the Australians' faith in the man sufficiently to weather the coming storm from Nippon.

Soon after the catalytic banquet, Parliament followed Curtin's lead, voting Mac the privileges of its floor, an unprecedented open invitation to speak before them at his whim. The commander in chief of the Australian Military Forces, Sir Thomas Blamey, who at this point was able to supply three of his own troops for every available American, soon joined the stampede, obligingly ceding all responsibility for overall strategy to MacArthur's handpicked staff. So it was that Douglas MacArthur endured the critical spring of 1942 with an entire country in his pocket.

For the Allies it was truly a portentous stretch. German forces were bearing down on Egypt, and in Russia had still yet to reach Stalingrad, while in the Pacific Japan's light losses in the first wave of conquest only whetted the appetite of the Rising Sun's naval planners. Most galling to MacArthur was Jonathan Wainwright's surrender in early May of all US troops in the Philippines, a capitulation he specifically forbade when he left him in command. But did Mac really expect the almost fifteen thousand Americans left to fight to the last man like the Japanese? The place was already lost; except for the use of Manila Bay, the Japanese gained nothing more from the final fall. How much MacArthur knew about the subsequent Bataan Death March and the vicious treatment of American prisoners at the time is hard to tell, but the savagery of the prior campaign was enough to convince him that the Japanese were waging what amounted to a war without mercy, and henceforth he would reciprocate.

But before he could spring into action, his immediate fate was being settled far out to sea between ships that never saw each other. In early May, the Japanese Navy sailed a carrier task force into the Coral Sea to cover an invasion force headed for Port Moresby, New Guinea, less than three hundred miles from the Australian coast. But a US Navy carrier task force combined with a joint Australian-American cruiser flotilla was waiting. In the resulting two days of air strikes the carrier *Lexington* was sunk and the *Yorktown* damaged, but they had inflicted sufficient injury on their opposite numbers that not only did the Port Moresby invasion fleet turn back, but the two Japanese flattops involved would be undergoing repairs and missing from a much more decisive clash the next month.

This time Admiral Isoroku Yamamoto, the architect of Pearl Harbor, sought to lure the remaining American carriers in the Pacific into defending Midway Island and then destroy them with a multipronged armada that included the cream of the Japanese carriers and naval aviators—essentially the same Kido Butai that had sunk the battleships in Hawaii. Instead, the US Navy set a trap of its own, one that over four days of air operations during the first week of June would carve the heart out of the Kido Butai, sinking four of its six big fleet carriers and providing the Allies some much-needed encouragement along with breathing room in the Pacific.

More specifically, the Coral Sea–Midway combination gave Douglas MacArthur time to set his plans in motion. Without the Japanese possession of Port Moresby, he believed the immediate danger to Australia had passed. But New Guinea still loomed just above the southern continent, shaped roughly like a fat bird of prey, and MacArthur understood it would be the key to his campaign, its center of gravity, and also that the place itself would largely dictate the when and where.

It wasn't a habitat to be possessed, but circumnavigated. That was because the equatorial interior was not simply inhospitable to humans, it was all but uninhabitable. Bisected by the 10,000-foot Owen Stanley Range, it was a wet, slippery vertical environment full of poisonous snakes, malarial mosquitoes, and just about every other existent tropical misery, often described with just two words: "green hell." Any enemy in his right mind, or at least one that wanted to win, would concentrate on the coast, MacArthur concluded, and that was what he would do, grabbing Milne Bay at the far southeastern corner and staging almost continuous amphibious landings along the north side of the island.

But to make this work he would need a precisely adapted force led by flexible and ruthless military opportunists, and that's what he built from the beginning. Despite its basic operational success, MacArthur's personal staff has been the focus of considerable negative attention, largely on the grounds of its corporate and individual sycophancy.

At the head of the list was Dick Sutherland, whom we last saw usurping Ike as Mac's chief of staff in the Philippines, the position in which he remained through the dark days in the tunnel, on the PT boat, and now Down Under. Sutherland's job was designed to make him unpopular, doing the boss's dirty work and enforcing his hard decisions so the mask above could remain unsullied and maximally convincing. He excelled at this, but he was also a misanthropic know-it-all who chronically overestimated his own abilities. Hence when Mac sent him to Washington to convince the Joint Chiefs of Staff that his plans were best and finished his presentation with the words "I would stake my reputation on the soundness of what I am proposing," George Marshall could not resist asking, "Would General Sutherland mind telling the Joint Chiefs of Staff exactly what this military reputation is?" He was that kind of person, but in his own hateful way he kept things moving, accurately translating MacArthur's ideas into orders and instructions; in that respect he was a functional functionary.

Another suspect sycophant was Mac's everlasting intelligence chief, Charles Willoughby, a Pantagruelian figure known to the staff as "Sir Charles," sarcastically so, since he was born in Germany and still spoke with a heavy accent. His saving grace for most of them was that he didn't get along with Sutherland, who in turn thought of Willoughby as a weak reed in a variable breeze, constantly changing his mind to suit the situation.

But in reality, Willoughby had a secret friend whispering in his ear, the same one who told the Navy that the Japanese were coming to the Coral Sea and Midway. Not only had their diplomatic codes been broken and disseminated as MAGIC; but their military cyphers were already falling prey to carelessness and relentless American cryptographers, and soon their plans would be distributed under the British tag ULTRA, sometimes faster than they got to Japanese commanders.

MacArthur's subsequent terpsichorean progress along the top side of New Guinea would in large part be choreographed by ULTRA. The oracle occasionally went silent, but for the most part Mac's brilliant leaps to "hit 'em where they ain't" and miraculous interceptions of Japanese reinforcements came courtesy of the dance master in Washington, or just outside of it, in Arlington Hall, the home of US Army cryptography.

That said, MacArthur still had to put together the people and the means to capitalize on his intelligence advantage. Whatever the shortcomings of his staff, the combat team he assembled was extraordinary. This would be an amphibious campaign, but once the troops hit the beach he needed ground pounders, and got two excellent ones. The first was a longtime henchman, Robert Eichelberger; tough and brave to the extreme, Mac regularly gave him the worst jobs, at one point telling him: "Bob take Buna or don't come back alive."

The other was a gift or at least a contribution from George Marshall. Walter Krueger was known to be among the best generals in the US Army; with Patton, he was one of the few elders the chief of staff hadn't purged after the Louisiana Maneuvers. But he still retained his German accent, and that, it seems, precluded assignment in Europe; yet it made absolutely no difference in the Southwest Pacific. Loath to waste a good combat soldier, Marshall sent him Mac's way to become the second bookend holding together his land component, doing the heavy combat lifting for the team.

But it was the movers more than the shakers that personified the improvisational brilliance of the force MacArthur assembled. Admiral Daniel Barbey ("Uncle Dan" to subordinates) was not simply a student of amphibious warfare, he was its mechanical engineer, having overseen the development and deployment of a veritable bestiary of vessels capable of bringing progressively larger quantities of troops and equipment, including tanks, right up to the beachhead. He had a landing craft for every occasion, and an operational approach just as varied, exactly what MacArthur needed along the harborless shores of New Guinea and the islands yet to come.

But probably even more useful—eventually Mac's equivalent of Archimedes's long lever—was the man who gave him the capacity to leap over or on top of the Japanese. Bitter about the Air Corps's performance in the Philippines and the flying wrecks sent to rescue him, MacArthur had already gotten rid of two air commanders when the latest replacement, George Kenney, arrived at the end of July 1942, also courtesy of Marshall.

Sutherland coldly ushered him into MacArthur's office, where he was treated to a half-hour monologue and pacing session: "I really heard about the shortcomings of the Air Force . . . They couldn't bomb, their staff work was poor and their commanders know nothing about leadership . . . he was even beginning to doubt their loyalty." Kenney took it all in silence and finally jumped in: "I'm here to take over the air show and I intend to run it . . . correct them and do a real job . . . produce results. . . . If it's a matter of loyalty . . . I'll tell you myself and do everything in my power to get relieved." MacArthur stared for a long moment, then smiled and put his hand on Kenney's shoulder. "I think we're going to get along all right."

Sutherland required further convincing. When Kenney found out the chief of staff was scheduling bombing missions on his own authority, he stormed into his office, picked up a pencil, and put a black mark on a piece of paper. "That is what you know about air power. The rest of the sheet is what I know about it." When Sutherland began fulminating, Kenney suggested they step in to see the general: "I want to find out who is supposed to run this air force." Instead, Sutherland backed down, and from that moment it was clear to everyone, including Mac, who ran the air component. "I don't care whether they raise the devil or what they do," he told Kenney, "as long as they will fight, shoot down Japs, and put bombs on target." Carte blanche, in other words.

If you wanted maximum performance, this was exactly the way to treat George Kenney. He was a relentless and truly inventive practitioner of airpower. Initially short on planes, he was infuriated that many needed repairs. Maintaining that an air force full of broken planes was worse than useless—it was a target—he attacked the maintenance problem with a ferocious enthusiasm that soon had sortie rates soaring.

Meanwhile, there was work to be done on the hardware. Initially short on new planes, the supply gradually increased, but it often consisted of fighters that had not done well in Europe, like P-38s and P-40s, or light bombers, particularly B-25s, of which there was a surplus. That didn't matter to Kenney and his aerial hot-rodders; they would turn them into what they wanted.

For example, P-38 "Lightnings" were fast and good divers, but like most two-engine fighters were thought to be inherently lacking in maneuverability. But in the Pacific pilots learned that because the plane's motors were so widely set apart, each in a separate minifuselage extending back to twin tails, they could literally steer the plane into sharp turns by throttling up on one and backing off on the other. Soon Kenney's pilots were holding their own in dogfights with even tight-turning Japanese Zeros, and then got an added performance boost when Charles Lindbergh arrived as an independent consultant and showed them that they could extend the plane's range and endurance by up to two hours simply by increasing manifold pressure and lowering engine revolutions. And he flew multiple lengthened combat missions to prove it.

But far more integral to Kenney's aviation chop shop was Paul Irvin "Pappy" Gunn, a former airline pilot turned military aviation innovator. If fast-flying attack planes were having trouble hitting targets, he wondered, why not give the bombs parachutes? Or at sea, why not skip them like stones? Both methods worked better, it turned out, and Kenney enthusiastically instituted them. Meanwhile, as his name suggests, Gunn's specialty was adding firepower, particularly to his two-engine A-20s and B-25s, stuffing them with .50-caliber machine guns positioned to fire from all angles, transforming them into strafing machines extraordinaire. So, operating in real time and using what amounted to spare parts, a group of black sheep aviators turned Douglas MacArthur's heretofore anemic Fifth Air Force into a prodigious killing instrument.

And George Kenney still had more to offer Mac's team, capabilities that had fundamental implications for waging war in a green hell. By the fall of 1942 his stock of C-47 transports (the military version of the DC-3) had grown large enough that he could tell both MacArthur and Australian general Blamey that wherever they fought in New Guinea, he could supply them from the air, preferably using an airstrip, but also with paradrops. Not only that, he could stage the invasion himself, flying the troops in, a proposition he demonstrated when he delivered, over Sutherland's objections, two full regiments in early October for what became the fated siege of Buna.

IN THE ERA OF INDUSTRIAL WARFARE, WHEN QUANTITY HAD A QUALITY ALL its own, attacking a country with nearly twice the population and five times the production capacity was a fundamentally dubious proposition, but not to the Japanese. Their military elite had convinced itself, based on a martial history largely fought out with swords, that they could make up the deficit with sheer audacity, dedication, and a willingness to sacrifice literally everything in pursuit of victory. It turned out to be a recipe for disaster, one that caused commanders to pursue impossible objectives, rigidly adhere to prior planning, and do it all with complete disregard for their men's lives, literally demanding they commit mass suicide in the face of failure. And inescapably locked into a duty-based society, the men not only obeyed orders, they did so with an enthusiasm almost guaranteed to turn their battles into bloodbaths, one way or the other. It was in this spirit that their campaign in New Guinea and the neighboring islands was waged.

Spellbound by its lightning descent down the rain forests and swamps of the Malay Peninsula, the Japanese high command came to believe that their army was an all-terrain vehicle that could go literally anywhere. This far-fetched assumption took the form of reality on August 8, 1942, when Australian scouts realized the Japanese column they had spotted was lugging mortars, heavy machine guns, and field artillery—all the tools of an invading force—up the Kokoda Trail, a slippery ribbon headed toward the 13,000-foot pass of the same name, all of it surrounded by green hell but undeniably the most direct route to what had become almost an end in itself, Port Moresby, directly facing Australia and the obvious jumping-off point for invasion. But it would prove the end of this army.

The Australians, supported by Kenney's bombers and C-47s dropping fifty thousand pounds of supplies a day, waged a brilliant wasting campaign all the way up and over, then took a stand on the last ridge preceding Port Moresby before driving the Japanese back and letting green hell do its work, decimating the force with disease, starvation, and jungle-induced pratfalls such as befell even its General Horii, who drowned in the treacherous Kumusi River. When it was over, all that was left were a few survivors of what had once been a proud but lethally overconfident force of nearly ten thousand men.

But the Japanese were far from finished. In late August, they had launched an amphibious landing at Milne Bay at the island's eastern tip, seeking to capture the Allied air bases there as a means of getting at Port Moresby. But

ULTRA told MacArthur they were coming, and he used the combat veterans of the Seventh Australian Division, now returned from the Middle East, to throw them back. Outnumbered almost four to one, the invaders held on for twelve days before, with their landing barges destroyed, they melted into the jungle, having sustained nearly 50 percent casualties and leaving as a calling card the horribly mutilated bodies of those they had captured. That too was just a prelude.

The two regiments Kenney paradropped in October headed toward Buna, two hundred miles up the coast from Milne Bay, assuming it would be lightly held, since it along with nearby Gona had served as staging areas for the now defunct Japanese thrust up the Kokoda Trail. But instead of the five hundred rearguard troops they expected, they were greeted by a hail of fire from thirteen thousand fresh combat veterans. For no particular reason except to hold on to it, the Japanese had reinforced the place.

They were dug in deep with extensive and interlocking fortifications, and met each American and Australian attack with an impenetrable wall of fire. After a few of these, officers could no longer get their men to go forward. Morale collapsed, and only Kenney's Fifth Air Force was keeping them in the fight with supplies, reinforcements, and bombs dropped on the Japanese. Back in Port Moresby, where Mac had moved his headquarters, he was furious. Remembering the heroics of the Rainbow in the First World War, but forgetting the effects of heat and tropical disease, he at this point told Bob Eichelberger to take Buna or die in the attempt.

It took him until the third week in January 1943, but with constant reinforcement Eichelberger got the troops fighting again and finally crushed the Buna-Gona stronghold. But it required the bitterest kind of no-quarter infantry combat, all of it taking place under nearly unbearable conditions. The Japanese lost the entire command, 13,000 men dead. Of the 14,500 Americans who saw action, 930 were killed, almost 2,000 wounded, and nearly 9,000 downed by disease, adding up to a nearly 80 percent attrition rate.

Those figures preyed on Douglas MacArthur like a bad hangover. He had never been about wasting his men and attrition, and given his theater's limited share of resources, this kind of war was absolutely unsustainable. But Mac didn't get into the starting lineup by being one-dimensional; he was in fact a brilliant switch-hitter. From this point, he began taking full advantage of the tools ULTRA, Barbey, and Kenney had given him to begin waging what he called "a new type of campaign—three-dimensional warfare—the triphibious concept," albeit borrowing the term from Churchill. It amounted to strategic hopscotch, played knowing most of the time what your opponent's next move would be, a winning concept and the one best calculated to get MacArthur where he always wanted to go.

Because of the split US commands in the Pacific—itself the product of a bureaucratic accommodation between George Marshall and the chief of naval operations, Admiral Ernest King—there were two separate American thrusts forward across the great ocean, the Navy's to the north and MacArthur's. In purely strategic terms they were beginning to form a giant pincer, eventually destined to converge on the Philippines. But that was really only incidental for the Navy; its key objective was to meet and obliterate the Japanese fleet, to avenge Pearl Harbor as much as anything. As for MacArthur, it's possible to conclude that everything he did after Corregidor was ultimately animated by his dream of getting back and fulfilling the muttered pledge he made as he stepped onto the PT boat.

Still, he did it brilliantly. On the way to the Philippines and once there, he would conduct a total of eighty-seven amphibious landings, all successful, and most serving to slash Japanese lines of communication and escape routes. And with an earful of ULTRA he was frequently able to "hit 'em where they ain't," keeping American casualties lower than any Allied theater commander—though the Australians claimed he did it by using them whenever possible. Yet the net effect was still parsimonious bloodletting. On our side, that is.

The Japanese entered the war with a distinct advantage in airpower. But it was a narrow one, focused on the tactical mission, and destined to be short-lived. Japanese combat pilots, particularly the carrier-based "Sea Eagles," were likely the best in the world. Their training was meticulous, and many had prior combat experience in China. But they were a small cadre, and their very excellence encouraged military planners to assign them exceedingly ambitious missions. Attrition was inevitable, but unlike the Americans, who could train pilots by the thousands, Japanese planners had failed to build the infrastructure necessary to replace them.

While they lasted, their lethality had a good deal to do with their planes, in particular the Mitsubishi A6M Zero. Developed in the late thirties based on combat lessons in China, the Zero design team took advantage of a special superlight duralumin and every possible weight-saving measure to create a fighter that had a combat range well over five hundred miles but still retained excellent tactical capabilities—exceptional low-speed turning, good acceleration and climbing, plus surprisingly effective armament. But light weight also proved the plane's undoing; there were no self-sealing fuel tanks or armor for the pilots, and its gossamer construction did not stand up well under fire. All of which only hastened the demise of Japan's elite pilots, not to mention the rookies who followed.

And as with the aviators, there was no provision for a second act. The Zero began the war and ended it as Japan's first-line fighter, though finally reduced to serving as a suicidal smart bomb. Meanwhile, American aircraft manufacturers served up two full generations of combat planes, each one containing a variety of different aircraft, amounting to something like an-all-you-can-fly buffet, one destined to sweep the skies of everything Japanese but kamikazes.

So it was that even in the far reaches of the Southwest Pacific, George Kenney managed to wrench control of the air from his adversaries and begin the real punishment.

To support the Kokoda operation, the Japanese had established bases at the small ports of Lae and Salamaua, about 150 miles up New Guinea's north coast from Buna. Now, perceiving a threat to them from the US Navy's operations in the Solomon Islands, army chief of staff Hajime Sugiyama decided to strengthen his local position by sending a convoy full of fresh troops to Lae in the first week of January 1943. But Kenney's reconnaissance planes were on the lookout and picked it up, touching off two days of American high-level bombing before US aircraft were driven off by Zeros with only a single disabled transport vessel to show for their efforts. But the clairvoyant Yank aviators were waiting when the convoy arrived at Lae, and attacked its ships continuously as the troops disembarked, killing at least six hundred, Mac learned from ULTRA. "Bombs came down like rain" remembered one soldier. "Several tens of planes coming in rapid succession made their intensive attacks which were utterly indescribable."

Rightly concerned about his air cover, Sugiyama responded by sending three thousand troops at the end of January to take the Australian airfield at Wau, on mountainous terrain about thirty miles southwest of Salamaua. But ULTRA told Mac, who had Kenney fly in a brigade of Aussies, who arrived in time to throw the Japanese back when they got within four hundred yards of the strip.

These setbacks left the Imperial General Headquarters sufficiently alarmed to order an eight-ship reinforcement convoy bound for Lae, carrying the entire elite Fifty-First Division, to proceed from the Japanese navy's central base at Rabaul on New Britain Island. On February 24, ULTRA in Washington confirmed the force would be landing at Lae on March 5. MacArthur spent an afternoon with his aerial henchman matching the ULTRA data with the map, whereupon Kenney concluded: "The weather will be bad along the west coast of New Britain. I'll bet they'll use it for cover." Mac responded: "Are you calling off all other air operations except this?" With the exception of supplying the Aussies at Wau, Kenney responded, everything he had available would be flown at the convoy.

During the interval before the Japanese ships sailed, Kenney worked his fliers to a razor's edge for what was to be his pièce de résistance, a precisely choreographed high- and low-level attack, staging a final rehearsal realistic enough to leave one plane crashed and two more badly damaged in the process of skip bombing. "The Japs are going to get the surprise of their lives," he told his young assassins.

On March 2, once reconnaissance planes reacquired the convoy and the weather cleared, Kenney sent his long-range, four-engine heavy bombers— B-17s and B-24s—out in two high-altitude waves, managing to sink two transports before dark.

The next morning dawned with the convoy now in range of his two-engine planes. With Zeros circling far above to protect against another high-altitude attack, Japanese lookouts aboard ship saw what looked and sounded like a cloud of angry bees heading for them just above the waves. Instead, the cloud consisted of over one hundred Allied aircraft: Australian Beaufighters, American A-20s, and thirty of Pappy Gunn's modified B-25s festooned with .50-caliber death-dealing. Skip bombs were soon bouncing into hulls, explosions tearing huge holes and scattering death among the soldiers packed below deck. And survivors who managed to get topside were cut down in droves by wave upon wave of machine gun fire, as were those who made it into the water, in rafts or just swimming for their lives. Soon the surface was whipped into a crimson foam, a combination of blood, oozing chunks of flesh, and oil from the sinking transports. But the planes kept coming, churning up this grotesque seascape in search of more victims, until finally when nothing could be seen moving the strafing stopped. Some of the Allied aircrews were said to have vomited as they got a low-level look at what they had done.

Not Kenney, who was in Brisbane and absolutely elated, but not more than his boss. Jean MacArthur remembered her own "Sir Boss" pacing in the dining room of their apartment, clutching the latest dispatches and muttering, "Mitchell! Mitchell!" Billy, the original American high priest of airpower, his dead friend from Milwaukee and court-martial victim, was now vindicated. When Kenney joined the MacArthurs at 3:00 a.m. and gave him what he thought was the final tally, Mac waxed ecstatic. "I have never seen such jubilation," Kenney remembered. Soon after Mac sent out a telegram: "Air Power has written some important history in the past three days. Tell the whole gang that I am so proud of them I am about to blow a fuse." Mac promptly announced to the press that twenty-two ships had been sunk and fifteen thousand Japanese troops were confirmed dead, a vast exaggeration. In fact over half of the Fifty-First Division survived, though only one thousand were in shape to fight and ever reached Lae.

Slaughter of combatants rendered helpless during and after battle has been part of warfare practically from the beginning if we are to believe the inscriptions of Sumerians and of subsequent Middle Eastern rulerships. Better documented was the epic bloodletting staged by Hannibal and his army at Cannae in 216 BC, putting to the sword nearly fifty thousand Romans in an afternoon. So it went through much of recorded history, not simply featuring characters like Attila and Genghis Khan, but also replete with tactical situations when it simply made no sense to take prisoners.

Nonetheless, in Europe at the end of the epically deadly religious wars of the sixteenth and seventeenth centuries, the emergence of a durable political balance of power along with a two-century freeze on weapons development did lead to war being waged in a somewhat more restricted fashion. Besides that, the mercenaries who did the waging took a long time to train, and were therefore a valuable commodity not to be unnecessarily wasted in battle, and worthy of a ransom if captured. What emerged in terms of behavioral psychology was an unwritten but still real code of conduct that came to hang about the practice of war-making, at least among the international class of professional officers who dominated the age's battlefields. It was far from exalted, basically consisting of offering quarter to the defeated, not mistreating or killing prisoners, and keeping the fighting away from civilians. It didn't matter that the French Revolution and Napoléon tore down the political edifice that engendered it; neither he nor his officers, even in the direst circumstances, entirely cast the code aside. Nor did their equivalents who waged World War I. Though the weapons the latter fought with vastly magnified the killing on and around the battlefield, these fundamental boundaries remained largely intact—even conceding that the eighty-mile reach of the German Paris Kanone and sporadic zeppelin and aircraft bombing raids on London did point the way to a different future.

It dawned almost the moment World War II began, most flagrantly on the Axis side. Abetted by separate but equivalent conceptions of their own racial superiority, both Germany and Japan waged war in a nakedly murderous fashion, their officers consistently disregarding military norms and their men primed for indiscriminate death-dealing. Even using the lowest-accepted figure of World War II dead, around sixty million, and subtracting the ten million Germans and Japanese who lost their lives, it becomes abundantly obvious which side was the most bloodthirsty. Yet these were very big numbers, and it should not be forgotten that in the last major wars both powers fought— particularly the Japanese in the 1905 conflict against Russia—each stayed within combat norms. Was racial hatred alone enough to make the difference?

If not, it seems logical to assign more of a role to the devices that actually did all the killing: the weapons themselves. After all, ten million dead Germans and Japanese was still a big number, and the Allies inflicted it using the same range of weapons with a similar murderous enthusiasm, and they were already working on something bigger—far bigger.

But back in Brisbane in the spring of 1943 Douglas MacArthur's joy at what has come to be known as the Battle of the Bismarck Sea could be unalloyed, without second thoughts. In the true spirit of Yankee ingenuity, he, George Kenney, and the self-described "Pappy" Gunn had put together a truly formidable killing instrument, never stopping to consider where chemical energy–based industrial warfare had taken them, a place that made death-dealing so easy it was almost reflexive. Among history's great soldiers, Douglas MacArthur would hardly be considered among the most sanguinary, but he was a military opportunist and used what he had.

It followed that Mac's war in the Southwest Pacific became both elegant and barbaric. Using his triphibious assets, he did consistently jump behind the Japanese, in particular ignoring their central base in Rabaul—though admittedly prompted from Washington by Allied leadership at the Trident Conference in late May 1943—then rendering it useless by cutting its access to points in need of supplies and reinforcement. Using ULTRA he deliberately sought out Japanese positions that were weak but also of strategic significance, then pounced.

Yet, in the end, many had to be taken by hard fighting. Using airpower, naval artillery, and vastly more equipment, the Americans always won in what increasingly became an island-hopping campaign, and in this theater they managed to keep their own casualties low. But the darker side of this equation, the war-without-mercy side, was that they had to kill literally all the Japanese. It was no accident that flamethrowers saw their greatest use and success in the Pacific war. Prompted by the behavior of the Japanese themselves, and driven by their own brand of racial hatred, American soldiers became as much exterminators as warriors. To most this must have been an excruciating experience, even if their own chances of being maimed or killed were low. And they had to do it over and over, since the chances of being relieved in the Southwest Pacific theater were slim. You were with Doug for the duration.

Given the circumstances, it's not surprising that he was unpopular with the troops. He was now a man well into his sixties leading postadolescents. His days as the "Fighting Dude" were long over, replaced by an update of "Dugout Doug," a far-off Wizard of Oz–like figure living a life of luxury in Brisbane. Nor did MacArthur exactly help his case. He seldom visited his men, and when he did he took ostentatious and theatrical risks that often impressed his now combat-hardened charges as unnecessary bordering on stupid. It was all part of the act.

With a string of victories separating him from Corregidor, MacArthur's ego was on the march. The man had become his mask. His battlefield communiqués might as well have been titled "Studies in MacArthur," since he was frequently the only one mentioned. That was good for his popularity at home, but infuriating to the troops, all the way up to Bob Eichelberger, who later wrote that Mac "not only wants to be a great theater commander but he also wants to be known as a great front-line fighting leader. This would be very difficult to put over if any of his particular leaders were publicized . . . He just wants it all for himself." Equally maddening was his tendency to announce victory when in fact there was still a good deal of fighting left to be done—it was almost like saying: "I'm moving on regardless of you laggards."

But you couldn't deny his success; together with the US Navy he had turned things around in the Pacific—exactly what Team America needed.

But before we give MacArthur too much credit, we should remember not only ULTRA, but also the detonator on the Mark 14 torpedo. Between the wars the US Navy had developed an excellent commerce-raiding submarine in the Tambor class, featuring advanced diesel engines developed in conjunction with the railroads and, critically in the South Pacific, air-conditioning. But the detonators on its otherwise advanced Mark 14 torpedoes had never been properly tested and proved to be utter duds. The Japanese first learned of this when they heard these fish without teeth bouncing off their hulls. The British had realized the detonators were faulty as early as 1940, but the US Navy's Bureau of Ordnance refused to perform the necessary live testing to reveal the problem until 1943, when it was quickly fixed. US submarines promptly went on a tear, ripping into Japan's commercial fleet like wolves into lambs, virtually eliminating the entire flock within two years and also adding another perspective on the strategic necessity of Mac's oeuvre.

Had the Mark 14 worked from the beginning, Japanese forward positions would have withered of their own accord, not only making his job a great deal easier but also a lot less glorious, elaborate, and eye-catching—a true combat backwater until, of course, Mac headed back to the Philippines, where we will pick him up later. But for now it's worth considering how in this era of warfare something so small as a detonator might have changed things so much.

CHAPTER 8

THE MAIN EVENT

The British invaded Washington for a second time in late December 1941. In 1814 they left it burning; now they came more or less hat in hand, but still intent on dominating their country cousins, who were now vital to their existence. To exert maximum pressure in the coming Arcadia Conference, their manager, Winston Churchill, moved into the White House while the rest of Team Great Britain took the field against their American counterparts for the first of many intersquad contests. The two managers, who actually were distant cousins, got along famously. But having Winston Churchill as a houseguest was a bit like entertaining a volcano—an alcohol-fueled one who spewed endless words and cigar smoke, erupting on a schedule that always extended deep into the night. The two jousted and felt each other out, enjoying every minute, while Eleanor Roosevelt worried about the impact on her husband's fragile health.

The real action, however, was actually taking place on Washington's field of conference tables. It was here that George Marshall pulled his Christmas surprise. One late afternoon, as everyone looked forward to turkey dinner, the American chief of staff grabbed history's bit between his teeth: "I am convinced that there must be one man in charge of entire theater(s)—air, ground and ships . . . operating under a controlled directive from here"— the "here" being Washington, where a joint British-American chiefs of staff element would maintain oversight. This was practically unprecedented for coalition warfare, and had been resorted to only in the desperate spring of

1918 when all Allied forces were consolidated under French general Ferdinand Foch. It worked and was Marshall's model, even though it had lasted for only a little more than seven months until war's end. Now he was proposing it at what amounted to the beginning of the alliance. The British cast a quick negative and left grumbling; even Marshall's fellow American chiefs were noncommittal.

But he was far from finished, and as they walked out he gave Ike a late Christmas gift, telling him to draft specific instructions to a proposed theater commander for the Far East to include Burma, the Dutch East Indies (Indonesia today), Australia, New Guinea, and the Philippines. It took him to midnight, but he cranked out what Marshall wanted, and the latter then used it alternately as a cudgel and a cookie. It took him two days but he finally knocked his fellow US chiefs into line, and then he enlisted the support of the president. Next came the sweet snack: he proposed and they agreed that the first supreme command go to General Sir Archibald Wavell. Much to his surprise the British chiefs gobbled it up, and suddenly Marshall had only Churchill to convince.

He found him late in the morning of December 27, propped up in bed, clad only in a Chinese dragon bathrobe smoking cigar and a scotch-and-soda at the ready—his natural habitat until lunch. He was not in a good mood, but that didn't deter Marshall, who rattled off his proposal pacing in front of the bed. When Churchill interrupted, asking what a general like Wavell knew about fighting a ship, which was a very special thing, Marshall shot back: "What the devil did a naval officer know about handling a tank." That was completely beside the point. As Marshall recalled, "I told him I was not interested in Drake and Frobisher, but I was interested in having a united front against Japan, an enemy which was fighting furiously. I said if we didn't do something right away we were finished in the war."

It was at that point that the prime minister ended the meeting and retreated to his toilette. The steam coming off the tub may not have been just from the hot water. Winston Churchill, the world's best arguer, had just been thrashed, and he must have known it. The innocent-sounding "unity of command" was loaded with significance. The Americans were giving it to Wavell now, but inevitably as their numbers and weight of arms grew, he would have to give it back, and the Americans would be the supremos.

Worse still, even though Churchill saw it coming, he would have to accede. George Marshall, though it was unsaid and unseen, held the diplomatic equivalent of a Louisville Slugger. From the British point of view the central achievement of the Arcadia Conference was to get the Americans to commit to a Europe-first game plan, thereby ensuring their own survival. But now this tough-talking American was demanding a quid pro quo—the bat's unmistakable implication being that if Marshall didn't get what he wanted, he would be willing to listen to the American people and turn toward the Pacific to seek the revenge they wanted first. Churchill, above all things, was a realist, and so it would be Europe-first and unity of command.

Although George Marshall had hit the equivalent of a walk-off homer, a game-winner against his best pitch, Churchill never held it against him. It's been said that one trait of greatness is being able to recognize it in others. Something like that seems to have transpired in this case; at any rate, from this moment Winston Churchill always treated George Marshall with the greatest of respect, which was not the case with his two key generals, chief of staff Alan Brooke and his best paladin, Bernard Montgomery, who both thought him a dullard. Had Marshall access to the correspondence between the two and their condescending military snobbery toward him, likely he would have smiled his thin smile and simply said: "All the better."

There was no institutional reason why George Marshall should have become a grand strategist; he simply saw a hole and filled it. At base, he did so by dominating the body that after the war would grow into the Joint Chiefs of Staff—at this point made up of Marshall, the Army Air Corps's Henry "Hap" Arnold, and the irascible chief of naval operations, Admiral Ernest King, described by his daughter as "the most even tempered man in the world . . . always in a rage." Marshall addressed that by deferring to him whenever possible, and in particular assuaging his fears of being outvoted two to one by proposing that Admiral William D. Leahy serve as chairman of the body, knowing full well that the role was simply that of a conduit, and that this longtime friend of FDR would happily serve as a bridge to the White House, simply transmitting the decisions of the chiefs. With Arnold, who had no interest in policy, he struck a sort of Faustian bargain: in return for his support in coming conference rooms, he gave the Air Corps what amounted to operational autonomy—the planes and the freedom to implement what aviation theorists believed would be a war-winning "precision" bombing campaign against Germany's war industries. This did not work out well for the Air Corps, or for the many hundreds of thousands of German civilians it killed, but it did provide a consistently solid base for Marshall to play with the other heavy hitters.

On the conference's last day, it also became clear that George Marshall understood that in this league ganging up was not only allowed, it was vital to winning. He and the chiefs endorsed the other side's proposal that "finished war materiel should be allocated in accordance with strategic needs," knowing full well that nobody had bigger needs than the British. It was an act of calculated generosity that would allow the combined team to take full advantage of the cornucopia of arms that the arsenal of democracy was just starting to produce, and that would one day bury Nazi Germany.

Meanwhile, at Arcadia, Ike had truly won his spurs with Marshall. It didn't matter that the product of his masterful order—what came to be known as the ABDA Command—had only a two-month life span before the Japanese would blow it to pieces at the Battle of the Java Sea. The concept and Eisenhower's blueprint had been established as policy, leaving MacArthur in Australia, very much an American, to pick up the pieces.

Ike's reward was to replace his pal Gee Gerow as chief of the War Plans Division in the middle of February 1942. As he left, Gee told him: "I got Pearl Harbor on the books; lost the Philippines, Singapore, Sumatra, and all of the Dutch East Indies north of the barrier. Let's see what you can do." Quite a bit better, actually.

Since War Plans had traditionally been about future wars, and now the US had plenty to fight in the present, George Marshall, using his own Occam's razor, turned it into Operations Planning Division (OPD) and had Ike run it for him "as his [Marshall's] own personal command post," Ike remembered.

One thing that surprised Eisenhower, especially after what happened at Pearl Harbor, was the lack of a formalized intelligence element at this level. But that was probably to be expected, since the source of the phrase "Gentlemen do not read each other's mail" was none other than Secretary of War Stimson—albeit more than a decade earlier, in a time of peace. The problem was soon addressed at FDR's insistence by the creation of the Office of Strategic Services (the CIA's forebear) under the Joint Chiefs, and through pouring more resources into fine-grained industrial and military analysis. Ike supported both, but he was not in the OPD for institution-building; he was there because, as with MacArthur, he had quickly become George Marshall's best-ever penman, and was rapidly developing into a true strategic wingman.

In late February Marshall asked Ike to address the pregnant question amounting to "What's next?" Ike responded with not just a mirror image of his boss's thinking, but also with a reasoned exploration of the implications. While he devoted some consideration to the Pacific, his emphasis was plainly on the Atlantic. But he gave short shrift to the British-favored objective, the southern coast of the Mediterranean, opting instead for a direct invasion across the English Channel, arguing that troops and supplies could be much better massed in England, and that once on the northwest coast of Europe, there would be a direct route to Germany largely without natural barriers.

He did make the invasion contingent on the Allies achieving complete air superiority, eliminating the U-boat menace, and provision of sufficient offshore naval gunfire and landing craft—thereby providing some strategic wiggle room. He also emphasized the vital role played by Russia, arguing that the USSR "must not be permitted to reach such a precarious position that she will accept a negotiated peace." Marshall had lots of questions as he listened to Ike's crystallized thoughts, but then responded: "This is it. I approve."

But Marshall wanted more specifics, something he could use to rev up the president and bludgeon the British. Ike responded with what is remembered as the "Marshall Memorandum"—an initial cross-Channel incursion as early as autumn 1942, labeled (with a bit of humor, perhaps) Sledgehammer, followed by a true second front in mid-1943, Roundup, all of it fed by a humongous buildup in England, to be called Bolero. Once again it exactly fit the bill, and Marshall and Stimson took it to the White House, where it got FDR's quick approval.

In early April 1942, Marshall and Harry Hopkins took it airborne to England. The British already knew all about the "Marshall Memorandum," one of their officers in Washington having had a short look and cabled home the main points. Even minus the element of surprise, Marshall was still in a formidable position as he presented his plan to Churchill, Alan Brooke, and his other service counterparts. As outlandish as Sledgehammer might sound and as premature as Roundup might be, the British desperately needed the vast infusion of men and supplies Bolero would bring to English soil, both in terms of securing the homeland and eventually winning the war.

Marshall certainly knew this and also what he could threaten—two weeks before, Ike had written a memorandum advising him that if the British would not solidly commit to first confronting Germany through a cross-Channel invasion, "the United States should turn its back on the Atlantic area and go full out against Japan."

The occasion didn't call for this big a bat, but Sledgehammer served as a handy substitute, a means of knocking the British back on the true path and keeping them there. Marshall operated from the beginning as if he fully understood how horribly scarred the English were by their World War I experience of sending troops to France, and how hard it would be to get them to do it again. Though he did maintain it would be used only as a means of distracting Germany so as to avoid a Russian collapse, Sledgehammer—a limited invasion of the Cherbourg Peninsula without hope of major reinforcement until the following year—was a suicide mission, and it's hard to believe a soldier as accomplished as Marshall didn't understand that from the beginning. But it did serve to register the degree of American impatience to get down to the business of defeating Nazi Germany. Perhaps Churchill saw it for what it likely was—a diplomatic marker—but Brooke missed the point, labeling Marshall "a good general at raising armies and providing the necessary links between the military and political worlds. But his strategical ability does not impress me at all!!" You might even say he missed it entirely.

Before leaving London and the British brooding over the specter of the cross-Channel invasion, Marshall took time to have a look at the lead element of what would soon become the US Army's European theater command and the epicenter of Bolero. He didn't like what he saw: there was no sense of urgency among US personnel; they were keeping banker's hours and spending a lot of time socializing with the Brits. He vowed to do something about that, and soon would.

Meanwhile Ike had been toiling faithfully in the vineyard of long hours, sometimes as many as eighteen a day for weeklong stretches, and in the process becoming an indispensable fellow. But if you were a people person like Ike, working for George Marshall was bound to be a frustrating experience. For example, given Marshall's dry-ice personality, it's a good bet that his recent reorganization that reduced the number of staff with direct access to him from sixty to six was in equal parts motivated by his mastery of delegation and his not wanting to deal with more people than was absolutely necessary.

Ike was one of the chosen few, but that was a problematic reward on a day-to-day basis. Characteristically, Marshall rode Ike with a very loose hand, but then jerked the reins to show him who was in charge, in this case with just one word. Ike liked Ike—the nickname, that is—and wanted to be called it always by friends and associates. He was forever "Eisenhower" to Marshall: the one time he slipped and called him "Ike," he made up for it by filling the air with so many "Eisenhowers" that their recipient could barely suppress laughter.

Then there was the matter of promotion. One day pivoting off the case of another officer, Marshall decided to acquaint Ike with his approach: "The men who are going to get the promotions in this war are the commanders in the field, not the staff officers who clutter up the administrative machinery," Ike remembered. "Take your case. I know that you were recommended by one general for division command and by another for corps command. That's all very well . . . but you are going to stay right here and fill your position, and that's that!" It was enough to flare Ike's temper, as had happened with MacArthur: "General . . . I don't give a damn about your promotion plans as far as I'm concerned. I came into this office from the field and I'm trying to do my duty. . . . If that locks me to a desk for the rest of the war, so be it." With that, Ike got up and took the long walk to the door. As he opened it he turned to Marshall, grinning his soon-to-be famous grin. In return: "A tiny smile quirked the corner of his face." Another test. Marshall had absolutely no intention of doing what he had just told Ike; in fact, he was now planning on the opposite. The man was all about indirection.

SO IT WAS THAT ON THE MORNING OF MAY 26, 1942, IKE FOUND HIMSELF IN London wearing two stars (Marshall, when questioned about violating his own policy on promoting staff officers, blandly labeled Ike his deputy commander) and in the company of Mark Clark, who had jumped out of Marshall's black book to also gather two stars. Both of them, dead tired from a long transatlantic flight, were arriving in Paddington station to begin their mission of assessing the London operation for their boss. Waiting for them was their driver, Kay Summersby, who would be the second romantic love of Ike's life.

She was an extraordinary woman, and a surprising choice for Kansas-bred Dwight Eisenhower. Her look was elegant enough for Paris couturier Worth to have employed her as a fashion model before the war. The daughter of a retired British officer, Kay received an upper-class upbringing, and, though Irish, she blended well with the British equivalent, and married a wealthy publisher, a union that had recently dissolved. This was a time of war, and Kay joined an ex-debutante corps of women drivers and was at the wheel of an ambulance through the height of the blitz. Now with an intimate knowledge of London's maze, she was carrying on as a chauffeur for high-ranking Americans, making sure they got where they needed to go. She was also an exceptionally good bridge player, and had a weakness for married men. It would take a while for Ike to find all this out, but he likely fell for Kay on that first ride.

But by this time Dwight Eisenhower had also grown into an awesome compartmentalizer, perfectly capable of filing away an entirely intriguing encounter for later reference, and then pursuing the main mission as if nothing had happened. So Kay drove Ike and Clark to practically every American military installation in England, as well as a lot else, including a meeting with Bernard Montgomery that went south before it began: "Who's smoking?"

"I am," replied Ike.

"Stop it. I don't permit smoking in my office."

Ike crushed the cigarette and, not knowing Monty had been almost fatally shot through the lung in World War I, continued to smolder even after the session was over. On the way back to London, chain-smoking Ike let loose with his initial opinion of Britain's greatest fighting general: "That son of a bitch!" Kay looked into the rearview mirror and saw Ike's "face was flaming red and the veins in his forehead looked like worms." Ike had reached sufficient rank that he could safely give vent to his blowtorch of a temper, and Kay was an early witness to what became a steady diet of explosions for those who worked for him. They passed quickly, but Ike had a higher gear of long-term grudges that he was much better at concealing.

On this trip it was reserved for Major General James Chaney, the US commander, and his entire operation. Ike saw everything Marshall had noticed—the banker's hours, the lack of urgency—and added a good deal more, including the wearing of civilian clothes and thinking of themselves as a "liaison group," not headquarters. "Specifically, they seemed to know nothing about the maturing plans that visualized the British Isles as the greatest operating military base of all time," Eisenhower later remembered. Mark Clark was fully on board, just not as angry as Ike, whom Kay heard say that the whole gang should be relieved and sent back to the United States "on a slow boat, without destroyer escort."

When he got back to Washington and spoke to Marshall, his message was less murderous, but no less damning. It was clear that Chaney had to go, and Marshall asked Ike what kind of replacement was needed. Ike responded that it had to be someone intimately familiar with Bolero and with the will to make it happen, recommending the general who had presided over the recent staff reorganization. Marshall remained noncommittal, further directing Ike to draft a document specifying the duties and responsibilities for a European theater commander, the second time he had been given such a task.

Ike turned in his directive on June 8. He must have been satisfied with the results, since he told Marshall to read it carefully, as it was likely to become a key set of orders in the war against Germany. "I certainly do want to read it. You may be the man who executes it. If that's the case, when can you leave?" Marshall made him wait three days, but that night when Mamie asked what his new job would be, he could reply, surely beaming his broadest grin: "I'm going to command the whole shebang." Marshall wanted Mark Clark as his deputy, which was fine with Ike, since he considered him a friend. Three weeks later he got his third star, the hardware that went with the shebang. In response to George Patton's immediate congratulations, Ike threw back what looked like a perfunctory pitch, but wasn't: "I particularly appreciate it, because you and I both know you should have been wearing additional stars long ago. It is entirely possible that I will need you sorely . . . As I have often told you, you are my idea of a battle commander . . . I would certainly want you as the lead horse in the team." No idle chatter here; these were prophetic words.

Ike flew into London on the night of June 24, 1942, met by Lord Louis Mountbatten, chief of Combined Operation Headquarters, who whisked him off to Claridge's, the city's toniest hotel, where he was told his orderly might find a bed at a nearby enlisted barracks. Eisenhower didn't like the reply, or the ambience. "My sergeant has had a long and trying trip like the rest of us. I prefer to have him stay here at the hotel." But nobody in the party was there long; Eisenhower saw to that. Glitz was exactly what he didn't want.

He had come to run a tight ship, and he found too many American officers at headquarters on Grosvenor Square in an entirely relaxed mood, failing to salute his car with its highly visible three-star flag, and arriving late and leaving early while he and his close aides were still buried in work. That didn't last long with Eisenhower around; almost instantly the entire crew was at their oars, doing their best to approximate his pace and endurance. He also grumbled about his social duties: "I don't think my blood pressure can take it if one more silly woman calls me 'My deeaaah general.' . . . I'm not fighting this war over teacups." When it came to soirees that mattered, like his weekly meals with Winston Churchill, Ike turned on the charm full blast, but the rest soon got only his regrets. Still, the place remained too full of distractions. Earlier, he had been heard to say darkly, "Some of these fellows have been in London too long"; now he seemed to have concluded that went for him too.

Ike's Mr. Fixit and designated buddy Lieutenant Commander Harry Butcher—a radio executive friend from Washington days, now on loan from the Navy—found his general exactly what was needed: Telegraph Cottage, an ultrabucolic English country house less than a thirty-minute drive to Grosvenor Square. Here Ike could and would retreat, and inevitably rebuild the Club Eisenhower—still boys and girls together, although everybody had left his spouse at home.

By this time Kay Summersby was very much a part of this rollicking but near suicidally dedicated and hardworking crew. One of the first things Ike did when he got back to London was to send one of his aides out to run her down. When he discovered she was driving for an Air Corps general, he blithely pulled rank and had her peremptorily perched behind the wheel of his new Army-olive Cadillac staff car.

But Kay grew into far more than that. An intuitive soul, she quickly grasped that you were Ike's friend to the degree you were useful to him, and she responded by becoming his personal Swiss Army knife. Most useful, perhaps, but often overlooked, Kay could not only seamlessly drive her general through London, she knew the even more intricate and unforgiving ways of the British upper classes just as well, and she would provide Ike with exactly the road map he needed when dealing with those at the top. Along with that came a dash of glamor and sexual mystery that made Ike look more the sophisticate than he actually was to this audience, particularly to the louche individual at the center of things, Winston Churchill. As noted, Kay was a whiz at contract bridge, Ike's new card passion. Poker is a strategic game, and Ike got so good that he had to stop for fear of winning too much from fellow officers. Bridge is a game of alliance and strategic communication, just what Eisenhower was presently doing, and the perceptive Kay quickly established herself as his preferred partner.

Sex, it seems, was the least satisfactory part of the relationship, described by Kay after Ike's death as a few rushed and fumbling attempts terminated when he lost his erection. Although Telegraph Cottage constituted extremely close quarters, Kay and the other distaff members had their own barracks elsewhere, so opportunities were rare. Compounded by Ike's long hours, three packs of cigarettes a day, endless cups of black coffee, an unhealthy diet, and practically no exercise, all the while being under tremendous pressure, it was a lifestyle bound to be tough on any libido. Yet you also get the feeling he was holding back; that, as with so many people who had grown close to him, Kay had a time and a place, and if his heart had strayed, his future still belonged to Mamie. Ike was torn, as we shall see, but ambition would call him back to her.

Another key individual joined the club and Eisenhower's future around this time. He was Walter Bedell Smith, the same "Beetle" who had summoned him to Washington for Marshall, and for whom Ike had used all his leverage to pry away and bring to England as his own chief of staff. He was Ike's Sutherland, only a lot more talented and perhaps even less likable. His job as the boss's enforcer was to say no, knock heads, and make sure orders from above were scrupulously followed, all the while doing his best to keep Ike's smiling mask spotless.

Although his preferred nickname was that of an insect—one who struck outsiders as about equally lovable—underneath was a remarkable man. Having joined the Indiana National Guard as a private, Smith had worked his way up the ranks on sheer ability, and on the way demonstrated just how effective the Army's educational and mentoring process could be, turning Smith not just into an extremely knowledgeable and competent officer, but something of a closet intellectual, one who read the likes of Joseph Conrad while Ike stuck to cowboy stories. Smith also demonstrated his depth by revealing a real flair for diplomacy, one sufficiently bright to lead him one day to become US ambassador to the Soviet Union. But for now, Eisenhower quickly learned he could trust "Beetle" to talk to anybody he didn't, and not suffer any unintended consequences—a precious commodity at this parlous moment as America got down to waging war where it really counted.

One of the sources of Smith's miserable personality was that he actually was miserable, wracked by recurring ulcers. One night, after he and his boss had dinner with Churchill, the prime minister took Ike aside and told him he thought Smith looked like a ghost, and was likely not far from becoming one, as paleness in an ulcer victim is inevitably a sign of bad internal bleeding. Ike ordered Beetle to bed forthwith, and told him to stay there until he felt better. Next, Ike told Harry Butcher to get Beetle a full-time nurse. Being the designated fun guy of the Club Eisenhower, Butcher came up with Ethel Westermann, considered by many the most beautiful nurse in the European theater. She was also bright and funny, and before long it became obvious to the others that they were sleeping together. That was okay with Ike, so long as they were discreet. The main point was that Beetle's health improved, and he was soon restored to his place at Ike's right hand, where he was needed. Meanwhile, like Kay, Ethel revealed herself as a lot more than eye candy, proving utterly fearless when things got deadly, all the while playing an excellent hand of bridge—the perfect combination for full membership in the Club Eisenhower.

It's also important to keep in mind that Ike's interludes at Telegraph Cottage were simply brief respites from the overwhelmingly complicated and time-consuming process of planning a war and putting America's legions in a position to come to grips with the Germans. But it's still worth considering that at this point the only combat veteran in Ike's military family was Kay Summersby, having driven an ambulance, at times stuffed with corpses and pieces of corpses, through London's blazing streets at the height of the blitz.

ENGLAND'S WAR WITH GERMANY HAD TURNED BLOODTHIRSTY IN SEPTEMBER 1940, nearly two years before Eisenhower came on the scene, when the Luftwaffe began a strategic bombing campaign in anticipation of Operation Sea Lion, an invasion of England. But after failing to knock out British warning radars and grossly underestimating the strength of the Royal Air Force's Fighter Command, the Luftwaffe allowed itself to be drawn away from attacks on vital air bases and aircraft plants to concentrate instead on the terror bombing of civilian targets. It was a thoroughly feckless decision, and, after the cancellation of Sea Lion, one that served no real military purpose, except the one that boiled down to killing for killing's sake.

As for breaking British morale, it seemed to have had the opposite effect. When Winston Churchill visited a bombed-out neighborhood in poor South London he heard—or thought he heard—not panic or recriminations but simply the angry cry "Give it 'em back."

And he had the means to do so. Unlike the Germans and also the Japanese, who had neglected to develop a true strategic bomber, the British had both the Halifax and the Lancaster, big, four-engine aircraft capable of hauling five tons of high explosives and incendiaries all the way to Germany and back, not to mention the equivalent B-17s and B-24s he would soon be getting in quantity from the Americans. And in February 1942 he settled on the man ready—actually itching—to lead the onslaught, Air Marshal Arthur "Bomber" Harris.

After a short run of ineffective and costly daylight raids, the RAF concluded that nighttime was the right time to bomb Germany. Losses dropped precipitously but so did accuracy, or any pretense of accuracy. While the Americans would insist on the efficacy of daylight "precision" raids, which they began without much success in the summer of 1942, the British developed another rationale: "dehousing." It was a fig leaf dreamed up by Lord Cherwell, Churchill's science adviser, who, after analyzing the effect of the blitz, concluded that what most bothered victims was losing their homes—"People seem to mind it more than having their friends or even relatives killed." Extrapolating to Germany, where twenty-two million civilians were concentrated in fifty-eight cities, Cherwell estimated that an eighteen-month bombing campaign could dehouse one-third of the population, and thereby "break the spirit of the people"—exactly what the Germans had failed to do during the blitz, but a perfect strategic excuse to set Harris loose.

The campaign began on March 28, 1942, when the RAF burned the wooden city of Lübeck with incendiaries, dehousing over fifteen thousand. Rostock, the venerable Hanseatic League port on the Baltic, followed in late April. But these were just warm-ups for the kind of attack Harris really wanted to stage. With losses down and heavy bomber production up, "Operation Millennium," a 1,000-plane assault, suddenly became a real option. He had to raid flying schools, training units, even the RAF's Coastal Command to gather the crews and planes, but on the night of May 30–31, 1,046 of his heavy bombers were headed for the cathedral city of Cologne with murderous intent. Guided by flares dropped by low-flying pathfinders, they came in waves, alternating high explosives with incendiaries, destroying six hundred acres of the city's center and igniting history's first man-made firestorm.

With great courage and also losses, the US Army Air Corps persisted in its daylight raids, but with "precision" measured at around a half-mile radius, US bombers were just as likely to engage in dehousing as they were to hit a factory. And when they did manage to do so, the machine tools inside proved a lot harder to destroy than the structures surrounding them. Hence, in spite of all the bombing, German war production not only ground on, it actually increased. Meanwhile, when American heavy bombers flew joint missions against Deutschland with the British, it was at night. It all added up to a lot of dehousing, to millions losing their homes, but more to the point, in 1956 the West German government would estimate that over 600,000 Germans had been killed by Allied air raids. Subsequent accountings have been somewhat lower but still don't change the point: hundreds of thousands of civilians were deliberately killed in a campaign that lasted to war's end by the side purported to be defending human rights. This is where industrialized warfare had taken us.

Strategically, if the bombing campaign had any effect, it was serendipitous—decimating the ranks of combat-experienced German pilots attempting to defend their homeland. This was especially the case after Allied bombers began arriving with a new generation of high-performance fighters with range-extending drop tanks. Ultimately, the Luftwaffe losses would leave the skies over Normandy almost exclusively friendly ones. But it took a while to get there, and it was hardly a direct itinerary.

IN THE BIG LEAGUES, IT'S EASY FOR EVEN A SUPERSTAR PLAYER TO OVERESTImate his own pull when dealing with management, a fact of life George Marshall would have demonstrated to him in the summer of 1942.

The war against the Euro-Axis was not going well. France and most of the rest of Europe had already fallen. In Russia, although the Soviets had beaten back the Wehrmacht on the outskirts of Moscow in late 1941, Leningrad was still under siege. Meanwhile, the Germans had crossed the Don, headed for Stalingrad, and the Caucasus oil fields to the south seemed within easy reach. In North Africa, the Italians and Germans, led by Erwin Rommel and his Afrika Korps, were besieging the British and Australians at Tobruk, beyond which lay Egypt and the vital Suez Canal. At sea, U-boat wolf packs prowled the Atlantic largely unmolested, while to the north a convoy bound for Archangel lost twenty-four of its thirty-three transports, causing the indefinite suspension of further shipments to the USSR's arctic ports.

It was in this context that Winston Churchill boarded a long-range seaplane on June 17, 1942, with General Brooke, and made the twenty-eight-hour jump to Washington, to be greeted there by George Marshall. Likely he underestimated the peripatetic prime minister's sheer stamina, for the next morning he discovered Churchill had already departed for FDR's home at Hyde Park, leaving both Marshall and Stimson fretting over what new scheme he might have concocted to lure their boss away from a cross-Channel invasion.

They had reason to be worried. Initially eschewing subtlety, Churchill took a sledgehammer to Sledgehammer, throwing the president a series of hard questions based on "who?" "what?" "when?" and "where?" and then capped these with a real stumper: "What shipping and, in particular, how many landing craft will be available?" With this he handed FDR a note stating that the British government and military believed a September cross-Channel attack was "certain to lead to disaster."

Next, Churchill threw Roosevelt his changeup. He knew FDR had recently "gone off the rails" by promising Stalin's foreign minister, Vyacheslav Molotov, a pressure-relieving European "Second Front" in 1942—knew it because Molotov told him so. But if Sledgehammer was suicidal, the British prime minister asked, "Can We afford to stand idle in the Atlantic Theater during the whole of 1942?" It was a leading question, and he answered it by again proposing the invasion of French North Africa, which had been put on the back burner months before. It may not have been in Europe, but it was definitely a "second front"—though a roundabout one—so the scheme had some merit to FDR, who was eager to fight Germans somewhere soon, if possible before the November presidential elections.

Churchill's pitch looked considerably harder to pass up the next day, June 20, when they reconvened in Washington with their military principals. In the midst of discussions an aide handed Roosevelt a telegram. He studied it, passed it to Hopkins and Marshall, then on to Churchill, who read aloud: "Tobruk has surrendered with twenty-five thousand men taken prisoner." This was staggering, "one of the heaviest blows that I can recall during the war," Churchill remembered, one that apparently opened the way for the Germans to the Nile delta, Suez, and the oil of the Middle East. Nor would he ever forget FDR's sincere sympathy, and more tangibly his offer to immediately send three hundred of the new Sherman tanks to help reinforce British forces now dug in at El Alamein, only sixty miles west of Alexandria.

George Marshall found himself pulled in two different directions, and in the process nearly became unhinged. Being a good soldier and wanting his side to win, he did his best to expedite the shipment of the Sherman tanks, which did arrive in time to help Montgomery save the day at El Alamein. But he was also prepared to up the ante considerably more by sending the British one of the Army's two fully equipped armored divisions. This wasn't just a suggestion to reassure Churchill and Brooke; Marshall took steps to make it happen, or seemed to at least.

He immediately reached out for George Patton at the Desert Training Center and brought him to Washington, ordering him to prepare for the deployment of the Second Armored Division to Egypt. No more, no less.

But Patton being Patton, he got overexcited and the next day sent the chief a memo proposing two divisions. Marshall's reaction was a swift jerk on Georgie's reins: "Order him back to Indio"—Georgie's purgatory in the desert. He even had one of his staff personally put him on the plane to California. "That's the way to handle Patton," he said to a colleague shortly after. Patton had hit on a sore point with Marshall: sending our entire armored strength to North Africa was exactly the wrong message to give the British when he was trying to pin them down to marshaling the resources in England for a cross-Channel invasion.

Instead, he wanted to outflank them, and also the president, which is where his scheme went awry. On July 20, he proposed that the Joint Chiefs present the president with a Pacific alternative to North Africa, which is what Admiral King and the Navy had wanted from the beginning. It went forward forcefully worded, to say the least. "If the United States is to engage in any other operation than the forceful adherence to Bolero plans, we are definitely of the opinion that we should turn to the Pacific and strike decisively against Japan."

Fighting words, and that's exactly how FDR reacted in calling the bluff. He replied with a one-two combination. The first was addressed to the chiefs: "I want you to know I do not approve the Pacific proposal"; the second was aimed at Marshall directly: "Exactly what Germany hoped the United States would do following Pearl Harbor." To make sure he was understood, this one was signed "Roosevelt C-in-C." This was about as stern a course correction as the hyperaffable and proverbially indirect Roosevelt was capable of giving. But there was only so much grand strategy to be made, and the manager wanted it completely clear that he would decide it. North Africa it would be, and he sent Hopkins and Marshall to London to make nice with the British and to make it happen: Operation Torch was scheduled for November.

Torch was an extremely ambitious and therefore risky concept: three simultaneous amphibious assaults, two of them combined US-British operations requiring passage through the potentially submarine-infested straits of Gibraltar and then separate landings at Oran and Algiers, while the third would be an exclusively American effort, a transatlantic assault on the western coast of Morocco. To further complicate matters, all this territory was under the ambiguous control of the defeated Vichy French, whose loyalties at this point were impossible to calculate, except that they liked the Americans better than the British, who had attacked their fleet in an effort to keep it out of German hands. Also, any apparent French accommodation with Axis forces logically could be presumed to be based purely on expedience. As to their own internal politics, that was anybody's guess, and this being the case, multiple players would give it a shot. It all added up to a hairball of complexity for whoever was put in charge.

While Hopkins and Marshall were in London they were in close contact with Eisenhower, and it was obvious to the chief of staff that he had tightened up the operation in the United Kingdom to what now looked like wartime status. It was also apparent that he had made an excellent impression on the British; here as elsewhere people liked Ike, and that translated into good working relationships. He had been sent basically to preside over the Bolero buildup, but at this point Marshall took the leap of faith and handed Ike the hairball that would soon catch fire as Torch.

Some at the time, along with a number of historians later, have wondered at his choice, given Eisenhower's lack of combat and command experience. Yet Ike was a veteran of what might have been the US Army's toughest torture test, six months of unremitting toil under the unblinking and unforgiving eye of George Marshall. That was what turned him into a starter.

Overall, the chief proved to be an excellent judge of talent and capabilities, but not as good as he thought, which was faultless. A place in Marshall's mental black book would prove to be a very secure perch, given his own coldhearted version of loyalty—"You must be good if I chose you." In several cases it left him hesitant or unwilling to remove obvious incompetents. But with Ike, Marshall's patience was well warranted. Torch was not just an order of magnitude more complex than anything he had yet attempted; it would transpire in the twin dimensions of battle and diplomacy, sometimes in both at the same time. Perhaps Hannibal or Napoléon might have shined from the outset, but it took Ike time to get used to all this complexity and ambiguity. George Marshall gave him that time with unremitting support and only helpful advice. Ike proved well worth the wait. And if the story did not turn out exactly as he had hoped from a personal perspective, Marshall, because he was a team player, would graciously endorse the outcome, which was in the end Ike over everybody.

Stuck back in California's "Little Libya," the Desert Training Center, George Patton might have found Eisenhower's treatment comforting had he known about it. George Marshall had become a black hole to him, swallowing his messages and phone calls and returning nothing. The only good news he got was the acceptance of his son at West Point, prompting some fatherly advice: "If some little fart hases [*sic*] you dont [*sic*] get mad do what he says and take it out on some one [*sic*] else next year. . . . We are really proud of you for the first time in your life see to it that we stay that way." With motivational skills like that, how could his luck not change?

It did. At the end of July 1942, Marshall jerked him back to Washington and promptly informed him he would command the Western Task Force of Torch, set to land on the Atlantic coast of Morocco. He was always in the chief of staff's holster, just in need of a less itchy trigger finger, one he would never develop.

Patton had a long time to prepare, around eleven weeks, and since this was an amphibious operation, that meant a lot of interaction with the Navy. With a revved-up, war-hungry George Patton, sparks were bound to fly. At one of his first meetings with Admiral Kent Hewitt, Torch's soft-spoken naval commander, Patton let loose a flaming broadside—"A torrent of his most Rabelaisian abuse"—causing a number of the admiral's staff to leave "convinced they could never work with a general so crude and rude as Patton."

It got only worse, and Hewitt complained to his chief, Admiral King. Even the explosive King thought Patton's conduct sufficiently outrageous to demand Marshall relieve him. At Eisenhower's insistence, he refused. "Patton is indispensable to 'Torch,'" Marshall maintained, arguing that the same qualities that made him miserable to work with made him an excellent battle commander—a line of reasoning calculated to make at least some sense to the ill-tempered King. At any rate, he stayed, and by September 24—a full month before the Western Task Force would leave Norfolk for North Africa—Patton had completed his operational plan.

Three days before their departure, FDR invited Hewitt and Patton to the White House, and, perhaps prompted by Eleanor's fondness for Beatrice Patton, received them in an effusive mood. "Come in Skipper and Old Cavalryman," Roosevelt greeted them, "and give me the good news" . . . "I asked him whether he had his old Cavalry saddle to mount on the turret of a tank and if he went into action on the side with his saber drawn. Patton is a joy. . . ." Not to be outdone when it came to hyperbole, Georgie replied, "Sir, all I want to tell you is this—I will leave the beaches either a conqueror or a corpse." A veritable love feast it was, the burned-out smell of the Bonus Expeditionary Force apparently long forgotten.

One of Patton's last acts before his departure was to visit Black Jack Pershing, now eighty-one and more or less in permanent residence at Walter Reed Hospital. At first, the old general didn't recognize him, but when he heard Patton's high, squeaky voice he knew him immediately. "I can always pick a fighting man and God knows there are few of them. I am happy they are sending you to the front at once. I like Generals so bold that they are dangerous." Pershing may have been old, but he knew Georgie.

It would take a hundred ships to transport Patton's 34,000 soldiers on the 4,500-mile journey to the Atlantic coast of Morocco. Observing the final embarkation of the Third Division put Patton in a reflective mood: "When I think of the greatness of my job . . . I am amazed, but on reflection, who is as good as I am? I know of no one." It must have been a blissful moment. He had a war, he had an army, and he was likely feeling he was on his own personal stairway to heaven, filled with reincarnated warrior Pattons.

Back in England, Dwight Eisenhower could only dream of such clarity. He was stuck with the whole hairball, its edges characterized by extreme fuzziness.

Ike had more than a flair for diplomacy, and if he could get what he wanted through talk rather than combat, he was always ready to do so, even in the middle of a war. Conveniently, when it came to North Africa he was of a like mind with his manager FDR, who in turn was beguiled by a very convincing diplomat of his own, Robert D. Murphy, who argued that the French in control there could be turned, convinced to switch sides, if the proper French leader could be supplied. Murphy's man was Henri Giraud, a general who had escaped from the Germans twice—once in each world war—earning him some admiration in metropolitan France, but not necessarily in North Africa.

To find out, an elaborate scheme was hatched. Beneath a heavy cloak of secrecy, Murphy flew into Britain on September 16 and was immediately spirited out to Telegraph Cottage. Here, in an all-night session, he spelled out his plan, part of which involved slipping a high-ranking American into Algeria for a clandestine meeting with key French generals, presumably to secure their support. Since Ike liked the idea, his deputy Mark Clark jumped at the chance to be the high-ranking American, and was duly injected via a British submarine.

The meeting proved a farce, and a dangerous one at that. Just one French general showed up, and he claimed only Marshal Pétain, head of the Vichy regime, could relieve him and his fellow officers of the responsibility to defend North Africa. Then a warning that the local police were coming caused Clark to bolt, later losing his pants and, more important, his wallet in the surf reboarding the submarine. The mission proved an utter snafu, and it should have been a warning, but Ike liked the concept and the participants, so he strung them all along for a while longer.

The potential magnitude of this sort of procrastination became apparent on November 1 when, with all three invasion fleets already at sea, Murphy cabled Eisenhower asking for a two-week postponement of the landings, because General Giraud was unable to leave France until then. Ike exploded to Marshall, who quickly got FDR to overrule Murphy's ridiculous suggestion. But for Ike this was far from the last of Murphy, Giraud, or other assorted French players of dubious reliability.

Ike had chosen as his initial command headquarters the Rock of Gibraltar, the great fortress and its associated airstrip being the last territory in western Europe controlled by the Allies. On November 4, Eisenhower and staff, split up among five B-17 bombers, took a night flight from London headed for the Rock. At this point in the war these were far from friendly skies, so the formation flew a few hundred feet above the waves through steady fog and rain.

Ike could rest easy, for at the controls of his plane was Paul Tibbets, already considered the best bomber pilot in the Air Corps. While both were undoubtedly preoccupied with their slice of the moment, it's still tempting to ask a rhetorical question about these two, an order of magnitude removed. Which one would ultimately be more important for the future of war? Ike, who would serve two terms as president of the United States during the height of the Cold War, or Tibbets, who flew the B-29 over Hiroshima named after his mother, Enola Gay?

We will leave that an open question, and return to when Tibbets finally caught sight of Gibraltar looming above the haze. "This is the first time I have ever had to climb to get into landing traffic at the end of a long trip!" Ike heard him say. All five bombers arrived safely; Tibbets's mission was accomplished, but Eisenhower's was just beginning.

For starters, he had barely begun setting up shop in his damp headquarters deep within the Rock when, courtesy of the ever-resourceful Murphy, arrived a most unexpected houseguest, Giraud. To compound matters, the French general was under the impression that he was no visitor, but instead the proprietor of the operation, its commander in chief. Also, he wanted to turn the entire force around and direct it to the southern coast of France. Probably Ike would have been better off casting Giraud immediately out of the Rock, but instead he chose the diplomatic route, explaining at great length exactly who was in charge and what the Frenchman's role would be: "Giraud will be a mere spectator in this affair." Events would prove that to be exactly the case.

Giraud was charismatically challenged, a figure with negligible support in either France or North Africa. Rather, he was Murphy's and FDR's straw man, a hollow substitute for Charles de Gaulle, a figure of true substance, but also arrogance—enough of the latter to convince the president that he was a Bonaparte wannabe and a danger to France once liberated. Churchill and the British were already betting on de Gaulle, but it would take Ike some time, and FDR forever, to realize he represented Gallic true grit and was the key to France's republican future. In the meantime, the Americans would stick with Giraud and whomever else they could dig up with some pretense of authority.

It's easy for things to get turned around and out of proportion when you are below several hundred feet of limestone. Gibraltar was not the place for Ike as commander in chief to be if he wanted to influence events. To make matters worse, his radio communications were in a state of near collapse, effectively cutting him off from all three landing zones and rendering him just as much a spectator as Giraud. When Torch was lit, both were in the dark.

Reportedly, the real Bonaparte maintained that "I would rather have a general who was lucky than one who was good." For a while, that probably best described Dwight Eisenhower's performance as commander in chief: for a mission as risky as Torch, you had to be lucky.

ON THE SAME DAY GIRAUD ARRIVED, A THRILL OF DREAD WENT THROUGH GIbraltar when a garbled transmission announced that a transport carrying a reinforced American battalion had been torpedoed and sunk with major loss of life in the Mediterranean, a day away from the Algiers landing site. But the casualties proved to be light, and it was to be the only troopship lost in the initial assault.

The Germans certainly were aware of the huge amount of Allied shipping passing through the Strait of Gibraltar. But simultaneously, Erwin Rommel's position was collapsing at El Alamein, and Bernard Montgomery and his Eighth Army were beginning to chase him westward across Libya. Hitler concluded that the Allied task force was headed there to cut him off, and ordered the rest of his submarines to the confined Strait of Sicily to ambush them. But they never arrived; the Torch landing sites were hundreds of miles to the east.

So the Allies landed on November 8, 1942, largely unmolested: 20,000 English and Americans on the beaches east and west of Algiers, and 18,500 on the coast near Oran, all Americans under the command of Lloyd Fredendall, a prominent alumnus of George Marshall's black book and one of his biggest mistakes. But for now things were stable.

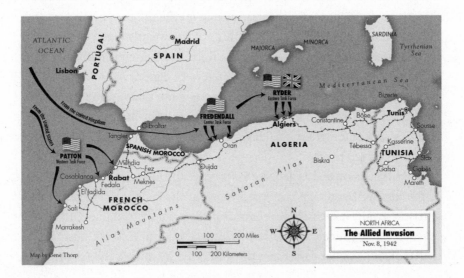

Meanwhile, out in the Atlantic, Fortuna gave George Patton at least a taste of adversity, but in the end proved still more bounteous. This was the most ambitious amphibious operation since the British arrived in New York Harbor to try to quell the American Revolution. And to the Navy participants it might well have seemed as ill starred. Particularly for those designated to operate the landing craft, this mission must have looked like an invitation to a demolition regatta, since almost without exception the surf is huge off the western coast of Morocco in November.

But when the Western Task Force got there on the morning of the eighth the seas at all three of the widely separated landing sites were flat. It was a major stroke of luck, but also accompanied by a bit of drama for Patton aboard Hewitt's flagship, the heavy cruiser *Augusta*, accompanying the main landing force to Fedala, fifteen miles north of Casablanca. Just as he was about to join his men storming the beaches against weak opposition, seven French cruisers swooped out of a smoke screen, guns blazing. The muzzle blast of *Augusta*'s first return salvo blew Patton's landing craft into small pieces, in the process destroying all his personal possessions except his emblematic ivory-handled twin pistols, which an aide had fetched moments before, and also dragging him back out to sea as Hewitt chased away what proved to be the half-hearted naval antagonists.

It was around one thirty when a soggy but still fuming Patton managed to hit the beach at Fedala, which his men had secured around five hours earlier. Still, there was a lot of disorganization and inertia to be seen. Because they had all been mislabeled, less than a fifth of the weapons and supplies had been unloaded, and wounded soldiers were not being properly attended. In the face of only desultory fire, the men were digging foxholes instead of taking care of business. These were the moments that George Patton was meant for. Suddenly, in the eyes of the men on that beach, he became gigantic, his mask ferocious, the idea of not doing what he said absurd. He got things organized pronto.

On this mission and in general, Patton was fortunate in his subordinates. At the southern landing point Ernest Harmon blocked the significant French garrison at Marrakech from reinforcing their compatriots at Casablanca. Meanwhile, fifty miles to the north of Patton, Lucian Truscott, destined for a memorable combat career, got things started by seizing the Port Lyautey airport, the only concrete strip in Morocco and fully capable of handling combat planes flown in from carriers at sea and Gibraltar.

Covered at the ocean side, from above, and to the south, a confident Patton led his now reasonably well-consolidated force toward Casablanca, and by November 10 was beginning to surround the city. He set the final assault for seven thirty the next morning so there would be plenty of light. Three hours before, his intelligence chief, Oscar Koch, woke him with the news that all indicators pointed to the French preparing to surrender. Oscar was a whiz at the dark art, and events would demonstrate his calls were not to be ignored. But Patton was playing his own game of bluff and intimidation; hence he rejected Koch's intimation that the attack be postponed. Two hours later the French folded.

It was at this moment of victory that George Patton revealed a new and entirely unexpected device in his professional tool kit—diplomacy. Patton the conqueror abruptly and very shrewdly turned into Patton the magnanimous, at least to the French and the locals holding things together. He fell back on his romantic code of the universal officer, treating his recent adversaries as indisposed colleagues and providing them equivalent peace terms. There would be a full prisoner exchange; they could keep their weapons but had to stay in their barracks for now, and in return the Americans would have free access to what they required to carry the fight to the Italians and Germans.

For these Frenchmen with no other option, this was a dream surrender, and they readily agreed. "Gentleman," Patton announced in more than passable French, "we have now settled everything, but there is one disagreeable formality which we should go through." The French peace negotiators visibly stiffened, but only temporarily, as they were assaulted by a phalanx of excellent champagne chilled to perfection. It was a love feast or at least a great drunk. You could almost hear the ghost of Humphrey Bogart tell the ghost of Claude Rains, "Louis, I think this is the beginning of a beautiful friendship."

But this taste of victory was hardly substantial. Patton's troops and the Americans at the other invasion points—around three-quarters of the total force—were still rookies in this deadly game of war. Here on the Atlantic coast they had just beaten French soldiers who would soon fight by their side, and who likely preferred the Americans to the Germans even while shooting at them.

The situation was much the same in Oran and Algiers, where the French fought harder but were never the central problem; the real enemies were geography and Axis troops—the Germans, really, but even the Italians had more combat experience than the Americans. If Torch was to light the way to quick victory, it was critical to consolidate rapidly and move in force into Tunisia to cut off the Axis supply and escape routes to Sicily. For Patton and his army this proved impossible when it was discovered that his sole rail link east through the Atlas Mountains was of such a narrow gauge that it couldn't accommodate his tanks; in the case of the others, it proved more a matter of will.

Ike, still separated from reality by Gibraltar's limestone, had sent Mark Clark to work on a cease-fire with Murphy's latest Frenchman, Admiral François Darlan, an archconservative and key man in the Vichy power structure. For the price of reaffirming his role as French high commissioner of North Africa, he did get his countrymen to stop shooting, but the cost to Ike and the Americans was considerably higher. While the talks dragged on, consuming precious time, the Germans were funneling troops and aircraft from Sicily into Tunisia, and Rommel and his Afrika Korps, after their defeat at El Alamein, moved steadily toward the same objective, soon to become a very formidable one.

Moreover, the atmospherics of a deal with an arch collaborationist proved nearly as unpleasant. "Not a pretty sight," observed Charles de Gaulle from London. In Washington, FDR was incensed that Ike had affirmed such a deal, and it took most of George Marshall's ability to spread oil on troubled waters to calm him down and see that any agreement that stopped the Franco-American fighting and allowed the US Army to get to the Germans in Tunisia had to be a good thing, or at least a tolerable one. But in the eyes of his manager Ike as commander in chief remained very much on trial, and once again the machinations of Murphy and Clark had not helped his situation. Only the very convenient and possibly OSS-inspired assassination of Darlan on Christmas Eve 1942 did that.

Meanwhile, it had taken Ike until the third week in November to move his headquarters forward to Algiers—joined there by Kay Summersby and Ethel Westermann, who were torpedoed, then rescued from lifeboats on the way. He was glad to see them, but showed little eagerness to go see what was happening at the front. Finally, on December 22, Marshall wrote him a cautionary note: "You are doing an excellent job and I want you to feel free to give your exclusive attention to the battles particularly as German intentions against your right flank seem evident."

The veteran British commander of the Eastern Task Force, Sir Kenneth Anderson, realizing both the dangers and opportunities to be found in Tunisia, had moved his force quickly inland; by November 17 he had reached the crest of the Dorsal Mountains, with the minarets of Tunis visible less than thirty miles away. But he was near the end of his supply tether, and at that moment his force was already seriously outnumbered and outgunned, with less than twelve thousand men and only light tanks available.

By this time the Germans gathered around Tunis numbered thirty-five thousand and had two full panzer divisions, some elements equipped with the new Tiger heavy tank mounting an 88 mm main armament—generally considered World War II's best and most versatile gun. Even worse, they were led by Field Marshal "Smiling Albert" Kesselring, a sort of human version of that gun—good at everything and always lethal. Probably the German military's best strategist and most resolute leader, whatever the circumstance, Kesselring, true to his nickname, remained forever optimistic.

Here in North Africa he saw only opportunity. He came from a Luftwaffe background—not an army one—and had become a master at mixing air and armored forces, World War II's deadliest and most effective tactical combination. In Tunis, he had plenty of the standard variety of German tactical aircraft (Bf. 109 fighters, Ju 88 fighter-bombers, and Stuka dive-bombers), not state-of-the-art but battle proven, and, more important, flown by veteran pilots used to fighting for their lives. It all added up to a significant combat multiplier in the hands of a commander who liked his odds—ones apparently growing better by the day as Hitler had reacted by pouring German and Italian troops into the Tunisian redoubt until they reached over a quarter million.

Under the circumstances Anderson did well just to hold on and establish a rickety stalemate, joining with the underarmed Free French to the south and Americans from the Oran task force to form a thin perimeter roughly along the line of the eastern Dorsal Mountains. Almost immediately the Germans grabbed not just air superiority but air supremacy, repeatedly sweeping the skies clean of American P-38s and P-40s flown by neophyte pilots from dirt fields hundreds of miles away, and then proceeding to bomb and strafe anything that looked useful to the invaders, along with their points of entry. In Algiers, the Luftwaffe kept Eisenhower awake with its night raids, one of which sank a ship full of air defense radar and communications equipment, compounding the unfriendly-skies problem.

Ike did manage to reinforce Anderson with troops sent forward in half-tracks rather than by narrow-gauge railway, the same choke point that back in Morocco made it impossible to send him the new Sherman medium tanks he had hoped to add to the battle. On November 28, Eisenhower did go forward with Clark to meet Anderson, but not anywhere near the combat zone. Finally, on December 9, Ike called George Patton in from Morocco and ordered him to the front to find out why Allied losses, particularly of tanks, were so high. In matériel terms the answer was obvious: the 37 mm main armament of the Grant tank didn't stand a chance against the 75 mm and 88 mm guns of its German opponents. But more disturbing to Patton was the message he got from the troops of the First Armored Division—they had been in combat for twenty-four days and he was the first general they had seen. Patton had recently been quoted by a newsmagazine stating that leadership was like "moving spaghetti—push it from behind it just buckles, but pull it from the front it will move forward." If he told this to Eisenhower it could not have registered, since he continued to try to lead from long range, and in the process missed a truly extreme example in Lloyd Fredendall, not far off. For now, though, Ike remained preoccupied with the politics of high politics.

January 14, 1943, marked the start of a major intersquad scrimmage, otherwise known as the Casablanca Conference, an event organized and secured by George Patton, who would not join the negotiations but made an exuberant greeter and majordomo. Team Great Britain steamed in with literally a ship full of experts, all their best players, and a quiver of persuasive schemes, one to be aimed directly at Ike.

There was little discord attached to agreeing that first priority still belonged to securing the Atlantic corridor by eradicating the U-boats. Nor were Roosevelt and Churchill the slightest bit out of synch when they announced that "unconditional surrender" would be their only terms for the

Axis powers, a declaration primarily aimed at reassuring Stalin there would be no separate peace. But from this point things got questionable, and the British had most of the answers.

Both de Gaulle and Giraud were at Casablanca, not as direct participants but hanging in the background. The British remained solidly behind de Gaulle, but FDR and Marshall continued to push their limp alternative Giraud, getting the two Frenchmen to sign a meaningless statement of unity that failed to disguise the vast difference in gravitas between the two.

Then there was the question of where to go next in terms of grand strategy. Here George Marshall was treated to a lesson in the gravity of propinquity. The immediate strategic objective of the North Africa campaign was now clearly Tunisia, whose Cape Bon Peninsula points like a finger directly at Sicily, less than a hundred miles away. By comparison the cross-Channel operation loomed off in the distance, exactly where Churchill and his very persuasive Army chief Alan Brooke wanted to keep it. Marshall was caught flat-footed as the British pitch whizzed by, leaving him flailing angrily against its overwhelming logic. Sicily was next, the cross-Channel operation pushed back to 1944.

Nor was this the last of Team Great Britain's confounding pitches. They let loose what looked at first like a softball, a proposal to unify the command of the entire Mediterranean and Middle East, and give it to Eisenhower while giving naval, air, and ground operations to British subordinates. It looked good to FDR and Marshall, until they realized it was aimed at Ike's head. The proposed ground commander was Sir Harold Alexander, one of Britain's most distinguished and aggressive combat generals, and a perfect headsman to lift those responsibilities from Ike's shoulders. But once Marshall saw the pitch for what it was, he used it to reinforce the principle of unity of command. Elevating Ike officially as "supreme" for the entire theater and clearly designating Alexander as one of his deputies would be treated as a good thing now, and even better for the future.

That was probably about the best George Marshall could get for Ike, who plainly crashed at Casablanca. Almost literally. As he flew in, his B-17 lost two engines, forcing Ike to don a parachute and be ready to jump before the pilot finally wrestled the plane to the ground. Safe but probably not in the best mood to brief the entire Allied high command convincingly, Ike found that his scheme to resume the offensive in North Africa, plan Satin, got anything but a smooth reception. His central stratagem involved using Lloyd Fredendall's II Corps to drive a wedge between the Germans already dug in to northern Tunisia—now under General Hans-Jürgen von Armin, a veteran of the eastern front and Hitler's personal choice to defend Tunis—and the newly arrived Afrika Korps, commanded by Erwin Rommel. But Eisenhower's pitch proved a hanging curveball to Brooke, who promptly knocked it into the stands by pointing out that Armin with eighty-five thousand veteran troops and Rommel's Afrika Korps, estimated at another eighty thousand, were much more likely to grind II Corps into dust. Ike attempted a rebuttal, but it was the last anybody heard of plan Satin.

But his bad day wasn't over. "Ike seems jittery," FDR told Harry Hopkins. Later that afternoon he got word the manager wanted to see him at his villa. Roosevelt went right to the Tunisian campaign: "How long is it going to take to finish the job?"

"With any kind of break in the weather, sir, we'll have them all either in the bag or in the sea by late spring."

"What's late spring? June?"

"Maybe as early as the middle of May." Events would show Ike called it almost exactly, but his manager remained far from convinced that he was the man for the job. Later, when Marshall recommended that Ike be given a fourth star on the grounds that his three British subordinates already had them, Hopkins recalled that "the President told General Marshall that he would not promote Eisenhower until there was some damn good reason to do it," and that "he was going to make it a rule that promotions should go to people who had done some fighting, and that while Eisenhower had done a good job, he hasn't knocked the Germans out of Tunisia."

Perhaps hedging his bets just a bit, but also trying to protect Ike, Marshall took him aside and proposed that if he wanted to continue to handle the politics, which were complex and important, maybe he should think about making George Patton his deputy for fighting. When it mattered, Ike had few equals at reading people, and he got together with Georgie that night.

"He and I talked until about 0130," Patton wrote. "He thinks his thread is about to be cut. I told him he had to go 'to the front.' He feels he cannot, due to politics." But when Ike popped the question, Georgie turned him down.

He wanted to kill Germans, and he had been in the Army long enough to know "deputy for combat affairs," or whatever such title, was an oxymoron that would land him behind a desk.

George Marshall was not about to leave for Washington without looking in on the state of Ike back at his headquarters in Algiers, which he did with the other chiefs on January 24. If there could be such a thing as a military mother hen, it would describe George Marshall's relationship to his protégé at this point and for a while longer.

Ike was working ridiculous hours and looked terrible, so Marshall ordered designated buddy Harry Butcher to get him outside as much as possible, preferably on the back of a horse but also with golf club in hand—wholesome activities, which implied he also sensed unwholesome overtones. He could hardly have missed Kay Summersby, especially after Telek, the Scotch terrier she and Ike shared, urinated on his bed. "I always had the distinct impression that General Marshall would have been just as happy if I did not exist," she remembered, likely understating the case. Marshall may have been suspicious, but Kay, Ethel, and the other distaff members of the Club Eisenhower all had valid roles and worked hard at them, so nothing was said, and Ike would persist with the old intelligence trick of hiding ambiguous assets right out in the open.

If he thought he had the better argument ("How can three stars be taken seriously by a bunch of fours?"), George Marshall was no man to turn down, even if you were the manager. It took him less than three weeks of convincing before FDR caved, and Ike's promotion went forward on February 11. It was about the only good news he would get in a month that had to rank as among the worst in Eisenhower's career.

ON HIS FIRST DAY AS A FOUR-STAR EISENHOWER WENT FORWARD WITH KAY AT the wheel, finally headed for the front, but first he was to pay a visit to Fredendall's II Corps headquarters. On the way, Kay told Ike, in case of a strafing attack, "Don't expect me to hold the door for you."

"Fine . . . Agreed"—his response reducing them both to recurring fits of laughter.

But that stopped as soon as they arrived at Fredendall's and saw what he had been up to, or down into, actually. He had detailed an entire battalion of engineers to tunnel deep into the side of a ravine to provide him and his staff a safe haven from the Luftwaffe's bombs, but one completely removed both physically and psychologically from the front and his men. Ike, especially after his own under-rock experience at Gibraltar, left hardly reassured that this key subordinate was in any sense ready for battle.

His worries only grew when he reached the recently occupied American positions at the Faïd Pass on February 14. Here he found the troops spread out and in a nonchalant mood, having been there for two days, and as yet they had made no effort to protect their positions with minefields. "Well, maybe you don't know it," Eisenhower told them, "but we've found in this war that once the Nazis have taken a position, they organized it for defense within two hours." Compounding his concerns, Ike found the US First Armored Division, which he had told General Anderson to keep together as a mobile reserve, scattered instead in what the British called "penny parcels" in an effort to stiffen the shaky line. All in all, it was an inspection tour that could only fill a commander with a sense of impending calamity. Eisenhower knew Kesselring, Armin, and Rommel by reputation alone, but that was enough to realize they were bound to attack if given the opportunity.

Three hours after Ike left, three of Armin's panzer divisions came crashing through the Faïd and Maizila Passes, then surrounded a regimental combat team assigned to defend Sidi Bou Zid and forbidden to retreat by Fredendall. During the night, First Armored Division's commander, Orlando Ward, managed to round up enough of his tanks to stage a counterattack in an attempt to relieve them, but he found himself caught between two panzer divisions, losing 46 out of 52 medium tanks, 9 self-propelled artillery pieces, and 130 vehicles before it was over. Among those captured was George Patton's son-in-law John Waters. On February 17 Anderson ordered a general retreat and was able to stabilize the line with better defensive positions just forward of Sbiba and Tébessa. But the Germans were not finished with these American rookies.

Two days later, Rommel's Afrika Korps, reinforced by one of Armin's panzer divisions, rolled into the Kasserine Pass to the south. Here, the American commander had committed the neophyte error of failing to occupy the heights, placing his troops instead on the valley floor "as if to halt a herd of cattle," Omar Bradley later observed. But these were battle-hardened humans in tanks, and those Americans who weren't cut down or captured in two valiant but futile counterattacks soon joined the stampede to the rear.

It was inevitable that this moment would come—for the infantrymen in particular, the weak link in the US Army during practically the entirety of World War II. But it was not their fault. Unlike the other branch specialties, they were almost exclusively conscripts, drafted and delivered to the front on a conveyor belt far beyond their control. They were newcomers who, under George Marshall's largely unsupervised delegate for training, Lesley McNair, received far too little in the way of a realistic introduction to combat. Now they were faced, many for the first time, with its unremitting reality.

In this case, their enemies consisted of some of the most lethal and merciless combatants in history, German veterans, many of whom had been in the business of bloodletting for nearly four years. No wonder the Americans ran and continued to do so until the line could be stabilized approximately fifty miles to the rear, but not before taking 3,000 casualties and losing 3,700 captured. The British Tommies, who had also been at war just as long, would take to calling them "Our Italians." But that would prove far from the case. While GIs might occasionally break, they always came back, and with a hail of firepower provided by brothers with lots of big guns and planes.

For the moment, however, things did look bad, particularly from Ike's perspective. He immediately went forward toward the front and, by projecting an aura of command karma, did manage to calm Fredendall—near nervous collapse at this point—along with a number of others sufficiently to keep the battle running.

Then, on February 22, to add a dash of personal chaos, *Life* magazine published a photo story, "Women in Lifeboats," featuring "the irrepressible Kay Summersby, Eisenhower's pretty Irish driver." The magazine had a circulation in the millions, and this would be the first Mamie had heard of Kay, or rather saw of Kay, since the photojournalist Margaret Bourke-White, who shared their lifeboat, made her and Ethel the focal points of the piece. Now as well as a war to run, Ike had some explaining to do on the domestic front.

But the lifeboat moment would pass, as would Eisenhower's strategic discomfiture. Rommel had run out the slack in his army's bungee cord. The reason he was in Tunisia had everything to do with being chased there from El Alamein along the North African coast by Bernard Montgomery and his

British Eighth Army. But Montgomery was an exceedingly methodical pursuer determined to give his opponent no tactical opportunities, and so he was easy to outrun. That had given Rommel and Armin the slack necessary to strike out against the Americans. But now his desert stalker was approaching, and Rommel had little choice but to be drawn back toward the original Axis perimeter, soon to become a noose.

On the way, he had a pair of malign revelations. One came in the form of a lethal rain of American artillery projectiles, not only from tubes better placed in the highlands, but in a profusion seldom experienced even by veterans of the Russian front. This was just the beginning. Detroit was now spewing out artillery shells by the millions, but critically they fed cannon and howitzers of only two basic varieties, 105 mm and 155 mm, thereby vastly simplifying logistics and opening the way for on-demand, wall-to-wall firepower that the Germans, with their mélange of captured and home-grown pieces, could not hope to match. Meanwhile, a shipment of 5,400 new trucks made sure that the shells and everything else GIs needed to fight kept on coming. Nor did their drivers any longer have to worry about getting attacked from above.

With stunning suddenness, the Allies seized control of the air, never again to relinquish it. The Royal Air Force was already highly capable, but limited in available planes and pilots. What changed the equation was the Americans, their sheer numbers and growing effectiveness as they gained experience. Early in the campaign the Air Corps was losing as many fighters in accidents as in combat. That soon changed, as did their familiarity with the fighting characteristics of their planes—the maneuverability of the P-40 and extremely high diving speed of the P-38. And more and more of them just kept coming, such that by the end of March the Air Corps was flying a thousand sorties a day to the Germans' sixty. That was enough so that by early spring 1943 they were seriously disrupting the Axis forces' vital links between Tunisia and Sicily, along with providing air cover for the Allied naval forces needed to truly cut them.

With these basics established, it all seemed to come together for Eisenhower. Although Harold Alexander shared his countrymen's suspicions of Americans as combatants, Ike's now deputy for ground forces did at least take immediate steps to integrate UK and US elements into something that began to look and act like a unified command. Meanwhile, he, Air Marshal Arthur Tedder, and Admiral Andrew Cunningham—the latter two highly competent but agreeable types who fell in naturally behind Ike's benign mask of command—all proved copacetic.

Emblematic of Ike's style was his deft removal of Fredendall. The man's military incompetence, not to mention his questionable valor, must have been obvious to Eisenhower, but he was also a distinguished graduate of George Marshall's black book and therefore had to be treated gently even in disgrace. Ike waited two weeks before relieving him, and rather than a reprimand Fredendall got Eisenhower's recommendation for a stateside training command, one at which Marshall could and would promote him to lieutenant general. Marshall had stood behind Ike at Casablanca, and now Ike returned the favor.

He wanted Mark Clark to take over II Corps. But Clark, now with a third star and in command of what had become the Fifth Army in Morocco, making him eligible for a fourth, declined to take the step down, something Ike never forgot or forgave. He next turned to Patton, who was also slated to command an army in the Sicily invasion; Georgie jumped at the opportunity of wielding a corps in combat and in the process whipping them into shape.

Patton arrived at II Corps headquarters on March 6 and immediately made his presence known with what Omar Bradley described as a "spit and polish" regime: "Each time a soldier knotted his necktie, threaded his leggings, and buckled on his heavy steel helmet, he was forcibly reminded that Patton had come to command the II Corps, that the pre-Kasserine days had ended, and that a tough new era had begun." This was what separated Patton from other generals and made him unique—his unrivaled ability to impose his uniquely ferocious mask over vast numbers, and in the process unify, or at least solidify, even the defeated and disheartened. One of his regimental armor commanders, Paul Robinett, caught his effect on II Corps almost cinematically: "He came with a Marsian speech and a song of hate; gross, vulgar, and profane, although touchingly beautiful and spiritual at times. . . . The old soldiers, who knew him as Gorgeous Georgie of Flash Gordon, rejoiced at this coming, even though they feared his rashness. They knew that there would be a pat on the back for every kick in the pants and that their interests would be his interests."

Patton's potion took effect almost immediately, reflected in II Corp's re-energized performance against the Germans facing them. Ike's appreciation came quickly in the form of Georgie's third star on March 12. Eisenhower knew he was doing a good job because Omar Bradley was already there as Ike's eyes and ears, a role Patton would not stand for, and got Beetle Smith to designate him his deputy commander. A minor incident, but it showed that Ike understood what Marshall understood about Patton—he had to be watched closely and like an attack dog kept on a leash.

Although their operational approaches were near polar opposites, George Patton and Bernard Montgomery were virtual bookends of eccentric behavior, egotism, and insubordination. But they were and would be Eisenhower's two best combat generals, so he would have to manage them, the bureaucratic equivalent of bull riding.

Strategically, the arrival of Montgomery's army in Tunisia had left the Axis forces surrounded by an ever-thickening cordon, a noose that could only mean their end. It took a while, in part because Alexander and Montgomery were reluctant to give Patton's revitalized II Corps a full role in squeezing the Germans and Italians up into the Cape Bon Peninsula, now largely cut off from Sicily. George Marshall didn't like the way this endgame was playing out and in mid-April warned Eisenhower about the "unfortunate results as to national prestige . . . and reaction of the public." Ike, in turn, flew to Alexander's headquarters and in effect ordered him to give the Americans their own sector and role in the compaction of Axis forces, a process Rommel personally escaped when he returned to Germany on sick leave.

The end came on May 13, 1943, two days short of Ike's prediction to FDR, with the surrender of around a quarter million Axis troops, about 125,000 being Germans, among them Armin, to whom Eisenhower pointedly refused to extend any military courtesies. Significant numbers did manage to escape, and German losses in North Africa were well below the 600,000 killed or captured by the Soviets at Stalingrad. But considering the geographical ambition of the campaign, its logistical challenges, and the fact that Americans were fighting for the first time against battle-hardened Axis troops, Torch and its Tunisian follow-up must be considered not just victories, but stunning ones.

Team America's two key players on the scene, Patton and Eisenhower, emerged with reputations enhanced. Almost from the moment he hit the beach in Morocco, George Patton proved the charismatic hypercombatant that Marshall and Ike believed him to be. Thus far he had only reinforced his position in the starting lineup. Ike's path had been more difficult, filled with slippery diplomats and a big rock, among other things, but as the campaign matured, it's safe to say his performance improved.

Shortly before the surrender Ike flew down to see Montgomery, their first meeting since Monty had told Ike to douse his cigarette in London. Each sized the other up with a jaundiced but penetrating eye. Montgomery wrote Alexander: "I should say he was good probably on the political line; but he obviously knows nothing whatever about fighting." Eisenhower, for his part, told Marshall: "He is so proud of his successes to date that he will never willingly make a single move until he is absolutely certain of success." That was the difference between the two: Monty would stay the same; Ike still had a lot of room for growth.

"LIKE FROGS ABOUT A POND" WAS THE WAY PLATO CHARACTERIZED THE Greeks. Much the same could be said of the Anglo-American military approach to this colossal war 2,500 years later, not just in the Mediterranean, but eventually with the cross-Channel operation, and of course in the entire Pacific campaign—by circumstance and tradition the English speakers were amphibious.

It followed that the next intersquad scrimmage, which took place in Washington for two weeks in late May 1943, would be dubbed "Trident," and that the issue of landing craft, their numbers and whereabouts, would hang over the proceedings like the morning mist in Foggy Bottom. Trident also marked the last time the two teams were evenly matched, each using a key advantage to get what was wanted.

The gravity of propinquity was the bat the British again wielded with undeniable leverage. Once Sicily was won, all those men and landing craft would be left just a few miles across the Strait of Messina from Italy, one of the three Axis powers and plainly teetering. How could an invasion of the Italian boot not make sense? It was a question Churchill and Brooke kept hurling, and one that FDR and Marshall had only matchsticks to swing against. So Italy would be the next Anglo-American target. The British had every reason to congratulate themselves, but only on a short-term basis.

Not only were the Americans growing stronger by the day, but Churchill understood that the potential could soon turn exponential. Because of Britain's early scientific contributions, he knew all about the American atomic bomb program. But Churchill's was an intellect informed by history, and he instinctively sought for his country full access to associated secret technologies and also a postwar nuclear alliance, prerogatives he believed necessary for Britain to remain a first-class power. While it still remained to be tested and therefore technically nonexistent, Roosevelt and Marshall were able to wield the projected bomb with something like the power of the real thing. In exchange for renewed access to the American nuclear program, they extracted a promise from Churchill that the cross-Channel invasion would definitely take place in the spring of 1944. To further hold British feet to the fire, it was agreed that come November four American and three British divisions would be transferred from the Mediterranean to England to further flesh out the Bolero buildup. Meanwhile, Ike would remain commander in chief in the southern tier, to plot further moves against Italy.

But Sicily, a.k.a. Operation Husky, came first. Barely remembered today, Husky at the time of its commencement on the night of July 9, 1943, was the largest amphibious assault ever attempted—175,000 troops would simultaneously land at twenty-six beach sites stretching over a hundred miles of coastline, to be followed by 300,000 more within two days. In an effort to ensure they were safe from above, Ike had staged a preinvasion invasion of the tiny island of Pantelleria, midway between Tunis and Sicily, seizing its airstrip and thereby cutting his fighters' flight time to the Husky invasion force in half.

But since this was war, the situation became inverted almost immediately, with those below slaughtering their brothers above. The worst incident took place as a regiment of Matthew Ridgway's Eighty-Second Airborne flew over the invasion fleet, causing a chain reaction among its antiaircraft gunners that left twenty-three C-47 transports shot down, another thirty-seven heavily damaged, and 1,400 of the regiment's 5,300 paratroopers killed or missing. Something similar happened to British gliders along the east coast of the island, and together they rank as the worst friendly-fire incident in history and a testimony to the blind and clumsy killing power of industrialized warfare. Nobody was safe here, even from your friends.

Otherwise, the landings went rather well. Just as would occur at Normandy, a storm rolled in the day before, but when his forecasters predicted it would pass, Ike told Admiral Cunningham, "Let's go," and the weather obligingly improved. Although a cloak-and-dagger attempt was made to convince the Germans the real target was Sardinia, it wasn't hard for them to figure out the Allies were headed for Sicily, and "Smiling Albert" Kesselring had a force of 300,000 waiting.

What they weren't prepared for was the breadth of the assault, along with the speed and ingenuity the Allies demonstrated getting a force of this size ashore. Most notably, the US Third Division arrived on the southern coast in a flotilla of new and highly specialized landing craft (LSTs, LCIs, and LCTs), which allowed them to proceed from port of embarkation to the beach fully equipped—tanks, vehicles, and artillery—all ready to fight, rather than long rides to shore in tiny Higgins boats and time wasted reuniting with weapons and equipment. There was even a six-wheel-drive cargo and personnel carrier shaped like a barge and able to propel itself to shore, and then continue on land with equal facility. Known officially as the DUKW but otherwise remembered as just "Duck," 21,000 would be produced before war's end—a kind of symbol of what the Germans were

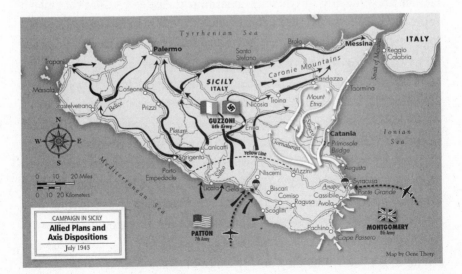

CAMPAIGN IN SICILY
Allied Plans and Axis Dispositions
July 1943

Map by Gene Thorp

facing, not simply innovation, but innovation in huge quantities courtesy of the arsenal of democracy.

Here in Sicily this profusion of men and equipment left the Allies with a firm foothold, the British on the east coast near the bottom of the island centered on Syracusa, and the Americans around the corner along the southern littoral bracketing Gela. The plan Harold Alexander had in mind for the rest of the campaign not surprisingly had Patton and his Americans in reserve, and centered on the British Eighth Army and its ineffable commander, Bernard Montgomery, racing northward along the coastal highway and grabbing Messina, the German army's vital link to the Italian mainland. But this was the wrong route, and the wrong man if you wanted to get anywhere quick.

Behind his "Smiling Albert" mask, Kesselring was a master of contingency planning; he knew from the beginning that if defeat were to be his lot in Sicily, the last thing he needed to hold on to was Messina, his troops' back door off the island. And it was not hard to predict what he had to defend first. Hence, as Montgomery and his compatriots proceeded north along the coastal road, they rather quickly found their way blocked by a hornet's nest of aroused Germans.

Montgomery's response was typical, logical militarily, but out of phase with everybody else. He would keep the pressure on the coastal road, but also try to continue his drive north on a second inland route, forcing the Germans to redeploy and thereby undermine their defense. But that inland route had been reserved for Patton and his Seventh Army. Montgomery, failing to tell either Patton or Alexander what he was doing, simply moved in with two of his own corps right in front of the Americans.

Blocking George Patton from combat was akin to checking water flowing downhill. The man was a geyser of operational possibilities, and by now thoroughly immersed in his role as überwarrior, one that definitely had minuses along with the pluses. Ike got to see that for himself two days after the landings, when he came forward just in time to witness a German armored attack on Omar Bradley's divisions around Gela, one that he eventually beat off with heavy casualties. But that was not good enough for Patton, who countermanded one of Bradley's tactical decisions, then complained to Eisenhower that he was simply "not aggressive enough"—poison words and bound to get back to Bradley, since Ike not only liked and trusted him, but he was a lion in George Marshall's black book and therefore somewhat of a sacred cow. And there was more that Ike didn't hear about, like Patton's June 27 speech to the Forty-Fifth Division. "We were told that General Patton said . . . 'Fuck them. No prisoners,'" one private remembered, as likely did a captain and a sergeant from the Forty-Fifth who massacred seventy-three Italian POWs shortly after. Any investigation of Patton's complicity was quickly short-circuited, yet there would soon be more outrages littering his path of victory.

But there was no denying his ingenuity in blazing that path. In this case, when blocked by a bunch of Brits, he told Omar Bradley to forget Messina and instead slip the Forty-Fifth and First Divisions and head for the northwest coast. Unlike Montgomery, Patton did remember to tell Alexander, but too late for him to do anything except approve the move, which would culminate in an all-out assault on Sicily's biggest city and port, Palermo. With a scratch force of the Third Infantry and Second Armor divisions Patton had given him, Geoffrey Keyes bolted forward one hundred miles in a few days, capturing fifty thousand Axis prisoners (almost all of them Italians) and the city itself on July 22, all at the cost of just three hundred casualties. At the same time, Bradley had also reached his objectives along the northern shore, leaving Messina a straight shot down the coastal road for the Americans. Meanwhile, the British and Montgomery were bogged down around Catania, still almost sixty miles from the same objective.

Unfortunately for Patton and the troops under him, this proved the briefest of glimmers, dimmed by the foresight of Kesselring, who had the rugged coast clotted and every possible ambush point covered. Patton responded by staging a series of three daring but costly amphibious landings that failed to catch anything German but hostile fire. And as he leapfrogged futilely forward, Patton's objective transmuted from beating the Germans to Messina, which was impossible, to arriving before Montgomery. "This is a horse race in which the prestige of the U.S. Army is at stake. We must take Messina before the British," he bellowed, as if it were simply a matter of prestige and not added casualties among his own troops.

Nonetheless, they did as they were told, and on the evening of August 16 a patrol of the US Third Division were the first Allied troops to move into Messina to find a wrecked city and no Germans: 100,000 troops, 10,000 vehicles, 200 big guns, and 47 tanks had been safely evacuated to the Italian mainland in what amounted to Smiling Albert's last laugh—in Sicily, at least.

Montgomery and his Eighth Army took it well, even granting a little more respect to Team America, plainly no longer the combat rookies they had first encountered in North Africa. However measured, Sicily constituted another overwhelming Allied victory, and this time a prompt one, taking five weeks instead of the more than six months in North Africa. And even if there was no equivalent bagging of captured Axis troops, they had still been driven from the island by weight of Allied arms. And no one did more of the heavy lifting than George Patton, the man of the hour. But being Patton and over-excited by his favorite activity, he almost immediately set about to destroy the good reputation he had earned on the battlefield.

Compared with the horrific figures coming out of Russia—almost simultaneously the Wehrmacht was suffering a half million dead and wounded along the Kursk salient—casualties during the Sicily campaign had been relatively light, around twenty thousand on each side. Still, that amounted to a lot of mangled bodies, both dead and alive, accumulated in just over a month in a relatively small place. It certainly had an effect on George Patton.

The Sicily campaign left him in a state of war-induced delirium, in the midst of a dream that had been his lifelong polestar. Now it was actually coming to fruition, the forever-reincarnated warrior in the midst of his own martial extravaganza, the presumed audience being all the other battle-dead Pattons looking down with admiration at his victories. His viewpoint came straight out of the *Iliad*, and either Hector or Achilles would have understood: gore was a thing to be celebrated, the juice of victory.

But there was another reality to the kind of warfare George Patton was fighting. "I could smell dead enemy while driving for at least six miles along the road," the aroma of victory he remembered in his war memoir. And dealing with the consequences of battle from the perspective of his own side was still more unsettling.

Patton plainly didn't like visiting the evacuation hospitals with the most recent and worst casualties, but he did so scrupulously, pinning Purple Hearts and giving each victim a word of encouragement. That is, until early August, when he came upon a patient with no visible wounds and asked him what hurt. "I guess I can't take it," was the soldier's mistaken reply, since in addition to what is known today as post-traumatic stress disorder (PTSD), he also was suffering from chronic dysentery and malaria. But that was enough to ignite Patton, who "slapped his face with a glove, raised him to his feet by the collar of his shirt and pushed him out of the tent with a final 'kick in the rear,'" according to horrified medical staff members.

He doubled down on this approach a week later in a different hospital, when he encountered a second PTSD victim, one who had resisted evacuation but now lay shivering in his cot. "It's my nerves. I can't stand the shelling anymore."

"Your nerves," roared Patton. "Hell, you are just a goddamned coward, you yellow son of a bitch. . . . You're a disgrace to the Army and . . . I ought to shoot you myself right now. God Damn you!" He then unholstered one of his famous pistols and began waving it in front of the traumatized soldier, finally punctuating his obscenity-filled diatribe by striking him multiple times. On the way out, he told the hospital commander: "I meant what I said about getting that coward out of here. . . . We'll probably have to shoot them some time anyway, or we'll raise a breed of morons."

Nor did any remorse creep in as the days passed; instead, he actually bragged about the episode to Bradley, who, though mortified, did his best to keep the hospital commander's report from climbing the chain of command to Ike. Yet a nurse had told her boyfriend and the news correspondents attached to the Seventh Army picked up the story, making it likely that someone at home would print it—eventually Drew Pearson, MacArthur's former tormentor. But by that time it was old news, picked up from servicemen and just a minor scandal.

Ike had fixed it. When he found out, though, the implications were chilling. His star battle commander, by striking an enlisted man, had incurred a court-martial and probably career-ending offense, since he had done it twice and in front of multiple witnesses. He refrained from sharing the news with George Marshall; instead, he met with the knowledgeable reporters, but flatly refused their offer to suppress the story in exchange for firing Patton. Instead, seeing an opening, he outlined for them a sort of stations of the cross mea culpa he had planned for his errant general: "They were flatly told to use their own judgment," which they did, and Eisenhower being their best source, there would be no leaks from their side.

Patton's only apparent regrets as he clowned and trudged his way through the extensive list of apologies demanded by Ike was what it had done to their personal relationship, writing abjectly of his "chagrin and grief at having given you, a man to whom I owe everything and for whom I would gladly lay down my life, cause for displeasure with me."

It was a whole lot more than displeasure: Eisenhower gave every sign of feeling betrayed. He knew now that Patton, if left to his own devices, could not be trusted to stay out of trouble, the kind of trouble that could bring them both down. Henceforth, he was more a problem than a pal. But even so Ike's reaction was severe, almost exquisite in its cruelty. He simply left Patton marooned on Sicily, watching helplessly as component by component of his Seventh Army was taken apart and shipped elsewhere, until all he had left to command was his headquarters and about five thousand men. Here he would sit for another five months before being written back into the starting lineup. But before we leave Patton forsaken in Sicily, it's worth noting that the move left the Germans, who executed around fifteen thousand of their own men during the war, mystified: "How could something so trivial sideline their best commander?" they wondered indiscreetly over the airwaves. It was information Ike and the Allies would find useful when they put Patton in charge of a phantom army, before giving him another chance with a real one.

VICTORY IN SICILY COINCIDED WITH YET ANOTHER INTERSQUAD MEETING, this one in Quebec, code-named Quadrant, continuing the nautical theme. Team America came prepared to wield their ever-growing preponderance in numbers and weapons supplied, to hammer home essentially what had already been agreed on. In a generous way, when it came to nuclear weapons, they reaffirmed Trident and a "complete interchange of information and ideas on all sections of the project." But otherwise they were determined to use their muscle to nail down their shifty ally, who always seemed open to negotiations, especially when it came to the cross-Channel invasion.

George Marshall and his fellow chiefs arrived with a fundamental complaint: US Army and Air Corps personnel dedicated to the Mediterranean theater had metastasized from a landing force of 80,000 in November 1942 to what was about to become a juggernaut of some 700,000, thereby starving the Bolero buildup. Determined to reverse the process, they again drove home the Trident decision to withdraw seven divisions from the Mediterranean, now simply in time to participate in the cross-Channel operation, a slightly more generous schedule. But otherwise little give was to be found among the Americans.

Churchill being Churchill had sought to preempt the invasion by visiting FDR at Hyde Park several days before the conference convened, full of alternative schemes other than the one leading across the English Channel, but doomed with the American president. Not only was FDR adamant about the landing taking place no later than the next spring, but he was also insisting that an American command it. Nor did Churchill's situation change when they reached Quebec, the highlight of the meeting being the confirmation of May 1, 1944, as the date for the cross-Channel operation, now termed Overlord. To further nail it down, Overlord was given first priority, the Far East and Pacific second, and the Mediterranean relegated to third, exactly where George Marshall wanted them.

Not where Douglas MacArthur wanted them, though; he felt perpetually shortchanged in the pecking order. He took particular offense at being told by Quadrant's players in far-off Quebec that he should bypass the Japanese megabase at Rabaul, even though that completely served his interests and was probably what he would have ended up doing anyway. Of course, out in the Southwest Pacific it was easy to get a myopic vision of how World War II was going, but the rest of Team America's starters were under no such illusions—they knew where the action was, and also where it would be the following spring.

But who would lead Overlord? It didn't take Winston Churchill long to realize the Americans were determined that it be one of their own, and so he decided to submit gracefully—though callously if you were Alan Brooke, who had commanded the Dunkirk evacuation and had hoped to lead the way back. It was hard to dispute that from the beginning the cross-Channel operation had been George Marshall's primary strategic objective—obsession, in British eyes—and there was a kind of unsaid assumption that he would be its commander in chief. Churchill thought so, and so apparently did Marshall, who told several conferees he was looking forward to working with them in London, while his wife, Katherine, had quietly begun vacating the chief of staff's quarters. Yet, true to form, he refused to actively lobby for the job with FDR, who equally true to form put off making a decision. Further clouding the issue was a letter the president received three weeks later from John J. Pershing, who had just broken all ties with Patton over the slapping incident, and now proceeded to undermine his star pupil's path to Overlord. "To transfer him to a tactical command . . . no matter how seemingly important, is to deprive ourselves of the benefit of his outstanding strategical ability and experience." So the issue and Marshall remained in limbo for a while longer.

Regrettably, that was not the case for the invasion of Italy, which like so many failed enterprises looked good initially. On July 25, 1943, recognizing the regime was finished, Italy's Grand Council and king took the hopeful but perilous step of deposing Benito Mussolini and arresting him. His successor, Marshal Pietro Badoglio, promptly dissolved the Fascist Party and began negotiating an armistice with the Allies, which was signed on September 3, the day scheduled for Montgomery's Eighth Army's two-division landing on the north and south side of Italy's toe. As he predicted, they would arrive without opposition.

Unfortunately, the missing element in all cases was the Germans, of whom there were plenty in Italy: the evacuees from Sicily, plus more veteran divisions Hitler sent pouring in from the north to flesh out Albert Kesselring's multitude of defensive schemes. Just looking at Italy's topography must have brought a smile to his lips, the rivers and highlands running across the boot forming a segmented carapace of defensive possibilities.

Little of this was apparent to either Winston Churchill or George Marshall, who were thinking geopolitically, not of Italy's endless ups and downs, when they seized on something much more ambitious than the initial landings on the toe: a surprise amphibious assault well up the boot at Salerno, a perfect position to grab the port of Naples and cut off the Germans to the south. As theater commander, Eisenhower was willing to go along with the scheme, but not enthusiastically, being very much aware that Salerno was at the absolute edge of the range of his Sicily-based fighter aircraft, and also a long way from Montgomery's divisions to be landed as a diversion.

He also had some questions about leadership. Two weeks before Salerno was scheduled to be set in motion, Ike sent Marshall a report comparing his generals as prospects for Overlord operational command. Patton—"He has qualities that we can't afford to lose unless he ruins himself"; Bradley—"Has never caused me one moment of worry"; Mark Clark, commander of the US Fifth Army, designated to storm the beaches at Salerno—"He will shortly have a chance to prove his worth"; the unsaid message being that in Ike's eyes he had not yet done so. A prescient paper it was; not only did Marshall shortly appoint Bradley to command the American portion of the Overlord landings, thereby providing another nail in Patton's cross of repentance, but events would show there was good reason to worry about Mark Clark as a combatant commander.

Operation Avalanche, which began on September 9, proved anything but at Salerno. To preserve surprise, no preparatory aerial or naval bombardment took place, just the invasion fleet and the men in landing craft heading toward shore greeted in English by loudspeakers on the beach: "Come on in and give up. We have you covered." Their veil of secrecy in tatters, they came anyway and did establish a beachhead at some cost. But their trials were just beginning. They were surrounded by enemy troops led by the veteran Sixteenth Panzer Division.

The counterattack, three days later, was ferocious, so much so that Ike was informed Clark was considering evacuating the beachhead, beginning with himself and his headquarters. Eisenhower was blunt with his wavering subordinate: "Headquarters should be moved last of all and the Commanding General should stay with his men to give them confidence. He should show the spirit of a naval captain, and, if necessary, go down with the ship." He also sprang into action himself, ordering his air deputy Tedder to throw everything, including his B-17 heavy bombers, into protecting the beachhead while directing Admiral Cunningham to do the same from the sea, resulting in the delivery of one thousand tons of bombs per square mile and more than eleven thousand tons of highly accurate naval projectiles. Meanwhile, on the ground, critical night drops of Matthew Ridgway's Eighty-Second Airborne, and the courage and tenacity of artillery units who kept firing in the face of attacking tanks, finally saved the situation, but just barely.

Ike made no move to relieve Mark Clark from command of the Fifth Army, which he probably should have done. But this was not his style, and Clark would make just sufficient future progress in Italy to retain George

Marshall's confidence. But Salerno was prophetic of both Clark and the campaign—the repeat misery of the second big landing at Anzio, the four-month bloodletting that Kesselring had prepared at Cassino, Clark's decision to capture Rome rather than retreating Germans, followed by continued resistance right up to the end of the war, Smiling Albert still hanging on in Italy for all but the last two months, the military version of the Cheshire cat. But by then Ike and even Patton would be long gone; only George Marshall would return, but mainly to visit the grave of his stepson, Lieutenant Allen Brown, killed near Rome in his tank, one of 119,000 American casualties eventually accumulated in a campaign that ultimately led nowhere.

CHAPTER 9

THE OVERLORD

On November 19, 1943, Ike was waiting for FDR as he pulled into Oran aboard the brand-new battleship *Iowa*, on his way to the Cairo Conference, the closest the Allies ever got to a full intersquad scrimmage. Two days before, in Malta, Eisenhower had met with Winston Churchill, who strongly hinted that the command of Overlord was still up in the air. "We British will be glad to accept either you or Marshall."

The plan was for FDR to spend the next day, Saturday, in Tunis receiving last-minute briefings before proceeding to Cairo the following morning.

"A night flight, Sunday night, would be better," Ike suggested.

"A night flight? Why?" asked his manager.

Daylight was too risky. "We don't want to have to run fighter escort all the way to Cairo. It would just be asking for trouble."

"Okay, Ike. You're the boss. But I get something in return."

"What's that, Sir?"

"If you're going to make me stay over at Carthage all Sunday, you've got to take me on a personally conducted tour of the battlefields—ancient and modern."

"That's a bargain, Sir."

But it was a deal struck on entirely insincere terms. Ike knew full well the nearest conceivable Luftwaffe threat was at least five hundred miles away, while Roosevelt had more than history on his mind in wanting to get to know Ike better. FDR loved sexual intrigue, and Kay Summersby, Ike's glamorous and now-famous driver, was at least partly on his mind when he agreed to extend his visit. He barely knew Ike, and had not seen him since his disastrous briefing in Casablanca. But subsequent success spoke for itself, and it was plainly time to give him some quality time before deciding who would lead Overlord.

It was the opening Ike needed and his response can be judged either unbelievably lucky or preternaturally clever. When the president flew into Tunis Ike had sent Kay to drive him from the airport, a mission foiled by the Secret Service. Almost as soon as FDR was settled in Ike's villa, he asked for Kay.

"I've heard quite a bit about you. Why didn't you drive me from the plane? I'd been looking forward to it."

"Mr. President, your Secret Service wouldn't let me drive."

"Would you like to drive me from now on?"

"It would be a privilege, Sir."

"Very well. You shall drive me then. I'm going on an inspection trip soon." So was forged the first link in Ike's tender trap, or picnic at least.

The next day found Kay at the wheel, Ike in the front seat regaling Roosevelt with his vast knowledge of ancient military history, FDR enjoying every minute in the back, soon to be joined there by a stroke of guest-list genius, Telek, their shared Scottish terrier and a near facsimile of the president's beloved Scottie, Fala. Around noon they came upon a eucalyptus grove in the desert. "That's an awfully nice place. Could you pull up there, Child, for our little picnic?" Except for the squads of infantry and half-tracks surrounding them, it was a bucolic scene.

They produced a basketful of sandwiches and Ike chose chicken for the president, claiming a special knowledge from a youth of Sunday-school outings. Before he bit, FDR patted the seat next to him. "Won't you come back here, Child, and have lunch with a dull old man?" Ike had the good sense to drift off, and Kay later remembered, "Roosevelt enjoyed himself immensely." No wonder; the couple had managed to concoct a desert version of one of FDR's favorite activities—long drives and picnics in the country with his distant cousin and intimate friend Daisy Suckley. Good luck obviously played a role, but the ever-ingratiating Ike also put himself into a position to be its beneficiary, and so did Kay.

In the back seat the manager wondered at her exact status, officially at least. "It seems to me that you might like to join the WACs," he mused.

"There's nothing I'd like more sir, . . . But it's impossible. I'm a British subject and I'd have to become an American citizen to be a WAC."

"Well, who knows? Stranger things have happened." FDR not only re-membered Kay, he remembered to WAC her with a commission as a second lieutenant. But as usual with Roosevelt, there was likely more than just flir-tation involved. On the advice of her very conservative parents, Mamie had recently turned down a private luncheon invitation from Eleanor, and what-ever the status of their marriage, the Roosevelts were a solid front against such affronts. So, it's likely that however much the wily manager may have been charmed by Kay, he also took a special pleasure in doing what he could to undermine Mamie.

But not Ike. Before FDR boarded his plane the following morning he opened up: "Ike you and I know who was Chief of Staff during the last years of the Civil War but practically no one else knows. . . . I hate to think that 50 years from now practically nobody will know who George Marshall was. That is one of the reasons I want George to have the big command." Then he added, as if thinking aloud, "But it's dangerous to monkey with a winning team."

The president was still weighing his options, perhaps. But Ike had put on a dazzling performance, and not simply the rolling version of the Club Eisenhower; his knowledge of the ancient battlefields around Carthage sur-prised the president and added a significant element of credibility in the eyes of America's first history buff. He hadn't yet made the decision, but from this point, presidential procrastination seems largely based on dread of breaking the bad news to George Marshall. Probably the best clue to his mindset was directly ordering Ike to attend and brief the Cairo Conference, which he had not been scheduled to do.

The meeting in Egypt was a segmented affair (November 22–26; Decem-ber 4–6, 1943), a prelude and epilogue of the first intersquad scrimmage with the Russians in Tehran, one with a great deal more significance. As such, both Anglos and Americans gave vent in Egypt to their strategic crotchets, those with little chance of acceptance by the other side.

In the case of Team Great Britain, it was Winston Churchill's latest

Mediterranean mania, a strategically nonsensical invasion of the island of Rhodes, and, as far as George Marshall was concerned, one more step away from Germany. He took him on in front of Roosevelt and the combined chiefs. As Marshall remembered: "It got hotter and hotter. Finally Churchill grabbed his lapels, his spit curls hung down, and he said, 'His Majesty's Government can't have its troops standing idle. Muskets must flame.' I said, 'God forbid if I should try to dictate but . . . not one American soldier is going to die on [that] goddamned beach.'" There could be no blunter statement of America's growing gravity in the partnership, and there would be no invasion of Rhodes. "I think Winston is beginning not to like George Marshall very much," FDR told his son Elliot, an incorrect assessment but a gauge of his mood. "I'll tell you one man who deserves a medal for being able to get along with Winston. And that's Ike Eisenhower."

The American counterpoint to Churchill's shaky colossus of Rhodes was FDR's ill-advised scheme to drag the ragtag squad from China led by Chiang Kai-shek into the alliance as full player. This was almost entirely a managerial initiative, and pretty much a nonstarter for the British squad, who still retained colonialist delusions, matched with absolutely no interest in promoting their Asian ally to the big leagues. It's hard to say there was much enthusiasm for the China theater even among Team America's military leadership. George Marshall's old friend from Tientsin days, Joseph Stilwell, had been both China theater commander and military adviser to Chiang Kai-shek since February 1942. As distributor of American Lend-Lease matériel, "Vinegar Joe" had a sharper eye for the regime's corruption than anyone; nor did anyone have a more trusted voice to the ear of George Marshall. It took Stilwell only around six months on the job to exactly capture the regime's strategy: "The probabilities are that the Chiang Kai-shek regime is playing the USA for a sucker; that it will stall and promise but not do anything; that it is looking for an Allied victory without making any further efforts on its part to secure it; and that it expects to have piled up at the end of the war a supply of munitions that will allow it to perpetuate itself indefinitely." Except for the "indefinitely," he nailed it. These were not team players, they were parasites, ticks on the war effort.

As November drew to a close the British and American first-stringers slipped out of Cairo and winged their way to Tehran for their first face-off with Stalin and his veteran war-scarred team. Being Stalin and this being Team USSR, he immediately warned Roosevelt of a plot on his life and the American legation's lack of security, kindly offering safer, and thoroughly bugged, quarters in the Soviet compound. They may well have heard every word and snore uttered by Roosevelt, Hopkins, Marshall, and the other chiefs, but they really didn't have to bother. Stalin already knew everything he needed to know.

In their first three-way encounter, after Churchill discoursed on the deep wisdom of the Mediterranean approach, Stalin parried by questioning the soundness of scattering assets into the Adriatic and Balkans. Then he got to the heart of the matter: Overlord should be the "basis for all 1944 operations." Not only that, he also suggested a diversionary attack in southern France to give Overlord a boost. And once Germany was defeated he even offered reinforcements against Japan.

Still more to the point, he was in a position to insist. It's important to remember that at this stage, among the Allies, the Soviet Union had endured by far the greatest share of land combat, and just that summer had ripped the heart out of the Wehrmacht at the Kursk salient, leaving it incapable of offensive warfare. When Stalin spoke strategy, people had to listen, particularly Winston Churchill, who increasingly found himself crushed between two giants.

But so did Roosevelt. At their second Big Three session, Stalin asked out loud the question on everybody's mind. "Who will command Overlord?" Taken aback, Roosevelt hedged with "it had not yet been decided." That was a response Stalin waved aside with a dismissive hand and a sarcastic tone: "Nothing [will] come out of the operation unless one man [is] made responsible not only for the preparation but for the execution of the operation." It was a gotcha moment; with one cold blast the Soviet leader had frozen in place the most liquid of politicians. While FDR wouldn't give the cagy Georgian the satisfaction of telling him, it was clear at that moment that he could no longer put off naming the Overlord commander.

In contemporary terms, you might say Stalin acted as a facilitator at Tehran, an apocalyptic war's version of an outside consultant, one who cut through the issues to reveal close to exactly what actually happened. Amid the endless toasts at Churchill's sixty-ninth birthday dinner, Stalin's stood out, a tribute to American "machines" without which "the war would have been lost." Before the conference adjourned, a firm date, May 1944, was set for Overlord's launching, and Stalin promised to set off a simultaneous offensive in the east to keep the Germans from transferring troops west. This probably marked the high point of Allied cooperation, and seemed to demonstrate that in this hellish league, even monsters can be heroes.

Back in Cairo, FDR came back to grips with reality and did what he had to do. Buccaneer—a plan for multiple amphibious landings in Burma to take pressure off China—had no support from the British, who wanted any extra landing craft for the Mediterranean. Consequently, less than two weeks after he had promised Chiang Kai-shek Buccaneer at the first Cairo meeting, Roosevelt snatched it away.

Telling George Marshall proved more complicated. Reverting to form, he tried letting Harry Hopkins do it, sending his aide to the general's villa with the message, as remembered by Marshall, that the president was "in some concern of mind" over the appointment. Did he really even want Overlord? Marshall refused to take the bait; he had made a career out of playing patriotically coy.

So the president had to do it the next day. Again they got into the same routine: "What do you want to do?"

"Feel free to act . . . in the best interests of the country . . . and not in any way to consider my feelings." The Zen-of-disinterest approach had served him well in the past, but it gave Roosevelt an opening, and he took it: "Well, I didn't feel that I could sleep at ease if you were out of Washington." On the way back FDR flew into Tunis, and at the airport, when he got into the car with Eisenhower, virtually his first words were: "Well, Ike, you are going to command Overlord."

Meanwhile, unbeknownst to any of the major players save one, George Marshall had decided to take the long way home. Specifically, he had quietly planned to head east and stop off in New Guinea, because, as Marshall recalled later, he "wanted to show MacArthur that he had not been forgotten." His presumed host must have known he was coming, since Sutherland and his four-engine C-54 was available to transport him. Still, it was a long and dangerous journey, highlighted when the plane was saved from running into a mountain by a fortuitous lightning strike, and much of the rest of the voyage was flown over enemy territory. But when they finally touched down in Port Moresby, Marshall was informed the general had moved on to Krueger's headquarters on Goodenough Island—stood up by a teammate who very apparently never forgot the "Chaumont clique" or that this was Black Jack Pershing's star protégé.

Marshall reacted by simply pursuing him to Goodenough Island, but before his arrival MacArthur had almost bolted again to supervise a small landing, arguing that a meeting would somehow embarrass the chief of staff. Instead, he stayed, and even in this steamy location the two managed to stage a very chilly luncheon. Marshall finally lost his cool when MacArthur introduced as an independent entity "my staff." "You don't have a staff, General," Marshall interrupted. "You have a court." Given the degree of Mac's provocation, this was a pretty mild rebuke. And that's about as far as it went. Even MacArthur had to admit that Marshall had listened to his pleas for more tactical air support, and as a result of the visit and his subsequent intercession with Arnold, "from that time on Washington became more generous to the SWPA."

George Marshall was, and always will be, a hard man to know, but this encounter cut to his core. Though he was still six months from the death of his stepson, this period probably marked an emotional-professional nadir. He had been Overlord's champion from the beginning, and not to have been chosen as its leader had to hurt. And if FDR made the right choice in Ike—who had already commanded three big operations—from another perspective it could be seen as a setup and act of betrayal.

But not to George Marshall. Even in his darkest hour he remained a consummate team player. He was an ambitious man, always had been, but his ambition perpetually remained in the context of the state and its well-being—the optimal outlook in an era of industrialized warfare: blind patriotism. If his fate was to be decided in the halls of Congress and the Pentagon, so be it; he would carry on exactly as before. When stepson Allen was killed, he was able to share his grief with Harry Hopkins, who also lost a son in combat, but elsewhere his demeanor never changed. Through thick and thin, honors and insults, the mask had become the man, one imperturbably determined to do the best for the US of A, and in the spirit of the age, regardless of casualties.

PERHAPS AS ANOTHER VERY MILD ACT OF RETRIBUTION, AND ALWAYS THE mother hen, Marshall had ordered Ike back to the States for a conjugal visit with Mamie before taking up Overlord. But their stay at the Greenbrier resort in West Virginia did not go well. "Every time I opened my mouth to say something to Mamie," he told Summersby later, "I'd call her Kay. She was furious." And with good reason. But on Ike's part, this was likely not guilt-driven alone; it also shed light on the way he related to people: at the moment, Kay was simply more useful to him than Mamie. Though this was not necessarily a permanent condition, Mamie had no way of knowing. Things only got worse when they returned to Washington, and FDR, pleading illness, invited Ike but not Mamie to the White House on his last night stateside. They chatted until ten p.m. FDR asked Ike how he liked his new title, "Supreme Commander." "I acknowledged it had a ring of importance, something like 'Sultan.'" Eleanor came in to be introduced; Mamie was not mentioned. But FDR not only remembered Kay, he told Ike to "give her my best wishes," and produced an inscribed photo of himself as a gift for her. Presumably, Ike did not show it to Mamie before he left town, but he did cable Omar Bradley to "please tell Colonel Lee to make proper arrangements for my driver to meet me at the station."

On January 15, 1944, Kay was waiting at the wheel of Ike's olive-drab Packard, and very soon in their official wanderings, at a meeting of Allied commanders at Norfolk House, they arrived to find the number one spot already occupied, preempted by a shiny black Rolls-Royce that, Summersby observed, "could belong to but one man in all of England: General Montgomery." From their very first cigarette-stifling meeting, Monty found ways to infuriate Eisenhower that rivaled in ingenuity his feats on the battlefield. He was a cross Ike would not only have to bear, but drag with him across Europe.

As before, Ike's three top deputies were British. Two of them, Air Marshal Arthur Tedder and Admiral Sir Bertram Ramsay—the organizer of the Dunkirk rescue fleet—were not only competent but agreeable. Ike had wanted to round out the trio with the equivalently copacetic Harold Alexander, but Churchill wanted him in Italy, so he got Montgomery. However disagreeable, he was by far Britain's most successful combat general and also one beloved by his men, in large part because he was careful with their lives—an unusual quality in World War II, but one that afflicted him in the eyes of many colleagues, including most of the Americans, with what Lincoln had called "the slows." In part, this proved a matter of circumstances and bad luck, but he was not a commander predisposed to race ahead if opportunity pointed in that direction.

Nonetheless, he definitely knew his amphibious operations, and could see that this one as presently conceived was too narrowly centered and seriously undermanned, a skinny intruder in a tough neighborhood. Rather than three divisions in Overlord's initial landings, he wanted five, and not at one site but five—three British-Canadian, and two American (Utah and Omaha)—along a sixty-mile front, to be followed by nine rather than five support and exploitation divisions. From the beginning Ike was of a like mind, and even postponed the operation for a month, to early June, in order to provide time to accumulate and accommodate the larger force.

Nobody knew better than Ike how colossally risky Overlord actually was. Never had Eisenhower overseen the invasion of a more obvious target: the Germans were perfectly aware of the huge Allied buildup in England; they knew it was intended for an impending cross-Channel invasion; and they had reinforced their already formidable defenses accordingly.

But that, almost miraculously, proved to be the Allies' edge, courtesy of one of military history's all-time strategic sleights of hand. Concentrating four of their vital panzer divisions around Pas-de-Calais, the port nearest to England, didn't just make good geographic sense to the Germans; the Allies were doing everything in their power to convince them that was their destination, including conjuring a phantom army poised menacingly in that direction, led by a general most feared. Normandy, of course, was the real objective, but if you were in German intelligence, Patton was not to be ignored.

Probably, Ike always intended to resurrect Georgie from exile; true fighting generals were simply too valuable to waste at this stage. But before giving him a real army, his penance would include looking menacing as commander of the fake force the Germans would obligingly label Army Group

Patton. While elaborate and obvious steps were taken to shield his presence in England, he was also provided assignments that would make sure the Germans knew he was there.

One of them was to speak to the "Welcome Club for American GIs" in Knutsford, home of Patton's real headquarters for what would soon become the fearsome US Third Army. After the appropriate intro—"General Patton is not here officially and is speaking in a purely friendly way"—a single member of the press (likely a plant) recorded Patton's seemingly innocuous words, ending with "since it is the evident destiny of the British and Americans, and, of course, the Russians to rule the world, the better we know each other, the better job we will do." The story did its job when printed in London—another piece in the German jigsaw puzzle that was Army Group Patton—but was reported in the US without including the Russians in the postwar triumvirate, and this omission and its potential diplomatic ramifications meant trouble for Georgie. After the slapping incident, he was on a very short leash with the public, many members of the Roosevelt administration, and even George Marshall, who left his fate in Ike's hands, though reminding him: "Patton is the only available Army Commander for his present assignment who has had actual experience fighting Rommel and in extensive landing operations followed by a rapid campaign of exploitation."

That went for both his real and unreal assignments. This was no slapping incident, this was a kerfuffle in a good cause, and Ike knew it. But the situation allowed him to let Patton believe his future hung in the balance, followed by a cruel if hilarious ceremony of contrition before Eisenhower's brand of forgiveness was granted. "When I gave him the verdict, tears streamed down his face and he tried to assure me of his gratitude. He put his head on my shoulder as he said it, and this caused his helmet to fall off—a gleaming helmet I sometimes thought he wore while in bed. . . . Without apology and without embarrassment, he walked over, picked up his helmet, adjusted it, saluted and said, 'Sir, could I now go back to my headquarters?'

"'Yes,' I said, 'after you have lunch with me in about an hour.'"

It was a cruel bluff worthy of Ike's poker skills, and had Patton been his equal he would have seen through it. However "Supreme," Eisenhower was in no position to remove the commander of Army Group Patton. What, after all, were the Germans to think? So Patton persisted in both unreal and real manifestations, and Ike had his pound of flesh, or a bouncing steel helmet at least.

George Patton, of course, was just a single facet of Eisenhower's vast set of problems. As before, he settled into Telegraph Cottage, but now in his own version of the mountain coming to Mohammed, he had the entire planning operation move from the distractions of London to a nearby former Eighth Air Force facility, "where his staff could buckle down to hard work" Ike told Kay.

He certainly did. A typical day in the events diary Kay kept for him spoke for itself: "Staged a meeting of . . . top commanders to discuss 'capability of air support of land operations.' . . . Lunched with Prime Minister. . . . Considered involved plans for invasion of southern France. . . . Held a long conference on the complicated Free French situation. . . . Talked with leading newspaper correspondents. . . . Approved plans for improving the American PX. . . . Ordered coordination of Special Service facilities among the Allies. . . . And with Kay Summersby, won eleven shillings in a bridge game at Telegraph Cottage."

A particularly valuable presence was Beetle Smith, smoothing the bumps in the Eisenhower day and, when necessary representing him, invariably making decisions exactly as he would have. He was equally gifted as a hatchet man and enforcer, making meticulously sure that what Ike wanted got done, and all without leaving bloodstains on the smiling mask. From now until the end of the war in Europe, there was no one more valuable to the supreme commander than Walter Bedell Smith.

Yet they didn't much like each other. Beetle didn't care to share meals regularly at Telegraph Cottage, but Ike insisted he do so for "professional" reasons. Actually, Kay and Ethel got along far better, providing at least some insulation between these two very hot wires, but not down deep. Of Eisenhower's closest associates, Smith was perhaps the most perceptive, and as such he came to understand the extent to which, and the calculation with which, Ike used people, and also that his ambition was big enough to eventually grow out of even his uniform. But, for now, the towering question marks surrounding Overlord drove them together, forging them into a solid leadership front equal to the tasks at hand.

First among them was Ike's struggle to get control of the air assets he needed, and this amounted to much more than controlling the skies over Normandy. He believed, and was right in assuming, that the success of Overlord was absolutely dependent on preventing the Germans from reinforcing the area surrounding the Normandy landing zone. The ruse might work for pinning down the local panzer divisions, but from farther out armies moved by rail, and stopping that necessarily entailed a massive bombing campaign against western France's entire transportation network.

This demanded heavy bombers, and a lot of them. There were already plenty based in England, but they were the assets of "Bomber" Harris and his US Eighth Air Force colleague Lieutenant General Carl "Tooey" Spaatz, and jealously hoarded for targets in Germany and the proposition that their destruction alone could win the war. But the real problem loomed above both: Winston Churchill remained favorably disposed to dehousing and equivalently reluctant to turn over Bomber Command to Ike, however temporarily. Pouring gas on Churchill's fire, Harris and Spaatz told him that French casualties incidental to the campaign against the rail net were likely to be as high as eighty thousand. Calling them both "prima donnas," and backed solidly by both Marshall and FDR, Ike went on the warpath. There could be no compromise: unless Normandy was sealed off from secondary reinforcement, it could become another frozen western front or even a suicide mission. Finally, in late March 1944, he told Churchill and backed this with a memorandum for the record stating that "if a satisfactory answer is not reached . . . I will request relief from this Command." In a long career Ike had threatened only Douglas MacArthur with quitting, and that time he proved to be dead serious. Fortunately for all involved (numbering in the millions), the prime minister did not try to call Ike's bluff or attempt to paper over the matter like MacArthur, especially after de Gaulle shrugged off the casualty estimates. Instead, he gave Ike the bombers.

To be fair to the "bomber barons," Eisenhower actually owed them his most precious asset in this risky endeavor, the prospect of absolutely controlling the skies over Normandy and the surrounding areas. Actually, it was a strategic regifting, coming about when the addition of drop tanks allowed the new generation of American fighters (P-47s and P-51s) to accompany the heavy bombers all the way to Germany, where they slaughtered an ever-younger and more inexperienced flock of fighter pilots flying the same

Bf 109s and Focke-Wulf 190s. The US bombers may not have come close to driving Germany from the war, but their protectors drained the lifeblood out of the Luftwaffe's aviators, and on the rebound left Ike with the promise of cleared and friendly skies, his ace in the hole in an otherwise problematic hand. And to further improve his odds, by D-Day British and American bombers had obligingly pulverized the French transportation net (along with fewer than ten thousand citizens) with 76,000 tons of high explosives, concentrating on marshaling yards, bridges, and tunnels—and, in the process, cutting overall rail traffic in half and, critically, thoroughly battering all approaches to Normandy.

As another means of taking pressure off Overlord, Eisenhower instinctively favored the diversionary landings in southern France suggested by Stalin at Tehran, now with the all-too-descriptive code name Anvil—soon changed to Dragoon. Churchill and his team wanted no part of the operation, however code-named; scuttle it, they argued, and split the landing craft between Overlord and Italy. Marshall and the other Joint Chiefs would have none of it, and over dinner at Chequers on a short transatlantic visit, he almost certainly told the prime minister what he had already assured Ike: unless Dragoon went forward, the landing craft were headed to the Pacific. Churchill tried Roosevelt, only to be reminded of pledges made to Stalin at Tehran and otherwise shown no inclination to call off the operation. Ike struck a one-sided bargain: use all the landing craft for Normandy, then redeploy them to southern France for an assault postponed to mid-August. Alan Brooke complained that "history will never forgive them for bargaining equipment against strategy and for trying to blackmail us into agreeing with them by holding a pistol of withdrawing craft at our heads." But that was now simply the nature of the relationship, and due to Team America's ability to flex its productive muscles when it counted. Churchill was more succinct and accurate when he conceded being "Dragooned."

Ike remained fixed on Dragoon, not simply as a diversion but also as an opportunity to introduce actual Free French combat units into the fray, yet in an environment less intense than projected for Normandy. This was symptomatic. His extended stay in North Africa and acute political antennae had led Eisenhower to some basic conclusions about the French, the most significant being that among the major players for power, Charles de

Gaulle stood head and shoulders above the rest—physically, at six foot five, and in every other respect. Ike desperately wanted de Gaulle's support for Overlord, and, reeking respect, called on him as provisional president of France.

"You were originally described to me in an unfavorable sense. Today I realize that judgment was in error. For the coming battle, I shall need not only the co-operation of your forces, but still more the assistance of your officials and the moral support of the French people. I must have your assistance, and I have come to ask for it."

"Splendid. You are a man!" exclaimed the haughty de Gaulle. "For you know how to say I was wrong."

That exchange turned the meeting into what Eisenhower later described as a "love fest," one that climaxed when he told de Gaulle that whatever "theoretical position" the management in Washington might take, "I will not recognize any other authority in France than yours." This was de Gaulle's postwar memory, but he had no reason to exaggerate Ike's words.

It was a stunner of a statement. Ike knew very well that FDR disliked de Gaulle, both viscerally and as a potential autocrat, but now he acted to force Roosevelt's hand by pushing all the red, white, and blue chips in the French general's direction. Not only that, but he subsequently succeeded in smoothing ruffled feathers in Washington, including Marshall's and most of all Roosevelt's, eventually getting them to accept his de facto installation of de Gaulle as "un fait accomplis."

In French or English, it was a brilliant maneuver. Ike realized de Gaulle was no autocrat, but was instead the tallest pole in the French tent, one capable of stabilizing it and governing while leaving him free to tend to the war and the Germans. Dwight D. Eisenhower from Abilene, Kansas, operated seamlessly at the level of princes, but, for now, this was just one more in a seemingly endless string of issues that filled his calendar as the days disappeared and June 1944 crept closer.

The sheer magnitude of the operation cannot be captured in the numbers, which sometimes can be deceptive. For instance, in terms of the humans involved, the initial landings in Normandy were not as big as those in Sicily—but that entirely understates the case. Stuffed in the south of England was a mechanized juggernaut of an army—a million and a half Americans alone—waiting to be brought ashore once the beaches were won, and then let loose in the French heartland. They were the reason "Mulberries"—entire portable artificial harbors capable of handling twelve thousand tons of supplies a day—would immediately be emplaced onshore, as would a gas pipeline be strung underneath the English Channel. For now, however, all of it accumulated in a ten-mile military-only staging area along the coast; in Ike's words, "the southern portion of England became one vast camp, dump, and airfield." All those amped-up humans—some itching to let loose their most savage instincts, others just terrified, and the rest oscillating in between—waiting for the word from a single man. From the first of February to the first of June, Eisenhower spoke to twenty-six divisions at twenty-four air bases, and at numerous other facilities, but now those men were waiting to hear just one thing.

It would be hard to overstate the amount of pressure Ike was under at this stage, but also pointless, since it's so obvious. He had moved from Telegraph Cottage and was now headquartered in the woods near Portsmouth on the Channel, smoking and drinking coffee more and more, while eating and sleeping less and less. And despite all his work and worry, he could do no more than the smallest insect about the most critical factors to success or failure—the moon and the weather.

The landings had to straddle low tide to expose German mines and obstacles, while sufficient moonlight was required for the paratroop and glider drops, which left only three possible days in June—5, 6, or 7. Otherwise, it meant sliding into July and not only perpetuating the Pas-de-Calais ruse, but also reversing the gigantic actual mechanism, a process that ultimately would include bringing back the many thousands of men already loaded into ships at sea.

That was exactly Ike's situation as June opened and bad weather moved in. Like a Roman general tagged with grim auspices, the supreme commander's climate augurs had only storm and bluster to report. Meanwhile, among the string of dignitaries who found their way to Ike's wooded headquarters during this critical period was Charles de Gaulle, who, with Roosevelt's reluctant okay, had been rushed to London. Ike gave him a half-hour briefing on Overlord, then pivoted to his worries about the weather and his twenty-four-hour go/no-go decision timetable. "What do you think I should do?"

Flattered and feigning a dash of reluctance, the Frenchman replied, "I will only tell you that in your place I should not delay." Ike did not ask just anybody such a question; likely it was a measure not just of an extraordinary rapport almost instantly forged, but also of the American's unerring perception that he was in the presence of a like mind.

On the evening of June 4, Ike left his trailer in the woods and headed for Southwick House to meet with his principal military commanders and their chiefs of staff—all in uniform except for Monty in corduroys and a turtleneck, and everybody smoking save him. All seated, three senior meteorological soothsayers entered, led by RAF group captain James Martin Stagg, who brought word from weather aircraft far out in the Atlantic of a high-pressure front rapidly sailing eastward. When it arrived, the weather on June 5 would begin to clear and the winds would drop; by nightfall conditions would be marginally acceptable for the airdrops, as would those for the sea landings the next morning, before the situation again deteriorated later in the day. There would be a twenty-four-hour window of time to jump through—that was all the augurs could deliver.

Ike next turned to his military commanders. The airmen, Arthur Tedder and especially Tafford Leigh-Mallory—who had already predicted 70 percent night drop casualties in perfect weather—thought it "chancy" and were leaning toward postponement. Admiral Ramsay, with all aboard and at sea but rapidly running out of fuel, sounded a note of urgency when he advised his colleagues that whatever they decided had to be signaled to the fleet within the next thirty minutes. Montgomery, for once resisting the temptation to lecture, had but one word of advice: "Go!"

Then all eyes turned to Ike, who sat silently, poker-faced. "I never realized before the loneliness and isolation of a commander at a time when such a momentous decision has to be taken, with the full knowledge that failure or success rests on his judgment alone," Beetle Smith remembered, his boss shuffling his options for what seemed all of five minutes. Finally, Eisenhower looked up and answered Monty and history: "Okay, we'll go."

But it was far from a sure thing and nobody knew that better than Ike. Sometime during the following afternoon he penned the following on a single sheet of paper, which he folded and slipped into his wallet: "Our landings in the Cherbourg-Havre area have failed to gain a satisfactory foothold and

I have withdrawn the troops. My decision to attack at this time and place was based upon the best information available. The troops, the air, and the Navy did all that bravery and devotion to duty could do. If any blame or fault attaches to the attempt it is mine alone.—June 5." As darkness fell he had Kay drive him to as many of the airborne units as possible before they flew off into the night. "It's very hard really to look a soldier in the eye, when you fear that you are sending him to his death," he told Kay later.

When they returned to the trailer, there were no more orders to give, nothing to do but wait. With Kay occasionally massaging the knotted muscles at the base of his neck, Ike was silent, alone with his thoughts, almost certainly focused on the monumental mission's chances for success. A lot had to go right. First, Utah Beach could be exited only by two thin cross-lagoon causeways, which, if held by the Germans, would trap the landing force. Ike's solution was the injection of two divisions of American paratroopers (the 82nd and 101st) the night before the landings to seize the causeways, a risky endeavor by any measure. If they failed, the Americans were down to one beach site, Omaha, backed by cliffs.

And even if they succeeded, the initial landings at Normandy would not relieve the necessity of perpetuating the Pas-de-Calais hoax. Now an even bigger deceit had to be sold to the Germans: that the assault at Normandy was simply a diversion from the real threat, the looming Army Group Patton. Ike's intelligence minions were reading ULTRA and believed the Germans remained duped, but this tissue of lies might rip at any moment, and could be expected to hold back reality for only so long—an interval measured by the time it took for British forces under Montgomery to capture Caen, the key communications and road link separating the German defenders at Normandy and their nearest reinforcements around Pas-de-Calais.

But that roadblock lay in the future. Ike's problem was the here and now, and on June 6 reports from the beaches were slow in coming. He was able to cable Marshall that the airborne landings and tactical bombing runs had gone well, and the approach channels were free of mines, but "I have as yet no information concerning the actual landings nor of our progress through the beach obstacles." He got that in the early afternoon, when reports of the crisis at Omaha trickled in, leaving Ike little to do beyond authorizing his

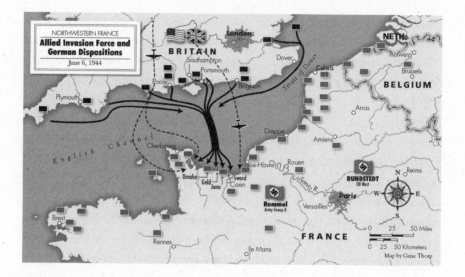

NORTHWESTERN FRANCE
Allied Invasion Force and German Dispositions
June 6, 1944

Map by Gene Thorp

tactical aircraft to drop their bombs through cloud cover, even at the risk of hitting Americans.

Initially transferred into Higgins boats twelve miles out at sea by orders of Bradley and Gee Gerow, and then confronted by an unexpected division of combat-hardened German infantry, the GIs at Omaha rallied in the face of heavy casualties, and by nightfall several elements had worked their way to the tops of the cliffs. By that time 156,000 Allied troops were ashore at the five beach sites, and although they cost 10,000 casualties, the landings at Normandy were clearly a success—or at least they hadn't failed.

THE ISSUE WAS FAR FROM OVER, BUT SECURING THOSE INITIAL SITES TRIPPED a compulsion among the leadership to set foot on the hallowed beaches, visitations that received mixed reviews from the actual combatants. Ike went first, landing at Omaha near midday on June 7, only shortly after Montgomery and Bradley had consolidated all five sites. He didn't have much to say, then promptly departed, leaving Bradley to conclude: "Ike's visit had been perhaps necessary for his own personal satisfaction, but from my point of view it was a pointless interruption and annoyance."

Next came George Marshall, who with the other chiefs in tow flew to England, where they were intercepted by Churchill and whisked on his personal train to Portsmouth, there joined by Ike. All were taken across the

Channel on June 12, landing in an amphibious DUKW; photographers documented George Marshall and his teammates, all in their midfifties and above, gingerly clambering over the side to exit. They visited an airstrip where wounded men were being evacuated, Marshall making sure the heroic were awarded medals, while Ike felt their presence was "heartening to the troops . . . possibly on the theory that the area is a safe one or the rank wouldn't be there." Bradley was less sanguine, cutting the tour short, concerned over what a sniper might accomplish if he caught this group in his sights. Meanwhile, Churchill, at a British beach, insisted on being brought forward so he could fire a gun in the general direction of the enemy. He then reunited with his fellow battle tourists aboard his train for an excellent dinner and plenty of champagne for triumphal toasts.

If Ike joined the festivities, it was only his smiling mask; beneath it he remained a deeply troubled man. "There was a terrible letdown after D-Day," Kay remembered. "Ike was tired as if he had run out of steam. And he was very much depressed." George Marshall knew Ike was looking forward to the graduation of his son John from West Point, and, perhaps glad to remind him of his familial status, pulled the strings necessary to allow the young man to spend his month's leave prior to advanced training with his father in England. The Eisenhower mood did brighten, particularly during the first week of the visit, but it didn't take Kay long to conclude "Ike was not a particularly doting father." His first night at Telegraph Cottage, John described having had to turn in all equipment including his mattress before graduation, leaving him to sleep on bare springs before departing West Point. Kay watched as Ike's face hardened, though he said nothing, and later asked him what he had been thinking. "What was going through my mind, was that across the Channel there are thousands of young men sleeping in foxholes—if they are lucky." Nothing could take his mind from that central reality, those men and their peril.

The focus of his worry and growing frustration was Caen, which Montgomery had promised to capture on the first day and the British and Canadians had almost delivered before defenses stiffened. The Germans, however misled, could still read maps that told them Caen occupied a critical spot between their own forces and began pouring in reinforcements accordingly. There ensued a battle that stretched through June, then sprawled over July.

Meanwhile, the Allied pocket in Normandy, at the cost of sixty thousand casualties during the first three weeks, was barely growing, expansion thwarted as much as anything by the bocage landscape, the venerable and formidable hedgerows that divided the countryside of individual farms like a three-dimensional chessboard and provided ready-made fortifications for the Germans. Consequently, American forces under J. Lawton Collins—only later known as "Lightning Joe"—made slow progress across the bocage of the Cotentin Peninsula, taking the vital deep-water port of Cherbourg only on June 29 and thus leaving the Germans plenty of time to meticulously mine and wreck the facilities, and in the process remove an alternative source of supply for the Allies at a particularly embarrassing moment.

Ten days earlier one of the worst storms ever to hit Normandy had wiped out the American Mulberry harbor off Omaha. Supplies would still flow, but not at the rate necessary to get an army across Europe; that demanded a major port with its facilities largely intact. And to further add to Ike's burden of worry, since the middle of June the Germans had been bombarding London and southern England with waves of V-1 cruise missiles—to be followed in September by the V-2, a true ballistic missile—a campaign that would eventually inflict 26,000 civilian casualties, and at this point had to be taken seriously as a threat to the Allies' nerve center and base of supply.

Meanwhile, the invading army would continue to accumulate in close quarters; near the end of July it numbered almost 1,600,000 troops and 333,000 vehicles. The Allies were all armored up, with nowhere to go. It was at this point that the supreme commander folded himself into the tiny back seat of a specially modified P-51 Mustang and ordered the US Ninth Air Force's tactical chief, Major General Pete Quesada, to take him on a close reconnaissance mission of the front, the spectacle of six stars sweeping low over Germans being a measure of Eisenhower's worry and desperation to find a way out. For he knew more than anyone that this rolling but now static behemoth of an army could be hidden from the Germans only so long, and once discovered, unless something changed, they had it in their power to put a stopper into the whole operation.

Actually, Ike could have chilled. From D-Day on, the Germans were either duped or fatally slow to react, undermined and nearly destroyed by the strategic antics of Adolf Hitler, or, for good measure, thrashed by bad luck, not to mention by Stalin.

"There can be no invasion within the next fortnight," the Germans' chief meteorological soothsayer had maintained on June 4. The miserable daylight weather of June 5 kept the Luftwaffe from flying, so they missed Ramsay's six-thousand-ship invasion fleet. That night the Allied airborne landings, though massive, were dismissed as simply local incursions by irregulars.

Erwin Rommel, now in charge of Channel defenses, also took the weather as an excuse for home leave in Germany to celebrate his wife's birthday. When finally informed of the Normandy landings, he did not think they were a diversion, but he was a long way off and had to return by car since the planes were socked in.

Meanwhile in Paris, his superior, Gerd von Rundstedt, commander of all German forces in the west, remained temporarily fixated on Pas-de-Calais, but as a precaution he placed two of the four panzer divisions stashed there on alert, only to be overruled by Alfred Jodl at the high command in Berlin on the grounds that Hitler had taken personal control of those divisions and he was asleep. It was midafternoon, ten hours later, before the Führer finally released the divisions, which were sent to reinforce Caen, thereby beginning the prolongation of the battle.

While Rundstedt quickly came to agree with Rommel that Normandy was no diversion, convincing the high command and particularly Hitler of that proved impossible. Nobody was a bigger believer in Pas-de-Calais, and since the Führer called the shots, the Fifteenth Army—nineteen divisions packing eight hundred tanks—remained stalled above the Seine, pointed at an army that did not exist and an invasion never to come.

But an assault that did was staged by Stalin and Team USSR, who, as promised, launched Operation Bagration on June 22, the third anniversary of Germany's invasion of Russia. It was designed to deliver a crushing blow—166 divisions numbering 2,400,000 troops, 5,000 tanks, and 5,000 warplanes aimed at the 700,000 troops of the German Army Group Center. It was a mismatch from the start, and by July 5 Hitler had lost half those soldiers, along with any possibility of sending reinforcements west with the Red Army closing in on East Prussia. Germany was now caught in an Allied nutcracker.

Back in Normandy, things only got worse on July 17 when Erwin Rommel's car was strafed by two Canadian Spitfires, leaving him with head injuries sufficient to end his military career. Three days later Hitler's career was almost but not quite terminated when a powerful bomb placed by military conspirators went off in a conference room. Defying the odds, he escaped injury with nothing more tangible than a perforated eardrum, but between the ears the blast seems to have thoroughly unhinged him. Before the interrogations were finished nearly five thousand were executed, and when Rommel's name came up Hitler forced his best general to commit suicide. But more important, his never-firm grasp on strategic reality was shattered; from this point, he would make one bad decision after another until Nazi Germany ceased to exist.

So, even if Normandy was beginning to look like a sandy, bocage-ridden version of the western front, the Allied team had a lot less to worry about than the other side. It was plainly time for plan B—actually a number of them—but the Allies rather than the enemy had that option, and on this team innovation sprang from below as well as from the top.

To wit, bocage busting became the personal crusade of an American sergeant named Curtis Cullin, who forged two large scythe-shaped blades out of German steel beach obstacles, then welded them to the front of a tank. He called it "Rhino," and it was that hard to stop, not simply cutting through earth and plant, but creating a giant divot with the blades, Eisenhower marveled, "actually allowed it to carry forward, for some distance a natural camouflage of amputated hedge," not to mention additional protection against antitank rounds. Bradley initially rejected the Rhinos, but when Collins experimented with them in the Cotentin and found they worked, Ike ordered them fitted to as many tanks as possible for the coming push, which in this case could be taken literally.

At the top, strategy being an ever-changing process, Allied leadership in late July turned a bad thing—the cacophonous traffic jam at Caen—into a good thing—the opportunity for a breakout far to the right at Saint-Lô, an operation aptly named Cobra. Due largely to the personalities involved, the origins and intent of Cobra would become an area of heated controversy in a lingering postwar battle of memoirs. Montgomery claimed he always intended that the strategic criticality of Caen would draw Germans magnet-

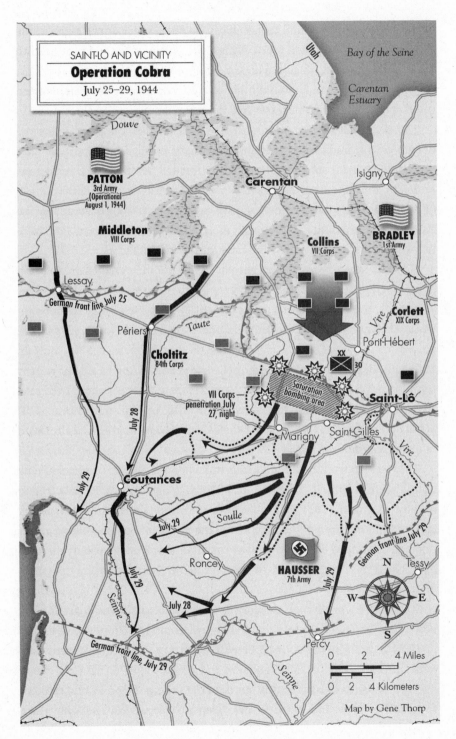

SAINT-LÔ AND VICINITY
Operation Cobra
July 25–29, 1944

Douve

Utah

Bay of the Seine

Carentan Estuary

Isigny

PATTON
3rd Army
(Operational
August 1, 1944)

Carentan

BRADLEY
1st Army

Middleton
VIII Corps

Collins
VII Corps

Corlett
XIX Corps

Lessay

German front line July 25

Taute

Vire

Pont-Hébert

Périers

Choltitz
84th Corps

XX
30

Saint-Lô

July 28

Saturation
bombing area

VII Corps
penetration July
27, night

Marigny

Saint-Gilles

Vire

July 29

Coutances

Soulle

July 29

HAUSSER
7th Army

July 29

German front line July 29

Roncey

Tessy

N

July 29

July 29

Semme

July 28

W   E

S

German front line, July 29

Semme

Percy

0   2   4 Miles

0   2   4 Kilometers

Map by Gene Thorp

like, thereby leaving a fatal vulnerability on the undermanned right, while Bradley, wanting full credit for devising Cobra, claimed Ike and Beetle Smith never understood it, which is a little ridiculous when you consider that only the supreme commander could orchestrate an operation this complex and the cast of characters it necessarily entailed.

The key to Cobra, its unprecedented venom, was a military-industrial update of the traditional preliminary artillery barrage, now termed "carpet bombing." On July 25, all of Pete Quesada's Ninth Air Force fighter-bombers, followed by 1,500 heavies, dropped nearly seven million pounds of high explosives on a seven-thousand-yard-wide corridor occupied by the Panzer Lehr Division, five thousand men, all combat instructors (the military equivalent of seed corn), and most headed for oblivion.

"I'm almost sorry for those German bastards," muttered George Patton as he watched hundreds of planes converge on the target areas. He was now exclusively the head of a real army, the US Third, which he had trained relentlessly while in Knutsford and had now joined in the Normandy pocket, poised for the breakout. Also observing the bombing was Lesley McNair, Marshall's deputy for training, whom he had recently sent as a hoax-worthy replacement for command of Army Group Patton. He was a short-lived one, though, as a disastrously short bombing error killed him and 110 other GIs. Word of his death was the first Marshall, back in Washington, actually heard of Cobra. But not the last.

The German front held for a while, but when the shell broke, there being nothing left behind, it commenced hemorrhaging Americans. Now free to become "Lightning Joe," Collins sent two armored divisions, led by a phalanx of Rhinos, through the gap, where they discovered almost no opposition and continued carving their way south to tank-friendly country around Avranches, soon to be joined by the rest of VII Corps, then by the whole First Army.

Following them through the corridor formed between the coast and the now-open German flank, George Patton squeezed the Third Army, and as in Sicily began sending it unexpectedly westward—this time toward Brittany and Brest. But Ike, Montgomery, and Bradley agreed that only one of his corps was necessary for this mission; the others they told him to point toward the Seine and Loire Rivers. That he happily did, sending them southeast, where, sweeping aside only scattered opposition, they reached Le Mans, seventy-five miles away, in less than a week before turning left to Alençon thirty miles north and back toward the Germans. In the meantime, the US First Army had moved east, then north, but at a slower pace, while the British Second Army and the Canadian First had also broken out and were headed south, in a direction roughly opposite the Third Army in Alençon. The combined Allied forces were beginning to outline a significant bulge in the German lines.

It was then that Hitler ordered the Wehrmacht's Seventh Army and Fifth Panzer Army to stage an all-out assault west toward Mortain and eventually the sea, the aim being to cleave the Allied Cobra. This might have made sense on a map, but it utterly disregarded the central reality of this campaign: the complete Allied control of the third dimension, the deadly skies above any German unit. Not only that, but ULTRA had told Bradley and Patton the attack was coming and where it was headed.

Consequently, from the moment it began early on August 8, the farther forward the German armored wedge drove west, the more vulnerable it became to being surrounded should Patton's Third Army and the Canadian First meet and close the escape route. Subjected to devastating air attacks once daylight broke, the German field commanders almost immediately wanted to call a halt and retreat, but Hitler insisted they continue toward the US First Army and Mortain. But they never made it, stopped by continuous enervation from above and the heroic stand of the US Thirtieth Division. Rommel's replacement, Günther von Kluge, ordered what was left of the Fifth and Seventh Armies to backtrack and hold the gap between the Americans and Canadians open long enough to escape.

At this point, at least as far as George Patton saw it, Bradley, on the verge of a modern Cannae, hesitated fatally, failing to give the orders to go all-out to close the gap, now only fifteen miles wide between Falaise and Argentan. As a result, it was not until August 21 that it was shut, trapping around 50,000 but allowing 100,000 to slip through on foot. They were the remains of the Nazi shield in northwest France that on D-Day had numbered forty divisions, 600,000 men, and 500 tanks. Now it was just small groups fleeing east.

Team America's two heavy hitters on the scene, George Patton and

Dwight Eisenhower, both reacted to this notable and devastating victory in characteristic ways. For Patton, the lessons were purely professional—"better ways to kill Germans," he might have phrased it. First, the preceding events made him a believer in ULTRA, and from this point he would pay close attention to it and the prognostications of his intelligence chief, Oscar Koch. Second, they had convinced him of the power and possibilities of tactical air support: "Just east of Le Mans," he wrote in his war memoir, "for about two miles the road was full of enemy motor transport and armor, many of which bore the unmistakable calling card of a P-47 fighter-bomber. . . . Armor can move fast enough to prevent the enemy having time to deploy off the roads, and so long as he stays on the roads the fighter bomber is one of his most deadly opponents." Hence, he concluded he could charge confidently ahead without concern for threats from either side: "If I had worried about flanks, I could never have fought the war. Also, I was convinced that our Air Service could locate any groups of enemy large enough to be a serious threat." And in this spirit he headed across the French homeland as fast and as far as his tanks would take him: "Touring France with an Army" was a chapter title in his memoir.

Ike's reaction to these spoils of war was decidedly less upbeat. "Forty-eight hours after closing the gap I was conducted through it on foot, to encounter scenes that could be described only by Dante. It was literally possible to walk for hundreds of yards at a time, stepping on nothing but dead and decaying flesh." Now, added to his burdens at the top, was Eisenhower's growing awareness of just how horrible war fought in this manner, with the tools technology and mass production provided, actually was. For seventy-five days, the fighting had blazed through Normandy, and when it was over the Germans were missing 500,000 men killed, wounded, or captured. But Allied losses totaled 200,000, two-thirds of them American, the price of victory demanded at this point by industrial warfare.

Eisenhower's postinferno tour coincided with his official move to France, allowing him to assume control of Allied ground forces from an increasingly obstreperous Montgomery, and also to direct his strategic energies away from the northern point of attack. Dragoon, the invasion of the South of France, was no longer needed to take pressure off Normandy, but Ike still made sure it went forward. On August 15, Allied landings, commanded

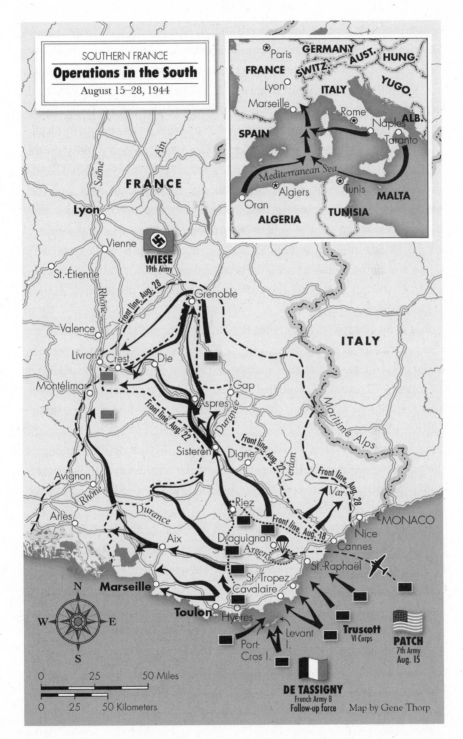

SOUTHERN FRANCE
Operations in the South
August 15–28, 1944

Map by Gene Thorp

by the very able Lucian Truscott but including a significant component of American-equipped French combatants, commenced along the Côte d'Azur.

Dragoon was wildly successful. By the end of the month, the much-needed ports of Marseille and Toulon were in Allied hands; only mildly damaged, they were soon to provide a high-capacity if somewhat far-flung supply point for Ike's ravenous force structure. Even better, German Army Group G, unable to establish a stable defense, was being chased rapidly up the Rhône, unaware that its withdrawal plans were being decrypted by ULTRA and exploited accordingly. By September 2, it had fallen back to Lyon, nearly two hundred miles upriver, and another two weeks found it past Dijon and heading for the Vosges Mountains, where it was finally able to dig in. Meanwhile, on September 10, forward elements of Truscott's IV Corps were able to link up with units of George Patton's Third Army coming from the east, marking the entirety of southern France being cleared of Germans.

But as any native knows, France is nothing without Paris. That was where the real action was taking place, and the Allied supreme commander already understood it was as much a matter of politics as fighting. A week after Normandy, Charles de Gaulle had paid a quick visit to Bayeux and received a tumultuous welcome, one sufficient to convince Ike and his civil affairs officers that their elaborate occupation manuals and script to replace the franc were unnecessary. France was utterly fractious, emotions boiling to the point they could take the country practically anywhere, but Eisenhower was henceforth committed to the proposition that de Gaulle could run the place, leaving him free to fight the war.

It was a proposition soon to be tested. On August 19, the Third Army's Seventy-Ninth Division reached the river Seine and established a bridgehead just thirty-five miles west of Paris, allowing Patton, after a visit, to brag to Bradley, "I pissed in the Seine this morning." Next, Georgie wanted nothing more than to liberate the City of Light.

But Ike wanted to steer clear, initially at least. An urban environment as big as Paris could swallow an army if the Germans chose to contest it street by street, which is exactly what Hitler had already ordered his forces to do, along with wrecking the museums, monuments, bridges, and other public structures. Walter Model, Kluge's replacement, was quick to realize this would also mean the encirclement and destruction of the remaining organized German forces in France; so, on August 20, he ordered his First Army and what little remained of the Fifth Panzer Army to evacuate their positions in front of Paris and head away from the city. "Tell the Fuhrer that I know what I'm doing," Model told an astonished Alfred Jodl.

That left Paris's fate in the hands of its military governor, General Dietrich von Choltitz, with orders to blow up a cultural heritage measured in millennia. But then came word that he didn't like how that might sound to posterity, and wanted instead to surrender the city. Yet Choltitz had at most forty-eight hours to get busy with the explosives or he would be replaced, likely executed, and the window of opportunity to save Paris from destruction would snap shut.

Before it did, Ike realized bypassing Paris was no longer an option. De Gaulle warned him of chaos if the Allies didn't move in quickly—a second Paris Commune, with the Resistance taking bloody revenge on collaborators and remaining Germans. He also had an excellent suggestion for how to do it, and also who to do it with. Rather than Americans, de Gaulle recommended sending first his own paladin, the indefatigable Philippe Leclerc, at the head of his own Second Armored Division. The French liberating the French—it was hard to argue against. With little more than a verbal shrug, Ike accepted the inevitable: "What the hell, Brad. I guess we'll have to go in. Tell Leclerc to saddle up."

As usual, luck was in Ike's stable. Things went splendidly. Leclerc arrived with sixteen thousand men in four thousand American-made vehicles on August 24, followed the next day by the US Fourth Division. Token resistance was swept aside, and late that afternoon Choltitz surrendered the city unharmed and avoided a very dark future in history. On Saturday, August 27, de Gaulle relit the "eternal flame" at the Arc de Triomphe, then led the Second Armored Division down the Champs-Élysées lined with two million Parisians. France was far from completely liberated, but its spirit and its center were free, and that meant everything.

So as not to draw attention from de Gaulle, Eisenhower waited until the next day to enter the city, whereupon he paid him a formal visit at the Élysée Palace, traditional residence of French presidents. "I did this very deliberately as a kind of de facto recognition of him as the provisional President of

France. He was very grateful—he never forgot it—looked upon it as a very definite recognition of his high position. That was of course what he wanted and what Roosevelt had never given him." One of Ike's better biographers calls him a "unique kind of American proconsul," but in Rome it was a rare proconsul who crossed the Senate or an emperor and survived the experience. Eisenhower in his inimitably unassuming way was operating as his own manager in the international arena, jumping off the field when necessary and taking command of the entire team.

More remarkably, he got away with it, and would continue doing so. Part of the credit went to the seamless nature of Ike's smiling mask, but the team captain also gave him a very long lead. When George Marshall had imposed unity of command on Allied planning, he really meant it. It was all about delegation; as far as he was concerned, a theater commander should be left to his own devices. If Marshall was initially surprised by Ike's de Gaulle connection, or learned about Cobra only when McNair was killed, he remained imperturbable George, backing his man on the ground to the hilt, an easy thing to do with things going this well.

IKE AND HIS ALLIED LEGIONS WERE NOW FREE TO MOVE ON. DE GAULLE NOT only restored order, he proved to be the greatest leader of France since Louis XIV, and a testimony to the Eisenhower eye for political talent. Otherwise, the future looked wide open for the American-made juggernaut, driving forward not unlike a really good pro football offensive line—the ends, tackles, guards, and center more or less representing the Army-size prongs in the Allied front. But sports analogies go only so far to explain this situation; while football players run on calories, an army in the industrial age no longer traveled just on its stomach. "We planned, following upon any breakout, to push forward on a broad front, with priority on the left. Thus we would gain, at the earliest possible date, use of the enormously important ports of Belgium," Eisenhower explained.

For the army he controlled was a gas-guzzling, ammunition-gobbling behemoth that required constant feeding and a steady gusher of hydrocarbons that only the nearby superport of Antwerp could begin to satiate. Meanwhile, his forces were burning through 800,000 gallons of fuel a day, Patton's Third Army accounting for 350,000 of that total. And still he wanted

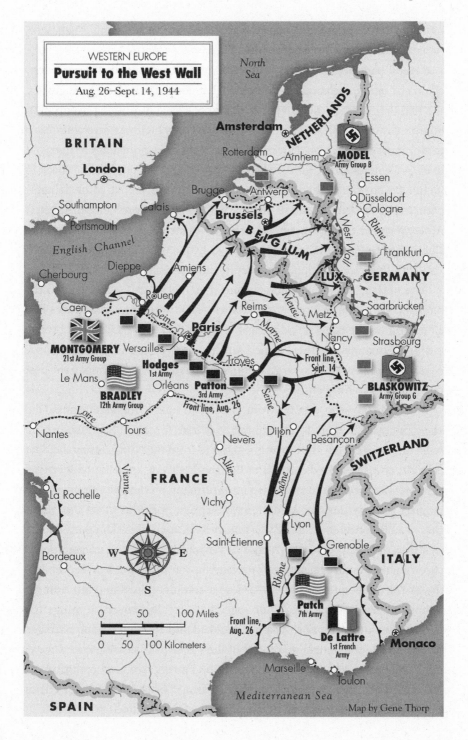

WESTERN EUROPE
Pursuit to the West Wall
Aug. 26–Sept. 14, 1944

North
Sea

BRITAIN

London

Southampton
Portsmouth

English Channel

Dieppe
Cherbourg

Caen

MONTGOMERY
21st Army Group

Le Mans

BRADLEY
12th Army Group

Nantes

La Rochelle

Bordeaux

SPAIN

Calais

Brugge

Amsterdam

Rotterdam

Arnhem

NETHERLANDS

Antwerp

Brussels

BELGIUM

Amiens

Rouen

Seine

Paris

Versailles

Reims

Marne

Meuse

Troyes

Front line,
Sept. 14

Hodges
1st Army

Orléans

Patton
3rd Army

Seine

Front line, Aug. 26

Loire

Tours

Nevers

Vienne

Allier

FRANCE

Vichy

Dijon

Saône

Besançon

Lyon

Saint-Étienne

Grenoble

Rhône

Front line,
Aug. 26

Patch
7th Army

De Lattre
1st French
Army

Marseille

Toulon

Mediterranean Sea

MODEL
Army Group B

Essen
Düsseldorf
Cologne

West Wall

Rhine

Frankfurt

LUX GERMANY

Saarbrücken

Metz

Nancy

Strasbourg

BLASKOWITZ
Army Group G

SWITZERLAND

ITALY

Monaco

0   50   100 Miles

0   50   100 Kilometers

N
W   E
S

Map by Gene Thorp

more: "I'll shoot the next man who brings me food," Georgie bellowed. "Give us gasoline; we can eat our belts."

Several half measures were instituted, and though massive in scale they could not satiate the voracious appetites that needed to be fed. On August 25, the Red Ball Express rolled—six thousand trucks kept running twenty-four hours a day on a dedicated system of one-way-only highways stretching back to Normandy—intent on delivering 82,000 tons of consumables to the front lines by September 2. Unfortunately, that proved little more than a snack to its hungry customers, and it wore out trucks and drivers at an unstainable pace, as can-do spirit goes only so far. That was not the problem with the US Army Corps of Engineers, which managed to string double pipelines from Cherbourg to Paris and from Marseille to Lyons with amazing alacrity, but also with the same disappointing results: not enough throughput.

Attempting to put his gallons where they mattered, Eisenhower, to the intense frustration of Bradley and Patton, sent Montgomery an extra fuel allocation on the basis of his critical location. When the Allies had broken out of Normandy and then turned left, the British and Canadians, whose beach locations were eastward, remained closest to the coast and in the best position to grab Antwerp.

Montgomery, who was far from convinced of its vital significance, did stage the operation, and Antwerp fell to British forces on September 4, almost undamaged. But the city was separated from the North Sea by the river Scheldt, and the next day Admiral Bertram Ramsay, knowing how critical the port was to sustaining the drive into Germany, warned Montgomery that his first priority must be to eliminate German positions at its mouth, or they would continue to mine the waterway and paralyze shipping. Monty was less than impressed by Ramsay's reasoning, and uncharacteristically chose instead a very daring leap forward in the third week of September, the ill-fated Operation Market Garden, a combined land-and-airborne attempt to spring across the Rhine, one that literally proved "a bridge too far." Meanwhile, Antwerp remained blocked and was soon being barraged by V-2s, while the hunger and thirst of Ike's army was slowing it to a crawl. The "October pause," as it has come to be known, would stretch through much of November as the slow work of clearing Germans out of the Scheldt estuary proceeded, and Antwerp could finally come on line.

George S. Patton, a.k.a. "The Boy," at an early age.
Public Domain

Georgie and Bea, a matrimonial moment for one tough couple, May 26, 1910.
Public Domain

Patton (*right*) wields a mean saber at the 1912 Olympics.
Public Domain

Patton with tank, and ready to rumble, summer 1918.

Public Domain

Back in the saddle—Patton at Fort Sheridan, circa 1923.

Heritage Image Partnership Ltd./Alamy Stock Photo

The Eternal Warrior in his element, Patton watches over his tanks.

Public Domain

Patton gets his third star from Ike—all smiles for the moment, March 12, 1943.

Public Domain

Patton near the end, wearing the spoils of war, 1945.

Public Domain

The MacArthur family unit, with one destined to shake the world—Doug, Arthur, Arthur Jr., and Pinky, circa 1886.

MacArthur Foundation

MacArthur at West Point—the picture-perfect cadet, already using
a uniform to maximum advantage. *MacArthur Foundation* RIGHT: Mac, sprouting a
corncob pipe and decked out in a prototype of his signature combat togs,
Veracruz, 1914. *MacArthur Foundation*

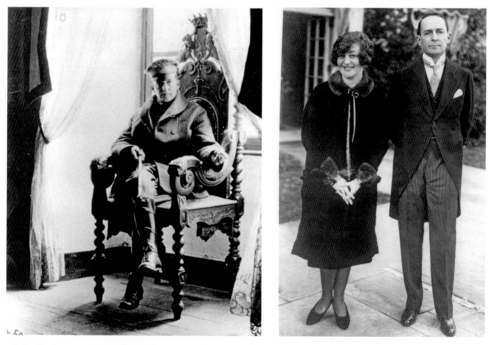

MacArthur, a man of destiny on a throne slightly too big for him, at his headquarters in
occupied Germany, 1919. *MacArthur Foundation* RIGHT: The MacArthurs as young marrieds—
Louise looking arch; Doug looking worried. *MacArthur Foundation*

MacArthur's arrival in the Philippines, with Ike in the background, exactly where his boss always wanted him to be, 1935.
MacArthur Foundation

MacArthur and Jean Faircloth—he found a woman who worshipped him, and she found her "Sir Boss"—shortly after their marriage, 1937.
MacArthur Foundation

MacArthur and General Richard Sutherland deep in the tunnel complex of Corregidor in March 1942: it looked like the end.
MacArthur Foundation

MacArthur, the face of
command in action,
directing a paradrop,
September 1943.
MacArthur Foundation

MacAthur hits the surf at
Leyte, and soon after onto
movie screens all over America,
October 1944.
MacArthur Foundation

MacArthur signs "a
solemn agreement
whereby peace
may be restored,"
September 2, 1945.
MacArthur Foundation

In all his glory, Douglas MacArthur views the Inchon landing unfolding before his eyes, September 15, 1950.
MacArthur Foundation

MacArthur addresses a joint session of Congress, marking the beginning of his own "Not Fade Away" tour, April 19, 1951.
MacArthur Foundation

An adolescent George Marshall, nicknamed "Flicker," during a vaporous boyhood in Uniontown, Pennsylvania.

George C. Marshall Research Foundation

The Marshall wedding party—George with his bride, Lily, and his brother, Stuart (*center*) looks on, "eyes wiped."

George C. Marshall Research Foundation

Marshall (*left*) laughing briefly on the Western Front. He wanted combat, but General Robert Bullard (*right*) sent him to American Expeditionary Forces (AEF) commander, General John "Black Jack" Pershing.

George C. Marshall Research Foundation

Beneath Black Jack Pershing's wing as his aide, Marshall (*right*) learned generalship.

George C. Marshall Research Foundation

Ike, looking lean and hungry, during June 1943 discussions with Marshall in Algiers.

George C. Marshall Research Foundation

George Marshall (*second from right*) flaunts a stiff upper lip after tracking down MacArthur on Goodenough Island, December 15, 1943.

George C. Marshall Research Foundation

Star-studded tourists—Ike leads an appreciative Marshall as they hit Omaha Beach, June 12, 1944.

George C. Marshall Research Foundation

President Harry Truman congratulates Marshall at ceremonies marking the end of his military career—retired today, rehired tomorrow.

George C. Marshall Research Foundation

Marshall, still clueless in China, meets Mao, March 5, 1946.

George C. Marshall Research Foundation

Marshall, marching toward greatness, in a Harvard commencement procession prior to his Marshall Plan speech, June 5, 1947.

George C. Marshall Research Foundation

Dwight Eisenhower (*left*) and his family, one fused by hard times and hard work, 1902.
Eisenhower Presidential Library

Ike (*front*) as a teen, as always, surrounded by friends.
Eisenhower Presidential Library

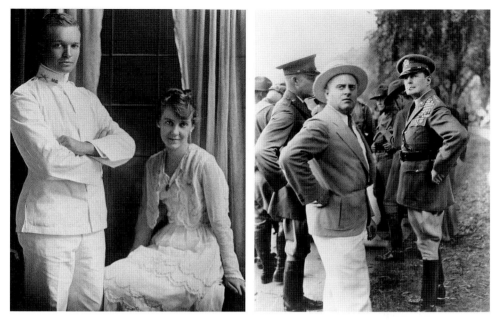

Ike and Mamie's white wedding, July 1, 1916. *Eisenhower Presidential Library*
RIGHT: Ike with MacArthur at the rout of the Bonus Expeditionary Force (BEF)—a bad day for Team America, July 28, 1932. *Eisenhower Presidential Library*

Ike with 101st Airborne before Normandy. "It's very hard really to look a soldier in the eye, when you fear you are sending him to his death."

Eisenhower Presidential Library

Club Eisenhower's zombie beach party—
Kay Summersby, Ike, and
Ethel Westermann. *Estate of Kay Summersby*

Eisenhower congratulates Patton
after the Battle of the Bulge, as General
Omar Bradley looks on.
Eisenhower Presidential Library

Ike and Mamie in Abilene, not a relationship entirely on the rocks, June 1945.
Eisenhower Presidential Library

Ike hears Truman has fired MacArthur; a mask at the crossroads.

Eisenhower Presidential Library

Ike takes the oath of office and becomes president, January 20, 1953.

Eisenhower Presidential Library

Ike and his longtime "batman,"
Sergeant John Moaney,
grilling steaks.
Eisenhower Presidential Library

John F. Kennedy soars at his
inauguration. Ike, though, was
weighed down by the military
industrial complex.
Everett/Shutterstock

The war, it seemed to George Marshall, had been put on hold. In an effort to add some explosive motivation, he tried to prevail upon the Air Corps's strategic vagabonds to stop bombing targets deep in Germany and concentrate on the actual battlefield—but without much success. In early October, he decided the situation warranted some face time with key players.

He flew first to Paris, then moved toward the front, where his force structure was stuck in neutral. He visited the Third Army, which despite their being out of gas and artillery shells he was delighted to find in excellent fighting condition; he personally pinned a Silver Star on one of the troops, only to be one-upped by Patton: "Georgie composed an inspiring citation of his own and presented the soldier with the DSC," remembered one of Patton's brigadiers. Given the strategic situation, Marshall found Patton's mood surprisingly buoyant and, seeking to reinforce it, gave him the latest news of Beatrice, not realizing that his upbeat condition had to do with Jean Gordon, who had arrived in England in July as a Red Cross volunteer, and was now installed as a "Donut Dollie" with the Third Army. However misinformed as to the true reason, Marshall did leave convinced his best gladiator was in excellent fighting trim.

He also went north toward the source of the problem, and paid a call on Bernard Montgomery at his headquarters outside Antwerp. What he got in return was a stream of vituperation, most of it directed at Ike, who according to Monty lacked "grip and operational direction." As a consequence, the whole effort had become "ragged and disjointed . . . a real mess." Being in didactic mode and hardly perceptive, Montgomery likely did not notice the hardening expression on Marshall's face; George later remembering that he "came pretty near to blowing off out of turn." This was more than a case of bad manners; at the heart of the matter was a very fundamental disagreement over strategy.

Marshall did place some blame for the slowdown on Ike's empire-building logistics chief, J. C. H. Lee, and wanted him replaced—something Ike never did—but otherwise his thinking was in lockstep with the man who had beaten him out for the job. Eisenhower's determination to advance on a broad front looked to Team Great Britain as an update of Black Jack Pershing's approach to World War I, yet they were wrong. Once the gasoline problem was fixed, Ike held a decisive hand. Years of strategic bombing aimed at Germany's fuel supply had finally taken hold, and at this point the Wehrmacht and the Luftwaffe were essentially running on empty. As Ike explained, "The combination of an overwhelming air force and the great mobility provided by the vehicular equipment of the Army enabled us to strike at any chosen point along a front of hundreds of miles." And without fuel, the Germans had only boots and horses with which to respond, giving every Allied thrust, should it break through, the potential of tearing at Germany's vitals. The thinking was sound, but at this lockdown stage you had to be a believer, and Team Great Britain wasn't.

The Brits argued all that mechanized firepower should be concentrated into a single sword thrust into Germany along a northern corridor that could ultimately be aimed at Berlin. Not only was it a sort of "Strategy 101" solution, it was entirely self-serving, since this was where the great bulk of British forces were located, and it was completely out of touch with reality. Put crassly, the British military was beyond stretching; the Americans were bringing most of the men and equipment, and they would call the strategic tune. Brooke, Montgomery, and no doubt others shared their gossipy contempt for American thinking, but it was they, far more scarred by the western front, who overreacted out of fear it might be reestablished.

It was also getting personal. When Eisenhower rejected the British approach as a "mere pencil-like thrust" vulnerable to a flank assault that might push it back against the North Sea, Montgomery responded by calling the broad-front approach "balls, sheer balls, rubbish." "Steady, Monty, you can't talk to me like that. I'm your boss," Ike reminded him, probably without his famous smile. Patton, who was attracted to the idea of a single thrust—his own, not Montgomery's—took to mocking the latter's offensive potential with an exaggerated English accent; the Germans, he said, "will be off their guard, and I shall pop out at them like an angry rabbit." Given their personalities, he got along with Montgomery surprisingly well early in the war, but now dismissed him as a windbag with a fatal dose of the slows. Yet between Ike and Monty there was true bad blood, soon destined to be shed in a showdown.

In reality, much of their anger and frustration was simply beyond their control. For one thing, northern Europe was settling into the worst winter in nearly forty years, one that numbed and slowed all military operations, particularly those of the Allied attackers, whom the weather frequently robbed of their critical tactical air advantage. Meanwhile, for those on the ground, memories of this time would forever be suffused with cold rain and mud and ice and snow, all of it stained red. For when fronts stabilized, it inevitably became feeding time for industrial warfare. Consequently, of the three-quarters of a million battle casualties the Allies accumulated in Europe, two-thirds would be suffered after the autumn slowdown.

The fleeing Germans had rallied, taken their place behind the "Westwall," the shield of fortifications Hitler and slave labor had built in front of the Rhine and beyond which the Reich lay. Just why the Wehrmacht's individual soldiers continued to fight so hard remains perplexing. When captured they seemed relieved, and there are numerous photographs of them marching unescorted to Allied POW camps. It was also true that within the ranks even the slightest sign of insubordination was frequently punished by execution. Still, such practices generally lead to passivity, and German troops remained anything but passive. Many had seen or participated in the terrible crimes the Wehrmacht had committed in the east, and though their behavior had been better against the Allies, they had good reason to fear defeat would bring truly harsh retribution against themselves and the German civilian population. And although Hitler had largely disappeared from public view, there were no signs of domestic revolt that might have undermined morale at the front, as had happened in World War I.

Meanwhile, to convince them he could stave off defeat, Hitler retreated to the realm of science fiction and strategic fantasy to unveil his final throws of the dice. In a last, desperate burst of creativity, the Third Reich's technologists forged virtually a complete prototype arsenal for the coming Cold War and new era of armaments. Not only did they manage to field the world's first cruise and ballistic missiles in the V-1 and V-2, they also energized the Luftwaffe's antibomber campaign by introducing the Waterfall surface-to-air missile, along with the first operational jet interceptor, the Me-262. And behind these was a veritable cornucopia of new armaments in late development or early fielding, including the the Hs 117 Butterfly air-to-air missile, and the Hs 293D television-guided bomb. On the ground, Hitler was mandating immediate production of Ferdinand Porsche's latest tank, the 188-ton Maus, almost four times the weight of the US M4 Sherman, and mounting a main gun nearly twice as large. Today, going to war in a Porsche sounds stylish but problematic, if for no other reason than there are not a lot of them, which proved exactly the problem in late 1944. To affect the actual course of the war, all these weapons required mass manufacture and fielding in significant numbers, and, with the exception of the V missiles and the Me 262, none were. And had they somehow multiplied in the Third Reich's battered war industries, they would have thirsted for fuel that did not exist, witnessed by Me 262s being towed from hangar to runway by oxen.

Meanwhile, calling nuclear science "Jewish physics," Hitler also missed the one weapon that would define the Cold War, and was at the time nearing completion deep in the Manhattan Project, primarily for use against Germany. Consequently, the longer the Third Reich held out, the greater the chance mushroom clouds would sprout over its cities.

Still, on the Allied side, combat entropy continued and even intensified. Montgomery's frustrations clearing the Antwerp approaches were paralleled by those of Americans to the south. Ike told Bradley to work his way eastward, and he in turn sent his First and Ninth Armies toward Cologne, getting within thirty miles before becoming locked in a bitter yard-by-yard struggle in the Hürtgen Forest. Even farther south, Patton's Third Army and Jacob Devers's Sixth Army Group (a post-Dragoon consolidation of American and French forces) did manage to bite a substantial chunk out of the German line, moving as much as seventy miles forward, and in the process taking Metz, Strasbourg, and Mulhouse, under ten miles from the Rhine. But from this point the going got very tough, gains shrinking to yards rather than miles a day. The net result was that by Thanksgiving Ike was quarterbacking a front line extending 450 miles from just east of Antwerp—which was finally up and spewing gas and supplies—down to the Swiss border at Basel. Even for the broad front's spread formation, it was a real stretch, and one that spelled danger.

But this was not apparent at home, where victory, if perhaps not imminent, was seen as inevitable. It was a vibe that Team America's manager imbibed to sooth his annoyance that his four stars were being outranked by not just British field marshals, but also those of other Allied services. The time had come to hand out some accolades to his triumphant starting lineup, and nothing worked for FDR better than adding an extra star to their individual constellations.

Henry Stimson was against it; so was the chief of staff, whom wags claimed did not relish the prospect of being called Marshal Marshall. He claimed he simply didn't need it to do his job, and remained characteristically mute on how he felt about being elevated to unprecedented rank. Anyway, FDR was the one doing the decorating and pushed a bill through Congress; on December 15, 1944, Marshall, Eisenhower, and MacArthur, along with Arnold, all became five-star generals of the Army. Yet it wasn't just George Marshall who greeted his apotheosis with an absence of cheer. Closer to the cutting edge, both Ike and Mac were in a better position to see rank coming at the cost of their men's lives. In November, Ike had written Mamie: "It's all so terrible, so awful, that I constantly wonder how 'civilization' can stand war at all?" Now he wore his new rank in almost a stealthy fashion—five tiny stars in a circle on each shoulder, no collar insignia unless in a combat jacket, and he abandoned his overseas cap, which required rank, for a visored one, which didn't. On the other side of the world, Mac also found little joy in his own elevation: "The old thrill of promotion and decoration was gone. Perhaps I had heard too often the death wail of mangled men—or perhaps the years were beginning to take their inexorable toll." Likely both.

OSCAR KOCH, GEORGE PATTON'S SAVVY INTELLIGENCE CHIEF, DIDN'T LIKE what he was hearing from ULTRA, that German ground force elements were on the move and calling for air cover—not the kind of thing you'd expect from static defenders. On December 9, at a briefing of the Third Army staff, he pointed worriedly to a particularly rough ninety-mile segment of the front, the Ardennes Forest, as the possible destination. It was occupied by VIII Corps, made up of two inexperienced divisions, and one worn out from fighting. George Marshall had inspected the area closely during his last visit, and he had been assured by Bradley and the local commanders that, though it was a bit of a calculated risk, the roughness of the terrain provided a healthy measure of safety. Koch didn't think so. This was the same Ardennes the Wehrmacht had crashed through in 1940, on their way to the conquest of France. Besides Patton, none of the other top commanders took Koch's threat scenario or the suspicious ULTRA intercepts seriously, and even he wasn't that worried. Then, at his 7:00 a.m. briefing on December 16, Koch told his boss that the Germans had slipped into radio silence. When Patton asked him what he thought that meant, Koch reminded him that when Americans went similarly blank, it inevitably meant an impending attack. Later that morning the Wehrmacht confirmed his assessment.

It was Hitler's strategic version of his fantasy arsenal, although with a lot more impact. His war industries czar, Albert Speer, had managed to squeeze out of the nearly dry German economy just enough fuel and land arms to stage one last great attack. And what better place than the site of the Wehrmacht's most brilliant strategic coup? Antwerp was just a hundred miles from the front—grab that, and the Allies would have to retreat to the Seine. When they did, he could race the force east and deal the Russians a similar lightning blow. In the eyes of the addled Führer at least, it looked like springtime for Hitler. Field Marshal Rundstedt was likely less optimistic, but he did put together in great secrecy a truly formidable attacking force: two fresh panzer armies, the rebuilt Fifth, under Sepp Dietrich, Hitler's ex-chauffeur, and the Sixth, commanded by the highly capable Hasso von Manteuffel, flanked by two additional armies, both reequipped. All this added up to twenty-eight divisions, almost 1,500 tanks and 2,000 pieces of artillery, all manned by more than 300,000 troops.

Deception and special operations were given particular attention. Along with actual captured Shermans, certain tanks were painted olive drab and disguised to look like them, an armored version of a malicious virus. To heap confusion atop surprise, Hitler's favorite commando, Otto Skorzeny—"a combination of thug and Nazi Lawrence of Arabia," according to one of Ike's biographers—put together a unit featuring English-speaking SS volunteers dressed in American uniforms and driving jeeps to sow confusion, including rumors of an element dedicated to assassinating Ike.

To maximize secrecy and minimize the intervention of the now-dreaded Allied fighter-bombers, Hitler waited until his own meteorological soothsayers came up with a prediction of at least five days of really bad weather. That period began on December 16.

Not surprisingly, the German attack began well, particularly against the ground held by the suspect VIII Corps, with the green 106th Division mostly

NORTHWESTERN FRANCE
Battle of the Bulge
December 16–25, 1944

Map by Gene Thorp

overrun, and the 28th and 99th Divisions hard hit and falling back. The Ardennes was starting to look like 1940, but only for a while.

To the north, the Allied flank refused to fold. The combat-grizzled Second Division dug in against the attack, while units of the First planted themselves firmly along Elsenborn Ridge, and V Corps brought up reinforcements, creating a barrier that forced the Germans to sideslip south. This not only narrowed their corridor of attack, but threw off their entire schedule sufficiently during the first twenty-four hours to preclude a truly catastrophic breakthrough.

But that doesn't mean the Allies, particularly the Americans, weren't caught flat-footed. Some have argued that Eisenhower was not surprised, that he already suspected there would be an attack through the Ardennes. But if so, why was he back in Verdun, far from the front, when it came?

Once it did, though, Ike moved promptly and decisively, taking direct command of the entire operation. Almost immediately, he addressed the worst case by ordering Lee back in Paris to send his logisticians and engineers forward to defend the river Meuse, which the Germans would have to cross to go farther. Next, and probably most important, he had the 101st Airborne trucked into tiny Bastogne, which, like the Gettysburg Ike knew, was a hub of numerous roads and therefore of momentous strategic import, particularly to the Germans if they expected to continue their rate of advance and ensure their supply line.

Finally, he sent the Eighty-Second Airborne north to Stavelot, an equally little place that suddenly assumed enormous importance. For Ike was enraged to learn that, in direct contravention of his order to keep all fuel dumps safely west of the Meuse, Bradley had stashed a million gallons in Stavelot—just what the fuel-starved German commanders needed, having requested 500 gallons for each tank and been told to roll on 150 instead.

It was the beginning of Ike's disenchantment with Bradley. Unlike our four he was no military superstar; his combat record was a mixed one, the brilliance of Cobra dimmed by his launching soldiers twelve miles off Normandy and his failure to close the gate on the fleeing Wehrmacht at Falaise. Now Bradley in Luxembourg was barely in touch with his forward commanders, and if he had been, he would have learned that for almost two days William Simpson, the head of the Ninth Army, had failed to do much of anything, while Courtney Hodges at the First Army had taken to his bed, suffering a collapse on the order of Fredendall's at Kasserine.

On the other hand, George Patton—who had only contempt for Bradley and called him "Omar the Tentmaker"—didn't just rise to the occasion, he slam-dunked it. Ike had called an emergency meeting of all the major players at Verdun on December 19 to formulate a strategic response. Patton not only had a plan, he had already worked on the details of implementation. "When it is considered . . . I left for Verdun at 0915 and that between 0800 and that hour we had had a Staff meeting, planned three possible lines of attack, and made a simple code in which I could telephone General [Hobart] Gay which two of the three lines we were to use, it is evident that war is not so difficult as people think," Patton recalled. He also looked to divine intervention, summoning his chief cleric, Pastor O'Neill: "Chaplain, I want you to publish a prayer for good weather. I'm tired of these soldiers having to fight mud and floods as well as Germans. See if we can't get God to work on our side."

As the man with a plan, George Patton dominated at Verdun. In addition to Ike and Georgie, Bradley, Devers, Tedder, and Beetle Smith were all in attendance, but not Montgomery, who sent his chief of staff, Freddie de Guingand, which struck all the Americans as a calculated insult aimed at them and the apparent vulnerability of the broad front. Eisenhower's intelligence chief was able to report that the German advance had slowed, but otherwise his opening briefing was suffused in gloom. Determined to flip the mood, Ike began briskly: "The present situation is to be regarded as one of opportunity for us and not of disaster," since the Germans were now out of Westwall and vulnerable. At which point Patton chimed in: "Hell, let's have the guts to let the bastards go all the way to Paris, then we'll really cut them off and chew them up" The room broke up and from this point Ike and Patton were practically finishing each other's sentences. "George, I want you to command this move—under Brad's supervision of course—making a strong counterattack with at least six divisions. . . . When can you attack?"

"On December 22, with three divisions"—a reply that set off a skeptical buzz in a room full of experience wondering how anybody could move that fast. But Eisenhower questioned only the force's size, allowing Patton to reassure him that the key to success was surprise, and that the time lost gathering the extra three divisions could come only at the cost of that vital element. Satisfied, Ike approved the plan, and Patton immediately put in the call to Gay that set it into motion. As they departed Ike mused, "Funny thing George, every time I get a new star, I get attacked." Remembering Kasserine Pass, Patton slipped him a final zinger: "And every time you get attacked, Ike, I pull you out." A fitting end to the best meeting in George Patton's life.

Bradley, prudently realizing it was an Ike-and-Georgie show, said almost nothing. He waited until the next day, and as Patton's direct superior reviewed his plan, which was to strike deep to the northeast to close off the entrance of the Germans' attack corridor. But Bradley realized this was too ambitious, that the key to everything was still Bastogne, and therefore sent him in that direction—his sole significant contribution to what history remembers as the Battle of the Bulge.

Meanwhile, Eisenhower was far from satisfied with Bradley's performance, and still felt in Luxembourg he was dangerously disengaged from his subordinates along the northern flank of the Wehrmacht's lunge forward. And to make matters worse, two SS divisions had just about surrounded Stavelot and its hydrocarbon bonanza.

His solution tells us a lot about the real Dwight D. Eisenhower; when pressed, he could be ruthless to his friends and reliant on those he despised—anything to get the job done. In this case, he turned over Bradley's First and Ninth Armies to Bernard Montgomery. He may have been his personal bête noire, but Monty's headquarters was already in the northern sector, and Eisenhower knew nobody was a better battlefield commander, the energizer needed to get Simpson and Hodges back in the fight. It worked like magic: Montgomery immediately began to "sort things out" and "tidy up the mess," scattering order over chaos almost like pixie dust. He certainly stabilized the north, but, being Montgomery, the blowback was his backstabbing campaign to also replace Ike. Then there had been the matter of telling Bradley.

"By God Ike, I cannot be responsible to the American people if you do this. I resign."

"Brad, I—not you—am responsible to the American people. Your resignation means absolutely nothing. . . . Well, Brad those are my orders." Hard words, ones Bradley would never forgive, especially considering that it was he who had sent George Patton to Bastogne.

By December 20, things there were beyond desperate, with the 101st surrounded by three German divisions already demanding surrender, though that simply prompted General Anthony McAuliffe's famed one-word reply: "Nuts!"

"Any man who is that eloquent deserves to be relieved," Patton roared when told. "We shall go right away." That, of course, was a relative term when you are coming to the rescue with anything as big as the Third Army. Nonetheless, within a week Patton had shepherded 250,000 men along with thousands of trucks, tanks, and artillery through sleet and snow over icy, treacherous roads, fifty to seventy miles, and just as he had promised at Verdun, on December 22 the three divisions of III Corps were ready to attack, now aimed directly at Bastogne.

Four days later elements of Patton's forces made contact with the 101st Airborne, and Bastogne was now a permanent obstruction to the German advance. And although the impact of Patton's prayer remains an open question, the weather did clear and US fighter-bombers were quickly tearing into armored columns. Meanwhile, the Germans, having never caught on to the fuel dump in Stavelot, moved on without a fill-up. The Germans impersonating Americans were tracked down and dealt with summarily. "One sentinel, reinforced, saw seventeen Germans in American uniforms. Fifteen were killed and two died suddenly," Patton chortled, citing a report from the Thirty-Fifth Division. By the time Manteuffel's panzers got within three miles of the Meuse they were out of gas, at which point Lightning Joe Collins and the Second Armored Division, sent by Montgomery to counterattack, virtually annihilated them.

All the generals involved asked Hitler's permission to withdraw, but as usual he refused. So the fighting dragged on through late January, and it became, in Rundstedt's words, "Stalingrad No. 2." The Germans lost between eighty thousand and a hundred thousand troops; they still had millions under arms, but critical was the destruction of most of their armored and motorized reserve. The Wehrmacht was now a force on foot against an army on wheels and treads. To make matters still worse, on January 12, 1945, on the other side of the Third Reich's nutcracker, the Russians launched their own massive winter offensive. There would be no spring respite for Hitler.

For the Americans at least, the Bulge was probably the signature combat event of World War II. By this time the US Army was a fully developed instrument of industrial warfare, long tested in battle and highly confident due to recent success. But being completely surprised by the toughest and most desperate of opponents had a revealing effect. While the weak performance of inexperienced units like the 106th Division was certainly not emblematic overall, the combat potential of American infantry remained a problem, particularly when stripped of its air cover.

George Marshall's policy of keeping divisions up to strength by backfilling with insufficiently trained replacements had infiltrated even experienced units with individuals who had never seen combat and were now expected to face the worst of it. It was a fate epitomized by the writer Paul Fussel, who as a twenty-year-old rifle platoon lieutenant looked to implement Marshall's Fort Benning–derived "fire and movement" doctrine: "But it had one signal defect, namely the difficulty, usually the impossibility, of knowing where your enemy's flank is. If you get up and go looking for it, you'll be killed." Nor was the inexperience limited to foot soldiers. One tank commander at the Bulge complained, "I spent long hours in the turret when I was literally showing men how to feed bullets to the gun. Could they shoot straight? They couldn't even hold the gun right!"

Such was the fate of the citizen soldier in the era of industrial war: one day, and however many before, a civilian; the next, given a full immersion experience in a death machine capable of squashing humans like ants, not to mention blowing them to shreds in an instant. It was telling that Eisenhower himself admitted that only convicts sentenced to at least fifteen years accepted the offer of a clean record to return to the front. By this time Marshall, Eisenhower, and Patton had all embraced a heretofore undemonstrated enthusiasm for integrating Black troops into combat units, though it would be hard to argue it was motivated by any sense of racial justice. They wanted bodies, and by now any color would do. It was cold, it was horrible, but somehow, like the rest, they endured, and in the process contributed to a crushing victory.

One key reason for that victory, as had been previously the case, was the unsung, or at least undersung, US artillery. Perhaps because the branch is traditionally divided into the many hauling and loading, and the few specialists aiming, it was better suited to absorbing inexperience. But, courtesy of the arsenal of democracy, it also possessed great quantities of guns and shells, all of two basic calibers, and at the Bulge, courtesy of Johns Hopkins University, a radio return-based proximity fuse. "The new shell with the funny fuse is devastating," Patton would write the War Department. Doubtless, he was referring to an incident he describes in his war memoirs: "On the night of December 25 and 26 we had used the new proximity fuse on a number of Germans near Echternach and actually killed seven-hundred of them." This was instantaneous death for the masses by means of shells cunningly designed to burst at exactly the right height every time, the latest progeny of technology and warfare's ill-starred mating, and one more example of the horrible direction in which military history gave every sign of heading.

Of course, this macabre drumbeat bothered George Patton barely at all. He gloried in the gore, it was by this time his natural element, like mud to a pig. One biographer notes that in letters he treated Beatrice to gruesome descriptions of dead Germans he photographed, once complaining about his lack of color film to capture their distinctive "claret color" resulting from being quick-frozen. He had become a connoisseur of corpses. The human detritus and the rest of the rubble of warfare, all of it on a gigantic scale, still struck Patton as somehow glorious. He was by now truly addicted to war; the rest of life would subsequently strike him as cold turkey.

But within his lethal sphere it's hard to imagine a better combat leader. One biographer, ignoring the bloodstains in the snow, called the Bulge Patton's "sublime moment," which in this context it undoubtedly was. "Shoot the works," he told the Third Army in his first order. "If those Hun bastards want war in the raw then that's the way we'll give it to them!" and they somehow believed him enough to carry roughly that message to the enemy. "Don't you love to see a man come rushing out of a doghouse?" George Marshall asked a colleague around Christmas. It was more than that. Already a military superstar, George Patton became a legend at the Bulge.

But if Patton saved the day, it was Eisenhower who held Team America together when things were most grim on the field. His mask of command never blinked, never showed any sign of the loss of nerve that signals defeat. Instead, he shifted seamlessly to Patton's plan B, and maintained the steadiest of hands at the wheel. And had there been no Patton and no plan

B, he still would have cruised to victory when the German armored columns ran out of gas and were savaged from the air. The key was radiating absolute confidence in the outcome—that was the ultimate role of commander in chief, and Ike never wavered.

Nor did home base, where the manager and team captain made a point of letting Ike direct the game on the field. George Marshall told his staff emphatically, "Don't bother him," and sent Ike a terse endorsement: "I shall merely say now that you have our complete confidence." Team America's manager was similarly tight-lipped. "Roosevelt didn't send a word to Eisenhower nor ask a question," Marshall remembered. "In great stress Roosevelt was a strong man." And under these circumstances silence was golden, the best message of support.

But it was also one the British refused to hear. Being tone-deaf, Montgomery took being given temporary command of all Allied troops north of the Bulge as reason for him to be given command of all Allied ground forces, and to implement his one-big-thrust strategy. Enraged, Ike sent Marshall a "him or me" cable and got a most decisive reply: "My feeling is this: under no circumstances make any concessions of any kind whatsoever.... You not only have our confidence but there would be a terrific resentment in this country following such action." Guingand, Monty's chief of staff, got word of Team America's top-down fury and rushed back to headquarters to persuade his boss to "eat humble pie" and send Ike a formal apology.

But that was not the end of it. With Churchill's okay, Montgomery held a rare press conference, supposedly intended to smooth ruffled eagle feathers but instead being yet another bash in the beak. Coming from Montgomery, his words oozed condescension, and they also contained downright inaccuracy. Calling it "a most interesting little battle," he went on to describe British units "fighting on both sides of American forces who have suffered a hard blow," when in fact the English were barely engaged, their casualties numbering less than 1,500, compared with 80,000 Americans. Almost two decades later, Ike was still boiling: "This incident caused me more distress and worry than did any similar one of the war." Patton was more succinct, calling Montgomery "a tired little fart."

But Team Great Britain refused to give up its assault on Ike. With Churchill's permission, his chief of staff, Brooke, tried to flank him in early February by floating the idea of bringing in not Montgomery but Harold Alexander as overall Allied ground forces commander. Ike responded like a man who knew his back was covered, telling Brooke he would be delighted to have his "great friend" Alexander as his deputy, but one without portfolio and likely to be dealing with quality-of-life issues in liberated areas: "There can be no question whatsoever of placing between me and Army Group Commanders any intermediate headquarters, either official or unofficial in character." Churchill, knowing a player with a heavy bat when he saw one, told Brooke to back off. Ike now had fifty-nine American divisions to wield; the English had eleven. So, whatever mistakes were made at the Bulge, Ike stood tall and Team Great Britain was reduced to the level of ankle biters.

This became even more apparent as the two squads met in Malta on the way to the big event at Yalta, the US-UK-USSR summit to be held in February. After visiting Ike in Marseille to assure him of his full support, George Marshall staged a showdown with Brooke, who loudly continued to question Ike and the broad front. Marshall responded in blowtorch mode, not failing in Brooke's words "to express his full dislike and antipathy for Monty," whom he saw as the ultimate source of the carping over what was inevitable: Ike and the broad front it would be. Later, once all German forces west of the Rhine had been destroyed and they both stood on its banks, Eisenhower claimed Brooke turned to him and said: "Thank God Ike, you stuck by your plan. You were completely right and I am sorry if my fear of dispersed effort added to your burden." Brooke being Brooke, he later denied it.

THE YALTA CONFERENCE, WHICH TOOK PLACE DURING THE FIRST TWO WEEKS of February 1945, remains the most remembered of the Allied intersquad scrimmages—but it's remembered more for what allegedly happened than for what actually did happen. It was perhaps suggestive that in keeping with the nautical theme, the code name was Argonaut, despite matters having turned intensely terrestrial. To compound the ambiguity, it was apparent to the participants and soon to be verified by events that Team America's manager was a dying man. Finally, Yalta took place in the Crimea, very much within the USSR, geography bound to stoke the paranoia of future participants in the Cold War trying to explain it in terms of betrayal.

Despite being perched on the brink of victory, George Marshall came to

Yalta a troubled man, torn between his obsession with maximum manpower and the implications of a science fiction weapon now virtually a reality.

At the end of December, General Groves had informed him, Stimson, and the president that an atomic bomb would be ready for dropping by August 1, and although his yield estimates were low and the device remained untested, he also assured them that it would almost certainly work. While at Malta, Marshall had breakfasted with the new secretary of state, Edward R. Stettinius, who had just been read into the Manhattan Project. Stettinius remembered: "As we watched the sun come up over the hills on this lonely island we discussed our hopes for the future and for the new age that was dawning"—a rare moment of philosophizing for George Marshall, but one that revealed his suspicion that something unprecedented was coming.

On a more practical note, Stettinius wanted to know what to say if Stalin brought up the bomb at Yalta, a question that prompted a dismissive "We'll deal with that one, if it comes up" from Marshall, apparently certain that America's greatest secret weapon was still secret. This was highly ironic, since Stalin was not only aware of the bomb, he had infiltrated both the American and British nuclear programs with spies, and he had an equivalent program of his own. It's hard to say how significant he thought an atomic weapon might be at this point, but he must have viewed Marshall's scarcely disguised enthusiasm to enlist the USSR's copious manpower for the war against Japan with a certain degree of bemusement.

Team USSR arrived at Yalta in a powerful position. They, not the Americans or the increasingly puny British, had broken the back of the Wehrmacht at Kursk, and now advanced units of Marshal Georgy Zhukov's armies were within a hundred miles of Berlin. Russian troops already occupied Bulgaria and Romania, over half of Hungary and Poland, along with most of the Baltic states. It was a Communist empire in the making, and however objectionable, those were the facts on the ground, and whatever the state of Roosevelt's health and Churchill's well-warranted suspicions, this was what they had to deal with. And deal they did.

Despite the howls of betrayal that cold warriors subsequently slung at him, contemporary historians agree George Marshall had limited skin in this game, in fact just a single objective. The bomb came with no guarantee. Hence, he reflexively fell back to what amounted to a lifetime preoccupation, the contingencies surrounding feeding industrial warfare with manpower in the millions. The larger part of the Japanese Army and some of its best units remained in Manchuria. He needed Russians to pin them down there while amphibious Americans invaded the Japanese home islands. And since the Soviet Union had never declared war on Japan, it was obvious concessions would have to be made. The other chiefs supported his view unanimously and without signs of arm-twisting; the necessity for mega-armies was everybody's frame of reference. So it was that on February 8 FDR and Stalin worked a deal for Soviet participation in the war against Japan at the cost of some minor territorial concessions—small change compared with what had to be conceded in Eastern Europe—and George Marshall had nothing to do with that. He got what he wanted at Yalta. History would reveal in a flash that he didn't need it and, for his trouble, eventually he'd be blamed for a lot more.

AS IF SYNCHRONIZED, YALTA'S END HERALDED RENEWED MOMENTUM ALONG Ike's broad front, studded with rolling rapiers. His plan called for a four-pronged attack beginning in the second week of February, along the northern tier, heading toward Germany's coast and in the general direction of Berlin. That would be followed two weeks later by a big push in the center between Cologne and Koblenz by the US First and Third Armies, along with two simultaneous attacks in the south by the US Seventh Army and the French First Army. Brooke's reaction to the plan was to complain to Ike that the northern mission being the harder one, he'd set up the Americans in the center for success. "I must tell you in my opinion there is no glory in battles worth the blood it costs," Ike shot back, expecting at that point the price to be high.

Since the Bulge, the Germans had been pushed back behind their last bastion, the Rhine, methodically obliterating all crossing bridges. The Allies had pontoon-based substitutes and could provide tactical air protection during their deployment, but Europe's second-longest river still presented a formidable obstacle, especially during the snowmelt of late winter. But then Fortuna intervened, not stopping the flooding, but definitely stanching the bleeding.

On March 7, the day Cologne fell, forward elements of Bradley's First

Army rolled into Remagen, south of Bonn, and found to their astonishment the Germans had botched blowing up the Ludendorff Bridge, which remained largely intact. Bradley immediately called Ike, who later recalled, "I fairly shouted into the telephone: 'How much have you got in that vicinity that you can throw across the river?' He said: 'I have more than four divisions but I called you to make sure that pushing them over would not interfere with your plans.' I replied, 'Well, Brad, we expected to have that many divisions tied up around Cologne and now those are free. Go ahead and shove at least five divisions instantly and anything else that is necessary to make certain of our hold.' The final defeat of Germany was suddenly in our minds, just around the corner." Despite furious but piecemeal counterattacks, by March 24 the Remagen bridgehead was ten miles long and three miles deep, packed with three entire corps primed for deep thrusts into the Reich.

The morning before, George Patton had called his boss: "'Brad, don't tell anyone, but I'm across' 'Well, I'll be damned. You mean across the Rhine?' 'Sure am, I sneaked a division over last night. But there are so few Krauts around there that they don't know it yet. So don't make any announcement.'" They soon found out, of course, and the Luftwaffe spent the day attacking the Third Army's growing strands of pontoon bridges, accomplishing not much more than providing targets for American antiaircraft gunners and their proximity fuses, shooting down thirty-three and enabling their commanding general to stroll safely across the next day. He stopped at the halfway point and unzipped his fly: "Time for a short halt to take a piss in the Rhine," the culminating moment of his urination tour of Europe's rivers. Flooding ensured that Monty was last across, much to the glee of the Americans, particularly Patton, who, as he had at Messina, reveled at beating the British anywhere. For His Majesty's Team, that did not dim the significance of crossing the Rhine, and being proverbial good sports, the event prompted Winston Churchill to label George Marshall "the true organizer of victory." But it was Ike's broad-front strategy, not Team Great Britain's one big push, that ensured the Allies would very soon be metastasizing through Germany top to bottom.

But Ike victorious was still beat. Fourteen months of twenty-hour days, punctuated by four packs of cigarettes daily, and energized mostly by as much coffee as he could drink, had plainly taken its toll. He looked like hell. Marshall noticed when he saw him on the way to Yalta, and was soon sending cables urging a break. In closer proximity, Beetle and Kay were worried he might be working himself to death, and were relieved when, with the finish line in sight, Ike conceded it might be time for some downtime. So, in the third week of March, the intrepid and overworked members of the Club Eisenhower headed for some fun in the sun at the Côte d'Azur villa of another convenient friend, Douglas Dillon, later one of Ike's political supporters. Bradley even joined the group the next day in Cannes.

But the trip proved a pale reenactment of the club's better days, more like a zombie beach party. There is a picture of Kay, the general, and Ethel Westermann in bathing suits arrayed on loungers and folding chairs—three very attractive people looking surprisingly modern, with Ike savoring a glass of red wine. But on closer examination, all share a completely disengaged look, almost a thousand-mile stare. Ike's was the worst. "For the first couple of days, all he did was sleep," Kay reported. "He would eat lunch on the terrace, with two or three glasses of wine, and shuffle back to bed again. . . . One afternoon, I suggested that we might play bridge that evening . . . 'I can't keep my mind on cards,' he said. 'All I want to do is sit here and not think.'" It may not have been much of a respite, but after five days of this Ike reflexively snapped his mask back in place and resumed being the steady hand of victory as if nothing else mattered.

George Patton also played true to form. By now thoroughly rehabilitated by a stunning combat record, a legend in his own time, but still the same Georgie, he was always ready to impulsively undermine himself.

In this case it came in the form of intelligence that his son-in-law John Waters, captured at Sidi Bou Zid in Tunisia, was now being held in a POW camp at Hammelburg, along with 5,000 other prisoners, including 1,500 Americans. Patton wanted to send XII Corps to rescue them, but its commander rightly maintained it was not worth the risk.

So he turned to plan B, a three-hundred-man motorized assault element—sixteen tanks, twenty-seven half-tracks, and three self-propelled assault guns—led by the large and very feisty Captain Abraham Baum, previously a combatant in New York's garment district. While Patton may have maintained, even convinced himself, that this was about all the American prisoners, nobody around him doubted it was really about just one, his son-in-law. This was why Task Force Baum was sent on a one-way mission to Hammelburg.

It went well initially; they chased away a few tanks and shot up a train full of antitank guns before reaching the camp, which promptly surrendered. But in the process, Waters was so seriously wounded he had to be left behind, his life later saved by a Serbian surgeon. Loading as many Americans as his vehicles would hold, Baum headed back, only to be intercepted by a much larger force of thoroughly aroused Germans. Surrounded and outnumbered, the little task force fought well before its inevitable capture, Baum having been wounded three times.

But, like Oscar Koch, Baum was a Jew, so he couldn't expect much credit from Patton, who would not even mention him in his own warped recounting of what happened. He had every reason to bend the facts, for had they reached the States several months earlier they would have finished his career, but Georgie caught a break as the episode was drowned in the accelerating rush of truly momentous events taking place during April 1945. Rather than disgrace, the month brought Patton his fourth star, thereby completing not only his personal constellation, but the martial galaxy of our four military superstars—twenty-four out of a possible twenty-five stars—but like the others he had reached the stage where the baubles of war no longer excited him. For Patton, at this point, only the real thing got him off. He was addicted to war, and as this one wound down, he also sensed the end of the road. "I will be out of a job," he informed Beatrice; peace "is going to be hell on me. I will probably be a great nuisance." Prophetic words.

Meanwhile, Ike's rolling rapiers, having penetrated the last barrier, moved quickly to the slice-and-dice stage. Most lethal was surrounding the Ruhr, Germany's still-throbbing industrial heart, with powerful elements of the US First Army and Montgomery's combined force. They trapped there the remains of Army Group B, the Wehrmacht's only significant fighting force left in the West, caught by Hitler's order not to retreat. After failing to break out to the north and south, Field Marshal Walter Model, rather than preside over the destruction of whatever industrial future Germany might have, disbanded his army, burned his papers, and shot himself on April 14. In addition to more than 300,000 prisoners, the Allies had the Ruhr, severing the base necessary to feed industrial warfare. Effectively, it was over.

One destination Dwight Eisenhower had no interest in reaching was Berlin, the final target of Team Great Britain's single big push, undertaken in the grand tradition of "take the capital and win the war." But to Ike Berlin had nothing to offer but bigwig Nazis. He knew the Germans would fight savagely to hold it, and that the Russians were vengefully intent on taking the city themselves. He had counted on Montgomery moving slowly in his deliberate way, but was surprised and enraged to learn he was heading rapidly toward the Elbe and Berlin, fait accompli evidently in mind. To head him off Ike instantly reassumed the role of prince and went directly to Stalin, working a deal to "make every effort to perfect the liaison between our advancing forces," thereby implicitly ceding Berlin to the Russians. He then told Monty he was coordinating Allied movements with the Soviet leader, and took the US Ninth Army away from him to make sure he was thwarted.

Team Great Britain was livid, and Churchill sent George Marshall a stinging protest, but Marshall, backed unanimously by the other chiefs, would have none of it: "The battle for Germany is now at a point when it is up to the Field Commander to judge the measures which should be taken." Instantly, Churchill raised the matter with FDR, but since his rejoinder arrived at the White House on April Fools' Day and the president was already on his terminal visit to Warm Springs, it was apparently Marshall who supplied the answer under his ailing manager's name, including the following: "I regret even more that at the moment of great victory by our combined forces we should become involved in such unfortunate reactions." That would end it. The Russians took Berlin, but at the cost of 350,000 casualties, and each of the Allies still received a slice of the fated city, though access was destined to become virtually the epicenter of the Cold War. Yet at the time Ike had the better part of the argument, one that spared American lives, and he made it stick.

Having penetrated what was fast becoming the cadaver of Nazi Germany, the Allied teams very quickly discovered just how cancerous its organs had grown—not simply evidence of the crimes that had already been committed, but also chilling and completely unexpected confirmation of what they might have been capable of inflicting.

Back in 1936 Dr. Gerhard Schrader, searching for new insecticides, discovered tabun and sarin, fantastically deadly chemical agents that blocked acetylcholinesterase, the neurotransmitter that allows muscles to relax. A factory was built at Dyhernfurth, near the Polish border, and twelve thousand tons of nerve agent were stockpiled, all of which the Soviets captured. American soldiers' first encounter with the agent came when they shelled a barge on the Danube in Bavaria and the German soldiers aboard began frantically waving white flags, knowing that their craft was packed with tabun-filled bombs and that one hit would produce a cloud capable of killing everybody in its path for many miles. It's worth considering what such munitions might have done at the Bulge or Normandy, or in V-2s aimed at London; but Hitler, a gas victim in World War I, never pressed for their use, a bit of ironic battlefield chivalry considering he had no trouble gassing millions of Jews.

Nor was that hard to miss, prefaced in our case with a monumental Nazi treasure trove. Advancing northeast of Frankfurt toward Gotha, elements of the Third Army had stumbled on a mind-bending cache stored deep in a salt mine at Merkers—basically the national gold reserves evacuated from Berlin after B-17s demolished the Reichsbank, along with the loot gathered meticulously by concentration camp administrators, and decorated with heaps of assorted art masterpieces. Hidden treasure being a powerful attractor, the discovery not only drew Patton's instant and avid attention, but also Bradley's and Eisenhower's. On April 12, they joined him to inspect and be dazzled by the accumulated wealth. It was at this point that Walton Walker, one of Third Army's key paladins, suggested they might want to look at things less pleasing to the eye by taking a short drive to Ohrdruf Nord, "the first horror camp any of us had ever seen," Patton remembered. The excruciating condition of the survivors, the multiple gallows and instruments of torture, made it obvious that the place amounted to an abattoir for humans, a shocking revelation to even these war-hardened Americans. "It later developed that [our guide] was not a prisoner at all, but one of the executioners. General Eisenhower must have suspected because he asked the man very pointedly how he could be so fat. He was found dead next morning, killed by some of the inmates," Patton wrote later in his war memoir.

The day's surreal events were not easily forgotten, and Patton, Bradley, and Ike stayed up until nearly midnight talking. Just before he went to bed, Patton realized he had not wound his watch and it had stopped. To get the time he turned on the BBC, where he learned that President Roosevelt had died at Warm Springs.

Back in Washington to organize her husband's last journey, Eleanor Roosevelt turned to the best planner she knew, George Marshall, asking him to take responsibility for bringing the body back from Warm Springs for the funeral service at the White House, and for the final trip to Hyde Park for burial. No surprise, he handled it all without a hitch, but also without any outward show of emotion—nor was there likely any inward. Marshall admired Roosevelt's fortitude, but his refusal to laugh at his jokes or visit Hyde Park was pretty clearly not just a matter of policy, probably more so after the president chose Ike to lead his armies in Europe. Likewise FDR, who was brilliant at using people, viewed Marshall essentially as a reliable team captain and strategist, but with no particular affection—an excellent functionary, nothing more.

But almost the opposite was true of the new manager. A waspish and direct ball of nervous energy, Harry Truman was a far different creature from his patrician predecessor. He sprung from the Missouri heartland and Kansas City machine politics, so much so that when he first arrived in Washington wags dubbed him the "Senator from Pendergast," referring to Tom Pendergast, the local political puppeteer and a notably corrupt one. But not Harry. Through all the dirty politics he was never on the take, largely because his self-image was formed elsewhere, in the farming community of Independence, then during the First World War.

He would be the only US president to see combat in World War I, as a captain in the artillery, a rank he achieved largely because his superiors instinctively believed he could be trusted with the lives and welfare of his men, a rough bunch drawn from Truman's home turf. He taught them how to fight and did not lose a single man in combat, a record that would form the foundation of his political career when everybody came home, especially since one of those soldiers was a Pendergast. The Army constituted Harry Truman's first, and for a long time only, career success; most important, it taught him he was a natural leader, a lesson he never forgot even in his lowest days.

It's also worth noting that, in a world of masterful dissimulation, Truman very openly held grudges and categorized politics largely in terms of friends and enemies. One incident that springs to mind as relevant was the fate of the Bonus Expeditionary Force. In 1938 he had excoriated Herbert Hoover for having "ordered the Regular Army out to shoot down the poor broken veterans of the World War," and it stands to reason he hadn't forgotten those regular army officers who did the ordering. In the case of MacArthur and Patton, the impression was definitely lasting, as witnessed later by his diary: "Don't see how a country can produce men such as Robert E. Lee, John J. Pershing, Eisenhower and Bradley and at the same time produce Custers, Pattons and MacArthurs." Ike caught a pass here, but their interaction was and remained prickly. Just one of our four was nowhere near Anacostia Flats that day, George Marshall. And there was nobody in uniform more revered by Harry S. Truman, as he was thrust so unexpectedly into the presidency.

The relationship had been nurtured in Truman's senatorial bailiwick, the Subcommittee on War Mobilization, where he had earned widespread respect for ferreting out waste, fraud, and assorted chicanery. In an utterly sectarian environment, Marshall, the master of the nonpartisan schtick in his testimony, struck Truman as something like the fabled "only honest man in Washington," and also a supercompetent one. Now the occupant of a position many, including himself, were not confident he was capable of filling, the new president clung to George Marshall like a drowning man to a life preserver.

Especially after the mysterious message he had received his first day on the job, when Henry Stimson took him aside "and told me that he wanted me to know about an immense project that was underway—a project looking to the development of a new explosive of almost unbelievable destructive power." But he would say no more. The next day his friend and mentor from the Senate, and most recently the director of war mobilization, Jimmy Byrnes, stopped by echoing the same message, but "even he told me few details, though with great solemnity he said that we were perfecting an explosive great enough to destroy the world."

This charade could not go on, but Marshall and Stimson were caught in a high-stakes balancing act. They understood that pressure was growing to get tough with the Soviets, particularly after their obvious suppression of democracy in Poland, yet their version of the Pacific endgame still needed those Russian bodies in Manchuria. But if the bomb was presented as an easy alternative to cooperating with the Soviets and our own inevitably blood-soaked invasion of Japan, the final piece of the grand strategy might be sidetracked by a still-untested widget, however grand its scale and implications.

Being master players, they sent the new manager off in exactly the preferred direction. However tough his rhetoric with Soviet foreign minister Molotov—a tongue-lashing "in words of one syllable," he reported—Truman followed Marshall's advice to stick with Stalin and FDR's deal at Yalta: "I was anxious to get the Russians into the war against Japan as soon as possible, thus saving the lives of countless Americans"—echoing the Team America–derived approach.

The details of the bomb were presented in a similarly modulated fashion. Marshall didn't even attend the meeting on April 25, just Stimson and Groves, whom they smuggled into the White House through a secret entrance. In addition to telling the president several bombs would be available in August 1945, Groves handed him a twenty-five-page synopsis of the entire program, nuclear physics included. "I don't like to read papers," was his immediate reply. After being persuaded to at least peruse it, Truman remembered, "Stimson seemed at least as much concerned with the role of the atomic bomb in the shaping of history as in its capacity to shorten the war"—exactly on message, in other words. Outside of setting up an interim committee for policy recommendations, nothing else was decided in terms of when, where, or even whether the bomb might be used.

One thing was certain: it would not be dropped on Nazi Germany, as that place was finished. In the first three weeks of April more than a million prisoners were taken by the Allies. On April 30, Adolf Hitler committed suicide, terminating the career of arguably the evilest ruler in human history—at least if you go by the numbers—one born in and consumed by industrial warfare.

Eisenhower refused to negotiate a surrender with the Germans; he sent Beetle Smith to do it. When Alfred Jodl and others sought to hedge, somehow work a deal surrendering to the Americans while continuing to fight the

Soviets, Ike told Smith to inform them they had forty-eight hours to accept his terms or he would seal his front and leave fleeing Germans to Russian retribution.

That brought Jodl to Ike's headquarters in Reims, and Smith escorted him into his office, where at 2:41 a.m. the German general signed the instrument of surrender. Ike then told him, "You will, officially and personally, be held responsible if the terms of this surrender are violated. . . . That is all." Some heard that as "Now get out!" At any rate, the terms were obeyed, but Jodl was still held responsible, and hanged for war crimes by the Nuremburg Tribunal in October 1946.

"Afterwards," Kay observed, "General Ike's face stretched into the broadest grin of his career. As the photographers milled around, he said, 'Come on, let's all have a picture!' Everyone gathered near the Boss as he held two of the signature pens in a V-sign." The historic photo would come to exist in two versions: one with Kay smiling over Ike's shoulder; the other, the official one, with her carefully airbrushed out—soon to be her destiny in Ike's life. While he claimed everybody went directly to sleep, Kay remembered ten or twelve members of the exhausted Club Eisenhower trudging back to Ike's chateau and drinking champagne until dawn, though without much enthusiasm. "It was a somber occasion," she concluded. Either way, it was over, at least in Europe.

ON THE OTHER SIDE OF THE GLOBE IT WAS STILL GOING ON. DOUGLAS MACAR-thur, in what he might have grandiloquently termed the "splendid isolation of the Southwest Pacific," had continued to methodically play his unique game of strategic hopscotch, coached by a very able kibitzer named ULTRA, but always with the same objective—not the homeland of the Japanese enemy, but the Philippines, from which he had been unceremoniously ejected in 1941 and ever since had provided the fire in his belly to get back. But in late July 1944, just as Cobra struck to break the stalemate at Normandy, his grand plan was suddenly at risk, and it had nothing to do with the enemy.

MacArthur had just received a cryptic message from George Marshall, whom he had treated so rudely just six months earlier in New Guinea. It simply instructed him to attend a conference at Pearl Harbor on July 26; neither the subject nor the attendees were stipulated. Long experience prompted Mac to intuit that one of them was probably the president, and also that this was likely a setup, though he could only guess at its nature.

Mac and FDR greeted each other in Honolulu like old friends, which in numerical terms they were, but as they rode, squeezed in the back of an open car with Admiral Chester Nimitz, each sizing the other up, appearances were initially paramount. The team manager was bemused by this star player's getup—Philippine field marshal's cap, a leather flight jacket George Kenney, his air chief, had given him, and a corncob pipe grown so large it looked like a portable stove—while Mac was startled by Roosevelt's pallor. "Doc," he told his physician later, "the mark of death is on him! In six months he'll be in his grave." He missed by a few, but at the moment the president had a lot more on his mind than his own mortality.

There was an election in November, and Roosevelt was intent on running for a fourth term no matter how bad his health. MacArthur had been making political noises, and though maladroit, he was vastly popular at home, and therefore worth further scrutiny, along with providing an excellent photo op. But his political agenda also dovetailed with a key strategic decision that had to be made.

This became clear after dinner, when the president led Mac, Nimitz, and Admiral Leahy into a conference room with one entire wall covered by a map of the Pacific. In his wheelchair with a bamboo pointer, FDR suddenly whirled and aimed it at Mac. "Well, Douglas, where do we go from here?" By now he must have known this was coming, another showdown with this antagonist of long standing. "Mindanao, Mr. President, then Leyte—and then Luzon."

There was, of course, an alternative. Rather than the Navy's and MacArthur's pincers converging on the Philippines, Nimitz detailed the option of one big strategic thrust all the way to Formosa. If there was ever a time for Douglas to do some fast thinking, this was it. The Navy was here in force, not only Nimitz but Leahy and a host of advisers, and besides, it was the team manager's favorite service. George Marshall's absence was also telling. Back in Washington, although Admiral King may have been the most vocal advocate of the Formosa invasion plan, Hap Arnold and the Air Force also thought it would make an excellent platform for the new B-29s to bomb Japan, an argument that also appealed to Marshall—at least enough to skip the trip and leave MacArthur to his own devices.

But he was made for such moments, pulling the strategic rug out from under his opponents by taking the moral high ground. This was Team America's responsibility. The Filipinos felt betrayed when the United States failed to protect them from the Japanese in 1941, and would not forgive a second betrayal if we failed to rescue them now—an argument calculated to penetrate the president's anticolonialist chink and subtly remind him of all those American soldiers also abandoned on Bataan and Corregidor.

Thrown back on his wheels, FDR attempted to parry: "Douglas, to take Luzon would demand heavier losses than we can stand." But that one bounced off MacArthur's strongest suit: "Mr. President, my losses would not be heavy, any more than they have been in the past. The days of the frontal attack should be over. Modern infantry weapons are too deadly. . . . Good commanders do not turn in heavy losses." Few generals in World War II could make such a claim, but his manager knew MacArthur was one, that his ULTRA-based "hit 'em where they ain't" strategic approach was inherently parsimonious of his men's lives. By the end even Nimitz showed signs of sympathy for MacArthur's arguments, and neither, unlike practically everybody else, was asking for reinforcements, just permission.

So Doug won: the Philippines it would be. The next day, before he flew back to his headquarters, the president asked MacArthur to join him for a tour of Pearl Harbor's facilities in the back of his convertible. "We talked of

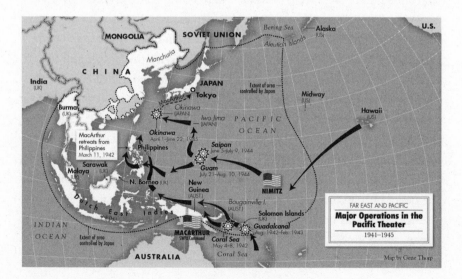

everything but the war—of our old carefree days when life was simpler and gentler, of many things that had disappeared in the mists of time," MacArthur remembered. It was their last meeting, these two frenemies, still telling each other lies and apparently enjoying every minute.

But in this context, it's important to add that one of the reasons Mac's arguments prevailed was that they were heartfelt. Among our four and a great many other Americans, MacArthur was unique in not harboring significant racial prejudices. Filipinos were not his "little brown brothers" but a promising nation of human beings deserving US protection. Granted, his utterances were never entirely free of condescending conventions, but it does seem his mind and actions truly were. This would be key to his remarkable success among Asians in the unlikeliest of roles, Team America's mikado. Yet the ego behind the mask that made it all so believable would, as it does in all good tragedies, eventually drag him down.

But not yet, or even soon. The momentum of events moved irresistibly behind Douglas MacArthur's version of the future, and it would continue to sweep him forward in just the right direction. Back in Hollandia (now Jayapura), his headquarters on the northern coast of New Guinea, the harbor steadily filled with warships, transports, and landing craft until they numbered almost five hundred vessels—the Philippine armada, a.k.a. the US Seventh Fleet, put together by the sure hand of Admiral Thomas Kinkaid. While it did incorporate some smaller escort carriers, its central problem, as Kenney kept pointing out, was air cover for a journey of over one thousand miles of open sea. And while Admiral William Halsey with the big carriers of the Third Fleet would be available for escort duty, Nimitz had also told him his primary objective was to destroy the Japanese fleet, not protect the beachhead once there.

But initially at least, Halsey, whose apt moniker was "Bull," amounted to MacArthur's chief instigator. On September 13, he signaled that his pilots had made multiple passes over the Philippines without encountering any air opposition, and on this basis he emphatically recommended that MacArthur speed up his agenda by forgetting about Mindanao and going directly to Leyte, a bullish proposition that not only delighted the recipient, but sounded right to Marshall and the chiefs. To help his own cause, on September 15 MacArthur began an operation that would result in the quick seizure of Morotai isle almost without casualties, once again leveraging ULTRA to leapfrog strongly held Japanese positions, and thereby acquiring a reliable air base over halfway to the Philippines. The operation was strategically an emphatic counterpoint to the Navy's Central Pacific island-hopping campaign even shorn of Formosa—a series of bloodstained lurches to Guam, Saipan, then the Marianas and Palau, to be followed by Iwo Jima and Okinawa: a veritable crescendo of casualties.

And to further bolster MacArthur's side of the strategic equation, beginning on October 10 Halsey staged another of his bullish endeavors, a ten-day destruction cruise off the Philippines, Formosa, Okinawa, and the Ryukyus, one that all but severed the wings of Japan's remaining naval air force, destroying over five hundred planes and obliterating two major bases. Mac was already aboard the heavy cruiser *Nashville*, leading his armada full of 175,000 veteran fighters toward Leyte, when Halsey's news arrived, likely clearing the last cloud from his strategic horizon. Yet it was a limited one. What lay beyond his view amounted to a darker, scarier, fundamentally more challenging picture composed by Japanese hands, not American ones.

Doubtless also on his mind as the hours passed aboard ship was betrayal. MacArthur, though relentlessly faithful to Jean, generally took a casual attitude toward his henchmen's sexual liaisons—Kenney, Sutherland, his chief of staff, and Willoughby, his head of intelligence, all had mistresses in Australia. But Sutherland's was different. She was the wife of a local steel magnate caught in Singapore and now languishing in a Japanese prison; also like Sutherland, she was obnoxious and inclined to overreach. MacArthur tolerated her as long as they were in Brisbane, but as they moved forward to Hollandia he had warned him, along with the others, that there would be no women included, not even Jean. As far as Mac was concerned that was the end of it, until one night rocking on his veranda, he turned to his physician and asked, "Doc, whatever happened to that woman?"

"She's ten miles down the coast. Larry just talked to her"—a response that prompted only two words: "Get Sutherland."

He actually had him arrested—a three-star general—the woman was sent back pronto, and from this point Mac treated Sutherland with calculated disdain. But he didn't get rid of him. While hardly a great chief of staff—key commanders such as Kenney, Krueger, and Kinkaid customarily dealt directly with MacArthur—Sutherland remained a functional functionary who got things done, so he stayed aboard. It was said that nothing openly enraged Mac more than this incident, but he would not allow personal feelings to in any way betray the mission. That said, it's worth comparing his inner staff to an analogous Club Eisenhower, where the mistresses had real responsibilities and everybody believed he or she was part of the team, not worshippers at the altar of the man in the mask. Sutherland's behavior was an act of adolescent rebellion, the equivalent of raising his middle finger to the Almighty. On the other hand, although Beetle didn't much like his boss, he always had his back. In the end that was the difference that mattered between Mac and Ike.

OCTOBER 20, 1944. THE ISSUE ON LEYTE AND IN THE GULF OF THE SAME NAME was far from settled, but MacArthur knew the value of a grand entrance, and was not above rushing it should he smell anything like the approach of victory. So, at one p.m., Mac, along with Kenney, Kinkaid, and Sutherland, all clambered aboard a landing craft, stopping briefly to pick up Philippine president Osmeña, whom he barely knew but needed for political purposes, then headed for Red Beach, where he was told the fighting was the hottest and the Twenty-Fourth Division remained pinned down. On the way, they passed four other landing craft burning and capsized. Kenney heard the zing of sniper bullets as they neared the beachhead, which the Twenty-Fourth had managed to push no more than three hundred yards inland.

Suddenly the engines reversed, the craft stopped, and the ramp dropped, still a ways from the shore in knee-deep water. Photographers and movie cameramen in place, Doug and his acolytes waded ashore and onto newsreels viewed by millions at home and ultimately finding a place in cinematic history—a stern visage in Ray-Bans and wet trousers symbolizing promises kept like none other. "People of the Philippines," he boomed into the mike of a radio transmitter, "I have returned! By the grace of Almighty God, our forces stand again on Philippine soil. . . . Rally to me! Let the indomitable spirit of Bataan and Corregidor lead on. . . . Rise and strike! For your homes and hearths strike!" Later that afternoon on Red Beach he came upon several Japanese corpses and kicked them over to examine their unit insignia: "The Sixteenth Division. They're the ones that did the dirty work on

Bataan." Had this been a war film of the forties, it could have ended right at this moment of retribution. But this was reality and the Sixteenth Division was far from alone.

The ULTRA team had looked long and hard at the Philippines, and were very much aware that the Japanese were building up their forces—Willoughby estimated to around a quarter of a million, but with only twenty thousand on Leyte—and also that perhaps Japan's best general, Tomoyuki Yamashita, "the Tiger of Malaya," had been placed in charge. But they missed a lot—the whole point, in fact.

The Japanese were planning not a holding action, but a showdown. Yamashita instinctively wanted to make his stand back on Luzon, but had been overruled by Tokyo. The larger plan demanded that he vigorously defend Leyte. And to help him do it, the high command had sent the crack First Division from Manchuria, a transfer ULTRA never heard about, leaving Mac and specifically Walter Krueger facing a considerably tougher foe than they expected. And that was the least of it, just one cog in the Japanese mechanism of destruction now being set in motion, without the slightest reverberation from ULTRA.

Since annihilating the Russian fleet in Tsushima Strait nearly forty years prior, the Japanese navy remained wedded to the notion of a crushing naval victory achieved primarily with surface ships wielding big guns; this, not the aerial coup against Pearl Harbor, remained the archetype. Now, though largely shorn of their air component—the Sea Eagles long gone, and more recently most of the Sea Sparrows—they grimly set about to spring their gun-toting trap on the Americans, aiming to break the back of US sea power, or at least the Seventh Fleet, in Leyte Gulf, thereby leaving Walter Krueger's Sixth Army stranded and inevitably snared on the island.

Events would reveal a lot of wishful thinking in this banzai pipe dream, but it did lead to the largest naval battle of the industrial-warfare era and one of the most decisive, but not in the way the Japanese intended.

They sailed up from Singapore with a battleship-heavy fleet, including *Yamato* and *Musashi*, the two largest in the world, intent on trapping Kinkaid's Seventh Fleet by first dividing their force, then slicing respectively through the San Bernardino and Surigao Straits, while using their remaining four big carriers as bait to draw Halsey's own carrier-heavy Third Fleet away from the action. Though largely devoid of planes and pilots, the Japanese flattops did serve as a red cape to Bull Halsey, who chased it hundreds of miles out to sea.

Other than that, just about nothing else worked. The larger force was quickly spotted on its way to San Bernardino Strait, and naval aircraft promptly crippled and sank the *Musashi*, leaving Admiral Takeo Kurita feeling vulnerable, vacillating, and unable to seize the moment when Halsey's departure left an undermanned element of the Seventh Fleet guarding the strait; Kurita chose to retreat instead. Meanwhile, on the Surigao side, the other flotilla had its "T" crossed as it entered the strait by five American battleships, all survivors of Pearl Harbor and now equipped with radar-guided guns, resulting in its near obliteration, the admiral drowned, and only one destroyer surviving. All told, the Imperial Japanese Navy lost four aircraft carriers, three battleships, six heavy cruisers, and eight destroyers at Leyte Gulf, effectively terminating it as a significant instrument of naval warfare.

But the battle hardly removed the danger to American sailors and warships; quite the opposite, it revealed a new and highly effective combination of technology and martial fanaticism: the kamikaze. Explosives-laden fighter aircraft flown with suicidal intent first crashed into the decks of US ships at Leyte, blowing up one escort carrier and heavily damaging two more. This was just the preface of what would become a frightening toll paid in American lives before war's end, and one more reason to do just about anything to the Japanese in return.

Back at Leyte the island, the merchant and other vessels anchored offshore and supporting US forces on the ground were no longer in danger of destruction; nor was MacArthur's grand scheme of liberation. But the real estate presently in dispute was already proving a disappointment. Intended to provide a secure air base for the subsequent invasion of Luzon, the ground quickly revealed itself as entirely too soggy to support any aircraft heavier than Kenney's P-38s. And the rains of November only made the situation worse. Meanwhile, by transferring air assets from Formosa and Japan to Luzon, the enemy was able to effectively, though at great cost, contest the skies above with land-based fighters. Mac needed solid ground for his own air operations, and he needed it quick.

The island of Mindoro, a hundred miles south of Manila, was his target, and he wanted the invasion set for December 5, a schedule that almost caused Thomas Kinkaid to resign on the grounds that, without Halsey's big fleet carriers to provide thorough cover, a single kamikaze hit could sink or disable one of his own escort carriers. The resulting argument had reached a fever pitch when a message from Nimitz established the basis for compromise: a postponement of ten days to give Halsey time to return. Mac was too amped to immediately accept, but signaled his intent by wrapping his arm around Kinkaid's shoulder: "Tommy I love you still. Let's go to dinner!" Halsey arrived early, and on December 13 GIs landed on Mindoro to find less than a thousand defenders. It took three days to clear them out and lay claim to four existing airstrips, all on dry, solid ground. Mac had his air base and the defenders of Leyte were basically surrounded. "What you have done on Leyte and are doing on Mindoro are masterpieces," George Marshall wired Mac, as if he didn't already know.

By this time Yamashita also understood the operation was a losing proposition, but his theater commander, Field Marshal Hisaichi Terauchi, ordered him instead to "muster all strength to totally destroy the enemy on Leyte." So he continued to reinforce, the futility of which was highlighted when Halsey's carrier aircraft caught one convoy off Ormoc, the main Japanese base on the island, and in the process drowned nearly a whole division, ten thousand men. Had they made it to shore they would have ended up just as dead; the situation was hopeless. It ended the day after Christmas with Yamashita and a few thousand survivors retreating to Luzon, leaving over fifty thousand dead compatriots, some of them among Japan's best troops. The grand plan, the showdown named Leyte, had come to nothing. All Nippon's warriors had left was suicidal enthusiasm—that in the face of Douglas MacArthur backed by the arsenal of democracy at full throttle. "The dark shadow of defeat was edging ever faster across the face of the rising sun of Japan," he remembered. "The hour of total eclipse was not far off."

IT WAS NO SECRET WHERE MAC WOULD LAND ON LUZON. ULTRA TOLD HIM that the Japanese were expecting him at Lingayen Gulf, exactly where General Masaharu Homma had disembarked almost precisely four years earlier, as it was the portal to the central plains leading south to Manila. Beyond that, the electronic oracle remained mute on how Yamashita planned to react.

The voyage of the invasion fleet did not provide grounds for optimism. It was kamikaze bait almost from the beginning, on January 2, 1945, and five days later entering Lingayen Gulf it took sixteen separate hits. But the fleet was huge, nearly a thousand warships and transports, and soon over 2,500 landing craft were headed to the beach, irresistibly so, since they found no Japanese.

"Running to Manila," GIs were told of the Japanese by Filipino locals. Actually Yamashita, perhaps remembering what happened to MacArthur when he took a stand at Lingayen and then Bataan, had withdrawn the bulk of his quarter-million-man army into the mountains occupying the top half of the island, where they could best prolong their defense.

But Walter Krueger didn't know that, nor did ULTRA; MacArthur, however, operating on instinct and adrenaline, sensed it. Krueger, a good general but a by-the-book commander, was haunted by the vision of thousands of Japanese piling down from the hills on his left flank to cut his six divisions' umbilical cord to Lingayen, and therefore urged caution. Mac's response was emphatic: "Get to Manila! Go around the Japs, bounce off the Japs, save your men but get to Manila! Free the Internees at Santo Tomas! Take Malacanan and the legislative buildings!" Despite all reservations, that's pretty much how it went down; speaking literally, at least. Back in Washington, in his official report to Henry Stimson, George Marshall more than conceded Mac's strategic chops: "Yamashita's inability to cope with MacArthur's swift moves" plus "his desired reaction to the [American] deception measures" fused "to place the Japanese in an impossible situation . . . forced into a piecemeal commitment of his troops." One by one Mac picked off his objectives: Clark Air Base was invested and taken, then the key port of Olongapo on the west coast, followed by the Bataan Peninsula in seven days, and finally Corregidor, where the Japanese commander blew up himself and the remains of his force in the Malinta Tunnel, site of Mac's own darkest days.

But what had been a strategist's dream turned into a nightmare in Manila. Yamashita believed the city was indefensible, but the senior officer in charge, Rear Admiral Sanji Iwabuchi, had every intention of trying with the sixteen thousand soldiers and sailors under his command. Yamashita not only acquiesced, but sent him three of his own battalions, making him complicit in what happened: mass murder in the guise of seppuku—a suicidal last stand.

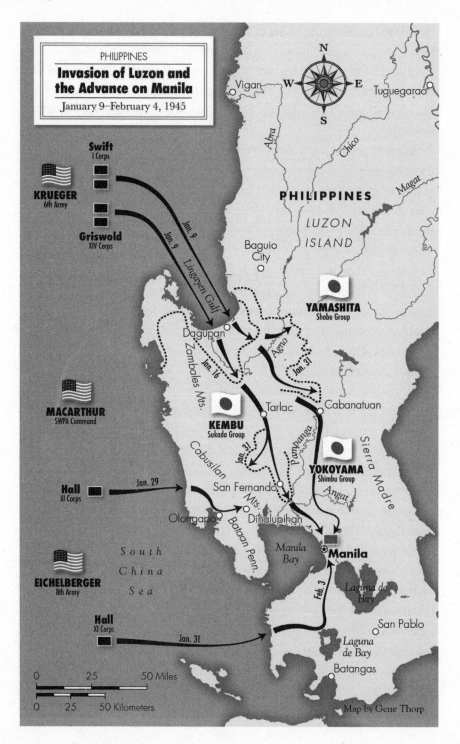

PHILIPPINES

Invasion of Luzon and the Advance on Manila

January 9–February 4, 1945

Vigar

Tuguegarao

PHILIPPINES

LUZON ISLAND

Baguio City

Swift
I Corps

KRUEGER
6th Army

Griswold
XIV Corps

Jan. 9

Jan. 9

Lingayen Gulf

Dagupan

Jan. 16

Zambales Mts.

Agno

Jan. 31

YAMASHITA
Shobu Group

MACARTHUR
SWPA Command

Tarlac

Cabanatuan

KEMBU
Sukada Group

Jan. 31

Cabusilan

Pampanga

YOKOYAMA
Shimbu Group

Hall
XI Corps

Jan. 29

San Fernando

Angat

Sierra Madre

Olongapo

Dinalupihan

Mts.

Bataan Penn.

Manila Bay

Manila

South China Sea

EICHELBERGER
8th Army

Laguna de Bay

Hall
XI Corps

Jan. 31

Feb. 3

San Pablo

Laguna de Bay

Batangas

| 0 | 25 | 50 Miles |

| 0 | 25 | 50 Kilometers |

Map by Gene Thorp

After the liberation of the three thousand American civilians at the University of Santo Tomas on February 4, fighting would rage through Manila for nearly a full month. It was urban warfare at its worst—house-to-house homicide with tanks breaching walls, then a bullet in the head to ensure every dead Japanese really was. When it was over, 80 percent of the city was rubble and 100,000 Filipinos lay dead, many simply murdered by the defenders, making Manila, besides Warsaw, the worst-hit Allied capital of the entire war. The general himself watched as his penthouse atop the Manila Hotel, apparently preserved like a museum up to that moment, went up in flames, incinerating several generations of MacArthur books and memorabilia—Pinky's legacy all gone.

Putting the best face on disaster, MacArthur staged an official ceremony with Osmeña on February 27, 1945, in the Malacañan Palace, virtually the only public building left standing. He declared: "My country has kept its faith. On behalf of our government I now solemnly declare, Mr. President, the full powers and responsibilities under the constitution restored to the Commonwealth. . . . Your country is again at liberty to pursue its destiny to an honored position in the family of free nations. Your capital city, cruelly punished though it be, has regained its rightful place—citadel of democracy in the East." It was a proclamation heard by Filipinos as an ironclad promise of sovereignty from the man they most trusted, and there was no going back on that.

Like Ike, Mac had no scruples about playing prince when the situation demanded. And unlike his tone-deaf approach to American politics, MacArthur's overwrought personality played just right to Asian ears tuned to authority figures—but this one was a prophet of democracy with the power to implement it, a potent combination here in the Philippines, later in Japan and what became South Korea.

With the end of fighting in Manila, George Marshall assumed that "Filipino guerillas and the newly activated Army of the Philippine Commonwealth" could "take care of the rest of their country." That was far from the case, but MacArthur, proverbially predisposed to declaring premature victory, was not about to advertise it in Washington. Instead, from February 28 to June 25, he quietly conducted fifty-two amphibious assaults on assorted Philippine islands, with Eichelberger running everything south of

Luzon and Krueger concentrating on exterminating the remaining Japanese on the island, and then presented it all to the Joint Chiefs as a done deal. Ordinarily, this would have provoked a stern rebuke at least, but by this time everybody's attention, including Mac's, had turned to a much bigger deal.

AS IN CHESS, THE ENDGAME IN WAR IS HARD TO APPLY, ESPECIALLY AGAINST A determined opponent. In the American Civil War, after years of battle failed to do it, the North required William Tecumseh Sherman to stomp across the face of the South in order to convince the Confederacy's citizens it was finished. Now analogous thinking was being applied to Japan in the form of an amphibious invasion of its home islands, albeit with an underlying sense of dread given the suicidal resistance already encountered. But there were alternatives.

The Navy, having extended the bloody footprints of its island-hopping campaign in the Central Pacific all the way to Okinawa, believed the combination of a total naval blockade and a heavy dose of strategic bombing could drive Japan from the war without necessitating an invasion.

Hap Arnold and the Air Corps, by now all but an independent service, were more than willing to give it a try, if for no other reason than to justify his faith in the troubled B-29 superbomber with a price tag exceeding the Manhattan Project's by 50 percent. Besides potentially carrying that project's still-untested progeny, the aircraft had done little to justify the money. The Japanese had overrun B-29 bases in China, and once the bombers were settled in the Mariana Islands, the results of their high-altitude daylight missions against Nippon's war industries had been disappointing. But with kamikazes as motivators, Arnold had the man and the means to transform the B-29 into a veritable fire engine of destruction.

That man was General Curtis LeMay, whom he put in charge of getting results. LeMay's lowest common denominator was always ruthlessness—that's what would make him so scary during the Cold War. In this case, he reasoned that since Japan's air defenses lacked radar, low-level night missions were the obvious recourse; he stripped the planes of defenses to increase bomb load, and told crews to proceed individually, not in formation, locating targets with their onboard radar. Actually, any old target would do, just as long as it was flammable.

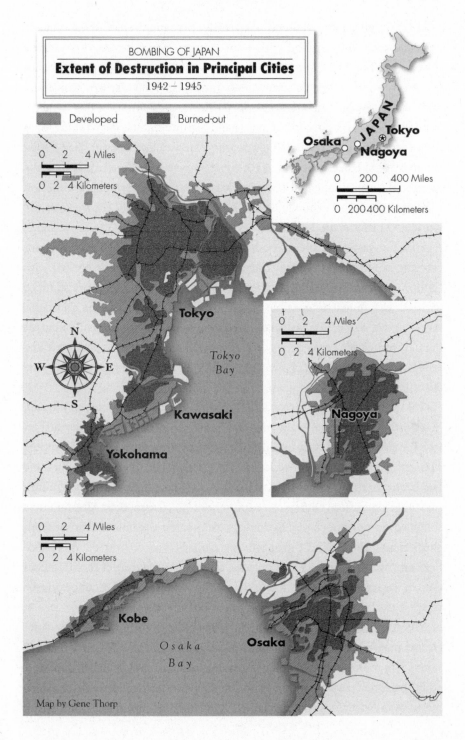

BOMBING OF JAPAN
Extent of Destruction in Principal Cities
1942 – 1945

Developed   Burned-out

Tokyo

Osaka   Nagoya

JAPAN

0   200   400 Miles

0   200 400 Kilometers

0   2   4 Miles

0   2   4 Kilometers

Tokyo

N
W   E
S

Tokyo Bay

Kawasaki

Yokohama

0   2   4 Miles

0   2   4 Kilometers

Nagoya

0   2   4 Miles

0   2   4 Kilometers

Kobe

Osaka Bay

Osaka

Map by Gene Thorp

On the night of March 9–10, 1945, nearly three hundred of LeMay's B-29s played torchbearer. Stuffed with E-46 cluster bombs designed to release thirty-eight bomblets filled with napalm—a jellied gasoline substance invented by a top secret research collaboration between Harvard University and the US government—they swooped in low over Tokyo. The first to arrive painted a fiery X on the city's teeming and highly combustible working-class district to guide the others. Winds nearing thirty miles per hour soon joined the individual blazes into one giant firestorm, an inferno that consumed a quarter of the city, killed or injured at least 125,000, and left another million homeless. "It was as though Tokyo had dropped through the floor of the world and into the mouth of hell," LeMay remembered.

Still, he wasn't satisfied. Soon his growing flock of B-29s would arrive carrying an even fiercer incendiary, the state-of-the-art M-74 packed with "goop," a combination of powdered magnesium and asphalt that, once ignited, was almost impossible to put out. Through the spring and summer, Nagoya, Kobe, Osaka, Yokohama, Kawasaki, and Toyama all had great chunks burned out of them by goop-initiated flames that cumulatively killed at least another 300,000 civilians. But all to no avail. The Japanese showed absolutely no signs of giving up.

This is why Team America's default position remained, almost to the end, invasion. Whatever the Navy and Air Force claimed, and however attractive the prospect of a war-ending science fiction weapon might be, George Marshall continued to believe Japan would have to be invaded. His planners had been working on it since November 1944, and had produced Downfall, a two-stage operation: Olympic, a September 1, 1945, assault on Kyushu, to be followed on December 1 by Coronet, a landing on the main island of Honshu. In the spring of 1945 the dual extravaganzas were postponed to November 1 and March 1, 1946, in part because of the huge manpower requirements demanding massive transfers of troops from Europe to the Pacific, and also to see if the blockade-and-bombing approach worked.

When it didn't, in mid-June Marshall pressed for Downfall at a White House meeting with Truman and his key advisers and the service secretaries. There he argued that the plan "offered the only way the Japanese could be forced into a feeling of utter helplessness." It would be difficult, but so was Normandy, and perhaps the Honshu invasion might not prove necessary. Since Truman was worried about bloodshed, Marshall sought to sugarcoat the whole proposition with absurdly low casualty figures—31,000 for the first thirty days of Olympic, and 70,000 for the whole operation, which was planned to include three-quarters of a million men. These figures were not just stunted, they directly contradicted the estimates of Admiral Leahy, the Joint Chiefs, and Henry Stimson, the lowest of which projected a quarter million casualties. But it was Marshall whom Truman trusted most to define military reality, so he grimly endorsed Olympic and put in place the men to execute it.

Douglas MacArthur, the exemplar of "hit 'em where they ain't," also knew it would be a bloodbath, but when presented the opportunity to "hit 'em where they live," he jumped at the chance. On March 20, two weeks before his death, FDR told a visiting George Kenney, "You might tell Douglas that I expect he will have a lot of work to do well north of the Philippines before very long."

The new manager and his captain, George Marshall, ratified Roosevelt's choice. It would be Mac leading Team America ashore, and in early April orders went out unifying all commands in the entire Pacific under service elements. That meant giving Nimitz everything afloat, giving Tooey Spaatz the strategic air forces, and making MacArthur commander in chief of the ground forces that would not only do the invading, but also, by implication, the occupying—a hidden proposition that would soon make him mikado and a good deal more.

But for now, he remained the team player. Consequently, when George Marshall worried, "'One school of thought is that much more preparation is necessary before the main operation . . .' I replied on April 20th strongly recommending a direct attack on the Japanese mainland at Kyushu for the purpose of securing airfields to cover the main assault on Honshu. . . . I recommended a target date of November 1." After the meeting with Truman, who remained worried about casualties and didn't much like the choice of commanders, Marshall asked MacArthur for his own estimate, prompting a reply that his figures of under 120,000 "were purely academic and routine," adding, "I regard this operation as the most economical one in effort and lives that is possible." It would be hard to imagine a man more on board with Marshall's agenda.

Like Ike, Mac professed an abhorrence of all the killing and a disenchant-ment with the stars and decorations it brought. Yet, when the opportunity arose to lead a mission he must have known was fated for apocalyptic death-dealing, he not only leaped at it, he was happy to dance to the tune of a master he was convinced was his adversary. Fortunately for all involved, this was a mission headed nowhere, an invasion that never came. For there was also the hidden alternative, the one that changed history and put an end to the curse of industrial warfare, but replaced it with something worse—at least potentially.

A REVOLUTION IS ALWAYS HARD TO SPOT, EVEN WHEN YOU ARE IN THE MIDST of it. Our four were right there, three in the know before the blinding flash, and it's worth considering what they thought the prospects might be.

George Marshall, under whose cloak the Manhattan Project mushroomed, was plainly conflicted. When in April 1945 Groves told him it was time to think about targets, he replied: "Is there any reason why you can't take this matter over and do it yourself?" A month later, in a meeting with Stimson, he warned that this revolutionary weapon involved "primordial consider-ations," and that its use to destroy an entire Japanese city would subject America to moral opprobrium for generations. Instead, the bomb should be applied to "a straight military [target] such as [a] large naval installation," and if they still didn't surrender, against a single "large manufacturing [area] from which the people would be warned to leave." Also, possibly because Groves was lowballing the bomb's blast and radiation potential, he thought it might be a useful opening for Olympic. But when asked days after if it should be used at all, he replied: "Don't ask *me* to make the decision." Later, when Stimson's chief aide John J. McCloy pressed Marshall, he insisted it "was a matter for the President to decide, not the Chief of Staff since it was not a military question." Finally, on June 1, when the interim committee tasked with policy guidance met and concluded that the bomb should be used as soon as possible without warning, Marshall raised no objections, like a man looking for cover.

Dwight Eisenhower first learned of the bomb when, in mid-July, the scene shifted to Potsdam, an intact suburb of wrecked Berlin, for the last major summit conference of the soon-to-be-no-longer Allied Team. Henry

Stimson took time to visit Ike's Frankfurt headquarters to inform him, only to be shocked and angered by his response: "I voiced to him my grave misgivings. . . . Japan was already defeated . . . dropping the bomb was completely unnecessary, and secondly . . . I disliked seeing the United States take the lead in introducing into war something as horrible and destructive as this new weapon."

Ike could afford to be frank, having already won his war, but he'd also seen enough of war. The astute but insecure Truman caught the general's gravitas in their initial meeting, Ike and Bradley having flown into Potsdam for lunch with the president, after which they took him on a tour of the shattered Nazi capital, at which point he caught Ike's attention: "General there is nothing that you may want that I won't try to help you get. That definitely and specifically includes the presidency in 1948." Likely knowing it would take him longer, Ike shot back a laughter-clad reply: "Mr. President, I don't know who will be your opponent for the presidency, but it will not be I." A little charade at the beginning of what was destined to be a tortured relationship.

In retrospect, the Potsdam Conference itself seems equivalently disingenuous. Present at the start was Winston Churchill, who was electorally removed by a war-weary British public during the conference and replaced by a diplomatic newcomer, Clement Attlee. Stalin was still Stalin, but with a lot more confidence, behaving much like the cat that swallowed the canary, in this case Eastern Europe. Truman's guard was up, but in comparison he was still a rookie.

The day before the conference opened, he had learned that the plutonium-based implosion bomb tested at Alamogordo in New Mexico had greatly exceeded expectations, exploding with a burst equivalent to forty million pounds of TNT. Seeking to gain leverage with the increasingly obstreperous Russians, a week into the conference he "casually mentioned to Stalin that we had a new weapon of unusual destructive force," Truman recalled. "All he said was that he was glad to hear it and hoped we would make good use of it against the Japanese." Given his espionage-enabled knowledge of the US program, Stalin may have already heard of the test; if not, the American president told him something critical—that the thing actually worked. Either way, according to Marshal Zhukov, that night Stalin ordered Molotov to "tell Kurchatov [the nuclear physicist in charge of the Soviet program] to hurry up the work."

Meanwhile George Marshall continued to hang his hat on the agreed-on Soviet belligerency against Japan in Manchuria as a companion piece for Downfall, and when Truman queried him whether "we could get along without them," like a patient pedagogue he replied: "Even if we went ahead in the war without the Russians . . . that would not prevent the Russians from marching into Manchuria anyhow." It was hard to argue with that logic, so Marshall got his Soviet cover for an operation that would never take place— not a great achievement.

Nor was the Potsdam Declaration, the conference's final ultimatum to Japan. It didn't just keep the demand for unconditional surrender that had been agreed on in early 1943, it also stipulated complete disarmament, the reduction of Japan's territory to basically the four home islands, and immediate Allied occupation. Moreover, and despite Henry Stimson's objections, the declaration failed to mention retaining the emperor or his position, thus making it likely it would fall on deaf ears. Nonetheless, the document also promised to eliminate the influence of militarists, install democracy, allow economic recovery, and to expect an end to the occupation once these demands were met—a wish list destined to form the core of Douglas MacArthur's agenda for Japan, one that utterly transformed its future. But to critics, this brighter prospect was overshadowed by the failure of the Potsdam Declaration to specify the means behind the promise of Japan's "prompt and utter destruction" should it fail to comply.

That or the perceived equivalent was now inevitable. In the end, the decision simply floated up to Truman, no one's having objected, and he did the same rather than issue a formal presidential order. It wasn't plausible deniability, but neither was it a firm affirmation of a very fiery swift sword.

On the way home from Potsdam, George Marshall met George Patton and Maxwell Taylor at Hitler's wrecked Berchtesgaden retreat in southern Germany and told them the secret. "Gentlemen, on the first moonlight night we will drop one of those bombs on the Japanese. I don't think we will need more than two." Patton was not inclined to agree, later maintaining: "So far as the atomic bomb is concerned . . . it is not as earth shaking as you might think. When man first began fighting man, he used his teeth. . . . Then one day a . . . genius picked up a rock. . . . Certainly the advent of the atomic bomb was not half as startling as the initial appearance of gunpowder. In my own lifetime, I remember two inventions or possibly three, which were supposed to stop war; namely the dynamite cruiser Vesuvius, the submarine, and the tank. Yet, wars go blithely on and will when our great-grandchildren are very old men." Doubtless, Georgie believed he had personally observed every stage, and was equally ready to wield this new cudgel in a brave new world of nuclear war. Fortunately, it seems both he and industrial warfare were headed instead for a dead end.

Douglas MacArthur first learned of the bomb on August 1, from Tooey Spaatz, now strategic air commander in the Pacific, and was soon heard grouching about Eisenhower's having been told first. But his understanding of the Japanese was also evident. "This will end the war," he privately told his physician. "It will seem superhuman—almost supernatural—and will give the Emperor and the Japanese people a face saving opportunity to surrender. You watch, they'll ask for it pretty quick." But on the same day he also advised Marshall, "In my opinion there should not be the slightest thought of changing the Olympic operation," demonstrating that, along with our other three military superstars, it's hard to be a prophet.

None of the four truly grasped the implications of atomic weapons for the future of warfare. Only Patton had a broad-enough knowledge of military history to understand that the course of arms had begun with the purely mechanical, then had become based on chemistry, and now had shifted to another power source entirely; but he then drew precisely the wrong conclusion, that nuclear weapons were just another brick in the wall of industrial war. MacArthur and Marshall seem to have grasped that this was something truly different, but only conditionally. Both had been molded by a lifetime of industrialized warfare—preparing for it, planning it, and waging it. The potential for massive mechanized conflict would remain their default position, both at the point of the bomb's inception and in the future.

News of the bomb caught Ike in flux. Besides Patton, who didn't much care, Eisenhower had seen more of World War II's destruction than his teammates, and it left him wondering about the entire future of civilization. More than the others, he was instinctively repelled by a new weapon of such power and wanted no part of it. Yet it would be Ike's destiny to lead America in an arms race that, step by irrefutable step, would build twin US and So-viet arsenals so potentially destructive that any major war between the two, or maybe anywhere, would be suicidal—the terrifying edifice of deterrence that keeps us all safe.

Meanwhile, the man chosen to actually deliver us into the new age was the same Paul Tibbets who had delivered Ike to Gibraltar to open the North African campaign back in 1943. Already known as Team America's best bomber pilot, Tibbets had been deeply involved in the development of the B-29 and had more hours at the stick of the finicky bomber than anyone else aloft—the perfect choice, one reflecting the lethal efficiency of an air force with four years of combat experience. The mission went off without a hitch. On August 6, 1945, Tibbets flew the B-29 named after his mother, Enola Gay, the two thousand miles from Tinian Island in the Marianas to Hiroshima, dropped the gun-type uranium-235–based bomb at 8:15 a.m. local time, then returned safely to base. Three days later a duplicate of the plutonium bomb tested at Alamogordo was dropped over the port city of Nagasaki. Though the circumstances made accurate counting impossible, it is estimated that the combined blasts killed more than 200,000 people. On August 15, a pre-recorded radio message from the emperor was transmitted across Japan—the first time all but a very few had ever heard his voice—advising that there was no alternative to "enduring the unendurable and suffering what is insuf-ferable": surrender on the Allies' terms.

WORLD WAR II'S END CAME WITH DOUGLAS MACARTHUR, STILL VERY MUCH the fair-haired (albeit thinning) boy of Clio, the muse of history, smack in the spotlight of his dreams. On August 12, he learned that his new manager, Harry Truman, had appointed him—with the approval of Attlee, Stalin, and Chiang Kai-shek, but without much other consideration—as supreme com-mander of the Allied powers in Japan (SCAP), to oversee its occupation. According to the instrument of surrender, this gave him absolute author-ity over all seventy-two million inhabitants, including the emperor, thereby making Douglas MacArthur the most powerful man in American history.

His first tasks as SCAP was to make an entrance and preside over the formal surrender ceremonies; he did both masterfully. No one knew what to expect when Mac and his staff boarded *Bataan II*, his C-54 transport, on

August 30 and took off from Okinawa on their way to Yokohama's Atsugi Airfield, a pair of worn-out B-17s their only escort. The general noticed that Kenney, Sutherland, and the rest were packing pistols in shoulder holsters. "Take them off. If they intended to kill us sidearms will be useless. And nothing will impress them like a show of absolute fearlessness. If they don't know they're licked, this will convince them."

Doug's bravado worked, but not in the way he and his companions thought. Actually, the Japanese were relieved, taking the Americans' disarmed condition as a sign they did not intend to harm them. On the way into Yokohama in a fleet of antiquated vehicles, the visiting team could not miss three thousand soldiers lined up stiffly with their backs to them, assuming it to be the best means to ensure nobody took a potshot at the convoy, and learning only later that it was a posture reserved for the emperor.

They reached Tokyo a day later and set up in the US embassy. It was basically in the same condition as when the American ambassador had left it four years before, and together with the granite-sheathed Dai-ichi Insurance building and Frank Lloyd Wright's Imperial Hotel, was among the few substantial structures left standing after LeMay's multiple firebombings. The next two weeks were spent largely preparing for the formal surrender, the details being worked out by Sutherland—his last assignment before Mac, now free to be vindictive, got rid of him. Not realizing the ax was about to fall, he did an excellent job.

Meanwhile, George Marshall, in a strategic act of kindness upon learning Jonathan Wainwright had been liberated by the Russians in Manchuria from a Japanese prison camp, where he had spent four years believing himself disgraced for his surrender in the Philippines, cabled Mac: "It seems to me that it would be most appropriate to have General Wainwright present at the signing of capitulation." MacArthur, who had been furious at Wainwright's surrender and vetoed Marshall's original effort to get him the Medal of Honor (he succeeded a month after the war ended), had little choice but to accept. But he seemed honestly moved when his former field commander arrived, reduced to a human scarecrow by his internment, and promised him his old corps and muttered his nickname "Jim" over and over, the only time he had called him anything but Jonathan.

At any rate, Wainwright made an instructive prop at the surrender ceremony, which took place on the morning of September 15, prominently lined up by Sutherland next to his fellow prisoner in Manchuria, British general Arthur Percival, who had surrendered Singapore to Yamashita, both of them living testimonies to Japanese wartime cruelty. MacArthur, remembering the dreadnoughts sunk at Pearl Harbor and as a tip of the cap to the new manager, insisted the ceremonies take place on the deck of the battleship *Missouri*, named for Harry Truman's home state. "The most startling weapon of war I have ever seen," Wainwright remembered. "I simply could not believe that anything could be so huge, so studded with guns." Soon he was joined by the American principals in the Pacific War: Nimitz, Halsey, Eichelberger, and Krueger, along with representatives from the other Allied teams—Great Britain, the Soviet Union, China, even the Netherlands—a potpourri of uniforms, joined by hundreds of the ship's sailors in white perched wherever they could find a space.

Mac seemed relaxed and in a backslapping mood, but actually he was strung tight enough to provoke one of the puking episodes that pockmarked his tensest moments, now in the captain's head of the USS *Missouri*. His pilot, who was guarding the door, asked him if he needed a doctor. "No, I'll be all right in a moment," and, as always, he was.

The Japanese arrived very much a pickup team. Absolutely no one of any significance wanted a place in this roster. Finally, the emperor had to twist arms, informing his foreign minister that any duty to his sovereign, no matter how humiliating, was an honor, an argument similarly applied to his chief of staff, who had threatened hara-kari. Joining them were six other officers and two civilians, similarly dragooned. It was not easy duty. As he stood listening to "The Star-Spangled Banner" being played, and beneath the gaze of all those sailors, one remembered: "There were a million eyes beating us in the million shafts of a rattling storm of arrows barbed with fire. I felt their keenness sink into my body with a sharp physical pain. Never have I realized that the glance of glaring eyes could hurt so much."

Then MacArthur stepped forward with a message more of hope than conquest: "We are gathered here, representatives of the major warring powers, to conclude a solemn agreement whereby peace may be restored." He never mentioned the unconditional surrender part. The Japanese squad was then ushered over to a green-felt-covered table to sign the document. When the first hesitated, a sterner Mac barked: "Sutherland, show him where to sign." That accomplished, MacArthur produced five pens to affix his own signature—the first two he gave to Wainwright and Percival, standing as witnesses, the next two were for West Point and Annapolis, and the last, a small red one inscribed JEAN, he kept. With that, MacArthur rose and addressed the assemblage: "Let us pray that peace be now restored to the world and that God will preserve it always. These proceedings are closed."

So ended the last world war in an era of unlimited conflict waged with conscripted millions armed by the industrial economies of entire states, one that arguably was on the way to bringing civilization itself to its knees. More than seventy million people died in the Second World War, over seven times more than its unprecedentedly bloody predecessor, and fourteen times the battle deaths of the American Civil War, the first major conflict in which technology and mass production were enlisted in the cause of death-dealing. This was a casualty curve headed only toward eventual societal collapse. But MacArthur was right: a new edifice of peace would be built.

CHAPTER 10

AFTERMATH

Team America as the organizing force in the nation's future was finished, and so was the infernal league whose gigantic struggles had come to dominate the globe—all of it over literally in a flash. But nuclear physics and human cognition take place in two separate realms, and in the case of the latter even obvious lessons take a while to sink in, particularly ones that render entire segments of societal endeavor irrelevant. So it would be with our military superstars and the team they had led to worldwide supremacy: success and institutional momentum would push them in roughly the same direction from which they had come; the rest—adjusting physics to politics—was a matter of fate and adaptation.

With the exception of Patton, all had the capacity to acclimate, but they did so in the context of a lifetime in the prior realm of military reality, as did their opponents in a new kind of conflict, dubbed, for want of a better term, the Cold War. That was the essential problem: old assumptions about a new reality, a discordance that could have led either side to stumble into apocalyptic combat. Providentially, it didn't happen. And that story, with all its perils, would still be told and perceived largely within the framework of Team America and its remaining superstars. Thrust forward by success but handcuffed by the past, they resumed operations on entirely new terrain and in the process, without wanting to, built a structure of deterrence based on mutually assured destruction, a situation epitomized by its unintentionally sardonic acronym, MAD.

Actually, the blinding flash and its implications were seen most clearly not through military sunglasses, but by civilian eyes conditioned to look for larger implications. Consequently, when the academic strategist Bernard Brodie picked up a copy of the *New York Times* and learned of the Hiroshima bomb he instantly told his wife, Fawn, "Everything that I have written is obsolete."

He made up for that quick. By early 1946, Brodie and his colleagues at Yale's Institute of International Studies had produced *The Absolute Weapon*, a slim volume that not only captured the central conundrum of warfare in a nuclear age, but also drew the essential policy conclusion: a terror-enforced peace. While several colleagues contributed sections, Brodie's thinking dominated.

Like the grim reaper turned prosecutor, he relentlessly built his case point by point: "The power of the present bomb is such that any city in the world can be effectively destroyed by one to ten bombs." "No adequate defense against the bomb exists, and the possibilities of its existence in the future are exceedingly remote." "The atomic bomb . . . places an extraordinary military premium upon the development of new types of carriers." "Superiority in numbers of bombs is not in itself a guarantee of strategic superiority in atomic bomb warfare." "Materials for its production must be considered abundant." "Regardless of American decisions . . . other powers besides Britain and Canada will be producing the bombs in quantity in a period of five to ten years." Together it added up to an airtight case, a cat's cradle of vectors leading to but one conclusion: only the insane would start a war under such conditions. The body of thinking for and against would eventually prove mountainous, but none of it substantially undermined *The Absolute Weapon*, and it still stands today as the Magna Carta of the nuclear age. But at the time, and for a long time, conventional wisdom prevailed, and our four protagonists got on with their lives.

FOR SIXTY-FIVE YEARS DOUGLAS MACARTHUR LED A CHARMED LIFE DOING dangerous things. It's a miracle he survived World War I, not to mention Corregidor and all the other bullet-ridden incidents that whizzed by without leaving a scratch. Then there was his swift rise in rank despite a towering and frequently self-defeating ego. Add to that a healthy dollop of being in the right place at the right time, and it already amounted to an improbably provident existence. But then things got even better.

Besotted by her boy, Lady Luck waxed Solonian in his next job description, leaving him absolutely in charge of both Japan and the Philippines, free to chart the political futures of almost a hundred million souls basically in any direction he chose. How else could you explain it? Had Harry Truman stopped to consider carefully the implications of making Doug supreme commander of Allied powers, he very probably would not have done so. Some suggest he wanted to keep him on the other side of the planet and out of US politics, but if so, his instincts failed him, Mac being no real threat in that quarter. More likely Truman, caught up in the rush of events, simply acquiesced to what was most convenient, and then most evident with MacArthur signing the surrender in Tokyo harbor, already the man on the scene.

So Luck prevailed and her choice was brilliant. MacArthur was a freakish example of the human capacity to integrate and exploit diametrically opposed concepts, in this case political visions. Viscerally and professionally he was an autocrat, his entire career having to do with ordering people around, something he was very good at. Yet his personal allegiance was unquestionably to democracy, his politics being those of a progressive Republican on the order of Theodore Roosevelt. Nor did he wrestle with the contradictions. MacArthur was serenely imperious in bringing democracy to Japan, while remaining distantly parental in fostering the American-inspired system he and his father helped build in the Philippines. In both cases his stance was impeccable.

Especially with Japan. Perhaps because of societal solidarity itself, reform has not come easily in the land of the rising sun. Instead, change arrived in earthshaking bursts, at least during what constitutes modern Japanese history, beginning in 1600 when Tokugawa Ieyasu consolidated the country, setting in place a rigid feudal shogunate that banned all firearms and contact with foreigners, leaving society crystallized until 1868. At this point, a group of young samurai, ashamed of how far their homeland had fallen behind Western powers, staged a coup d'état known as the Meiji Restoration, one dedicated not only to reviving the emperor's power, but also to turning their country into Team Japan, one featuring a modern, industrialized economy, a conscript-based army, a massive steam navy, and wolfish imperial ambitions. Here matters stood until 1945 and Doug arrived for another big change.

His opening moves were brilliant. Food shortages in the home islands stretched back to 1941, and by late 1945 the country was suffering from mass malnutrition. Nor was there money or a merchant marine left to import food. To compound matters, tradition and their own behavior as conquerors dictated that the vanquished feed the American victors. Mass starvation beckoned—it was part of losing.

But MacArthur wrong-footed tradition and took off in another direction, the one captured earlier by his countryman John Adams: "The shortest road to men's hearts is down their throats." Rather than starve the Japanese, he would feed them. He quickly ordered that his occupation troops be supplied externally, while seizing over three million tons of the food the US Army had stockpiled in the Pacific, then directing all of it into Japanese mouths, an act that not only saw the population through the winter, but warmed the hearts of former enemies who took obligations, particularly debts of gratitude, very seriously. When there was carping from Washington MacArthur neatly summarized his position: "Give me bread or give me bullets."

Having established this base of goodwill, he could proceed with an agenda of retribution without poisoning the possibilities for major social and political reform later. Specifically, he was charged by the Joint Chiefs of Staff with disarming and demobilizing the nearly seven million men left in Japan's armed forces; disbanding the military and secret police; and trying those accused of war crimes. The first two were fairly straightforward military procedures, and not likely to meet much opposition, though getting troops back from China, particularly from the Russians in Manchuria, was protracted by logistics and growing political tensions—the poor treatment returning soldiers received from their Soviet captors being a factor in the country's later unpopularity in Japan.

The war crimes, on the other hand, required a deft, or at least not too sanguinary, touch. Eleven hundred suspects were almost immediately arrested, among whom were two prime ministers, guards of American POWs, and Tokyo Rose (actually several women propaganda broadcasters), all jailed to await trail. MacArthur stood well away from these proceedings as an eleven-judge tribunal listened to almost fifty thousand pages of testimony before condemning 174 men to death—a list later shortened to just seven, including prime ministers Tōjō and Hirota. MacArthur could have commuted their sentences, but didn't. Nor did he attend the hangings, or allow any photographs. In effect, he washed his hands of the whole procedure, which was in direct contrast to his machinations in the Philippines to ensure generals Yamashita and Homma were promptly executed.

Then there was the matter of the emperor. His fate was completely in Mac's hands. He could have publicly humiliated him, even included him with the others on the gallows. He did nothing of the sort. Instead, he co-opted him, neutered him into a constitutional monarch, then set him in place as the cornerstone of Doug's new order. And while his instincts were mostly in synch with his domain, he also benefited from the advice of Bonner Fellers, a controversial figure in the intelligence quarter, but one well grounded in things Japanese. It was his idea to steer the emperor and his family clear of the war crimes tribunal, and also to exhibit the patience of a confident hegemon.

The decorum of defeat demanded that Emperor Hirohito pay his respects to MacArthur, part of "enduring the unendurable," and although the American embassy was just a short distance from the Imperial Palace, it was likely the longest journey Hirohito ever contemplated. Mac could have simply summoned him, but he agreed with Fellers that it was best to wait imperturbably. After three weeks, on September 27, 1945, an unprecedented motorcade was seen crossing the moat and heading for the lair of the conquerors, a Daimler limousine carrying a slim, somber-looking gentleman dressed in morning clothes.

MacArthur knew he was coming but kept the visit secret from the press so there would be no photographers except the one he wanted. Fellers, not the general, was at the entrance for the greeting, a calculated and unsettling breach of protocol. "He looked frightened to death," Fellers remembered.

Upstairs, MacArthur awaited him in his office clothes—khakis, open collar, and only his five stars signifying rank—a sartorial slur worn to show exactly who was in charge. Otherwise, Mac's approach was affable, reminding the emperor that he had met him earlier as a five-year-old prince, introduced on a visit to his father, Arthur, then in Tokyo at the end of the Russo-Japanese War. But Hirohito refused the bait; he was on a self-imposed mission that transcended small talk: "I come to you General MacArthur to submit myself to the judgment of the powers you represent as the one to bear sole responsibility for every political and military move taken by my people." Imperial suicide, seppuku in the offing.

Mac was initially taken aback, since he was planning exactly the opposite; but as the implications sunk in he was elated: "He was an Emperor by inherent birth, but in that instant I knew I faced the First Gentleman of Japan in his own right"—Citizen Hirohito, in other words. Doug outlined his democratic plans and found his audience not just acquiescent but knowledgeable and apparently enthusiastic. Suitably encouraged, the general brought up the sticky issue of Hirohito renouncing his own divinity, one of the Allies' nonnegotiable demands. He didn't press, but this too was inevitable, the constitutional monarch in Doug's new order. There was also a photo op— Mac in his khakis towering above the formally dressed Hirohito. The Japanese press was horrified, but the photograph's seismic impact across the country was that it exactly caught the true nature of the relationship.

MacArthur formally opened his reform campaign a week after the imperial visit with his October 4 civil liberties directive, which annulled all restrictions on freedom of speech, the press, assembly, and religion, while simultaneously firing most of the policing infrastructure responsible for the previous suppression of these rights. This was enough to cause the sitting prime minister and his cabinet to resign en masse, but Mac was just getting started as his replacement, Baron Kijūrō Shidehara, discovered when summoned to the general's office a week later. Mac read him a seven-point memorandum of reforms he wanted, including independent unions, an end to child labor, school reform based on "more liberal education," a breakup of the zaibatsu-based economic oligarchy, plus equal rights for women in an utterly male-dominated society. And, he added, he was expecting a new constitution from Shidehara.

What he got five months later in early February 1946 was classically Japanese, passive-aggressive and committee written: "It turned out to be nothing more than a rewording of the old Meiji constitution," MacArthur remembered. His response was decisive. The Japanese side having failed, it was Team America's turn. He assigned the job to his twenty-person Governance Section and gave them ten days. To guide them he provided a State

Department document stipulating a truly representative legislature based on popular (male and female) sovereignty, an end to titles and feudal privileges, a "budget to be patterned after the British system," and an emperor subordinate in all his powers to the constitution. Finally, he slipped in one of his own provisions, one with profound implications: "No Japanese Army, Navy, or Air Force will ever be authorized and no rights of belligerency will ever be conferred upon any Japanese force"—in other words, a renunciation of war.

Suitably guided, the exhausted members of the Governance Section completed their draft in just nine days, and General Courtney Whitney, Sutherland's replacement as blunt instrument, delivered it to the Japanese the next day, where it had the approximate effect of a billy club, leaving them stunned to the point of incomprehension, knowing only that they had to submit. In a personal meeting on February 21, Shidehara made one last attempt to get MacArthur to back off some of the constitution's most radical provisos, especially article 9, renouncing war. But Mac stonewalled him, leaving Shidehara and his colleagues no choice but to submit it unamended for his final approval on March 4. He unveiled it to the public two days later, along with a statement from Hirohito confirming "my desire that the constitution of our empire be revised drastically upon the basis of the general will of the people and the principle of respect for the fundamental human rights." That was followed on April 10 with the first open election in Japan's history, one that chose the legislators who would on November 3 ratify the full constitution with just a few changes, set to take effect on May 3, 1947.

It all worked better than anyone, except perhaps MacArthur, expected. As a nation, the people of Japan were profoundly disheartened by recent events, but the society had not lost its social cohesion, or the vast network of obligations that held it together. This would provide the stability on which the disorderly but liberating routines of democracy would rest. So the Japanese went to the polls in droves, casting over fifty-five million valid votes in a turnout that included over eleven million women. When Mac was told by a horrified Japanese legislator that a prostitute had been elected to the House of Representatives with over a quarter million votes, he replied dryly: "Then I should say there must have been more than her dubious occupation involved."

A lot more. The Japanese people were eager for change, and in accordance with historical precedent were capable of absorbing it in huge quantities, even though it was brought by an outsider. Stern but decent treatment flipped the entire society; suddenly the American conquerors became exemplars in practically all things, including baseball, which Mac also promoted. Through their eyes he now personified the gatekeeper to a better future, the mikado of the modern era, an admittedly temporary condition, but still an astonishing development in anyone's life.

MacArthur showed no signs of surprise; he was that immersed in the role. Holed up in the Dai-ichi building and the embassy, he kept himself rigidly apart from the Japanese, in his almost six-year stay traveling exactly nowhere in the country—a separateness the people expected from the figure at the top. Twice in late 1945 Truman invited him home to receive the accolades of a grateful America, only to be refused on the grounds of "the delicate and difficult situation which prevails here." It's an excuse that failed to satisfy either the president or George Marshall, but then, what leading actor leaves a hit show even temporarily? He also turned down the administration's proposal to keep an occupation force of up to 400,000 in Japan, part of Marshall's push for a big peacetime army; Mac effectively halved that figure. "I'm going to do something about that fellow," Harry Truman was heard muttering. But that would be a long time coming. Meanwhile, through a combination of guile, manifest success, and a central position, MacArthur prevailed in dispute after dispute with Washington, the man on the scene proceeding according to his own plan, camouflaged by nearly seven thousand miles.

But there were limits to reform, ones defined by traditions he either underestimated or simply did not understand. Consequently, the results of his efforts to change fundamentally the Japanese economy skewed between spectacular success and ultimate failure.

In the plus column was certainly land reform. "Most farmers in Japan were either out-and-out serfs, or they worked under an arrangement through which the landowners exorbited [*sic*] a high percentage of each year's crop," MacArthur observed, all funneled to some 160,000 absentee proprietors. Traditions die hard in Japan, and it took him two tries before he got what he wanted from the Diet. All absentee-owned land was now subject to compulsory sale to the government for what amounted to the price of a carton of black-market cigarettes per acre, the same rate offered to tenants to buy it with money lent them at low interest and payable over thirty years. More than five million acres changed hands, eventually leaving 90 percent of the land owned by the people farming it, and in the process immunizing the countryside from the Communist version of rural reform—no small thing as the Cold War relentlessly heated up. "I don't think that since the Gracchi effort of land reform in the days of the Roman Empire there has been anything quite so successful of that nature," Mac maintained, barely exaggerating.

Far less fruitful were his efforts at the other end of the economy, the oligopoly of corporate zaibatsu—Mitsui, Mitsubishi, Sumitomo, and Yasuda being the biggest—that dominated business, banking, and industry. Since war damage had reduced GDP by almost half, while industrial production was around 15 percent of prewar levels, MacArthur took this nadir as an opening to restructure the entire system, and, in the spirit of the Republican trust busters of his youth, set about breaking up the zaibatsu.

In August 1946, he ordered that within three months all cartels and trade control leagues be abolished; at the same time, he established the Holding Company Liquidation Commission to supervise the breakup of the interlocking directorates that were key to the zaibatsu's power and effectiveness. And for good measure, 2,200 top zaibatsu stakeholders were banished from their corporate realms. It would seem MacArthur cleaned house.

But unlike land reform, which required only a change of ownership, his antizaibatsu campaign struck at a much more sophisticated and adaptive mechanism. Like a hammer pounding a blob of mercury, the droplets scattered in all directions, but always with the potential to be reassembled. MacArthur missed how meticulously adjusted these corporate entities were to Japanese society and business practices, as natural in this context as birds flocking. Thus, in September 1947, a year into the program, one SCAP economist noted that only two of sixty holding companies designated for dismantling had actually been dissolved. Instead, they evolved into enterprise groups, more loosely organized, less coordinated, with limited financial interdependency—but still filling the same niche and becoming wildly successful in the process.

Ironically, MacArthur's most enduring role in economic reform was likely his invitation in 1947 to quality-control prophet W. Edwards Deming to help plan the census. Deming soon took business circles by storm, presenting his methodologies to hundreds of engineers, managers, and top industrialists, who in turn spread them across the gamut of Nippon's factories to work a revolution in product development and quality control. That was more the way things actually changed in postwar Japan, less a matter of diktat than of prompting, a process subtler than the one the general likely thought he was invoking.

Douglas MacArthur believed that his own and his father's long experience in Asia had left him with a special understanding of what he called "the Oriental mind." He certainly did have some significant insights, but not X-ray vision. His success in changing Japan depended as much on his imperious personality and good timing as it did on his knowledge of the people, and he showed absolutely no signs of getting to know them better, hermetically sealing himself off instead.

He lived in the embassy's moderately splendid isolation with Jean and little Arthur, whose only appropriate playmate was Crown Prince Akihito. Now that her husband really was "Sir Boss," Jean took to her new role almost instinctively—southern belle morphed into mistress of ceremonies. For when it came to Americans, the MacArthur enclave was a semipermeable membrane selectively open to visiting delegations of officials, politicians, and members of the press.

The social routine seldom varied. Those expecting cocktails did not get them. Instead, Jean spent the next half hour circulating, finding out who they were and what they wanted. Once his Cadillac was heard in the driveway, she invariably sang out, "The General is coming!" followed by, as he entered the reception room, "Why, it's the General! Hi, General!" This sort of mating call he invariably responded to by striding across the room to kiss her, only then pivoting to address his guests: "You must be hungry, I know I am."

If they were, they didn't get much to eat, MacArthur's idea of an evening repast being soup, salad, and a cup of coffee. Instead, he filled the void, if not their stomachs, with one of his patented monologues, a combination of impromptu grandiloquence, capacious facts and figures, and presumed sagacity as to the mysterious ways of the East. Then, as suddenly as it began, it ended with MacArthur slipping upstairs and Jean escorting the guests out the door, presumably satisfied in mind if not in body.

Not everybody was, most significantly George Kennan, the head of the State Department's Policy Planning Staff. A much-revered figure even today, the prime architect of the containment strategy that dominated US policy in the Cold War, Kennan in action was temperamental and thin-skinned, and had a vindictive streak. When he arrived on March 1, 1948, he was also on a mission to find out for the Truman administration what exactly the effects of Doug's fifty-two-card pickup in Japan were.

Pretty plainly, Kennan brought along a briefcase full of preconceptions—his area of expertise was Russia, not Japan; he was unabashedly Eurocentric; and his commitment to his own containment strategy led him to believe it could be applied to the Far East. Hence, MacArthur's antimilitary provisions were obsolete and dangerous: the "idea of eliminating Japan as a military power for all time is changing," read a September 1947 Policy Planning Staff document. "Now, because of Russia's conduct, [the] goal is to develop Hirohito's islands as a buffer state." As with postwar Germany, to Kennan this meant bringing Japan's economy back as fast as possible, a recovery he believed was being hampered by the antizaibatsu campaign.

Given their policy differences, the key one-on-one meeting went surprisingly well. MacArthur was at that moment concerned about the multinational Far Eastern Commission, a body that included Russia, and the possibility it might try to curb his power and reforms—contingencies Kennan had no trouble reassuring the general would not happen. Otherwise, though the discussion remained cordial, MacArthur's perception that "we parted with a common feeling, I believe, of having reached a general meeting of the minds" proved far from the case.

Upon return, Kennan authored a scalding forty-two-page denunciation of SCAP, claiming that MacArthur's reforms had caused "a high degree of instability in Japanese life generally," and more specifically charging that his trust-busting was holding back economic recovery while his disarmament provisions had left Japan effectively defenseless. Kennan doubtless made some good points, particularly if you were trying to build a bulwark against the Communist menace rather than eliminating it, which would prove to be

MacArthur's instinctive reaction as a soldier. In any case, Kennan in Washington was on his home court, and his ideas found their way into NSC 13/2, a policy document signed by Harry Truman on October 8, 1948, ordering new reforms to cease and old ones dialed back.

This marked the end of Doug the Lawgiver, unquestionably the most benign and influential period in his long life. Today in Japan it's not exactly obvious that he was ever there, the immediate postwar period being a time most would rather forget. Still, among those willing to remember, few would deny that, more than any other individual, Douglas MacArthur was responsible for launching Team Nippon on its democratic way, a liberating voyage chockablock with prosperity and, thus far, free from the curse of war.

UNLIKE DOUGLAS MACARTHUR'S EXPERIENCE, WAR'S END FOUND DWIGHT EI-senhower completely exhausted and uncertain about what to do next. As it turned out, the exhaustion had a lot to do with the uncertainty. So as time passed, his true path became steadily clearer, the one that would lead him to the Oval Office. But early on, the evidence points to some serious consideration of chucking it all save Kay Summersby. That certainly seemed the case when, according to Kay, he insisted they fly to London and celebrate V-E Day, a very public episode captured on film with Ike and Kay joined in their theater box by son John, Kay's mother, and Omar Bradley, followed by dinner and even a bit of dancing at the American nightclub Ciro's. One of Ike's signature moves was to hide things right out in the open, but this amounted to spotlighting. All eyes—not to mention cameras—were on Eisenhower, so he must have known this would get back to Mamie fast. Did he care? At this point, maybe not.

In 1973, in the process of recording his oral biography, *Plain Speaking*, Harry Truman maintained that "right after the war was over, he [Ike] wrote to General Marshall saying that he wanted to be relieved of duty" in order to divorce Mamie and marry Kay. To which, in Truman's version, Marshall replied he would "bust him out of the Army." Truman also maintained that before leaving office he "got those letters from [Eisenhower's] file in the Pentagon and I destroyed them."

Some maintain this was the fabrication of a very old man bitter over his relationship with Ike. Besides, Marshall never would have threatened to end

Ike's Army career; they were both five stars, a rank from which they could not be retired, except voluntarily. Yet Truman's tale makes more sense when told from the perspective of the historian Garrett Mattingly, the prizewinning author of *The Armada* and at that time a Naval Intelligence officer whose duty it was to read and censor all outgoing high command cables. He didn't forget this one, and in the early fifties told colleagues at Columbia University, where Ike was then president, essentially the same story, except he had Marshall threatening to relieve Eisenhower as supreme commander in Europe, which was in his power to do.

It's important to remember that Marshall was just as exhausted as Eisenhower and a decade older; he desperately wanted to retire. To that end an earlier daisy chain deal had been struck among Bradley, Ike, and Marshall, one that had the former returning home immediately to take over the Veterans Administration, followed by Ike once Japan was defeated, to replace Marshall as chief of staff for two years, to then be succeeded by Bradley. Now Ike was threatening to throw a spanner in the works by dumping Mamie and not coming home—fighting words to George Marshall, who was in no mood to compromise his retirement to accommodate Ike's love life.

Instead, he jerked him home in the middle of June 1945 for the triumphal tour MacArthur had dodged but was now mandatory for Ike, along with the conjugal visit that went with it.

The trip proved instructive and not necessarily unpleasant. Truman sent the presidential plane, the *Sacred Cow*, to fetch him across the Atlantic and tote him from parade to parade. As far as the eye could see, everybody liked Ike; he drew crowds bigger than anybody since Charles Lindbergh in New York and Washington, and in Abilene a multitude four times the town's population greeted the local boy made good, who also got to see his mother, Ida, having missed his father's funeral in 1942 due to the war. Back in the nation's capital, Marshall supplied him with a prepared text to address a joint session of the House and Senate, one that Ike promptly chucked to speak off the cuff, a switch that earned him the longest standing ovation in congressional history.

The reception he got from Mamie was likely to have been less positive, though not necessarily negative either. There is a picture of them together in the back of a convertible in Abilene, Mamie with her hair piled high, still with the bloom of youth, both smiling at each other and their shared good fortune—a private moment in a crowd of gawkers; it doesn't look like a relationship entirely on the rocks.

Once again, George Marshall put them up at the same Greenbrier they hadn't liked on Ike's last visit. But this time they were joined by Mamie's parents, around whom Ike always felt comfortable, along with their son John—the familiar roster of his adult family. There were likely to have been some uncomfortable moments with Mamie knowing after this visit he was returning to Kay; nonetheless, he had corresponded with her lovingly and continuously throughout the war. There had been no clear break on this side, nor would there be. Domesticity suited Ike. His attenuated sexual relationship with Kay may well have been more than exhaustion; he was and always would be a family man. And now that he had seen the crowds for himself, he understood that his way to the very top was a real possibility. But he also knew no one divorced had ever been elected president of the United States. To reach that goal he needed Mamie; she was a vital ingredient to his future. He must have known that as he returned to Germany, George Marshall's mandatory domestic leave having accomplished exactly what was intended.

Ike's remaining time in Europe and then as chief of staff was all about burden shedding—not pleasant duty, but necessary to line himself up correctly for the big jump.

Critical to that process was Lucius Clay, a prime example of Ike's brimming reservoir of useful friends. Son of a US senator who taught him the world of politics, Clay chose a military career and first met Eisenhower in the Philippines, where their wives also bonded. He was a natural administrator, and Ike had thought of replacing his logistics chief J. C. H. Lee with Clay when Lee's lifestyle in Paris drew fire from the press and Marshall. This didn't happen, but Ike saw Clay as perfect for helping him set up the military government in the American zone of occupied Germany, and also as an eventual successor after he left. Clay slid into the role as if he were made for it, which says a lot, since he would soon enough find himself at the center of the Cold War, then a little later at the heart of Ike's political future—a handy man to have on your side.

This was plainly no longer true of George Patton. He had written Marshall asking for "any type of combat command from a division up against the Japanese," but came up cold: MacArthur didn't want him; neither did Eichelberger. So he stayed in Germany, still head of the Third Army, parked in eastern Bavaria, which he was now to administer. Mostly, he spent his time hunting, inspecting troops, and participating in military ceremonies. His state of mind can be surmised from his remark to reporters at a temporary military cemetery at Nuremberg: "That man died so bastards like you can continue to breed!"

In June 1945, he was pulled back to the States for a monthlong bond-raising tour. Jean Gordon was still hovering nearby, and Patton also apparently established a sexual liaison in Knutsford, the scene of his misquoted speech and original headquarters of the Third Army, so reuniting with Beatrice was a touchy proposition. He certainly generated a lot of attention, huge audiences wherever he went, but there were also the usual gaffes. He told an outdoor crowd in Boston that soldiers killed in action were not necessarily heroes and frequently fools. It always put Patton on edge to visit the wounded, but his daughter Ruth Ellen was working as an occupational therapist and insisted he visit the amputee ward at Walter Reed Army Hospital. As he arrived in full uniform Patton let loose at the reporters who had gathered: "I'll bet you goddam buzzards are just following me to see if I'll slap another soldier, aren't you? You're all hoping I will!" It's hard to imagine what Patton thought press like this would do to bond sales. Before he left for Germany he told his daughters this was the last time they would see him alive, though it would not be true of their mother—a clairvoyant bull's-eye.

But if his days were numbered, he didn't make much use of them scrupulously administrating governance in Bavaria. So long as the war was on, Nazis were Patton's mortal enemies; but now, having held the reins of government, they could be useful in running the place, particularly since he had so little interest in doing it himself.

Eisenhower was back at Douglas Dillon's Riviera villa with Kay when Bedell Smith's message caught up with him, announcing that Patton was way off the ranch in Bavaria, and hinting that he might want to return to Germany pronto to deal with the problem. Beetle was no alarmist. Patton was mouthing off about the Russians being the real enemy and Germans Americans' natural allies in an inevitable war. His antisemitism was also back in force, taking the wretched condition of the Jewish survivors of the death camps as indicative of their true natures.

Ike visited Patton on September 16 and tried to talk some sense into him until three a.m., but more outrageous statements and rumors of bad conditions in the displaced persons (DP) camps brought him back a week later. At one huge DP facility Ike was mortified to find the guards were German, some former SS. When Patton tried to blame the squalor on the inmates "pissing and crapping all over the place," Ike responded angrily: "Shut up George." But Patton kept it up: "There's a German village not far from here, deserted. I'm planning to make it into a concentration camp for some of these goddamn Jews." Needless to say, that never happened. Around the same time, in a telephone conversation taped secretly without Eisenhower's knowledge, Patton told one of Ike's deputies, "We are going to have to fight them sooner or later. . . . Why not do it now while our Army is intact and we can have their hind end kicked back into Russia in three months? We can do it easily with the help of the German troops we have, if we just arm them and take them with us. They hate the bastards." This diatribe prompted a response similar to that of his boss: "Shut up, Georgie, you fool!"

That he was, and it was just days before he said the unsayable. At a press conference, in response to a question on the state of denazification, he was quoted as saying: "The Nazi thing is just like a Democrat-Republican election fight," a statement that soon swept America's newspapers and yet again put Patton in the public bull's-eye.

Ike's rage upon hearing it was reportedly monumental, but by the time Patton entered his office on September 28, he was just coldly furious. He and Beetle took turns questioning Georgie, trying to get at exactly what was said. They also marched before him Patton's chief accusers, spewing a wealth of credible evidence of his mismanagement. Finally, Ike got to the point of the meeting, his "good idea" that Patton be transferred from command of the Third Army to the Fifteenth Army, a headquarters without troops, charged with writing a history of the war in Europe. Given the degree of provocation, it was probably the kindest thing Eisenhower could have done, but Patton would never forgive him or Beetle. Nor was Ike eager to make amends: "If you are spending the night of course you will stay with me, but since I feel you should get back to Bad Tolz as rapidly as possible, I have my train set up to take you and it leaves at 7:00 . . . in a half hour."

"I took the train," Patton wrote in his diary. It marked the end of anything resembling friendship.

He didn't last long with the Fifteenth Army. Military history did interest him, but as a commander of men without any troops, he simply puttered at his new job. He hunted just about anything warm-blooded, and rode the best horses obliging German aristocrats could provide. His relationship with Jean Gordon had been complicated by her earlier having fallen for another married officer, who then wanted to return to his wife, causing Patton to assign him to a combat zone. Did this leave Georgie and Jean a couple, or at least coupling? It's hard to say.

He was scheduled to fly home on December 10 for Christmas, and didn't plan to return; either the Army offered him something better than military history, or he'd retire. Beyond that he had no plans.

On the morning before he was set to leave, he headed out to hunt pheasant around Manheim—Patton in a chauffeur-driven Cadillac with an aide, followed by a jeep full of guns and a hunting dog. Forty-five minutes into the expedition Patton's driver—distracted by the general's "Look at all those derelict vehicles!"—T-boned a two-and-a-half-ton truck. His speed was likely to have been below twenty miles an hour when they hit; the damage to the Cadillac was limited to a smashed radiator and front fender; neither driver nor aide was injured, only George Patton, who was dealt a death blow. It was as if the hand of fate threw him like a fastball into the front and rear seat partition, gashing his scalp and crushing the vertebrae in his neck. Veteran of countless pratfalls, Patton knew instantly he was in bad shape: "I believe I am paralyzed. I am having trouble breathing." He told his aide to rub his finger and arms, and as they waited for the ambulance he concluded, "This is a helluva way to die."

Certainly, he would have preferred a bullet, but otherwise his demise could hardly have been better timed. He lasted another eleven days, as predicted to his daughters, a duration long enough for Beatrice to arrive and for them to make something like amends—for better or worse, a couple now and forever. A few friends were allowed to see him, but not Bedell Smith, who arrived only to be firmly rebuffed. Patton gave Beatrice explicit instructions that both Beetle and Ike be excluded from his funeral, his final judgment being: "I hope he makes a better President than he was a General." Besides keeping him in traction and on heavy medication there was little his doctors could do except wait for the inevitable, when his lungs started to shut down. He told his favorite nurse on December 21 he was going to die that day, and true to his word slipped away in his sleep at around six p.m., the end of sixty eventful years.

In retrospect, it was an excellent exit strategy. While far from being America's greatest general, he was arguably the best combat leader in the country's history. Put simply, he never lost a battle, and he terrified his enemies sufficiently that it was possible to build an entire false army around his reputation alone. As an operational commander, he took full advantage of mechanized armor and tactical airpower, racing forward when others hesitated, certain that P-47s would guard his flanks like Rottweilers. But the climax of industrial warfare that became Patton's stomping ground had not only passed, it had become with stunning suddenness a thing of the past. George Patton's gigantic personality was useful and necessary because it could be draped over hundreds of thousands of disoriented, frightened, and barely motivated souls, to convince them they were an army. Every aspect of the image he sought to project was exaggerated to play to the GI multitudes, to get them to fight and do it together.

Huge, conscription-based armies would lurch along for a while, but the future belonged to much smaller, highly technical forces, ones commanded by a different breed than George Patton—electronics-empowered paladins capable of moving seamlessly from the battlefield to the ranks of TV talking heads. It could be that the lasting allure of the 1970 film *Patton* has something to do with how absurdly anachronistic the character portrayed by George C. Scott—in a rasp the real general could only dream about—actually seems to generations of us never thrust into situations this desperate: black humor but also comic relief. At any rate, the movie ensures he won't be forgotten.

Meanwhile, once he was gone, Beatrice continued to settle scores. She made sure that Beetle and Ike not only didn't attend the funeral, but weren't even listed as honorary pallbearers. Then there was Jean Gordon. She had returned to the United States in November and was staying at an apartment in New York. When she heard of Patton's demise, she was sad, but thought it for the best. "There is no place for him anymore."

Shortly after, Beatrice, back in Boston, asked her own brother Fred to arrange a meeting with Jean at a local hotel. Fred booked the room and extended the invitation, he and Jean arriving with no idea what was planned. After a few minutes, there was a soft knock on the door; it was Beatrice. She studiously removed her coat and hat, then whirled, pointing a finger at Jean while chanting the Hawaiian death curse: "May the Great Worm gnaw your vitals and may your bones rot joint by little joint." The sheer malevolence in her voice caused Fred to flee, so what more was said remains unknown. But just days later Jean stuck her head in the gas oven of the New York apartment and committed suicide. Family legend has it that she left a note: "I will be with Uncle Georgie in heaven and have him all to myself before Beatrice arrives." That would not be until 1953, when, horseback riding with brother Fred, Beatrice suffered an aneurysm and died before she hit the turf, certainly a fitting end for a Patton.

PATTON OFF HIS PLATE, THERE WAS STILL A LOT OF HOUSECLEANING FOR IKE to accomplish before he could take on his new, or at least modified, persona. Much of this was unpleasant and fraught with difficulty, yet characteristically Ike proceeded purposefully and barely raised a ripple of protest.

No general likes to take his army apart. In Europe and then as chief of staff that duty largely fell to Ike, and he handled it expeditiously enough to quell the impatience of both soldiers and public alike. It was Eisenhower's headquarters that was saddled with transporting more than a million not-very-happy-about-it soldiers to the Pacific for the invasion that would never come. They at least were in no hurry. But the millions designated to go home were in no mood to wait around in their tents. "As I gnawed at the problem the answer occurred—to double the passenger load in the transports, sleeping and feeding the men in shifts," he later recalled. Without consulting his staff, Ike tried the idea out on a sea of thousands of very impatient former POWs. "Do you want to go back home conveniently—or would you rather double up, be uncomfortable, and get home quickly?" What he got in return was a roar of approval, and soon enough transports were teeming with cramped but otherwise happy troops itching to become civilians or at least get stateside.

Unlike many Cold War zealots, Ike instinctively tried to address his future adversaries as fellow human beings. He got to know and like Marshal Georgy Zhukov, a relationship that would pay dividends a decade later. He had extensive talks with Stalin in Moscow, even stood with him on Lenin's tomb for an endless parade. He also received the usual propaganda-laced tour of the shining marble metro, the GUM megastore, and Sturmovik airplane factory, but it was on the flight out to the Kremlin that he saw what most impressed him: almost uninterrupted war damage. The Soviet Union may have looked like a military giant at war's end, but Ike learned it was also a badly wounded one. Time well spent. Before he left Europe, he had managed to acquire considerable firsthand knowledge of the USSR he would one day have to stare down—and, fatefully, an insatiable thirst for more.

That was not the case with Kay. As with others in his past, she drew his favor because she was useful to him, the linchpin of the wartime version of the Club Eisenhower, a perfect amalgam of skills he needed—driver, gofer, social adviser, confidante, bridge partner, romantic interest, even personal trainer on horseback. Now those days were irrevocably over. Likely, Ike knew it when he returned from America. In a guilty gesture, he tried to give her a platinum and gold cigarette case de Gaulle had specially made for him, his five stars outlined in sapphires. It was a gift Kay had the good sense to decline, especially since she could no longer see her future in that tiny constellation.

It came in November, a week after Ike left for Washington to replace George Marshall as chief of staff, when a telegram arrived dropping Lieutenant Summersby from the general's personal staff scheduled to join him in Washington. That was followed by a typewritten letter telling Kay, "I am terribly distressed, first because it has become impossible to keep you as a member of my personal official family, and secondly because I cannot come back and give you a detailed account of the reasons." At the bottom was a handwritten PS: "Take care of yourself—and retain your optimism." A little on the cold side? One of Ike's best biographers, Jean Edward Smith, maintains, "George Patton would have said a warmer good-bye to his horse."

That may be true, but Ike didn't entirely abandon her. Not long after, orders to go to work for Lucius Clay, Ike's key handyman in occupied Berlin, and a promotion to captain came through for Kay. She settled in with

Telek, the Scottie they had used to charm FDR, and took up her role running the VIP guesthouse for an endless succession of dignitaries. She had seen enough rank during the war to not be impressed with more, and quickly grew bored.

Within a year she managed to wangle orders to the United States without Clay's noticing, and she arrived at the chief of staff's office in the Pentagon with Telek, greeted by a very red-faced Ike. "I soon got my new orders, I was sent as far from Washington, D.C. as one could get and still be within the limits of the continental United States: California." Kay took the hint and soon resigned from the Army. She would never see Ike again, but did move to New York and secured a book contract for a very domesticated telling of her days with the general, *Eisenhower Was My Boss*. She bounced through several more marriages, and saved the unexpurgated version, *Past Forgetting*, for 1976, seven years after Ike's death and a year after her own. Through it all she took care of herself and retained her optimism; she was a brave woman and deserved better.

WITHIN A MONTH OF TAKING OVER FOR GEORGE MARSHALL AS CHIEF OF STAFF in mid-November 1945, Ike was wondering about his own future, and also that of his profession: "This job is as bad as I always thought it would be," he complained in his diary, but then added food for thought: "I'm astounded and appalled at the size and scope of plans the staff sees as necessary to maintain our security position now and in the future. The cost is terrific. We'll be merely tilting at windmills unless we can develop something more in line with financial possibilities."

True to form, George Marshall badly wanted a permanently enlarged Army with a peacetime draft to support it as part of his legacy, and Ike obligingly carried the ball before Congress, testifying to that effect. But it put him in an ambiguous position, since job one as chief of staff had been to demobilize the prior army. He accomplished that "dreary business," as he called it, with dispatch, turning five million soldiers into civilians by the end of 1945, but it left him grumbling over the statistical techniques and electronic devices it took, things representing the next generation of technology beyond him. "Since then, I've always mistrusted, a little, even the most handsome, most intricate and guaranteed computer."

But he was also awed and fascinated by what it represented, and nowhere was this truer than with nuclear weapons. It was obvious he had done some initial exploration of this paradoxical terrain when asked about the American nuclear monopoly during congressional testimony just before becoming chief of staff: "Let's be realistic. The scientists say other nations will get the secret anyway. There is some point in making a virtue out of necessity."

Then, in late March 1946, Eisenhower obtained a preliminary draft of *The Absolute Weapon* and appears to have read Brodie's essays carefully, underlining and annotating key passages. Exposure to the Magna Carta of the nuclear age almost at its inception seems to have had an indelible effect on his thinking from this point forward: on one hand, holding out the promise of a future free of the kind of destruction that had so appalled him in Europe; on the other, the possibility of something much more horrible, but also the key to safety. Ike would never escape the embrace of this paradox, nor would we.

It's important to point out that this was highly contrarian thinking when the conventional wisdom entailed trying to find a path to nuclear disarmament. In October 1945, his manager, Harry Truman, had called for an international "renunciation of the use and development of the atomic bomb," then sent Bernard Baruch to the United Nations the following spring to negotiate such a deal. After fourteen months of negotiations George Kennan concluded they had gone nowhere: the Americans would not destroy their weapons without a Soviet guarantee of security, which the Russians refused to give without the bombs' prior destruction—all the while gaining time for their own atomic equivalent.

In the interim, Ike never wavered, criticizing the Marshall-guided Joint Chiefs study of the impact of nuclear weapons for not addressing the transitory nature of the US monopoly, and opposing the policies of maximum secrecy and continued emphasis on conventional arms. Then, in his own report to the chiefs in late January 1946, while warning against "excessive reliance," he urged that "all possible methods of delivery of atomic weapons . . . be studied and developed." It was a position from which he seldom strayed until, much later and like the sorcerer's apprentice, he stepped back to gape in astonishment at what had been created, as well as at the military-industrial complex that went with it.

But as chief, Ike didn't yet allow adjusting to the nuclear future to dominate his attention; it couldn't. By its nature the position demanded almost constant interaction and travel; plus, Dwight Eisenhower had good reason to see and be seen by as many Americans as possible. In barely over two years, he gave seventy-six major speeches, frequently at universities; testified before Congress twenty-six times; and visited every state in the union at least once. He became in the process not simply a transitory war hero, but a familiar and ingratiating presence to America at large, particularly among movers and shakers.

He even flew fourteen thousand miles to visit Douglas MacArthur, the man widely perceived to be his prime competitor for the nation's highest office. "He came and told the soldiers he would get them home to mother and they gave him, 'Three cheers and a Tiger. Hip, Hip, hooray,'" MacArthur remembered sourly. But both were more interested in presidential politics than they were in troop levels, with Mac pressing Ike on whether he was going to run in 1948, and Eisenhower wondering if he was being enlisted by his former boss for a run of his own.

Meanwhile, his manager, Harry Truman, a political animal to his core, had recognized Ike's presidential possibilities from their first meeting in Potsdam, and nothing subsequent had changed that assessment, complicating the relationship and leading to an atmosphere of mutual misunderstanding, then distrust, and ultimately true dislike on Eisenhower's part.

While he was chief, interactions with Truman remained cordial but distant. Still, a continuing irritant was likely to have been Truman's obvious and extreme admiration for George Marshall—calling him "the greatest military man this country has ever produced" at his retirement ceremony. It was a line likely to have gotten under Ike's smiling facade; after all, he, not Marshall, had been chosen Overlord and become victor in Europe. Nor were Eisenhower's military views shown the deference that Marshall's had been by FDR and then by Truman; never was he drawn into the president's inner decision-making circle. Hence Ike, a master bureaucratic player himself, did his job scrupulously but at arm's length—a burden to be endured on the way to his future.

Another relationship in need of mending was with Mamie: here he did better. With the exception of two brief interludes, they had been separated for three and a half years when they resumed domesticity in Fort Myer's spacious quarters no. 1. Both had changed considerably, grown more willful and independent. Ike, of course, was used to being obeyed by everybody in sight plus millions more, but Mamie too, as wife of a big-time general, had developed her own coterie and become a social arbiter in her own right. She was not about to be steamrolled or marginalized by an overbearing husband.

Yet there was plenty of room for compromise between these two. Ike wanted things his own way, but he was no domestic tyrant. He was happy to leave lifestyle decisions to a partner who had long since shown herself capable at making them. Meanwhile, Mamie loved the service and luxury her husband's career was bringing them. She was an indoor girl, she craved the soft life, one full of Dorothy Draper "modern baroque" interiors with plenty of servants, and that's what she got. Then there was the force of habit and familiarity; this couple had been together for a long time and been through a lot. Kay may have been stylish and stimulating, but Mamie played to Ike's comfort zone like no one else. So they settled in again together, if not redis-covering what they had before, glad to share its luxurious successor.

Having served his agreed-on time as chief, Ike was ready to recast his image—the same smiling mask, but this time clad in civilian garb; academic gowns, no less. On February 6, 1948, he put his uniform aside (five-star rank allowed him to avoid full retirement) and began the big makeover.

First, he would write a book, as might be expected a war memoir, but one guaranteed to provide him with a very comfortable chunk of change through sale of all rights for a lump sum of $635,000. That allowed Ike to claim it as a capital gain taxed at 25 percent, not income, where the rate was 82 percent, thus netting him $500,000, or $5.4 million in today's money. But if Double-day and William Robinson of the *New York Herald Tribune* put together a sweetheart deal for Ike, they absolutely got their money's worth too.

He told them he planned to produce the book during his three-month terminal leave, a promise bound to engender skepticism if not outright laughter among veteran publishers. But he was serious, working sixteen-hour days, dictating up to five thousand words a session to a tag team of three secretaries backed by a researcher, who checked them against Eisen-hower's own carefully assembled diaries, orders, and records. He had risen to prominence writing for Pershing, MacArthur, and Marshall; World War II was a nonstop gabfest; by this time, syntactic coherence was second nature to Ike. Using Grant's much-praised memoirs as a model of simplicity and clarity, Eisenhower produced a gem of his own, unpretentious, understandable, and uninterested in settling wartime scores. But unlike Grant, who does not even mention and may not have understood the huge impact rifled small arms had on the Civil War, Ike did not fail to reflect on the bomb:

> Henceforth, it would seem, the purpose of an aggressor nation would be to stock atomic bombs in quantity and to employ them by surprise . . . while the defense would strive to prevent such delivery and in turn launch its store of atom bombs against the attacker's homeland. . . . With the evidence of the most destructive war yet waged . . . about me, I gained increased hope that this development . . . the ultimate in destruction would drive men, in self-preservation, to find a way of eliminating war. Maybe it was only wishful thinking to believe that fear, universal fear, might possibly succeed where statesmanship and religion had not yet won success.

It wasn't wishful thinking, it was the new future, and Ike grasped it almost from the beginning.

Upon receiving the manuscript in record time, Ike's very pleased publisher, William Robinson, whisked him and Mamie off for a two-week vacation at the Augusta National Golf Club in Georgia. Ike had played irregularly in the Philippines, but he had gotten serious on the course behind Telegraph Cottage, where he'd hacked away with growing intensity and dedication. The original Ike was an excellent athlete, but his knee injury at West Point, aggravated by an incident with a small plane during the war, left him with a stiff swing prone to slicing and a furious incomprehension at his merely above-average results. That only drove him to more golf, which along with bridge became his chief recreational activity.

Robinson aimed to please, and had waiting for Ike at Augusta a select group of golfers and bridge players to keep him company. They turned into quite a bit more. All his life useful people showed up just when he needed them. Now he was in the market for an updated Club Eisenhower, a civilian version that reflected Ike's new interests and image. These guys were it. All of them were rich businessmen and Republicans, save George Allen, a Democrat and corporate lawyer Ike had known in London. This being an all-male iteration, the name was changed to "the Gang," and the roster was fleshed out by Robert Woodruff, board chairman of Coca-Cola; W. Alton Jones, CEO of what would become Citgo Petroleum; Ellis Slater, president of Frankfort Distilleries; and Clifford Roberts, an investment banker, who with the golfer Bobby Jones had founded Augusta National—in other words, a pack of plutocrats. They made Ike a club member, built him and Mamie a cottage off the tenth tee, and subsequently were instantly available when he needed a weekend of bridge and time on the links. Also, like compass needles, they pointed in the direction of Dwight D. Eisenhower's political comfort zone.

But that was not necessarily the image he wanted to project, so his next move he played like a hand of poker. After speaking at the Metropolitan Museum of Art, Ike had drinks with Thomas Watson, the CEO of IBM and also a member of Columbia University's board of trustees. Watson got right to the point: Nicholas Murray Butler, Columbia's president of forty years, was stepping down. Was Ike interested in the job? What about Milton, Ike wondered; he was the Eisenhower brother with the academic record. That instantly forced Watson to raise the stakes, admitting that Columbia needed someone with an international reputation, thereby conceding Ike's value. Still he kept Watson sweetening the pot—zero academic responsibilities, minimal entertaining and fundraising—until Ike laid down his cards, having bluffed his way to a winning hand and an academic future.

His timing could not have been better. Four months after he was installed as Columbia University's thirteenth president, in June 1948, *Crusade in Europe* was released to transatlantic critical acclaim and huge sales, giving Ike just the credential needed among potential naysayers on the Morningside Heights campus. Columbia was the perfect strategic retreat, but also one calculated to burnish his reputation as a man above the fray. Ike may have been enormously popular, but 1948 was the wrong presidential race for him. On the Republican side, the one he wanted to be on, Thomas Dewey, governor of New York and GOP presidential nominee in 1944, had all but sewed up the nomination, and Ike was wise to discourage the inevitable nibbles he received from both Republicans and Democrats.

Instead, he and Mamie settled into their sumptuous Dorothy Draper–redecorated quarters and, riding a wave of approval and popularity, the new president set about running the place more like a modern manager than prince of an academic fiefdom—getting Columbia's finances in order, persuading New York City's mayor to turn West 116th Street into a pedestrian mall where it bisected the campus, and talking the football coach and a Nobel Prize–winning physicist out of defecting to Yale and Princeton respectively. But he never really engaged with the faculty, at one point declining to host the traditional reception for honorary degree recipients in favor of a private dinner with Lucius Clay and his wife. Also, Ike's academic innovations, his American Assembly, and particularly the Citizenship Education Project, struck some as superficial and lowbrow—"jejune" was likely a word applied. His reputation on campus also faded in part because he was there so little, only for ten of his first twenty-seven months as president.

It hadn't taken him long to realize that the reason academic politics were so vicious, as Henry Kissinger used to say, was because the stakes were so small. After running a war and then the Army, Columbia must have seemed like a penny ante game to Ike. There were also political calculations. He had listened to the returns of the '48 election over a night of bridge with his new friends Cliff Roberts and Bill Robinson, fully expecting Dewey to win, and when he didn't, he was "just as disappointed as Robinson and I were," Roberts remembered, and also "indicated quite clearly to me that he was having second thoughts about his decision to stay clear of political involvement." Dewey had lost twice, and the 1952 Republican nomination was suddenly a target of opportunity. But to occupy that ground Ike realized that Morningside Heights wouldn't do as a jumping-off point; he needed to get back in the national spotlight, and that meant Harry Truman.

In short order, he wrote the reelected team manager about rejoining the lineup: "I always stand ready to attempt the performance of any professional duty for which my constitutional superiors believe I might be specially suited."

That was pretty unctuous, but Truman did need help at a key position. The National Security Act of 1947 had unified the Army, Navy, and the now independent Air Force under a single Department of Defense, and his first pick for secretary, James Forrestal, was over his head and floundering emotionally, while the three services were increasingly at each other's throats. In late December Ike agreed to become acting chairman of the Joint Chiefs and institutional hand-holder for Forrestal, his idea of a part-time job, and add to that commuting back and forth between Washington and Morningside Heights. This lasted until February 1949, when the service tussles over the future of naval air and the gigantic B-36 bomber, followed by Forrestal's collapse and resignation, demanded Ike's full-time presence. Columbia's trustees gave him a leave of absence, but Mamie remained on campus, with Ike marooned at Washington's Hotel Statler—that is, when he could get away from the Pentagon. Piling more on, the new secretary, Louis Johnson, told Ike he expected him to stay for the "next six months *at least*. He says he told the Pres. he'd take job only if I stayed on!"

So he kept it up—the long hours, the four packs a day, the heavy hotel food—until it exacted its toll. On March 21, 1949, at a luncheon with a group of movie moguls, Ike doubled up like a jackknife, and by the time they got him back to his hotel room "his stomach was bloated and getting larger. It seemed like a balloon," remembered one aide. Eisenhower knew he was seriously ill, later recalling that "General Howard Snyder, my physician, treated me as though I were at the edge of the precipice and teetering a bit. For days, my head was not off the pillow."

The diagnosis was acute gastroenteritis, and Harry Truman deployed the *Sacred Cow* to fly him to Key West for further tests, two week's bed rest, and no solid food or cigarettes. The latter he quit permanently—"I gave myself an order," he liked to maintain—while the former gradually restored his vigor over the next month at Augusta National, reunited with Mamie and his golf clubs. "I feel stronger every day," he told Secretary Johnson in late April, "though I must admit the 'drives' are somewhat short of their expected destination." It had been an instructive episode—he'd had several prior gastrointestinal attacks but this was the first time mortality got Ike's attention, at least enough to persuade him he could no longer ignore his health on his way to the top, a lesson well worth learning at a time of personal renovation. His recuperation also forced Ike into the longest period of true relaxation within memory, one certainly deserved.

ON NOVEMBER 29, 1945, JUST OVER A WEEK AFTER THE VERY TIRED GEORGE Marshall finally shed the burden of chief of staff, he learned that there was such a thing as being too impressive to a very important person. The phone rang at Dodona Manor, the columned mansion Katherine had purchased for their retirement in the Virginia countryside. It was Harry Truman. "Without any preparation, I told him: 'General, I want you to go to China for me.' Marshall said only, 'Yes Mr. President,' and hung up abruptly."

Whether Marshall saw this coming is hard to tell. Truman himself had been blindsided when his ambassador to China, Patrick Hurley—the same Hurley who as Hoover's secretary of war thought MacArthur had done such a great job with the BEF—had come home for instructions, then abruptly and loudly resigned, denouncing the State Department, administration foreign policy in general, and the president in particular. Even from Truman's perspective China was a hornet's nest, but he still couldn't see why Chiang Kai-shek's Nationalists and Mao Tse-tung's Communists couldn't put aside their differences for the greater good and give democracy a try. Who better to catalyze that process than the quintessential nonpartisan and his favorite general, one whose days in Tientsin gave him at least the veneer of a China hand. So George Marshall became his special ambassadorial envoy, scheduled to leave immediately.

He flew by way of Tokyo, spending the night at the US embassy, where his mood and demeanor immediately caught Douglas MacArthur's attention: "Mentally, he had aged immeasurably since his visit to New Guinea. The former incisiveness and virility were gone. The war had apparently worn him down into a shadow of his former self." Yet Mac missed that the remaining shadow was still a very dutiful and determined one, and also a master at delegation; thus, he would persist in this fool's errand and later in much more promising endeavors.

His apparitional fortitude was revealed the next day when he reached his destination and met with Stilwell's replacement, Brigadier General Albert Wedemeyer, who told Marshall "he would never be able to effect a working arrangement between the Communists and the Nationalists. . . . [Chiang] would not relinquish one iota . . . [while the Communists] were equally determined to seize all power with the aid of the Soviet Union." To this admonition Marshall gave short shrift: "I am going to accomplish my mission and you are going to help me."

From our own perspective, it was an almost laughably naive comeback, but not atypical, especially if you were one of Team America's heavy hitters.

Although communism may have been on the move, the United States had never been stronger—no country had. With the country possessing over half the GDP of a war-devastated world and having suffered virtually no physical damage itself, and being the champion of democracy, victorious and respected everywhere—plus being the sole possessor of the atomic bomb—it was easy to see the future with a tinge of megalomania. Wedemeyer, though pro-Chiang and no fan of Marshall's, was not without insight into his approach: "His reputation was so great and his political influence . . . so overpowering that he thought he could accomplish the impossible."

Specifically, Marshall was charged in a cover letter signed by Truman with finishing the evacuation of Japanese troops from Manchuria, followed by the departure of the fifty thousand US Marines supervising that process; negotiating a cease-fire among the warring Chinese; and finally, convincing Chiang to convene a conference of all parties, including the Communists, aimed at national political unification. Hercules might have had it easier, given that the withdrawal of the Marines in the north was bound to create a power vacuum and further violence, while Marshall's only power lever was a promise of US dollars in a pacified future, a diplomatic non sequitur bound to lead nowhere.

Nonetheless, duteous George plowed ahead with his tenacious two-step. He had been warned about Chiang's duplicity by his friend Stilwell, yet Marshall still reassured him in their first meeting that it was the Communists who would have to "relinquish autonomy" over their army in order to earn "sympathy in the United States." This was exactly what Chiang Kai-shek wanted to hear from his American patron, and he reciprocated with an apparent willingness to go forward with the US agenda.

The next day Marshall continued his charm offensive, meeting with Zhou Enlai, Mao's deft and sophisticated mouthpiece, and hearing similarly encouraging words. Not only were the Communists ready for a "cease firing," but as a basis for further negotiations they would concede Chiang and his party "would be in first place" in the new power structure. Zhou even concluded the meeting by telling Marshall the Communists wanted "American style" democracy.

Almost magically the cease-fire agreement materialized in early January 1946, and the Political Consultative Council (PCC) began meetings on what appeared to be a constructive note. By February 4, Marshall could report back to Truman that the council and its deliberations resembled our own "Constitutional Convention," and that it was "doing its job well." Shortly after, Marshall could tell Truman the "prospects are favorable for a solution to this most difficult of all problems," the integration of the Communist and Nationalist military force structures. That too was obligingly accomplished after several extended rounds of negotiations among Marshall, Chiang, and Zhou, the three signing a formal agreement on February 25. "I can only trust that its pages will not be soiled by a small group of irreconcilables," George Marshall concluded at the ceremony.

Actually, there were plenty of irreconcilables and only a small amount of trust to be had. On the Nationalist side, they were the "CC Clique," the far-right wing of the overarching Kuomintang Party, a key element of Chiang's grip on power and unalterably opposed to unification with Communists. As far as they were concerned, Marshall was just a meddler.

Then there was Mao. Marshall wanted to meet him, take the measure of the man. Consequently, as part of a larger public relations tour to talk up the agreements, his C-54 touched down in Yan'an, Communist Party headquarters, where Mao and his associates lived in thousands of caves carved into the sides of hills. Pear shaped, round faced, dressed in rumpled peasant garb and a Red Army cap, he was nothing like the elegant, ingratiating Zhou. Enigmatic and beyond measuring, he "remained a mystery," Marshall later conceded. Nonetheless, he did "assure" him that "the Chinese Communist Party would abide wholeheartedly by the terms" of the cease-fire and military consolidation. That was good enough for George, though he would have been interested to learn that, twelve days earlier, Mao had told the Yan'an politburo that "the United States and Chiang Kai-shek intend to eliminate us by way of nationwide military unification."

Instead, Marshall wrote Harry Truman that it was time for a trip back to Washington to lobby Congress and the bureaucracy for the funding to support a reconciled China. Before he left, at Jinan on March 2, he described America's role in China in baseball terms: "We have not the authority of an umpire, but we endeavor to interpret the rules and agreements that have been arrived at in Chungking. And baseball goes along with American democracy." The problem was these Chinese were interested in neither. "The

Americans tend to be naïve and trusting," Chiang confided. "This is true even with so experienced a man as Marshall."

It's no surprise that his paper-only agreements started to come apart almost as soon as he left. The removal of the American Marines and the rush to reoccupy Manchuria doomed the cease-fire, and his diplomatic staff urged him to return immediately. But Marshall was exhausted and wanted to bring Katherine, so he did not get back to Chongqing (Chungking) until April 18. One could say George dropped the ball here, but this presumes it was a real ball in the first place. The fact was, the Nationalists and Communists were mortal enemies, and a showdown between them was a long time coming and all but inevitable no matter what George Marshall said or did.

But being who he was, he grimly set about trying to bring the sides back together, a doomed and thankless task. On May 6, he told Truman he had reached "an impasse" and that "the outlook is not promising," not for China and not for George Marshall, Team America's man in the field, stuck on base without hope of scoring.

But then three days later Ike arrived with good news from the manager. Now chief of staff and planning an inspection tour of the Far East, Eisenhower had been taken aside by Harry Truman and asked to deliver a message to Marshall: Secretary of State Byrnes intended to resign; was he interested in the job? When Ike popped the question, Marshall greeted it with uncharacteristic enthusiasm: "Great goodness, Eisenhower, I would take any job in the world to get out of this place. I'd even enlist in the Army." But not until September, he told Ike; by that time, he hoped to have a cease-fire and both sides back on the road to unification. "But this gives me a wonderful ace in the hole because I have been terribly worried."

While Madame Chiang swept Katherine up to the couple's luxurious retreat in the mountains, with her own stone cottage overlooking the pool, George trudged through the steamy summer in dogged and futile pursuit of agreement. His self-imposed deadline of September came and went without progress, but still he would not give up. In his last policy discussion with Chiang he warned him that although he had won several recent battles, the Communists were too widely dispersed for him to ever fully defeat them with the economic resources available from China's economy; his only alternative was compromise. After this prompted a one-hour intransigent rant from Chiang, Marshall turned to Madame Chiang, who was acting as translator, and told her he had a final message for her husband: "You have broken agreements, you have gone counter to plans. People have said you are a modern George Washington, but after these things they will never say it again."

Once Truman read his report of the meeting, he told Marshall he could terminate the mission and come home. But Marshall hung on for a few more weeks to watch over the passage of a meaningless constitution by a National Assembly devoid of Communists and most other political parties.

LATER, AROUND THREE HOURS INTO HIS FLIGHT BACK TO WASHINGTON, MAR-shall's pilot walked back from the cockpit to deliver a radio transmission: "Congratulations, Mr. Secretary Marshall." In his absence, Truman had submitted the nomination to the Senate Foreign Relations Committee, chaired by Republican Arthur Vandenberg, and, without hearings or opposition, he passed it to the larger body, which voted unanimously on January 8, 1947, to make George Catlett Marshall the nation's fiftieth secretary of state. It clearly helped that the China mission had taken place on the other side of the planet, but this process was an accurate measure of Marshall's prestige. While one member of Vandenberg's committee did worry that the appointment might look like "the military is taking over our foreign policy," the senator himself wondered if the general might soon be swept into the White House on a wave of esteem.

Characteristically, Marshall moved immediately to pour nonpartisanship on even these not very troubled waters: "I am assuming that the office of Secretary of State, at least under present conditions, is non-political and I am going to govern myself accordingly. I will never become involved in political matters and therefore I cannot be considered a candidate for any political office." That this statement was taken seriously was a testimony to the degree Marshall had Washington mesmerized at this point.

That plainly started at the top. Barely weeks after Marshall was sworn in, Harry Truman gushed, "The more I see and talk to him the more certain I am that he's the great one of the age." And it extended to his new colleagues at State, though they were very much upper-crust civilians and already feeling under assault by a wave of military or former military now occupying key posts. "The moment General Marshall entered a room everyone in it felt his presence," observed his undersecretary Dean Acheson. "It was a striking and communicated force. His figure conveyed intensity, which his voice, low, staccato, and incisive, reinforced. It compelled respect. It spread a sense of authority and of calm. There was no military glamour about him and nothing of the martinet."

Of course there was. It was the same mask of command, just dressed in a suit. He would not change, because he could not change; but he could still overawe, and that's how he operated. "He said to me," Acheson later wrote, "that I was to be the chief of staff and that I was to run the Department and that all matters between the Department and the Secretary would come to the Secretary through me with my recommendations, whatever it might be, attached to the action proposed." Same as it ever was—everything that could be delegated, would be. He was still a very tired man in his late sixties, and it remained his best weapon against overwork—an average day began at nine a.m. and ended at four thirty p.m., including an early-afternoon nap. And while his mental black book lacked a section for diplomats, Marshall retained his eye for talent and used it to gather a superb roster around him—not only Acheson, but also George Kennan, Chip Bohlen, and Robert Lovett, with his old staff officer Beetle Smith already in Moscow as ambassador.

But the most lasting mark of Marshall on the State Department was his almost reflexive streamlining of administrative procedures in what had been a jumble of review and reassessment with practically everybody free to join in; that was replaced by a direct line of authority and a specifically designated Policy Planning Staff. It added up to the same administrative architecture Marshall had been perfecting for decades, but applied courteously and with the gravitas to make it work at State.

Most fortuitous was his think tank at Policy Planning. The world was in the midst of tectonic political change: not only had America been thrust above all, but Europe lay in ruins, its overseas empires doomed, and looming was the dark cloud of communism threatening to inundate everything. It was a time for big thinking by people who knew what they were talking about. George Kennan certainly fit the bill.

As we saw with his dealings with MacArthur, Kennan could be vindictive and mandarin as a bureaucrat, but as a formulator of fundamental policy he stands with very few other Americans, most of them presidents. A Russian expert of long standing, Kennan was the author of the fabled "Long Telegram" sent from Moscow in February 1946, which not only incisively stated the magnitude and irreconcilable nature of the Soviet threat, but also proposed means of thwarting it through what he called "adroit and vigilant application of counter-force"—coercive means short of all-out war, employed over the long term to allow the internal contradictions of the system to destroy it from within. Sound familiar? That is what actually happened four decades in the future. Kennan had written the libretto for the Cold War: containment, a strategy to meet and defeat the Soviet Union without resort to nuclear weapons. It kept us alive.

But at the outset Kennan and his seminal policy planners still had a lot of thinking and convincing to do. The number one skeptic was Marshall himself, who was the last of the major World War II players to conclude the Soviet Union was the new adversary. While in China he remained unconvinced that the Soviets were an important force, much less a dictating one, behind Mao and the Communists. Now, as secretary, his conversion took place over only a matter of months.

It began with startling news. On Friday, February 21, the British ambassador phoned Acheson that he was about to deliver a "blue piece of paper" to the secretary—dip-speak for an urgent message. Put simply, His Majesty's Government was broke and could no longer provide financial and military aid to Greece and Turkey, both in imminent danger of falling under the sway of the Soviet Union. Would Team America pick up the tab?

Over the weekend Acheson and his staffers put together a position paper arguing that the situation was dire and required that the United States step up to the plate and pinch-hit for the Brits immediately. When briefed on Monday, Marshall responded that he was inclined to agree, but then characteristically delegated responsibility for developing aid packages and moving them forward to his undersecretary. Next, the two went to the White House, where convincing Harry Truman proved easy.

What was not going to be easy was convincing Congress, where both houses were controlled by traditionally isolationist Republicans. This took place in two steps. The first was a February 27 bipartisan gathering of congressional leaders from both houses, plus respective ranking members of the foreign affairs committees, at the White House, called by Truman "to advise them of the gravity of the situation and the nature of the decision I had to make." He then passed the ball to George Marshall, the supremely nonpartisan man, trusted by both sides of the aisle. While all agree he did make it "quite plain that our choice was either to act or lose by default," Marshall's dry presentation and lack of passion did not seem to be lighting a spark of urgency. Acheson thought his boss "flubbed his opening statement," and felt compelled to speak up in more lurid terms, arguing that unless Greece and Turkey got aid, the "infection" would spread to "Iran and all to the east," to "Africa through Asia Minor and Egypt," and "to Europe through Italy and France." Depending on the source, the key man in the room to convince, Arthur Vandenberg, may or may not have encouraged Truman to say exactly that sort of thing before Congress, but the meeting left the president sufficiently encouraged to move to step two.

He too decided to go lurid in his short but hard-hitting address to a joint session of Congress on March 12. Prior to delivery, Kennan was shown the text and didn't like it, while Marshall objected to the tone, saying there was "too much flamboyant anti-Communism in the speech." Truman not only requested aid for Greece and Turkey, he also invited all countries threatened by the Soviet Union to call on the United States for help. A startling, open-ended agenda was unveiled before hitherto isolationists, but it had the desired effect. Within days it was being called the "Truman Doctrine," and by early May the Greek-Turkish Aid Act was approved by big Republican majorities in both houses. The Cold War was on.

MEANWHILE, THE CONVERSION OF GEORGE MARSHALL INTO SOMETHING RE-sembling a cold warrior continued in April 1947 at the Moscow conference of foreign ministers aimed at resolving the postwar fate of Germany. Described by one journalist as being "as rigid as the Washington Monument" and holding himself aloof from other delegates, Marshall nevertheless interacted enough to find out that not just Germany, but all of Western Europe was in parlous political and economic shape.

Then, after forty-three progress-free meetings, Marshall decided to go see Stalin. The Soviet leader seemed affable enough, doodling wolves' heads on a blue pad, but as Bohlen, acting as translator, reeled off his answers to Marshall's questions, the secretary realized he cared not a whit about progress at this conference or any other. The present conditions in Europe were exactly those in which communism thrived, low-hanging fruit ripening through inaction. "We thought they could be negotiated with. . . . I decided finally at Moscow . . . that they could not be," he remembered. "It was the Moscow conference, I believe, that really rang down the Iron Curtain."

Back in Washington he met with Kennan and gave him the task of formulating a set of comprehensive recommendations aimed at helping Europe get back on its feet through large injections of American cash and other assistance. What he got back in the famous May 23 PPS/1 memorandum appears to have crystallized Marshall's emerging viewpoint:

> The Policy Planning Staff does not see communist activities as the root of the difficulties of Western Europe. It believes that the present crisis results in large part from the disruptive effect of the war. . . . American effort in aid of Europe should be directed not to the combating of communism as such but the restoration of the economic health and vigor of European society. . . .
>
> The formal initiative must come from Europe, the program must be evolved in Europe; and the Europeans must bear the basic responsibility for it. The role of this country should consist of . . . the later support of such a program."

It was an agenda that exactly meshed with George Marshall's basic bureaucratic instinct—when in doubt, delegate—so he ran with it. Acheson gave the prototype a test run before a business group in Mississippi, and when the reception was positive Marshall decided it was ready for a more public viewing, choosing Harvard as an appropriate unveiling venue. Bohlen did the drafting for what he told the university's president would be a few remarks and maybe "a little more."

The morning of June 5, 1947, was beautiful and it was commencement, but as the *Harvard Crimson* explained, "a crowd of 15,000 showed up in the Yard not so much in expectation of seeing history made, as simply in awe of the man." Harvard's president put the vibe further into context by calling

Marshall a "soldier and statesman whose ability and character brook only one comparison in the history of the nation."

Being compared to George Washington is a hard intro to follow, but since the first George wasn't a great public speaker, not impossible. Marshall's flat, measured delivery was belied by the cogency of his words. "Our policy is directed not against any country or doctrine but against hunger, poverty, desperation and chaos. Its purpose should be the revival of a working economy in the world so as to permit the emergence of political and social conditions in which free institutions can exist." He also made it clear that he was not excluding the Soviet Union or its satellites. "Any country that is willing to assist in the task of recovery will find full cooperation, I am sure, on the part of the United States Government." Nor did he fail to mention the delegation clause—"The initiative, I think, must come from Europe"—or the urgency of action in his parting words: "The whole world's future . . . hangs on the realization by the American people of what can best be done, of what must be done."

While the secretary's words were little noted by the American public at large, they reverberated across the Atlantic and back to the US policy community to change history in the form of the Marshall Plan, an unprecedented act of national generosity that achieved exactly what it promised—the restoration of Europe's economic, political, and social fabric—plus a whole lot more. Arguably that short speech in Harvard Yard marked the most significant few moments in George Catlett Marshall's life, which is interesting in light of Douglas MacArthur's reformation of Japan and Ike's future presidency. Of our four megasoldiers, only Patton's greatest deeds came in warfare alone.

The Marshall Plan's aspirations and magnitude were extraordinary. Without saying so, it promised a peaceful alternative to the Truman Doctrine—investment in recovery rather than open-ended confrontation. It was "adroit counterpressure," and the money involved was huge. The Europeans originally asked for $22 billion ($256 billion in 2022), and Congress eventually authorized a total of $13.3 billion ($155 billion) in grants and loans; actual transfers amounted to more than $12 billion ($130 billion) between 1948 and 1951.

Such international generosity from a country with a long isolationist tradition was sufficiently sui generis to beg further explanation. It was not just
a matter of beating back communism; a great deal of self-interest was involved. At this point the United States was the world's greatest producer of
just about everything, one with production capacity well in excess of domestic demand; the lure of vast overseas markets beckoned, especially Europe's.
But to buy American they needed money: to ensure future cash flow, the
pump needed to be primed. This was exactly how it worked out, at least for
a while, and helps explain why the Marshall Plan floated.

The other factor was Marshall himself. He seems to have known from the
outset that it would take a huge effort to put this one across. His health was
not good—a vastly enlarged kidney would require removal in a year—and to
compensate he went the next step in delegation by bringing an alter ego on
board as undersecretary. Robert Lovett, another banker bored by business,
had come to Washington in 1941 to administer the expansion of the Army Air
Force, no small task and done Marshall's way. Assigned the job of procuring
tens of thousands of aircraft, Lovett proceeded without pestering him. But
Marshall didn't just trust him to perform; they thought alike and almost always reached the same conclusions. Administratively speaking, they might
as well have been the same person, and that's how Marshall deployed him.

Suitably reinforced, he set about to campaign for the plan. In mid-October,
it was Lovett who testified to Congress on the need for an interim aid pack-
age, which was passed in late December and cost $522 million ($6 billion in
2022). At this point Harry Truman submitted the full European Recovery
Program: "I've decided to give the whole thing to General Marshall. The
worst Republican on the Hill can vote for it if we name it after the General."

Marshall got the message: with naming rights came the responsibility of
getting it through Congress, and he subsequently let loose the full range
of his persuasive modes. With the Senate Foreign Relations Committee he
was the soul of reasoned argumentation, capitalizing on his copacetic relationship with its chairman. "We couldn't have gotten much closer together
unless I sat in Vandenberg's lap or he sat in mine," he later recalled. But
before the House Foreign Affairs Committee he let loose fire and brim-
stone: "Left to their own resources there will be no escape from economic
distress so intense, social discontents so violent, political confusion so

widespread, and hope for the future so shattered that the historic base of Western civilization . . . will take a new form in the image of the tyranny we fought to destroy in Germany."

Having pointed Congress in the right direction, Marshall took the plan on the road in mid-January 1948 to rally public support, leaving Lovett in Washington to cover for him. "Oh Lord I traveled all over the country," he later recalled. "I worked on that as hard as though I was running for the Senate or the Presidency. That's what I'm proud of, that part of it. . . . It was just a struggle from start to finish."

As always, Marshall remained a team player. He took no credit for the formulation of the plan, but he got it done, more than earning naming rights. In April, the Marshall Plan passed both houses of Congress by big margins and was delivered to Harry Truman's desk. Mission accomplished. But it hadn't been easy: "God bless democracy!" he wrote an acquaintance. "I approve of it highly but suffer from it extremely."

Nor was the solution to his and America's basic problem, the Soviet threat, simply a matter of cash flow. All the economic aid in the world couldn't erase the glowering Russian military presence already occupying most of central Europe and pointed west. In 1947 the United States had 200,000 troops stationed in Europe, compared with 1.1 million Soviets in a tank-heavy juggernaut designed for offensive operations. The Truman Doctrine was a temporary stopgap, and it did not address this central threat in any institutional manner. The United States, of course, did have the bomb, but still not many, and it had little in the way of doctrine as to when and how it might be used, and, more important, no clear understanding of its power as a deterrent. So Western Europe was left reeking vulnerability.

The seed of a solution was planted when George Marshall dropped by the office of British foreign minister Ernest Bevin in mid-December 1947, after another unsuccessful conference with the Russians in London. Bevin told Marshall he believed the power to resist could come from some sort of alliance of democracies, perhaps not formal but "a spiritual federation of the West." He followed this up with something more tangible, a paper delivered to Marshall in mid-January 1948 proposing, under the sponsorship of the United States and Canada, a coalition of Europe's democracies, including Britain, France, the Low Countries, Scandinavia, Greece, Italy, and eventually Germany.

Kennan worried such an alliance would mark "a final militarization of the present dividing-line through Europe." Marshall, though favorably disposed, hedged in a characteristic manner. It was only after Bevin had introduced his proposal before the House of Commons that he learned from Lovett in London that the US approach hinged on delegation—Europe should take the initiative with a regional alliance that, once formed, could be joined by the democracies of North America. Propelled by a Soviet-backed Communist coup in Czechoslovakia in February, the regional alliance fell together like a self-organizing entity, beginning with an initiative among Benelux countries, soon joined by Britain, France, and then, feeling very pressured by the Soviets, Norway.

But back in Washington, US participation was clouded by Arthur Vandenberg's worries over additional guarantees to Europe. But after the international situation continued to heat up and Marshall testified before his committee, he reversed course and put forward a resolution favoring regional agreements. Then, on June 23, Marshall alerted Canada and the Europeans that Team America was ready for exploratory talks on forming the new league.

Marshall sent Lovett for the initial rounds, but he took the *Sacred Cow* to Paris in late September 1948 when talks grew more serious. While there, he and Katherine flew to Italy, low over Allen's last battlefield, then on to Rome for an audience with the pope. During their talk Marshall brought up the atomic bomb and its perceived role in deterring Soviet aggression. He also voiced his concerns over the morality of threatening to use a weapon capable of instant incineration of human beings by the hundreds of thousands. But the Holy Father said nothing critical and "seemed to indicate general approval," Marshall noted in a memorandum of conversation.

Atomic weapons were plainly on his mind. Brought up in a world without them, Marshall was obviously aware of the vast disparity in conventional forces between the looming Soviet land armada and the still just contemplated Atlantic alliance. But by November "I felt the Soviet leaders must now realize that the use of this instrument [by the United States] would be possible and hence the deterrent influence." On this basis he reassured the Europeans that, although it would take time to build a conventional force capable of giving the Soviets pause, it could be done safely under the wing of the US nuclear monopoly. At this point, George Marshall certainly still believed in the primacy of conventional arms, but circumstances backed him into the corner of nuclear deterrence, as they would, eventually, everybody else.

So had his right kidney backed him into a corner, and after Thanksgiving he entered Walter Reed to have it removed. He didn't leave the hospital until the end of December, and Harry Truman sent him to Puerto Rico to recover where it was warm. In November, his manager had unexpectedly won reelection, but there would be no second term for George Marshall. On Inauguration Day, January 20, 1949, his resignation became effective. Blessed retirement, at least for a while.

MEANWHILE, IN MARSHALL'S ABSENCE, LOVETT HAD CONTINUED THE NEGOTIations with the about-to-be allies in Washington, formalizing the North Atlantic Treaty Organization (NATO) in April 1949. Fueling that process had been a year of steadily rising tensions with the Soviets, centered on Berlin, which they blockaded, and Ike's man Lucius Clay, along with an inspired airlift, managed to break the blockade in May. But the basic strategic problem and the growing confrontational mood remained.

And the year was far from finished as a harbinger of bad news. In China, the Mandate of Heaven had continued to slip through Chiang Kai-shek's fingers and head irresistibly toward Mao. On the same day that the North Atlantic Treaty creating NATO was signed, a million of his troops crossed the Yangtze and pushed Chiang and 300,000 of his most reliable soldiers into a pocket around Shanghai. By early May the Communists were at the gates, and Chiang along with as many Nationalists as he could take fled to the island of Formosa. To a generation of American Republicans, it was Team America, not that estimable anti-Communist Chiang Kai-shek, who lost China. And when they turned to whom specifically to blame, they found among the State Department's China specialists not expertise but Chicom sympathizers, and eventually George Marshall.

Then there was the matter of the US nuclear monopoly. The deterrent effect of having sole possession of the bomb was largely psychological, but not entirely. Practically speaking, one bomb, or just a few, wasn't going to stop the Russians. And they were really hard to make. The vast Manhattan

Project had generated only three in the three years before Hiroshima, and by April 1947 the US arsenal consisted of precisely seven. Inevitably, this led to a priority effort to get more, fast. The Sandstone atomic test series in the spring of 1948 succeeded far beyond expectations to produce a bomb design so much more efficient that by June 1949 the US arsenal topped the century mark, enough to stop any army. But just when Team America and its manager might have thought they could rest easy, on September 3 a B-29 reconnaissance aircraft picked up a higher-than-normal radiation count in the northeastern Pacific, irrefutable evidence of a recent nuclear test. There could be no other conclusion: the Soviets had the bomb. It was a whole new ball game.

CHAPTER 11

FIRE ON ICE

Just before the sun rose in Tokyo on Sunday, June 25, 1950, the telephone clattered in Douglas MacArthur's dark bedroom at the US embassy. It was the headquarters duty officer: "General, we have just received a dispatch from Seoul, advising that the North Koreans have struck in great strength south across the 38th Parallel at four o'clock this morning."

"I had an uncanny feeling of a nightmare. It had been nine years before, on a Sunday morning, at the same hour that a telephone call with the same note of urgency had awakened me in the penthouse atop the Manila hotel." Doug tried to convince himself he was still asleep, dreaming, until the voice of his chief of staff, Ned Almond, cut in: "Any orders, General?"

This had been some time in coming, though the warning signs were easy and convenient to ignore. The Korean Peninsula was a strategic backwater, but one attached at the top to Chinese Manchuria and Soviet Siberia. A poor and oppressed colony of imperial Japan since 1910, Korea had been under temporary Allied occupation—Soviets in the north and American troops in the south—since the end of World War II, separated by a line of demarcation drawn in a very ad hoc manner along the Thirty-Eighth Parallel by a young colonel, Dean Rusk, the future US secretary of state, having no idea of its potential significance.

As the Cold War heated up, the geographic division hardened into two separate political entities—a Communist regime centered in Pyongyang in the north led by Kim Il-sung, and in the south the Republic of Korea (ROK),

dominated by strongman Syngman Rhee in its capital, Seoul. While the Soviets concentrated on building Kim a 200,000-man army and equipping it
with heavy artillery, modern T-34 tanks, and contemporary tactical aircraft,
the Americans in the south focused on repatriating Japanese soldiers and
trying to restore local order sufficient for their own withdrawal. MacArthur
spoke glowingly of South Korea's freedom at Rhee's inauguration in July
1948, but showed little interest in defending it beyond expanding its constabulary to fifty thousand men, a trick he had tried in the Philippines without any success. Actually, he thought the peninsula indefensible: "I wouldn't
put my foot in Korea. It belongs to the State Department. . . . They wanted
it and they got it. I wouldn't touch it with a ten-foot pole." Nor apparently
would Dean Acheson, now secretary of state. In a January 1950 address before the National Press Club, he stated that the US "defensive perimeter
runs along the Aleutians to Japan and then goes to the Ryukyus (primarily
Okinawa) . . . to the Philippine Islands," excluding both Taiwan and Korea,
and adding that in case of an attack, "the initial reliance must be on the
people attacked to resist it." These were words best not said in public, much
less in front of reporters. Did he expect they would go unheard by the other
side?

For, on that other side, Kim Il-sung was doing his best to become the tail
that wagged the Communist dog. Specifically, he wanted to use his newly
acquired military clout to conquer the South and unite the peninsula, betting it would not provoke more than a token American response. Mao was
certainly of a like mind, having just consolidated China without any US
military reaction. Stalin, though now nuclear armed, didn't want a direct
confrontation with the Americans, but this was a matter of client states,
and the Americans had apparently signaled they would not intervene. So
he reluctantly gave Kim the go-ahead, reasoning that this was after all the
minor leagues.

There was plenty of warning. Charles Willoughby, Mac's proverbially pre-
varicating intelligence chief, had developed a peninsula-wide network that
generated almost twelve hundred reports pointing to a relentless North Korean buildup above the Thirty-Eighth Parallel. But he also convinced himself
that Stalin would make the final call, and that he would wait. His boss was
even more optimistic; a month before the attack Mac waxed philosophical

to a *New York Times* correspondent: "I don't believe that war is imminent because the people of the world would neither desire it nor would they be willing to permit it. That goes for both sides."

It didn't. For Kim that goes without saying, but back home in Missouri, Team America's manager, Harry Truman, was in an equivalently bellicose mood, determined to respond, however far away and strategically unimportant Korea might be. Throughout his presidency Communists had been rounding the bases with embarrassing regularity, culminating in the loss of the China game, leaving Truman to face a string of beanballs from right-wing Republican supporters of the recently deposed Chiang Kai-shek. Now this naked aggression had to be resisted, a view his team of advisers shared to a man when he returned to Washington. So naked, in fact, that US diplomats at the newly formed United Nations were able, courtesy of a Soviet walkout, to enlist the organization in what Truman called a "police action," thirteen member states promising troops to fight alongside the Americans if they committed significant ground forces.

This was a question that still had to be answered, since the conflict came at an embarrassing moment for Truman, who had just slashed the Pentagon's resources to the bone as part of his "economy budget." He had already approved provisions for far higher future defense funding in the historic internal document NSC 68, but he needed a public scapegoat fast, and Louis Johnson, his secretary of defense and a staunch supporter of "holding the line" on budgets, fit the role perfectly. Combative and unpopular with the services, few save Johnson were sorry to see him go, but this left a gaping hole at the top of the US defense establishment just as it was thrust into war.

George Marshall was vacationing with Katherine in remote Huron Mountain, Michigan, when the local country store got a phone call for the general from the president of the United States, who then held the line until they found him. The conversation was short: "Drop by the White House the next time you are in Washington," followed by the now familiar but ever reassuring, "Yes, Mr. President."

At their September 6 meeting Truman came right to the point: He had to get rid of Johnson. Would Marshall "act as Secretary of Defense through the crisis if I could get Congressional approval"? Marshall agreed to serve up to one year, his sole condition being that Robert Lovett, his interchangeable man, be brought on board as deputy secretary. But he warned the president: "They are charging me with the downfall of Chiang's government in China. I want to help, not hurt you." "Can you think of anyone else saying that?" Truman wrote his wife, Bess. "I can't, and he's of the *great*."

Clearly the manager was sold on his unimpeachable general, but that proved far from the case among Senate Republicans during confirmation hearings. Robert Taft, the leading contender for the 1952 GOP presidential nomination, thought if he voted for Marshall he would be approving a policy favorable to communism in Asia. But that was mild in comparison with the extended rant of Indiana's William Jenner, climaxed by the only enduring words he would ever utter: "General Marshall is not only willing, he is eager to play the role of a front man for traitors. The truth is this is no new role for him, for George C. Marshall is a living lie." Senator Vandenberg, who was dying of cancer, managed to hold the line, and on September 17, the general was confirmed by a vote of 57 to 11.

But Jenner's outrageous statement was an accurate measure of the vehemence with which Communist conspiracy theories were held by his wing of the party, soon to be revealed in full by Senator Joseph McCarthy. Marshall might react contemptuously when told of the attack—"Jenner? Jenner? I do not believe I know the man"—but there was a new reality in Washington and he would have to deal with it. If only two words were available to summarize George Marshall's career, they would be "Europe first." These Republicans focused on Asia, and just as they had in World War II, they looked to Douglas MacArthur as their oracle of strategic wisdom.

So yet again this military odd couple was reunited, as always Marshall in command but never in control of his meteoric antagonist, both now fated to fight an entirely unexpected war together. Actually, Truman would have loved to get rid of Mac, and had been advised by Omar Bradley to do so at the outset of the conflict. But propinquity is power, and MacArthur was the man on the scene, or at least one country over, so like it or not he had been put in charge of the war, adding the new title commander in chief Far East to his SCAP appellation.

It was in this role that he forced everybody's hand in Washington, flying to Korea on June 29, the day after Seoul fell, and cabling home: "The South Korean forces are in confusion . . . the enemy advance threatens the over-running of all of Korea. . . . The only assurance for holding the present line and the ability to regain later the lost ground is through the introduction of United States ground combat forces into the Korean battle area." Twenty-four hours later Truman authorized their use, a momentous decision, but one unlikely to save the situation given the status of the US troops available.

The first to arrive were a tiny combat team equipped with obsolete anti-tank weapons; Task Force Smith was promptly overrun and annihilated by North Korean T-34s at Osan. Next came the Eighth Army's three combat divisions in Japan, the Twenty-Fourth, Twenty-Fifth, and First Cavalry. It may have been a tribute to MacArthur as proconsul that these elements could be safely withdrawn, but they had grown soft as occupiers and were in no condition to fight the highly aggressive and surprisingly flexible North Koreans. US B-29s slowed the onslaught, but unit after unit kept getting chewed up. When MacArthur requested five more divisions from the Joint Chiefs, he was enraged to learn that NATO had first call on available troops and he would have to wait. Finally, as the North Koreans neared the southwest coast of South Korea they wheeled left and, flanking the Americans, forcing them by early August into a tiny perimeter surrounding the port city of Pusan (Busan) at the bottom of the peninsula. Here the US troops fought furiously for the next month, desperately trying to stop the place from becoming an American Dunkirk. It was a grim situation.

But one not without hope. Mac's appeals for reinforcements were at last being heard, the Joint Chiefs agreeing to bring the Eighth Army to full combat readiness while shipping him two more Army divisions, the First and Second, plus the First Marines and two infantry regiments. Welcome relief, but MacArthur had no intention of pouring them into the scrum at Pusan. As far back as early July he told his operations chief, "Let's think of a couple of end runs." Later, when contemplating how to employ the First Marines, MacArthur turned to a map and pointed his corncob pipe: "I'd land 'em here, at Inchon," a port 150 miles northwest of Pusan and just thirty miles from Seoul—a true Hail Mary amphibious operation.

Not only did Inchon involve a long sea passage, but the inlet leading to the city possessed some of the biggest tides in the world, providing only two short periods a day when landing craft could get close enough to shore to off-load men and equipment. If the timing was off, everybody might end up

stuck in the mudflats, sitting ducks. But if it worked it would trap the entire North Korean invasion force, caught between MacArthur's hammer in
the North and the anvil of the Eighth Army commander Lieutenant General
Walton Walker pushing out of Pusan.

Only Douglas MacArthur could have thought up and sold such a scheme.
Commanders like Bradley and "Lightning Joe" Collins remembered European amphibious operations as bloody slogs, but MacArthur's experience,
courtesy of ULTRA, had been much better, and in the process he learned
to strike where least expected, and "hit 'em where they ain't." On these
grounds, or at least mudflats, and given Douglas MacArthur's supreme confidence in himself to accomplish practically any military feat, Inchon seemed
like a good bet.

Or so it appeared to a band of officers of the highest rank, sent to Tokyo
to lay critical eyes on Mac's plan, now code-named Chromite. They arrived
on August 23 skeptical and left spellbound. The general spoke for forty-five
minutes without notes, finishing with:

> The only alternative to a stroke such as I propose will be the continua-
> tion of the savage sacrifice we are making at Pusan. . . . Are you content
> to let our troops stay in that bloody perimeter like beef cattle in the
> slaughterhouse? . . . The prestige of the Western world hangs in the
> balance. . . . It is plainly apparent that here in Asia is where the Commu-
> nist conspirators have elected to make their play for global conquest.

Much as he had convinced FDR to invade the Philippines in Hawaii, he
rolled over these potential naysayers with grandiloquence. Chromite was
a go.

And it went brilliantly. Landing on September 15, 1950, having transited
around a typhoon, the 260-ship armada so surprised the North Koreans that
they hadn't bothered to mine the channel or even turn off the navigation
lights. The fighting was a rollover, Mac's troops led by the Marines defeating between 30,000 and 40,000 defenders while losing only around 500
dead. Meanwhile, Walker not only broke out of Pusan, but the North Koreans began shedding their uniforms and heading for the hills. By the end of
the month Seoul had been retaken, along with 7,000 prisoners, the fighting

adding 14,000 more North Korean casualties to the 50,000 troops already lost in the campaign—all at the cost of around 3,500 Americans killed and wounded. Despite the Joint Chiefs' objection that he lacked the authority, Mac restored Rhee to his capital, thereby reestablishing the status quo antebellum. By any measure, Inchon was a spectacular success, probably MacArthur's most stunning and lopsided triumph in a career full of them.

Congratulations poured in from everywhere. Churchill was definitive: "MacArthur did a perfect job." Bull Halsey went even further: "The Inchon landing is the most masterful and audacious strategic stroke in all history." Ike chimed in academically from Columbia: "I cannot stay the impulse to express the conviction that you have given us a brilliant example of professional leadership." George Marshall, just settling into his new post, was more sedate and also pointed: "Accept my personal tribute to the courageous campaign you directed in Korea and the daring and perfect strategical operation which virtually terminated the struggle." MacArthur's reply was doubtless read with a sarcastic grin by Marshall: "Thanks George for your fine message. It brings back vividly the memory of past wars and the complete coordination and perfect unity of cooperation which has always existed in our mutual relationships and martial endeavors." Its perfect insincerity was in itself a warning to the new secretary of defense that Douglas MacArthur was living in his own reality, one that went well beyond Marshall's assumption that the struggle was "virtually terminated."

The very audacity of the Inchon campaign should have been a tip-off: this was not the work of a man who would settle for a tie on a remote peninsula. There were earlier obvious signs he wanted a broader stage. At the end of July, when things were desperate, MacArthur, miffed that Truman had declined Chiang Kai-shek's offer to send two divisions to fight in Korea, had resolved to pay him a visit in Taiwan, and even to send three squadrons of F-80 jet fighters to bolster his defenses against China, a decision furiously rejected by the administration as well above his pay grade. Then there were his remarks at the Chromite review in Tokyo that Asia, not just Korea, was the primary Communist objective for global conquest. Unlike his Washington-based interlocutors, MacArthur for half a decade had been living the life of an absolute monarch, a veritable Rising Sun King, and it had plainly left him with a magisterial view of his own power and purview,

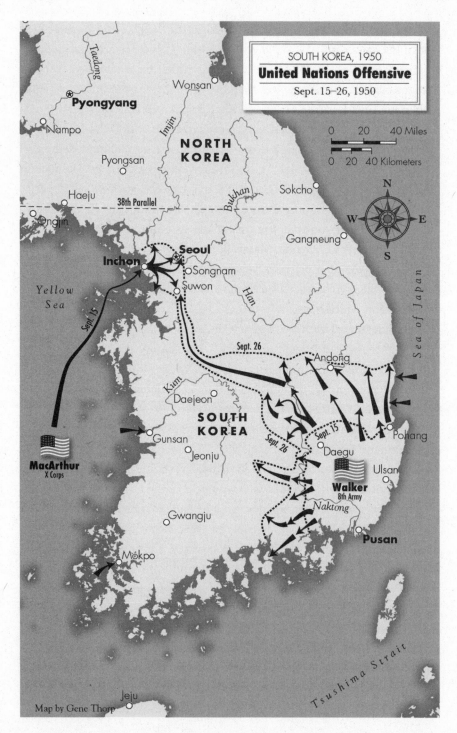

Map by Gene Thorp

one bound to put him on a collision course with those presumed less strategically enlightened.

ON THE OTHER SIDE OF THE GLOBE THE UNEXPECTED GOOD NEWS FROM KOrea found Team America's principles in a far less transcendent mood. A lot was happening elsewhere, the implications of which had to be balanced against the effort on the peninsula. Given that the Soviets now had the bomb, a new degree of caution was an obvious, almost visceral mandate. Just how much caution remained to be seen, but the heart of Team America's batting order—Truman, Acheson, and Marshall—all instinctively grasped that with Korea sharing a border with the Soviet Union, this conflict could potentially escalate beyond control.

More than prudence argued it was better to end this now that the situation had been restored. Harry Truman's "police action" had grown very unpopular at home, so much so that he realized he could no longer get a declaration of war from Congress—a precedent with major implications for the future. Given the massive and now active nature of the Communist threat, the administration had activated the peacetime draft and reflexively expanded Team America's ground forces to 634,000 in early July 1950, then to a million in August, and then, under George Marshall's auspices, to a million and a quarter for fiscal year 1951. As tired as he was, at least George found himself doing what he did best, building and arming a behemoth force structure; but this sort of competence didn't make a conscription-fed Army, or the Korean War, or him, any more popular with the American public.

Then there was the alternative to be considered. New designs made nuclear devices vastly more efficient in terms of yield and the fissile material they consumed, and this, along with no clear military doctrine for their use beyond "firepower is good," caused the US nuclear arsenal to reach nearly three hundred devices by the time the North Koreans invaded. But that hadn't stopped them, nor could atomic weapons be effectively employed once the attack had begun.

Yet there was something else in the offing that might truly deter: a weapon of virtually unlimited power. Nuclear physicists quickly realized that the hundred-million-degree temperatures generated by a fission reaction were sufficient to fuse hydrogen into helium, the same process that fuels the sun, and one that would release extraordinary amounts of energy here on earth if engineered into a bomb. It was so much energy that many of the scientists involved in the Manhattan Project were repelled by the concept. Not Edward Teller, a brilliant Hungarian immigrant physicist for whom a thermonuclear device became both an obsession and a crusade. For as he obsessed over H-bomb design, he also waged a political campaign that was soon getting attention in Congress.

Meanwhile, with the news of the Soviet atomic test, Truman took the question to the Atomic Energy Commission, whose advisory board was unanimously against developing a hydrogen bomb. Not satisfied, he went to his own National Security Council (NSC), which gave him a cautious recommendation to go ahead. And when one member, David Lilienthal, expressed "grave reservations," Truman cut to the chase: "Can the Russians do it?" All heads nodded affirmatively. "In that case we have no choice. We'll go ahead." Later, remembering the meeting and his efforts to block the H-bomb, Lilienthal wrote it was akin to saying "no to a steamroller."

George Marshall's experience was not much different. There at the inception, he had been swept along by nuclear events. He was perfectly aware of the danger these revolutionary arms posed to civilization. But now, as secretary of defense, he was also lured by the attractions of tactical nuclear weapons in sufficient numbers to hold off the Soviets in Europe, at least long enough to backfill with conventional arms. He was not happy about the decision to go forward with the thermonuclear project, but he did not object; he was drawn along by the siren's song of technological possibility and fear that the other side would sing it first. Temptation and trepidation—together they jerk history along its rocky path.

And so it was dealing with Douglas MacArthur back at the Thirty-Eighth Parallel. George Kennan; the UN allies, including the United Kingdom; and Lady Luck's archenemy, Prudence, argued this was enough, that aggression had been reversed, the essential mission completed. But more than just the strategic temptation of an apparently open way up the peninsula argued the opposite. During the invasion, Kim's secret police had killed more than twenty thousand civilians labeled "class enemies," while Walker's men discovered mass graves full of American prisoners, most shot in the head with their hands bound.

A number of key Washington players argued that Kim's was a particularly noxious regime that now deserved to be stamped out. At State, Dean Rusk maintained that the Thirty-Eighth Parallel was a completely artificial barrier, having drawn it himself. Across the Potomac at the Pentagon, the Joint Chiefs had already told MacArthur if there was no sign of Chinese or Russian intervention he was to "make plans for the occupation of North Korea."

Then, on September 27, with the okay of Acheson, Marshall, and Truman, the chiefs specifically authorized MacArthur to "conduct military operations . . . north of the 38th Parallel," subject to just three conditions: he must not cross into Soviet or Chinese territory, send aircraft over either country's territory, or allow any but ROK troops to approach the Yalu River bordering on China. When the general wanted something firmer, Marshall sent him an "eyes only" message: "We want you to feel unhampered tactically and strategically to proceed north of the 38th Parallel." As if to drive the point home, on October 2 Marshall cabled Mac: "We desire you to proceed with your operation without any further explanation or any announcement and let action determine the matter. Our government desires to avoid having to make an issue of the 38th Parallel until we have accomplished our mission." As if to accommodate him, five days later the UN General Assembly overwhelmingly endorsed a US proposal broadening the mission objective to "a unified, independent and democratic government" for all Korea. Marshall had been dealing with MacArthur for upward of thirty years and knew he read every directive in light of his own version of reality. If this didn't look like a carte blanche in his eyes, it's hard to imagine what would have.

A word of warning did come from Zhou Enlai that China wouldn't "supinely tolerate" breaching the parallel, even telling the Indian ambassador in Beijing that his country "would send troops to the Korean frontier to defend North Korea." Word got back to Washington, but Truman dismissed it as a bluff: "A bald attempt to blackmail the United Nations by threats of intervention in Korea." So the counterinvasion swept north practically unopposed, and rather than considering the words of Zhou, the man Harry Truman wanted to talk to was his star player in the field, Doug MacArthur.

Marshall cabled Mac that the manager would like to meet with him somewhere in the Pacific, preferably Oahu, on October 15. Mac, claiming to be in the middle of a war, insisted on Wake Island, almost 2,300 miles farther west. "I am sure the President would be glad to go on and meet you at Wake Island," Marshall replied sweetly, but neither he nor Acheson were about to be strung along that far and begged off the junket. But with congressional elections weeks away and his popularity sagging, Harry Truman needed a photo op with this military magician who seemed on the verge of turning a walkover into a solid win, so he swallowed his pride and went the extra miles.

Mac arrived first and had time to shave, shower, and breakfast before roaring up the runway to greet Truman's plane as if in a great hurry. His garb—the battered field marshal cap, the open-collar khaki shirt—immediately offended Truman, who later told his biographer: "If he'd been a lieutenant in my outfit going around dressed like that, I'd have busted him." Rather than salute, he held out his hand to the commander in chief.

Truman took it all in stride, and their first and only meeting proceeded with a kind of false bonhomie, a supposedly off-the-record chat belied by a note taker stationed just out of the general's sight. Mac began by apologizing for a widely publicized letter he had recently sent to the Veterans of Foreign Wars talking up Formosa as an "unsinkable carrier-tender," and Truman reciprocated by telling him the matter was closed, which it wasn't.

Soon they got to the issue at hand, Korea. All present were of a like mind that it would be fully occupied by Thanksgiving. Then Truman raised the question that loomed above everything: "What are the chances of Chinese or Soviet interference?"

"Very little," Mac shot back, puffing on his corncob. "Only 50,000 or 60,000 could be gotten across the Yalu River. They have no air force. . . . If the Chinese tried to get down to Pyongyang there would be the greatest slaughter." When asked if the Soviets might lend a hand, or rather a wing, he replied: "It just wouldn't work with Chinese Communist ground and Russian air. I believe Russian air would bomb the Chinese as often as they would bomb us." One of Truman's advisers marveled: "He was the most persuasive fellow I ever heard. I believed every word of it." The problem was, they were all wrong. Instead, the roof was about to cave in.

THE AMERICAN-LED UN INVASION OF NORTH KOREA CONSISTED OF TWO SEPARATE thrusts, necessarily bisected by the mountainous spine of the Hwangnyong

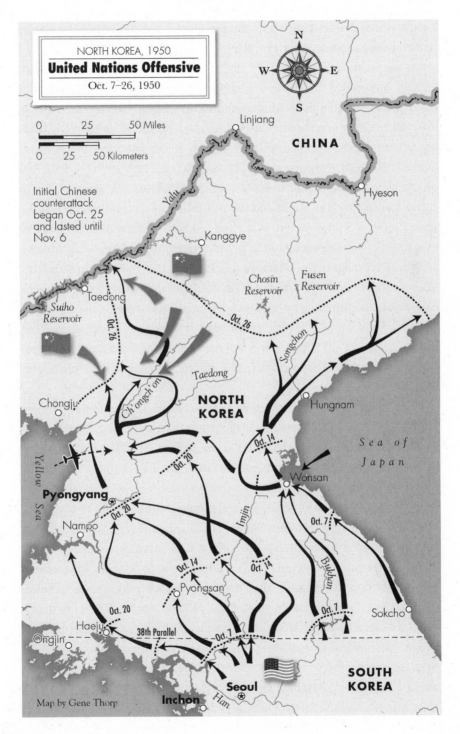

NORTH KOREA, 1950
United Nations Offensive
Oct. 7–26, 1950

0   25   50 Miles

0   25   50 Kilometers

Initial Chinese
counterattack
began Oct. 25
and lasted until
Nov. 6

Linjiang

CHINA

Hyeson

Yalu

Kanggye

*Chosin
Reservoir*

*Fusen
Reservoir*

*Suiho
Reservoir*

Taedong

Oct. 26

Oct. 26

Chongju

Ch'ongch'on

Taedong

**NORTH
KOREA**

Songchon

Hungnam

*Sea of
Japan*

Oct. 14

Oct. 20

Wonsan

*Yellow
Sea*

Pyongyang

Oct. 20

Oct. 20

Oct. 7

Nampo

Imjin

Oct. 14

Oct. 14

Bukhan

Oct. 7

Pyongsan

Oct. 20

Sokcho

Oct. 7

Haeju

38th Parallel

Oct. 7

Ongjin

Seoul

**SOUTH
KOREA**

Inchon

Han

Map by Gene Thorp

Range, which divides the top of the peninsula, a situation similar to the one on Bataan and no more propitious.

To the west, Walton Walker's Eighth Army crossed the Thirty-Eighth Parallel and headed for Pyongyang, capturing the capital after three days of fighting, even setting up headquarters in Kim's office. On the eastern side, the X Corps, now commanded by Ned Almond, staged an amphibious land-ing at Wonsan and established itself directly across from the Eighth Army. Between them ROK troops were supposed to lead the way to the Yalu, but on October 24, MacArthur ordered both X Corps and the Eighth Army to "drive forward with all speed and full utilization of their forces." When the Joint Chiefs of Staff objected, Mac cited Marshall's message about wanting him to feel "unhampered tactically and strategically." They backed down and the two prongs of the UN force continued to race north for another day, when suddenly near the Yalu the Sixth ROK Division found itself fighting a new enemy, one directed by bugles and drums and firing old Japanese rifles, with deadly effect as it nearly annihilated the division in two days. They were members of the Chinese Fortieth Army, and when two more ROK divi-sions ran into them, they too collapsed, leaving the flank of the Eighth Army naked.

The Chinese weren't supposed to be there. Persistent rumors of Chinese infiltration had certainly reached MacArthur, but he chose to first ignore, then downplay them. Instead of the more pessimistic CIA projections, he listened to Willoughby, his intelligence chief, the man with the sources on the ground, who told him confidently that no more than 16,000 Chinese had slipped into North Korea, when the real number was in excess of 200,000.

That became apparent after Walker sent his tank-heavy motorized First Cavalry Division to cover his right flank and infantry elements of the Eighth Army got within twenty miles of the Yalu. Suddenly Chinese were everywhere. On November 1, they hit the First Cav hard, virtually annihi-lating one of its regiments in what some call the worst US Cavalry defeat since the Little Bighorn. Meanwhile, on the other side of the Hwangnyong Range, Almond's X Corps also found itself confronted with lots of Chi-nese. After breaking up the ROK Third Division on November 2, they tore into the Seventh US Marine Regiment, fighting furiously for five days with little regard for casualties. Then, as suddenly as they had arrived, they vanished like a ghost army, an apparition with bullets. The Americans were flummoxed.

MacArthur's answer was that Chinese troops were pouring into Korea from Manchuria over six bridges crossing the Yalu, and he ordered a massive strike of ninety B-29s to eliminate them. The chiefs hesitated, and when Acheson, Marshall, and Truman seemed ready to approve, Dean Rusk took the proposal to the Allies, who unanimously objected to this potential widening of the war, so vehemently that the raids were canceled, leaving MacArthur furious. Then, when he finally prevailed and staged those raids, they not only failed to knock out three of the bridges, but brought another unpleasant surprise.

Cloaked in as much secrecy as they could muster, on November 1, 1950, the Sixty-Fourth Fighter Aviation Corps of the Soviet Air Forces began operating over Korea. But as ominous as the political implications of their presence was what they were flying—the MiG-15, at that moment the best fighter aircraft in the world by a substantial margin. Small, highly maneuverable, but heavily armed, it had a top speed of nearly 700 miles per hour and a ceiling of almost 51,000 feet, both far in excess of anything the United States had in the skies above Korea. The keys to the MiG-15's performance was a rugged yet powerful jet engine and its revolutionary swept wing, both obtained from British and German designs, but at the time entirely unexpected from the Soviet Union. Americans took their fighters seriously, and soon had a swept-wing model of their own, the F-86 Sabre, with capabilities much closer to the MiG's. But coming on top of the Soviet atomic bomb, the MiG-15 served to crystallize the impression of the USSR as a credible high-tech adversary, one that would serve to fuel the great Cold War arms race. Meanwhile, Russians and Americans were killing each other in the stratosphere, and the Chinese had their air cover, not that they needed it.

As days turned into weeks in November with no more Chinese attacks, "I decided to reconnoiter, try to see what was going on," Mac recalled. He boarded his personal transport, a Lockheed Constellation, flew from Tokyo to Korea, then ordered his pilot to traverse the entire Yalu at five thousand feet without fighter escort. It took four hours, assumed to be their last by the crew, but they had Douglas MacArthur aboard so they arrived back safely, but with the general still clueless, having seen nothing of the missing army.

And so Mac remained, in the words of then Army chief of staff "Lightning Joe" Collins, "like a Greek hero [marching] to an unkind and inexorable fate."

The Chinese were already there, lurking unseen in the Hwangnyong Range, and now that an icy Siberian front had frozen the Yalu thick enough for even motor vehicles, they were joined by still more infiltrators until their numbers reached in excess of 300,000. On November 25, the day after Mac's futile flight, they struck. Two Chinese armies hit ROK II Corps like a tsunami, inundating and surrounding it, allowing but one regiment to escape. Next it was the Americans' turn, and by the morning of the twenty-seventh the Eighth was being driven back by four Chinese armies, while two more circled south with the intention of cutting the road back to Pyongyang. That same night, on the other side 150 miles to the east, X Corps took its share of abuse: 120,000 Chinese came crashing down on the First Marines and the Twenty-Seventh Infantry Division west of the Chosin Reservoir. Everywhere the tactics were the same, a human battering ram of infantry regiment after infantry regiment, surging forward along a narrow front. The Americans as always fell back on firepower and inflicting horrific casualties, but also simply falling back, in the case of the Eighth Army fast enough to begin to look like a rout.

Back in Tokyo, Douglas MacArthur was plainly having trouble adjusting to being caught on the other end of surprise. For four days after the initial Chinese surge he continued to urge Walker and Almond forward, giving them pullback orders only when the Chinese were lapping around the inner flanks of both. The plan was for the Eighth Army to retreat to Pyongyang and for X Corps to establish a perimeter at Hungnam, on the east coast about eighty-five miles above Wonsan. But given the slow start, this was a great deal easier said than done.

For the Marines, getting back from the Chosin Reservoir required breaking through upward of thirty roadblocks and ambushes in frigid weather, but also buying time for the rest of X Corps to secure Hungnam, all to be evacuated by the Seventh Fleet beginning in mid-December—105,000 men, 91,000 Korean refugees, 17,000 vehicles, and over 90,000 tons of munitions and supplies. In the words of one MacArthur biographer, "It was Inchon in reverse."

But the retreat of the Eighth Army was worse. In particular, the Second Division blundered into a mountain pass lined with Chinese on both sides and by December 1 had lost five thousand men along with all of its equipment. So bad was the debacle that Walker decided to abandon Pyongyang and head south as fast as possible—so fast that it was not until December 20 that advanced units of the Chinese caught up with his new front line just north of Seoul.

Three days later Walker headed off to tour his command, but instead his jeep crashed into a ROK truck and he was killed instantly. To replace him MacArthur had but one name in mind, Matthew Ridgway, the fabled World War II airborne commander and athletic instructor at West Point when he was superintendent. It was an excellent choice. More than anyone, Ridgway saved the situation in Korea, but in doing so he became an alternative to Mac himself.

"The Eighth Army is yours, Matt," the general had told him in their meeting at the Dai-ichi building the day after Christmas. "Do what you think best." That was a given with Ridgway, Mac's first truly independent-thinking subordinate since Ike. Known as "Iron Tits" for the two hand grenades he wore strapped to his chest, he was quick to anger and hated reporters and publicity. But his men loved him for jumping out of planes on missions and sharing the dangers of war. He was a soldier's general, his mask shaped like one of them.

And he would need to put the best face on things, since the situation was continuing to go south. On New Year's Eve, six Chinese divisions rolled over ROK positions, forcing Ridgway to abandon Seoul, its airfield, and nine thousand tons of supplies on the way to establishing new lines seventy miles below the Thirty-Eighth Parallel. But by mid-January he was no longer interested in simply holding the line; he was ready to counterattack with a series of engagements he and his troops called "the meat grinder," aimed at decimating Chinese. Operations Wolfhound (January 15), Thunderbolt (January 25), Punch (February 9), Killer (February 21), and Ripper (March 7) did exactly that to the ever-aggressive Chinese, until their army was a hollowed-out shell and they were forced to give up Seoul and retreat just north of the Thirty-Eighth Parallel to a complex of forts and tunnels dug out of solid rock by North Koreans in the tens of thousands.

As far as Washington and the UN Allies were concerned, that was where they could stay. The administration, the State Department, and the Joint Chiefs of Staff all entered 1951 having staged a rapid retreat in war objectives and were now in solid agreement that the Thirty-Eighth Parallel would constitute the border of an independent South Korea, and that the war should be liquidated through a negotiated cease-fire such as the one the British were proposing.

Korea remained without strategic value, and the situation there was inherently dangerous. US intelligence was monitoring Russian broadcasts from supposedly Chinese and North Korean MiG-15s engaged in combat with Americans. But an attack on their bases, either in China or the Soviet Union, would constitute another step in an escalatory process that could plausibly lead to all-out war, presumably tripping a Soviet onslaught in Europe and leaving the United States with the stark choice of using nuclear weapons to stop it. Truman had spoken in late November of considering meeting the great Chinese surprise with a response that "includes every weapon we have." Despite his adding, "I don't want to use it. It is a terrible weapon," even this veiled threat prompted a quick trip to Washington on the part of the British prime minister and a clear signal that loose lips on this subject were subject to practically unlimited amplification. As for the future, the manager and his teammates were clear in their judgment: they wanted the game halted, not expanded into extra innings.

But that did not go for Douglas MacArthur, still stationed way out there in left field. Despite all signals to the contrary, he convinced himself it was still game on. In the grim days of December, he told Collins the situation could be saved only by expanding it to a naval blockade of China, by a free hand to hit targets in Manchuria, and by reinforcements from the Nationalist Chinese on Formosa. A month later George Marshall was more than annoyed with MacArthur for telling the Joint Chiefs that his men's morale would "become a serious threat to their battle efficiency unless the political basis upon which they are asked to trade life for time is clearly delineated," commenting to Dean Rusk that when a commander starts questioning his troops' morale, it's time to look into his own.

Next came his February 11 cable to the Joint Chiefs fleshing out his suggestions to Collins. Mac now called for twenty to thirty nuclear weapons to eliminate air and supply bases in China and create a belt of radioactive waste to seal Korea's northern border, followed by amphibious landings on both sides of the peninsula by two Marine divisions and a half million of Chiang Kai-shek's troops from Formosa to create a giant pincer to trap the remaining Chinese massed at the Thirty-Eighth Parallel. MacArthur estimated the entire campaign would take ten days, and might even cause the fall of Mao.

To put it mildly, Mac's missive was not well received. The idea of using nuclear weapons in this campaign knocked chairman Omar Bradley back on his heels: "I've never heard anything so preposterous in my life." The chiefs curtly replied that his entire strategy was out of the question; the war was to be cauterized, not expanded to China. That would be, as Bradley told Congress three months later, "the wrong war, at the wrong place, at the wrong time, and with the wrong enemy." MacArthur had drifted completely out of phase with reality. His towering ego had allowed him to confuse his own interests with those of his country. He could not bear to see his last war end in a stalemate, so he wanted to blow it up and, in the process, rescue his military reputation. If there was ever a loose cannon on board the ship of state, it was Douglas MacArthur at this point.

Back in December Truman had ordered Marshall and Acheson to direct diplomats and military commanders in the field to clear all speeches and press releases with Washington and in general to "exercise extreme caution in public statements." In the Senate, Robert Taft complained that the manager's intent was to gag MacArthur, which it was.

Not that it had much effect. As had George Patton, Mac chose to use the fourth estate to stage his final campaign of rebellion and self-destruction. It began on March 7, 1951, when he called a press conference to announce that unless he received "major additions . . . the battle lines in the end will reach a point of theoretical stalemate." He followed that up eight days later by telling the United Press the Eighth Army should not be stopped short of the "accomplishment of our mission in the unification of Korea."

The administration chose to greet both statements with silence, so Mac raised the ante with what Acheson called "a major act of sabotage." On March 20 the chiefs informed MacArthur that the administration would shortly announce that the United States was ready to enter negotiations with North Korea and China. Four days later, Dai-ichi issued a communiqué threatening to expand the war in a way that "would doom Red China to the risk of imminent military collapse." Consequently, he, Douglas MacArthur, was "ready at any time to confer in the field with the commander-in-chief of enemy forces in the earnest effort to find any military means whereby realization of the political objectives of the United Nations in Korea . . . might be accomplished without further bloodshed."

Mac had cut off his manager at the knees. "I couldn't send a message to the Chinese," Truman told his daughter. "He prevented a cease fire proposition right there." Also that same day, Mac had given Ridgeway permission to move north of the Thirty-Eighth Parallel without bothering to tell Washington. At this point it became clear up and down the chain of command that MacArthur had to go. Marshall later maintained it was the proclamation of March 24 that had done it. "It created a very serious situation with our allies . . . [T]he leader in the field comes forward with a proposition that terminates that endeavor of the Chief Executive. . . . It created, I think specifically, a loss of confidence in the leadership of the government."

Yet at the top they hesitated. Ironically, MacArthur's offer of a peace negotiation was getting favorable press coverage, and firing the general now would leave his manager "on the side of sin," in the words of Lovett, standing in for Marshall. Instead, he, Rusk, and Acheson convinced the fuming Truman to settle for a stern reprimand, in part because they all had faith in MacArthur's determination to say or do something for which there would be no escape.

They didn't have long to wait. On April 5, one of Truman's bitterest critics, Republican House minority leader Joe Martin, read into the record a letter from Douglas MacArthur agreeing with Martin that Chiang should be encouraged to attack mainland China, denouncing the "Europe first" focus of US foreign policy, and closing with the inflammatory words "There is no substitute for victory."

"This look[s] like the last straw," Truman wrote in his diary. "I've come to the conclusion that our Big General in the Far East must be recalled." But he wanted his key advisers, particularly Marshall, in lockstep behind him. Lovett claimed Marshall was "revolted" by such a letter "to the leader of the opposition," but Lovett, his alter ego, still continued to prevaricate, suggesting that MacArthur be brought to Washington "for consultations and reaching a final decision after that." When that failed to fly, Marshall backed away from it.

But over the weekend, during the process of getting the Joint Chiefs' approval, he continued to search for alternatives. Bradley, in his backbiting way, later suggested that Marshall had personal reasons: that, having already suffered the taunts of Jenner, and it being well known that these two five-stars were hardly simpatico, he thought the rage on the right engendered by Mac's removal would largely come down on his own shoulders, and wanted to avoid it. This makes sense, actually. Marshall had spent a lifetime fashioning a mask made to be viewed the same from the right or the left, and in the process worked his way into a bipartisan sweet spot that was key to his power and reputation. The Republican assault was not only disorienting, it had to have hurt his self-esteem. Why suffer more? George Marshall was not a vindictive man. With the exception of his brother, he had never taken any particular pleasure in getting even. Professionally, no one had treated him worse than Douglas MacArthur—suspicious to the point of paranoia, epically rude, and subtly contemptuous of his noncombat record—but it was seldom if ever reciprocated. Nor was it evident in this process of dumping Doug; Marshall was just bowing to the inevitable.

"So it was with Douglas MacArthur," wrote biographer William Manchester in what may be the best line ever on the man. "Brave, brilliant, and majestic, he was a colossus bestriding Korea until the nemesis of his hubris overtook him." From the perspective of seventy years passed, the Korean conflict can plausibly be seen as a stillborn World War III. All the ingredients were there—diametrically opposed ideologies, aggressive leaders and commanders, huge force structures—twice before; they had combined into all-out societal war. Now Mac wanted to take the next, quite possibly culminating step.

But the specter of nuclear war defused both him and the conflict. At this point the United States had almost all of them, but the moral disinclination to use them, combined with the fear of what just a few Soviet bombs could do to American cities, placed tremendous pressure on the Truman administration to de-escalate, and this in turn provided an opening for the Communist trio to accept the stalemate and eventually liquidate the fighting through negotiations. Nuclear deterrence was in its infancy, but still proved strong enough to fend off this first, very dangerous challenge to its barely established hegemony.

TRUMAN WANTED MARSHALL, WITH ACHESON'S ADVICE, TO DRAFT THE TERMI-nation orders, which he could have written with an icicle. After a single sentence informing MacArthur he was no longer SCAP, UN commander, or commander in chief Far East, it went on: "You will turn over your com-mands, effective at once, to Lieutenant General Matthew B. Ridgway. You are authorized to have issued such orders as are necessary to complete de-sired travel to such place as you select. My reasons for your replacement will be made public concurrently with the delivery to you of the foregoing message." It was supposed to be brought personally by the secretary of the army, who was in the Far East touring the front, but due to a transmission failure, it didn't reach him. Instead, an aide heard the news on the radio and called Jean, who then delivered it to "Sir Boss" in a reverent whisper. "Jean-nie, we're going home at last," he responded in his usual upbeat tone, not sounding like he was surprised at all. And when he finally did see his written orders, their language convinced him: "George Marshall pulled the trigger."

And so he departed, leaving an enormous shadow behind. When asked how it felt to take MacArthur's place, Ridgway replied: "Nobody takes the place of a man like that. You just follow him." Back home, Dwight Eisen-hower was more skeptical of his former boss, telling reporters: "When you put on a uniform there are certain inhibitions you accept." But for now Ike's comment went against the grain. This was Doug's time for accolades and sympathy.

Both the Japanese Diet and the Korean National Assembly passed unan-imous resolutions praising his achievements. Japanese prime minister Shigeru Yoshida and Syngman Rhee each sent personal letters expressing regret at his departure and thanks for saving their respective countries. As their Cadillac limousine headed out of Tokyo toward the airport shortly be-fore dawn, Doug, Jean, little Arthur, and his nanny, Ah Chue, gradually be-came aware of a vast crowd lining the highway—over half a million ordinary citizens had come out to bid him goodbye and thank him for the changes he had wrought.

It only got better as his Constellation touched down in San Francisco on

April 18, 1951, greeted by tens of thousands who had broken through police barriers to catch a glimpse of the conquering hero unjustly forbidden to conquer—in their view and also the opinion of two out of three Americans, according to a Gallup poll. It had been nearly sixteen years since he set foot on his native soil, gone but very apparently not forgotten. Just as in Tokyo, half a million San Franciscans lined the road from the airport to the Saint Francis Hotel, where Arthur found a bonanza of candy and milkshakes, and a student visa awaited Ah Chue. The next morning another half million showed up for an abbreviated ticker tape parade, a short address in front of city hall, then back to the airport for the long flight east.

The plane set down at National Airport outside Washington just past midnight. No surprise, Harry Truman stayed put in the White House, nor did Acheson attend. Besides the Republican congressional leadership, the official greeters did include Jonathan Wainwright, the Joint Chiefs, and, leading the military contingent, George Marshall. MacArthur, still eagle-eyed at seventy-one, spotted the perceived triggerman instantly. "Hello George!" he roared. "How are you?" They shook hands for photographers before a crowd of twelve thousand surged forward to completely interrupt the presentation of a silver tea service; it required a half hour to push the general through to his car.

Twelve hours later, dressed in a uniform without decorations, only his five stars, Douglas MacArthur stood poised at the rostrum of the House of Representatives, about to deliver one of the most famous valedictories in American history. Bathed in wave after wave of applause, as his eyes scanned the crowd he must have noticed that Marshall and the chiefs had now joined the rest of Truman's cabinet among the missing. No matter. He had been working on this speech since leaving Tokyo, he knew he had a good one, and he had been told there would be multiple millions watching on televisions like the one he had first seen at his hotel in San Francisco.

Finally, the clapping stopped and MacArthur began, filling the chamber with thirty-four minutes of dulcet oratory, the sum of which amounted to an antithetical rebuke of administration foreign policy. Asia, not Europe, should be the focus. As for Korea, "Once war is forced upon us, there is no alternative than to apply every available means to bring it to a swift end. War's very object is victory—not prolonged indecisions. . . . In War, indeed there can be no substitute for victory." His recipe included bombing north of the Yalu, along with a naval blockade of China's coast, and unleashing Chiang for raids on the mainland—a full-court press that only the right wing of the right wing truly wanted, but one superbly delivered.

He had already been interrupted twenty-nine times by applause before he set up a final thunderous eruption with a note of pure pathos: "The world has turned over many times since I took the oath on the Plain at West Point, and the hopes and dreams have long since vanished. But I still remember the refrain of one of the most popular barracks ballads of that day, which proclaimed, most proudly, that 'Old soldiers never die. They just fade away.' And like the soldier of the ballad, I now close my military career and just fade away—an old soldier who tried to do his duty as God gave him the light to see that duty. . . . Good-bye."

It was, of course, balderdash; Doug had no intention of fading away. He wanted to be president, to have the exquisite pleasure of evicting Harry Truman from the White House. He immediately embarked on a ticker tape tour that crisscrossed the nation with stops in not only New York, Chicago, Boston, Cleveland, Detroit, Houston, Dallas, Miami, Seattle, Portland, and Los Angeles, but also in San Antonio, Fort Worth, Little Rock, Norfolk, Manchester, and Murfreesboro. They turned out by the millions, mainly just to welcome him home, but the antievanescence tour as it unfolded looked like a presidential omen.

There was a showdown of sorts at the joint congressional hearings on Korea in early May. MacArthur got three days and three hundred pages of testimony to argue for not substituting victory in Asia, and proved less than convincing, particularly in his cavalier disregard for the risks posed by the Soviets. For its turn at bat, the administration brought the heart of its lineup—Acheson, Omar Bradley with the other chiefs, and in the cleanup spot, George Marshall. Comparing performances, one observer thought that "MacArthur had hoped to breach the walls with dynamite; Marshall with greater cunning . . . breached the walls with the slow prodding of a battering ram." It's likely Mac left his part of the hearings satisfied with his performance, but he had really been preaching to the right wing of the choir, convincing the already convinced.

He went back to New York, settled with Jean and Arthur into a jumbo ten-room penthouse in the Waldorf Towers just a few floors away from Herbert Hoover's, and essentially awaited being anointed for the position his new neighbor had once occupied. But Mac was even worse at politics, and the right-wing friends he expected to help him were politicians themselves, with ambitions of their own. Meanwhile, as he continued to crisscross the nation with speaking engagements, he did so in full uniform, a serious miscalculation in a nation where many were beginning to see the unprecedented peacetime influence of military men as a threat to democracy. Add to that a decade-and-a-half absence from the home field. Americans had changed; Douglas MacArthur's orotund oratory and noblesse oblige approach to politics came from another era, and no longer resonated with the public. It would be a long wait at the Waldorf; endless, in fact. Meanwhile, across town on Morningside Heights, the people's real choice, Dwight Eisenhower, continued to play his cards carefully, drawing into his hand more useful friends, placing his bets strategically, and setting himself up to pick up the big pot.

Back in the nation's capital George Marshall was getting ready to finally retire from the game, but before he left he wanted one more turn out in the field, ostensibly to get a better feel for the Korean battlefront. He quietly left on June 5. He and Ridgway flew in from Tokyo, only to find the weather horrible. When the new Far East commander suggested they switch to jeeps and scrub the rest of the tour, Marshall asked if he would do that if he were alone and, when the answer was negative, insisted they resume flying. "The pilots shuddered," remembered Ridgway's public relations officer. "It was so dangerous it was silly but Generals are like that. They think they always have to be braver than anyone else and they get away with it." They all did, landing safely, but the secretary of defense can't have learned much flying over the Korean front in that kind of weather.

And the forecast got only worse as soon as he returned to Washington. George Marshall was being blamed for Mac's removal—that and a whole lot more. On June 14, Senator Joe McCarthy, the Republican Party's über-anti-Communist, took the floor and delivered a three-hour excoriation of Marshall's entire career since the beginning of World War II. Opening with the paranoid proposition that "our present situation" could only be "the product of a conspiracy . . . on a scale so immense as to dwarf any previous venture in the history of man," he then proceeded to detail the misdeeds of the archconspirator, ranging from advocacy for a premature invasion of Europe, to "making common cause with Stalin" at Tehran, to creating (with Acheson) "the China policy which, destroying China, robbed us of a great and friendly ally," and finally was responsible for the strategy that turned Korea into "a pointless slaughter." And his motive for all this mayhem and misfortune? McCarthy maintained it was "to diminish the United States in world affairs, to weaken us militarily, to confuse our spirit with talk of surrender in the Far East, and impair our will to resist." In other words, the devil made him do it.

McCarthy delivered his diatribe to a mostly empty chamber but a packed gallery, the latter foretelling the widespread attention the speech received, if only for its brickbat iconoclasm. But when asked to respond to seeing his career turned upside down, Marshall was ready: "If I have to explain at this point that I am not a traitor to the United States, I hardly think it's worth it." It was the reply in keeping with the main lesson of his career: less is more.

But also that of a very tired man, sick of the game at last. At a staff meeting on September 12, nine days short of a year in office, he casually noted: "At eleven o'clock I cease to be Secretary of Defense." A car drove him back to Dodona Manor and Katherine, where, basically, he stayed. There would be future controversy and residual honors, but he was out of the game for good. Unlike our other three all-stars, there would be no memoirs. He said he didn't want to stir up old squabbles, but he also remembered Pershing's negative experience writing his own. Rumor had it that he had been offered a million dollars for an autobiography, but George Marshall chose privacy and Virginia.

CHAPTER 12

IKE

As he sauntered into Italy in 1494, Charles VIII, feckless king of France, had no idea history's gears were changing, only that he came well armed. For he rode at the head of a new kind of army, truly gun based, with a particularly fearsome new weapon. Guns had been known in Europe for more than two centuries, but primarily as huge and barely movable siege pieces. Now Charles came equipped with bronze cannon firing iron balls, compact enough to be horse-drawn, making them available on battlefields as well as devastating against the stone curtain walls that surrounded Italy's cities. Add infantry armed with arquebuses, along with pistol-packing cavalry, and the result was a revolutionary instrument of destruction. But not necessarily a decisive one. Instead, Charles lit a firestorm of war—sixty-four years of almost continuous fighting in Italy, followed by a leap northward to ignite the Low Countries, France, and eventually Germany. Mercenary armies wandered unchecked, preying on populations like giant parasites. Before it finally burned out in 1648 with the Peace of Westphalia, tens of millions had died from the fighting along with the famine and disease it spread. Henceforth, monarchs and heads of state would tightly control Europe's armies, but it had taken a century and a half to exert that grip.

Now, in the middle of the twentieth century, humanity faced what looked like a far more dangerous, utterly unforgiving challenge. Nuclear weapons didn't simply have to be controlled, they had to be integrated into an international system plainly prone to societal-scale warfare, and in a manner that would ensure that they would never be used again, an almost oxymoronic proposition. There could be no more slip-ups; a single failure once these weapons proliferated could plausibly lead to the incineration of civilization itself, or at least its urban version. Every weapon humans had ever invented had been used on other humans, this one twice already. And they arrived at a time when arms technology in general was cresting. Things did not look good for *sapiens*.

To provide some measure of mutual security from annihilation, subsequent events would have to boil down to keeping the two teams perpetually off the field and in training camp—all the time building nuclear muscle and practicing using it, but never, never for real, relying instead on threat behavior, minor-league teams, selective dirty deeds, and keeping the best eye possible on the other guy. This was no easy game to keep from going radioactive, and it largely fell to the Americans to dictate the pace and magnitude of the action, leaving Team USSR to react. Harry Truman in the company of George Marshall set the initial conditions, but the game was shaped and dominated by the last man standing in our murderers' row, Dwight David Eisenhower. Ike's turn as manager was not without foibles—he was fooled in a number of contests, even by allied teams, and the results he produced were not the final ones. But he kept the base runners advancing, the folks in the stands alive, and in the process became the greatest of our four Hall of Famers.

IKE WAS EVERYBODY'S PICK TO BECOME THE FIRST NATO COMMANDER; HIS record in World War II had established him as coalition commander par excellence. It took some time, though. Truman, who would do the asking, and Eisenhower the accepting, got along like porcupines at this point, so the mating proceeded cautiously. Meanwhile, Taft and the Republican right wing were ardently opposed to NATO and the concept of "collective security." But the North Atlantic Council unanimously recommended that Truman appoint Ike, and on April 2, 1951, he took over as Supreme Allied Commander in Europe.

There wasn't much to command. Facing seventy-five Soviet divisions and equivalent numbers from satellite states, the United States and the United Kingdom each had one division in Germany, not much more than a trip wire.

Both countries were in favor of rearming the Germans, but the French were not, despite most of their army being deployed to sit on colonies in Southeast Asia. To compound matters, de Gaulle was long gone, having stepped down in 1946, and subsequent French politics had reverted to its proverbial prewar confusion, with Communists making up the country's second-largest party. Much the same situation existed in Italy too.

But just outside Paris, ensconced with Mamie in a palace built for Napoléon III, Ike was less worried than he was resolute. On his flights to and from Moscow shortly after the war, he had seen with his own eyes the enormity of the damage the Wehrmacht had inflicted on the Soviet Union. He was fully aware of the US nuclear preponderance, and was fairly sure the Soviets were also. In his eyes, they wouldn't be attacking anytime soon.

Meanwhile, he had his own agenda. Ike had always operated comfortably at the level of princes, and now he became the primary voice of America to twelve separate states. He spoke confidently, but in perfect cadence with George Marshall's original message: America has your back, but NATO is ultimately about self-defense. During his year in Europe, Ike bounced from capital to capital delivering that message from behind the same smiling mask that proved yet again both convincing and trustworthy to military and civilians alike. Almost single-handedly he was stitching the alliance together with sheer personality, or so it seemed.

Just about everybody liked Ike, and back at the palace he remained very much aware of the implications for his future. He was great at playing coy, and when he put on his uniform again, convenient rumors arose that he was no longer interested in the presidency. But if certain photographs can be revealing, there is one of Ike just after he arrived in Europe being told of MacArthur's firing; it is the face of a man plainly weighing his options. Meanwhile, members of the Gang flew in steadily from the States for golf and bridge, and to talk politics. On the subject of the presidency he couldn't have been more two-faced, firmly denying any interest in the nomination while in uniform, but then telling Gang member Bill Robinson, "The seeker is never so popular as the sought. People want what they cannot get."

Lucius Clay even seems to have believed it: "Ike was no fool. . . . He knew that he had tremendous standing in America, but that if he entered into a political contest he could lose the nomination, in which case his standing would be greatly lowered. Or he could lose the election, in which case it would be lowered even more." Ike didn't intend to do either, but his continued reticence acted to motivate Clay, his chief liaison and facilitator in his subtle quest for America's top job.

After staring down the Russians with the help of the Berlin Airlift, Clay had left the Army in 1949; his savvy and business connections opened the way to his quickly becoming chairman of the Continental Can conglomerate. Son of a three-term US senator, Clay not only knew the mechanics of politics, but also was plugged into the internationalist wing of the Republican Party that was Ike's logical jumping-off point. Eisenhower had not yet said whether he was a Republican or Democrat, and once again in November 1951, Truman impulsively offered to help Ike gain his party's nomination. But the choice of Clay as go-between was a clear indicator of the direction Ike was headed, much as a golfer's choice of clubs is a tip-off to his or her approach to the hole.

A key moment came in late September 1951, when New York governor Thomas Dewey—twice-failed Republican candidate but still head of the GOP's internationalists—invited Clay to dinner and told him that although Taft was almost certain to be the nominee, he couldn't possibly win the general election. Only Ike could prevent a Taft nomination from happening. Message sent and received; Clay promised Dewey he would do everything he could to convince the general to throw his eagle-crested cap into the ring.

Ike did, but only reluctantly, it seemed. The New Hampshire primary was coming up in March 1952, and the state's governor and new political ally Sherman Adams thought it was an ideal opportunity for Eisenhower to demonstrate his popularity. But to do so he had to appear on the ballot with an affirmation that he was a Republican, and as the filing deadline approached in early January there was no word from Ike. Finally, on January 4, just under the wire, Clay authorized another instant ally, Senator Henry Cabot Lodge, to formally announce that Eisenhower would accept the Republican nomination if offered, followed three days later by a tepid affirmation that the prior statement "gives an accurate account of the general tenor of my political convictions and of my voting record." But then the general added a pregnant qualifier: "Under no circumstances will I ask for relief from this assignment in order to seek nomination for political office."

Even Ike's closest allies were flummoxed. Over there in Europe, did he realize that Taft had been planning his nomination for four years? Or how many IOUs Taft already held? Or, alternatively, how many people stateside really did like Ike, were wild about Ike? Plainly, this military Cheshire cat needed further convincing.

In early February, the burgeoning network of Ike allies staged a public demonstration of affection, a midnight rally in New York's Madison Square Garden. Eighteen thousand packed the arena, plus thousands milling around outside, all of them chanting, "We like Ike!" Tex McRary, the inventor of talk radio, and his tennis-star wife, Jinx Falkenburg, organized a spectacular gala that included Humphrey Bogart, Lauren Bacall, and Clark Gable. Mary Martin, star of the smash-hit musical *South Pacific*, sang "I'm in Love with a Wonderful Guy . . . Ike!" Irving Berlin closed the show by leading the crowd in a rousing version of "God Bless America."

Another husband-and-wife team—Floyd Odlum, head of RKO Pictures, and aviator Jacqueline Cochran—took it from there. Odlum had a Hollywood-grade two-hour movie made of the proceedings, while Cochran, known to be among the best pilots in America, male or female, personally flew it across the Atlantic and narrated a screening for Ike and Mamie at the palace. "Tears were just running out of his eyes, he was so overwhelmed," Cochran remembered. He began to "talk about his mother and father, but mostly about his mother." Finally, prompted by Cochran's "Taft will get the nomination if you don't declare," Ike responded: "Tell General Clay to come over for a talk. And tell Bill Robinson that I am going to run."

"I thought it was a lot of damn foolishness," Clay remembered, "but it did have a real effect in persuading General Eisenhower to announce." Mamie also thought watching the film made up Ike's mind to run. Yet the thought lingers that the tears were of the crocodile variety; beneath them was actually chuckling Ike the poker player, scrupulously building the pot. He had read *The Odyssey*, knew Penelope's perpetual reticence only drew more suitors. It was the same ploy he had set out in front of Bill Robinson: the sought-after always trumps the seeker. Perhaps it was some of each, strategy intertwined with emotion, or vice versa. Whatever it was, it had whipped America into a frenzy of "I like Ike." On March 11, 1952, in his first election, Eisenhower took the New Hampshire primary in fine style, with 50 percent of the vote to Taft's 38. A week later in Minnesota, home of favorite son Harold Stassen, Ike took second with over 100,000 votes, all of them write-ins, since his name wasn't on the ballot.

Self-organizing atop this building wave of popularity was a political superstructure intent on channeling its momentum against the better-established Fortress Taft. Dewey, Adams, and Lodge were all elected officials with substantial duties, while Clay was eager to get back to business; consequently, as the campaign progressed, more junior types moved in to play key roles. Three in particular stood out—each Ike came to like and trust, and eventually for each he carved out a substantial role in his presidency.

Bobby Cutler, bachelor and Boston banker, was brought in on the recommendation of his wartime boss, George Marshall, who described him as "a rose among cabbages." Cutler was a perfect sounding board for Ike. Considered brilliant since his days as a Harvard undergraduate, he was also very funny and a ferociously hard worker. All of this was fine with Ike, and Cutler found himself quickly engaged in deep conversations with the candidate; but as frank and sincere as Ike seemed, the new sounding board sensed an unapproachable side to Eisenhower where only his own counsel was forever kept. For his part, Ike almost immediately sized up Cutler as a man who could keep a secret and also build on it; he reserved for him the task of constructing and managing the coming administration's national security apparatus.

Guiding the campaign through the shoals of the legal and political environment was Herbert Brownell. He had managed both of Dewey's losing presidential efforts, but few blamed him. Brownell's knowledge of the mechanics of politics was vast, as was his knowledge of the law. Although periodically tethered to his firm in New York, Brownell was still able to fashion a winning strategy and play a key role at the party's convention, one that left Ike dazzled enough to make him attorney general when the dust settled.

Eisenhower had always enjoyed good relations with the press. During the war, on the grounds that they were loyal Americans, he had confided in them more than any other Allied general, and they responded with glowing stories that were instrumental to his popularity at home. Now, as presidential candidate and for years after, his run of good luck with the fourth estate would continue, largely due to the ministrations of a witty, likable, but also scrupulously honest press secretary, James Hagerty. He had been performing the same function for Dewey and simply slid over to Ike, ever the talent magnet. They hit it off from the beginning, and needed to. Newspapers still remained the amphetamines of presidential campaigns, and this one was shaping up to be a nasty one, played out on their front pages. On a number of fronts Ike would have some explaining to do.

When Clay came to Europe in February he spoke like a Dutch uncle, reminding Ike that getting the nomination would be far harder than winning the general election, Taft having already accumulated 450 of the 604 delegates needed. Still, Ike procrastinated, reluctant to formally relinquish his command and begin campaigning, but he had told reporters that one thing that had caused him to "reexamine" his "political position" was a "fairly considerable dislike for MacArthur's politics and policies." Hence, when Douglas MacArthur was picked to give the Republican keynote address, Clay and Dewey saw their opening, concocting a scenario for Ike of a convention deadlocked between him and Taft and turning to his nemesis, Mac. That did it; Ike wrote Truman, asking to be relieved of his command and retired from active duty on June 1. But meanwhile Taft had swept primaries across his Midwestern base in Illinois, Nebraska, and Wisconsin—expected, but still deflating to Ike's adherents.

It got worse as he kicked off his campaign with a nationally televised speech in Abilene on June 4, and the hometown heavens opened up. Dressed in a plastic rain hat and see-through slicker, the candidate laboriously plowed through his standard-issue Republican-speak diatribe with all the enthusiasm of a drenched eagle. Watching from New York, Dewey, Brownell, and Clay were stunned. "Absolutely dismal . . . about as disappointing an opening campaign speech as I've ever experienced," Brownell grumbled. This was not the Ike they knew, or one anybody else would particularly want to know—just a tired old man stumbling over platitudes. They brought him back to Morningside Heights for a weeklong reboot—not a makeover, but a restoration. There would be no more personality-blocking prepared texts; Ike was far better speaking off the cuff, allowing not only his folksy charm but also his rock-solid policy grounding to roll out unencumbered. Next, he set up headquarters at the Brown Palace Hotel in Denver. Near the Douds and on familiar ground, Ike rocked a packed crowd of locals ten days after the Abilene disaster sufficiently for Dewey to wire him: "I hope you never use a text again." In practically no time, Ike was ready for prime time, but there was practically no time left—just over three weeks before the convention.

Short as it was, the Eisenhower charm offensive did demonstrate substantial traction, and by the time the GOP convened in Chicago on July 6, he had captured around 500 committed delegates, while Taft, whose momentum was clearly flagging, had worked his way up to 525. But it was all downhill from that point. After losing three credentials fights, Taft's forces were thrown back on Douglas MacArthur's keynote address to inject some energy into their deflating cause.

The next afternoon, wearing a civilian suit for the first time in public, the general flew into Chicago from New York, a man on a mission. Two, actually. The first was, through sheer force of oratory, to obliterate Dwight Eisenhower's path to the nomination; the second was to slide into that slot himself, dark horse Doug. Instead, as his words poured forth over the packed arena and eighteen million television viewers, they amounted to less a keynote than a sour note, perhaps his worst speech in memory, both strident and banal enough to have the delegates talking among themselves before it was half over. Both missions not accomplished, so ended the ill-starred political career of Douglas MacArthur. His mask was an outdated one, and only ever really convincing as that of a warrior.

Ike's was not only suitable for mixed company, his political guise was about to pick up the big pot. On July 11, the balloting for president began. In the Eisenhower suite at the Blackstone Hotel, Clay and Brownell were parked in front of the TV intently counting votes, the strategy being to stay even on the first ballot, then begin picking up defectors on the second. "Eisenhower was certainly the calmest person in the room," Brownell remembered. That was because the process largely eluded him, especially when New York delivered ninety-two votes for him and only four to Taft, Dewey having strong-armed fourteen unexpected delegates. Brownell was instantly on the phone to Sherman Adams, their floor manager and chief arm-twister, telling him to go for broke, call in all chits, push Ike over the top on the first ballot. In the end, over the objections of favorite son Harold Stassen, the Minnesota delegation switched all twenty-eight of its votes to Ike, and with that, suddenly he had the nomination. Back at the Blackstone, Ike was flab-bergasted. "He'd never seen anything like it. He didn't know how Clay and I knew what was going to happen," Brownell remembered. Ike was no rube, nor did Brownell think him one, and politics being more like poker than the Army, he proved a fast learner—but in the choice of a vice president, not fast enough, since it would take place the next day.

Actually, he didn't know it was his choice to make; he had to have it explained to him by Brownell and Clay that the delegations on the floor, though technically the deciding body, would vote for whomever he chose. But when he presented to Brownell and Clay a list of those he considered to have the necessary experience—largely business executives—they gently explained that the delegates were political animals and needed to hear the name of one of their own kind as his running mate. Since Ike didn't know many politicians, he looked to his handlers, and they looked to Richard Nixon, junior senator from California. Nixon had played a key role ensuring friendly behavior by the California delegation in the opening credentials fights; he was also young to Ike's old, political to his core, and as a result of the disclosures he helped generate in the Alger Hiss Soviet espionage case, popular with the right wing of the party.

In the eyes of Clay and Brownell he was the obvious choice, but not in Ike's. He had met Nixon once, and it's probably safe to assume that he already didn't like him, never would. When it came to evaluating people, though Eisenhower had vast experience, he largely relied on instinct, and there was something about Nixon that never passed muster. Possibly he sensed a moral void at the center of the man, but whatever it was, it ensured that Richard Nixon never penetrated Ike's inner circle—a very significant waste of talent; presidential timber, actually. At any rate, Nixon was on board, just not in a first-class cabin—he was more like a barnacle.

IRONICALLY BUT INEVITABLY GIVEN THE TIMES, THE ONLY BAD WEATHER EN-countered by the USS *Eisenhower* on its cruise to the presidency blew in from the starboard wing of his own party. Taft people rumored that he was Jewish—"Ike the Kike"—sounding ridiculous in the process. Kay Summersby was dragged up, but Truman had destroyed Marshall's fierce letter to Ike on the subject, and Kay wasn't talking. So that went nowhere. Then there were insinuations that Mamie was an alcoholic. She did enjoy a drink, more so when Ike was away at war; but she also had a balance problem, later diagnosed as Ménière's disease, that left her staggering. Ike lived with her and knew the truth, so he gave the stories what they deserved, dead silence, and they too passed.

More troubling to Eisenhower personally and enduring well past the general election was the storm whirling around George Marshall, energized chiefly by the vitriolic windbag senators Jenner and McCarthy. Apparently understanding that his name was beginning to sound like the infamous Dr. Mudd's, Marshall wrote Ike a somewhat belated letter of congratulations, excusing himself with "I felt because of the vigorous attacks on me by various Republicans any communication with you might be . . . detrimental to your cause." Ike was quick to thank him, but didn't say anything about Republican attacks.

A chance to say something public came in early August, when back in Denver Eisenhower held his first press conference since the convention. A journalist from *The New Yorker* tipped off Hagerty that he planned to ask Ike about the Republican assault on Marshall, giving him plenty of time to prepare. When the softball was lofted, Ike was completely in character. His face turned scarlet, veins stood out on his neck and forehead, he jumped up and pointed a finger at the reporter: "How dare anyone say such a thing about General Marshall, who was a perfect example of patriotism and loyal service to the United States. I have no patience with anyone who can find in his record of service for this country anything to criticize." Impressive fulmination, no doubt; yet Ike conveniently failed to name names and pin the tail on Marshall's accusers, and thus he failed to escape from this tar pit of a dilemma.

The plot got only stickier when candidate Eisenhower learned that sharing the stage with him for his big speech in Indianapolis on September 9 would be Mr. "Living Lie" himself, Indiana's senator William Jenner. Looking for a way out, Ike turned to his handlers, only to be told unanimously by Brownell, Dewey, and Clay that attendance was mandatory.

It was a hands-on experience for Ike, literally. Appropriately enough, the candidate delivered one of his most partisan speeches of the campaign, but even that backfired, since with every sustained burst of applause Jenner grabbed Ike's arm and raised it like a prizefighter's. Not naming names again, he finished by imploring the crowd to support the Indiana GOP ticket "from top to bottom," presumably including Jenner, and was rewarded with a prolonged bear hug from the thoroughly aroused senator, one caught by photographers for the morning papers. After it was over Ike left so fast, nobody on the platform initially realized he was gone. "I felt dirty at the touch of the man," he told his speechwriter Emmet Hughes.

He vowed to come clean in Wisconsin, Senator Joe McCarthy's home state. For his major speech in Milwaukee Eisenhower inserted a stout defense of Marshall, calling attacks on him "a sobering lesson in the way freedom must not defend itself." But after reading an advance copy, party leaders begged Ike to take it out, claiming it would split the state GOP and give the Democrats the advantage in the fall. Sherman Adams agreed, reminding Ike that this was, after all, politics. So he caved, deleting the paragraph defending Marshall when he spoke. Unfortunately for Ike, the press had received the original text, and zeroed in on the omission as an abject surrender to McCarthyism, a story that made every front page in the country.

The episode left Eisenhower not just feeling further begrimed, but, it seems, truly abashed. Back in Virginia George Marshall greeted the incident with a shrug, telling Rose Wilson, over the decades still his confidante: "Eisenhower was forced into a compromise, that's all it was." This is how it struck the pols around Ike, but they also realized nothing that happened during the campaign upset him more than his failure to stand up for Marshall.

The two were never exactly friends—Marshall still called him "Eisenhower"—neither, really, were any of our four, except for Patton and Ike at the beginning. But all shared adult lives spent almost exclusively in the military, an environment where trust and loyalty are paramount. It was on this basis that Marshall had plucked Ike out of relative obscurity and fostered his rise to the highest levels. Even in the pursuit of Overlord, they remained completely principled, though each desperately wanted the job. George was hard on Ike over Kay Summersby, but he was also trying to save his career and his future. Up to this point, each always had the other's back. But this episode not only showed Ike how corrosive politics can be to even one's most firmly held principles, it also shamed him. Years later, Katherine

Marshall told a biographer: "Don't attack President Eisenhower about the McCarthy thing. He did everything in the world to make it up to George and me." Everything except directly attack McCarthy. Ike's strategy was, in effect, "ignore McCarthy and he will disappear." That eventually held true, but did not prevent him from being further slimed.

Ike found himself on firmer ground against Democrats. With Taft removed from the race, grizzled manager Harry Truman decided to call it quits; he had served almost two terms, and the senator from Ohio was the only Republican he could and wanted to beat. Adlai Stevenson, the governor of Illinois, was slated by the Democrats to be his replacement. Witty, urbane, and capable, he would cast but a pale shadow across Eisenhower's smiling mask as the campaign moved toward its inevitable conclusion. As the days ticked down, Ike seemed to have developed a genial contempt for Stevenson and his droll ripostes, but also an animosity that bordered on hatred for Truman.

One incident in particular did it. With the Democratic National Convention over and the candidates set, Truman wanted them both to have an intelligence briefing, trusting Omar Bradley, as chairman of the Joint Chiefs of Staff, to handle the invitations. Since Stevenson was to be briefed first, Bradley sent him his invitation without an equivalent going to Ike. Consequently, when the press reported that on August 12 Governor Stevenson had been briefed at the White House by his friend Omar Bradley and former right-hand man Beetle Smith, now head of the CIA, Eisenhower was livid and made no secret of it.

Some have noted it served his interests politically, letting him spotlight Democratic partisanship, but there was a lot of Ike's temper involved, the part suggesting Truman set up the incident, using his own close associates to humiliate him. The outgoing manager tried to make it right, personally inviting him to his own briefing plus lunch, and when Eisenhower stonewalled, Truman sent him another handwritten note: "I am extremely sorry that you have allowed a bunch of screwballs to come between us," closing with "From a man who has always been your friend and who always wanted to be." He received only silence in return. Like Kay, Harry Truman was about to learn that once it was over with Ike, there was no reclama.

Compared with this incident, or his own perceived failure to defend Marshall sufficiently, the campaign's most famous episode, the Checkers saga, was more like a badly stubbed toe to Ike, since he didn't much care for, or care what happened to, Richard Nixon. As scandals go, it was pretty thin gruel. The story broke over not much—the transfer of some political money, $16,000, most of it used to cover senatorial office expenses, in a perfectly legal manner—but when confronted, Nixon claimed it was a left-wing conspiracy to smear him because of his sterling record as an anti-Communist. Instantly, the story was splattered across front pages from coast to coast.

Hagerty and Sherman Adams, key men on the campaign train, along with Brownell, Dewey, and Clay in New York, were all appalled by how a supposedly seasoned politician could have let this easily removed tumor metastasize, and consequently wanted him gone, to be replaced by California senator William Knowland. Eisenhower himself issued the verdict. "Of what avail is it for us to carry on this Crusade," he told reporters, "if we ourselves aren't as clean as a hound's tooth?" Dewey strongly suggested to Nixon that he quaff the proffered tankard of Kool-Aid—go on national TV, explain himself, offer to resign, and leave the decision to Ike. Just before airtime this escalated by phone into "at the close of the broadcast tonight you should submit your resignation to Eisenhower." But nobody reckoned on Tricky Dick, who ended the call with: "Just tell them that I haven't the slightest idea what I am going to do. . . . And tell them I know something about politics too."

For a man supposedly not much good on TV, Richard Nixon delivered one of the great tearjerkers in American history that night. What was supposed to be a mea culpa Nixon transformed into a lachrymose tale of six-year-old daughter Tricia's adorable cocker spaniel, Checkers, and his wife Pat's "respectable Republican cloth coat." As Ike watched in Cleveland with Mamie, he looked over in dismay to see his wife dabbing tears from her cheeks. And Nixon closed with even worse news. "Wire and write the Republican National Committee whether you think I should stay or whether I should get off; and whatever their decision is, I will abide by it," deftly removing his own fate from Ike's hands. Having provided an altogether dazzling lesson in survivalist politics, Nixon stayed on the ticket, but at the cost of being forever banished from Eisenhower's trust and affection. The vice presidency is not necessarily a bad place to serve an eight-year term, yet for Nixon it was always a cold and unwelcome one, with scant hope of parole.

Meanwhile, candidate Eisenhower and his newfound gang of pols quickly got back on message and on track, with the campaign train crisscrossing the nation through September and October, Ike doing whistle-stops from dawn to dusk, and in the process putting to rest the notion that at sixty-two he lacked the energy and endurance to be president.

Then, on October 16, Truman, speaking in Hartford, challenged the Republican candidate to come up with a plan to end the Korean War, which by this time had degenerated into a grinding stalemate, and Ike saw his opening to administer the coup de grâce. A week later he responded at Detroit's Masonic Temple, linked to nationwide TV, pledging to go himself: "That job requires a personal trip to Korea. I shall make that trip. . . . I shall go to Korea." Ike's promise was an instant hit, a reassuring balm to a war-weary nation. "For all practical purposes, the contest ended that night," the Associated Press reported subsequently. Later, Omar Bradley would sourly maintain that the pledge was "pure showbiz. Ike was well informed on all aspects of the Korean War and the delicacy of the armistice negotiations. He knew very well that he could achieve nothing by going to Korea." Perhaps, but also irrelevant. This was politics, and Ike was giving the country what it wanted.

And its citizens returned the favor on November 4 with a landslide, giving him all but nine of the forty-eight states, which translated into 442 electoral votes to Stevenson's 89. Looked at retrospectively, our four superstars had commanded a massive number of organizations and people, but only Ike managed to become commander in chief. As a matter of context, this is worth noting, since it was largely how he looked at his new job. Domestic issues, with the exception of money matters, he wanted others to handle. And while multiple presidents have entered office with grand domestic agendas, only to end focused on foreign policy, Eisenhower saw national security as paramount from the beginning.

It was on these grounds that on November 29, 1952—less than a month after the election and almost two before inauguration—the president-elect departed for Korea. Before he left he was briefed by the chiefs to the effect that there were but two options—continue the stalemate, or renew the offensive north—and Ike didn't like either. "Dad could have seen there were hills by looking at a map," remembered son John, now a major and stationed along the front. But besides providing a father-and-son reunion, this trip was for Ike about the ritual of decision-making—feet on the ground, talking to the participants, getting a feel for the environment, then forging a new consensus. To foster buy-in and spread the fingerprints, Ike also brought along a mixed pickup squad of military and civilians; in addition to Bradley, they included Hagerty, Brownell, defense secretary–designate Charlie Wilson, and Admiral Arthur Radford, commander in chief in the Pacific.

The trip lasted only three days, riskwise highlighted by Ike's Mac-like flight along the Thirty-Eighth Parallel in an Army light plane. But he met everybody he needed to form sound conclusions. After conferring with Syngman Rhee, Ike came away seeing an Asian Charles de Gaulle, and one just as likely to stray off the ranch.

The key commanders Ike knew well—James Van Fleet, head of the Eighth Army, was a West Point classmate, and leading all UN forces was Mark Clark. As Bradley told it, "Clark and Van Fleet had cooked up a victory plan that could only be described as MacArthuresque," featuring Chiang's troops from Formosa and the use of atomic weapons being given "serious consideration." "I know just how you feel militarily," Ike told them, "but I have a mandate from the people to stop the fighting. That is my decision." Ironically, he would make it stick with the threat of nukes, the difference being Eisenhower was bluffing. And on July 27, 1953, barely six months into his presidency, UN and Communist forces agreed to an armistice, and the fighting did stop.

But if Bradley was wrong about Ike's trip to Korea achieving nothing, the passage home compounded his error, since it literally turned into a shakedown cruise for his coming administration. Ike flew from Seoul to Guam, where he boarded the heavy cruiser *Helena* headed for Hawaii, at Wake Island picking up the designated secretaries of state and treasury, John Foster Dulles and George Humphrey, along with Clay and the budget director Joseph Dodge. Eisenhower was planning his presidency like a military campaign, and he intended this session at sea to chart the grand strategy for the coming four years.

Unlike the political neophyte of the campaign, Eisenhower guided the proceedings with a firm and confident hand. Clear priorities were set and principles enunciated. The emphasis would be external, the focus to remain on Europe and containing the Soviet Union as NATO matured. Domestically, there would be no savaging of New Deal programs: Social Security and agricultural price supports would continue.

But Ike was a deficit hawk and deeply worried about the burgeoning national debt. Hence, tax cuts would be deferred until the budget could be balanced. Since wars are the most costly things governments do, the liquidation of the Korean conflict and concomitant force drawdowns offered massive savings, the security shortfall to be made up with an enhanced reliance on nuclear weapons, subsequently to be labeled the "New Look." That it was: Dwight Eisenhower was ready to bet the nation's future on Bernard Brodie's proposition that nuclear weapons had made all-out war, or in Ike's view just about any war, impossible; that their only use was to deter—and, of course, to bluff.

Bluffing being one of Ike's primary acquired life skills—an impenetrable mask that made him, beyond a certain point, unreadable—the motives behind his behavior at some junctures simply escape satisfactory explanation. And he was nowhere more opaque than in the contrasting manner he treated Douglas MacArthur and Harry Truman as his time in office neared.

Shortly after boarding the *Helena* Ike cabled MacArthur: "I am looking forward to informal meeting in which my associates and I may obtain the full benefits of your thinking and experience." Not only that, but it was he who came to the Waldorf on December 17, with Dulles tagging along to imbibe some of that wisdom, or perchance to separate the parties should things go awry.

They didn't. Instead the two sat through a two-hour recitation of the MacArthur plan for world peace, one that included an immediate one-on-one with Stalin, where Ike would demand the prompt unification of Germany and Korea under democratically elected governments; the subsequent neutralization and withdrawal of all foreign troops from not only these two countries, but also Austria and Japan, all capped by provisions outlawing war inserted into the respective Soviet and US constitutions. Should Stalin fail to comply, nukes were to be wielded, not necessarily at him, but in Korea and China.

Ike let his new mouthpiece do the talking, and Dulles proved up to the task: "Your present plan is a bold and imaginative one and could well succeed. I believe, however, that Eisenhower should first consolidate his position as President before attempting so ambitious and comprehensive a program." As they got up to leave, MacArthur took a fourteen-point memorandum of his plan, carefully folded it, and placed it in the breast pocket of Eisenhower's suit, murmuring, "God bless you."

It was the last anybody around Ike saw of the document; it was also the last meeting between these two giants. By all prior indications Ike had no interest in Mac's advice and never sought it subsequently. The meeting could be seen as politically expedient, Ike covering his right flank, except it went unreported for a decade until MacArthur published his memoirs. As far as personal motivation, the relationship between these two had reached the stage of exquisite dislike. Objectively speaking, this meeting made no sense, but it does seem to have some ritual and symbolic meaning—two old soldiers who, though they might never bury the hatchet, still found it necessary to pass the torch.

DWIGHT EISENHOWER'S PRESIDENCY BEGAN ON A SOUR NOTE, ONE THAT should have left him contrite, but instead failed to penetrate in the slightest degree his now glacial anger at Harry Truman. It was customary prior to the inauguration for the president-elect and his wife to stop at the White House and join the outgoing first couple for coffee. This time was different. Ike and Mamie stayed in their limousine until the Trumans got the message and belatedly joined them for the ride down Pennsylvania Avenue in icy silence.

But Ike could not contain himself. He had obviously wanted his son John to attend the ceremony, but for appearances' sake had refused to pull the strings necessary to bring him back from Korea. Then, unannounced, John arrived. Now he asked Truman who had ordered him home. "I did," he snapped back, done personally as a surprise for the Eisenhowers—enough to melt most men's hearts. But not Ike's. The two men rode on in a silence not broken until John Kennedy's funeral ten years later.

It was another example of Eisenhower behavior that not only defies expectations but also explanation. Objectively speaking, as a manager Harry Truman had always been favorably disposed to Ike, his only possible transgressions being his preference for George Marshall, or rumors that he might release the Summersby letter. But for the Eisenhower we have come to know, preference and rumors should not have amounted to rage and enmity. Those around him quickly became aware of Ike's torrid temper, but less obvious were grudges revealed on occasion as being as cold and massive as an icecap. In terrifying times Americans had elected a figure of reassuring competence and steadiness, and for the most part that's who Dwight David Eisenhower was; but there was also a darker, more unpredictable side that rendered him a far more complicated figure than all but a few suspected. That other side probably helped, rather than hindered, in the ultimately dangerous game that came with his new job.

In a moment of false pity, Harry Truman famously sympathized with his successor: "He'll sit here, and he'll say, 'Do this! Do that!' And nothing will happen. Poor Ike—it won't be a bit like the Army. He'll find it very frustrating." It's a statement most revealing of how poorly he understood his star player.

Eisenhower arrived in office knowing exactly what he wanted to do—keep America safe in unprecedentedly dangerous times—and with a roster meticulously chosen to carry out his strategic agenda. Being a general to the core, Ike instinctively relied on orchestrated meetings and panels to help him make decisions, along with carefully articulated staffing to ensure they were carried out—both mechanisms designed to keep him free to focus on the larger strategic picture. The henchmen he picked for key roles slid into place like gears in a carefully designed but still somewhat mysterious mechanism.

At the center, keeping everything on time and running, was Sherman Adams, White House chief of staff. He was the new Beetle, a politicized version and no more likable; he got things done through the same combination of intimidation and hard work, taking literally hundreds of phone calls a day and always too busy to say either "Hello" or "Goodbye." With few exceptions—Hagerty seems to have been one—the only way to Ike was through Adams. Critics at the time thought this isolated the president and left Adams with too much power, but it now seems clear that's the way Ike wanted it.

The focus would be foreign policy, and Eisenhower's choices reveal a species of personnel engineering that was absent from those picked to fill key domestic posts, the names here being largely supplied by Clay. To handle the mechanics of statecraft, both aboveboard and under the table, Ike brought in a sort of doppelgänger in the form of the brothers Dulles.

Of the two, John Foster was superficially the least attractive. "Dull, duller, Dulles," was the phrase often attached to the new secretary of state. British prime minister Harold Macmillan quipped that "his speech was slow, but it easily kept pace with his thoughts." Yet he was better than that. Beneath Dulles's pomposity and bloviating anti-Communist rhetoric, Ike found a rich vein of international expertise, and within weeks of taking office he began meeting Dulles nightly at six o'clock in his White House study to talk about the world. Dulles's grandfather, John Foster Sr., had been secretary of state, and Ike used to joke that his namesake had been preparing for the post "since he was five years old." As such he proved to be an excellent cat's paw—glad to articulate the harsher side of policy, talking "brinksmanship" and massive retaliation—while always allowing Ike to walk the rhetoric back with the voice of reason.

Less reliable as stalking horse was Allen Dulles, representing the spook side of the fraternal tag team as director of the Central Intelligence Agency. Master runner of spies in Switzerland (including Carl Jung) during the war, Allen Dulles belied the covert image by playing the hail-fellow-well-met. Yet as Louis Auchincloss, the novelist-attorney who worked with both Dulleses at a Wall Street firm, and got to know them well enough to see beneath the respective facades, later recalled, "I thought of Foster as maladroit but, beneath it all, warm." As for his brother, "he was a hail fellow, but his laugh was humorless. . . . Allen was shrewd—but cold as ice."

Beetle Smith, who had run the CIA prior to the election, treated Allen, his operations chief, with his standard-issue ill humor: "Dulles, goddamnit, Dulles, get in here!" But rudeness came to border on disdain as he got to know him better. Dulles was not only a notoriously sloppy administrator, but also one afflicted with tunnel vision. Smith recognized that the technical side of intelligence held great promise in areas like electronic surveillance, decryption, and surreptitiously acquired imagery; yet Allen Dulles instinctively clung to his spies, to the point where his boss labeled him the "Great White Case Officer." But he was also a survivor, and good enough at Washington's postelection version of musical chairs to find himself planted in the director's seat. The new president, it seems, had big plans for the kinds of dirty deeds that were Allen Dulles's specialty.

As a fail-safe mechanism and minder, Ike installed Beetle as number two at the State Department with a direct line to his desk, his charge being to keep the Dulles boys—Foster's rhetoric and Allen's schemes—securely on the Eisenhower policy ranch. "Ike always had to have a prat boy, someone who'd do the dirty work for him," Smith complained to Richard Nixon, who knew the feeling. The scheme was not only cynical; ultimately it would not work. The Dulleses' connections were simply too deep (e.g., sister Elea-nor occupied the German desk at State) for them not to gradually outflank Smith and cut him out of the key decisions. Yet the maneuver still testifies to the care with which the new president constructed Task Force Eisen-hower, the squad that would implement his grand strategy, the next national project on Team America's diminishing schedule of big events—walking on the moon and bringing the Soviet Union down peacefully through its own internal contradictions being among the last.

On an even more general note, back in the fifth century AD, the Roman military author Vegetius advised, "Therefore let him who desires peace pre-pare for war," a prescription for deterrence that subsequently stood the test of time for 1,500 years. But the fact was that in the era of industrialized warfare, having a big, well-trained military no longer appeared much of a guarantee against attack, only an imperative for someone else to build a bigger, more capable one. This was a path the Soviets were following and would continue to follow: fashioning a massive force structure designed to overwhelm through numbers and armored momentum, with high-tech weaponry and nuclear weapons being integrated into what was otherwise a conventional battle plan.

Just after the election, Ike was on a golfing vacation at Augusta National when he was briefed that on November 1 an American thermonuclear device five hundred times more powerful than the atomic bomb had been success-fully tested. "It's a boy!" the H-bomb's chief progenitor, Edward Teller, had exclaimed, but Ike wondered aloud to his briefer at the reason "for us to build enough destructive power to destroy everything."

Yet a month later, aboard the *Helena*, Eisenhower's contradictory deter-mination to rely increasingly on nuclear weapons indicates that he had rethought the issue, that it was less a matter of dollars saved than a new way of looking at deterrence, one first suggested by Bernard Brodie and now amplified by the H-bomb into an entirely new proposition: "Therefore let him who desires peace, make war entirely too horrible to fight." This radical alternative to Vegetius was not simply a matter of one man's decision to pursue it; it was propelled by a massive wave of technological possibility that would push us, perhaps irresistibly, toward mutually assured destruction, the madness that is our present salvation. But it would seem that Ike led the way with the "New Look" announced in May 1953, and it was manifested by his channeling 40 percent of the defense budget to the Air Force, much of it going to its nuclear bomber element, the Strategic Air Command (SAC).

At the edge of the state of the art, Eisenhower would rely on the scientific community and, despite his cost-cutting gestures, a rapidly maturing military-industrial complex, one he would later decry. Ike was what we would call a techie, had been all his life, fascinated by automobiles and aviation, learning to fly himself. Inherently, he respected and admired scientists and engineers, but he would not simply value their breakthroughs and advice; they would become his foot soldiers in the grand strategy to make war impossible to fight. It's important to add, however, that Ike was never happy about what was being created—from beginning to end the process terrified him, just as it did practically all the other players on both sides—but as with those other players he was drawn like a moth to a flame by technological possibility and the logic of the situation to build a nuclear arsenal beyond all reason except the ultimate one—to deter.

This fundamental strategic approach and accompanying fears would have a number of significant offshoots during the coming eight years. While more than ready to bluff, even brandishing the nuclear joker on occasion, Dwight Eisenhower remained extraordinarily cautious about committing US forces anywhere they might actually have to fight. He really does seem to have feared that in the bipolar world he had to deal with, absolutely any war had the capacity to suck us into the nuclear vortex. And as a corollary, he would place heavy emphasis on covert operations, coercive measures applied secretly as a substitute for overt military engagement, seeking war's objectives without war—not an easy thing to do consistently.

Finally, there was the necessity of tending to the strategic balance. The days of America's nuclear monopoly were not just years past—on August 12, 1953, the Soviet Union successfully tested its first thermonuclear device, barely nine months after the US equivalent. Although it would never become so, the nuclear balance suddenly gave signs of being poised on a razor's edge, and keeping track of the Soviet side abruptly became of paramount importance to Dwight Eisenhower and concomitantly to the Central Intelligence Agency in what proved to be a symbiotic and unsustainable relationship, one ultimately causing the biggest embarrassment of his presidency. Still, these misperceptions were excusable on the grounds that Ike's vison/nightmare of nuclear deterrence was an early version, the ultimate architecture being heftier than he had feared and easily capable of bearing the weight of a variety of armed conflicts, just not total war.

THE SUCCESS OF THE EISENHOWER ADMINISTRATION WAS FOUNDED ON ITS principal's kaleidoscopic experience at running things. Had he gone to sea, Ike would have captained a tight and tidy ship, one with everybody responsible primarily for doing his or her own job. This was particularly true on the domestic front. For example, Herbert Brownell was amazed at the latitude the president gave him to set legal policy, just so long as he got the procedure right.

Being Ike, this had an institutional counterpart in the form and function of his cabinet meetings. He didn't really intend for them to decide anything, but rather to simply act as his primary policy sounding board. But they were far from bull sessions. The president insisted they be held with metronomic regularity on Friday mornings, always according to a strict agenda. Ezra Taft Benson, secretary of agriculture and an apostle in the Mormon church, early on had suggested that each meeting begin with a moment of devotion, and that too became part of the ritual—so much so that when reminded on one occasion that it had been accidentally skipped, Ike erupted: "Oh, goddamnit, we forgot the silent prayer."

It was much the same at Eisenhower's real instrument of policy and governance, the National Security Council (NSC)—hardly a secret panel, but conveniently operating a level below the more visible cabinet. Established as part of the National Security Act of 1947, the council had been treated by the previous manager, Harry Truman, simply as an advisory board, but now

Ike put it on bureaucratic steroids, expanding its membership to include Treasury Secretary Humphrey and Budget Director Dodge, along with other experts as required.

It too occupied an unvarying time slot, on Thursday mornings, with now national security assistant Bobby Cutler opening the proceeding with "Gentlemen, the president!" in a tone so liturgical that Ike eventually had to tell him to dial it back. But as an indicator of its importance, during the first 115 weeks of his presidency the NSC met 115 times. Eisenhower presided over these meetings much like the national bird presides over its nest, adding some nourishment from time to time, but mostly keeping an eagle eye on the participants. He demanded that they be well informed and had full grasp of the contingencies. Ike was famously infuriated when on March 3, 1953, Stalin died, and nobody in Washington seemed to know what to do or the implications for Russia: "What we found was the result of seven years of yapping was exactly zero. We had no plan," Ike complained. That was not to happen again on his watch, subordinates were forewarned. Yet he seldom rushed the discussion. "Let's not make our mistakes in a hurry," was a standard admonition. At times, he pushed matters in a desired direction; occasionally he took an amoral position calculated to see what his advisers might say and where it might lead. Should the conversation reach closure, Ike might well announce a decision. But more frequently, he took off to the Oval Office or the family quarters to ruminate further with only himself as counsel.

Ike and Mamie took to the White House like it was a really comfortable pair of slippers. This was government housing deluxe, and as seasoned veterans of a long line of lesser quarters they took full advantage. Tradition prevented a Dorothy Draper interior makeover, but Mamie did make sure the residence area was well padded. As part of her cushioned existence and also because of her balance problems, she habitually spent the first part of her day still in bed, playing queen bee to her staff—issuing directives on social events, keeping a close eye on accounts, and generally playing the part of a pretty tough supervisor. She was definitely more than the dowdy grandmother of *TV Guide* and *Reader's Digest* America, but like Ike's, the role was good cover and allowed them some freedom to pursue their respective agendas.

One reason it was so convincing was that it accurately represented an important part of both, the domestic side. Like many long-married couples Ike and Mamie had reached a final accommodation, near amalgamation. When asked why she avoided public causes, she always answered: "I have but one career, and its name is Ike." Unlike most first couples they shared a bed—so she could "pat Ike on his bald head anytime I want to," Mamie told reporters. While drinking was kept purposely moderate during formal occasions at the Eisenhower White House, upstairs Ike continued to enjoy multiple scotches and Mamie her old-fashioneds. Most nights, supper took place on trays, both of them parked comfortably in front of the television—TV dinners, only better cooked. At this point, both were basically addicted to servants, Mamie daily primped by her personal maid, Rose Woods, and Ike spending a half hour each morning being meticulously dressed by his long-time African American batman, Sergeant John Moaney—duty that included holding the presidential underpants as he stepped into them.

The co-ed days of the Club Eisenhower were long gone. Mamie had her women friends, and Ike had the Gang, for which the White House proved an excellent gathering place and stamping ground. When Ike needed a break from world leadership, he had merely to say the word and Gang members would sweep down in their private planes to join him in some frivolity. He had a putting green installed at the White House and there was plenty of golf available at congressional and other local courses, plus on rainy days or otherwise, there was always recourse to full-contact bridge. Though his definition of it had evolved, through it all Ike still knew how to have a good time.

Ike and Mamie had occupied a long string of residences, the obvious trend being bigger and better, but it was not until 1950 that they managed to lasso that part of the American dream that involved owning a home of their own. Their choice of location, if not predictable, was understandable: Gettysburg was near the original Eisenhower family jumping-off point on the way to Abilene, the mother of all American battlefields, and the site of Ike's first command. Accommodating Gang member George Allen, who owned a farm nearby, obligingly came up with an available property—a two-story brick farmhouse on 189 acres well equipped with Holsteins, egg-laying chickens, a big barn, and appropriate agricultural machinery. Actually, the farm was in a lot better shape than the house. "I must have this place," was the wifely bottom line. "Well, Mamie, if you like it, buy it." Flush with cash from *Crusade in Europe,* Ike in fact did the buying, writing out a check for $40,000 ($437,000 in 2022 dollars).

As so many other first-time homeowners have learned, this was only the beginning. The restoration involved a major rebuild by contractors from Baltimore (Ike insisted on union labor), lots of heavy equipment, and a subsequent interior makeover by none other than Dorothy Draper, a process that stretched over most of the Eisenhower presidency and left him $215,000 ($1,910,000 in 2022) less prosperous.

But if inside was Mamie's, the outside remained Ike's, presiding over the farm like a benevolent monarch, reaching back to his rural roots to re-create Grandfather Jacob's rich spread, the one his father walked away from. At Mamie's insistence, "Shangri-La," FDR's retreat in Frederick County, Maryland, had been fixed up and renamed "Camp David" after their grandson. But that was for weekends; Gettysburg was the ground Ike intended for his last stand in retirement. But in the meantime there was no escaping what was turning out to be the toughest job in the world.

This being Ike, our temptation might be to compare his presidency to a really tough round of golf, in the driving rain, on a long, devilishly difficult course pockmarked with sand traps; but this only trivializes the obstacles he faced. More appropriate would be to add lightning and an 8.5-magnitude earthquake, but it still falls short. Given the nuclear stakes involved, the Cold War, by its very nature, consisted of a string of recurring crises, each one solving very little but generating enormous tension. Actually, both sides were locked in a perpetual peace grip, but only a tiny few understood this, so fear instead was the pervasive international emotion.

Compounding the tumult was the simultaneous collapse of Europe's world of colonial empires, replaced by a legion of South Asian and African states all struggling to find new identities and futures in an environment bracketed by a Manichean struggle between communism and democracy—a veritable tinderbox of armed possibilities. And all too often Dwight Eisenhower was stuck in the middle, no friend or defender of colonialism but also pressured by alliance loyalties to go easy on foot-draggers, or in the case of the French, those attempting to reassert themselves. Through all the ups and downs, Ike survived the roller coaster of events not just by holding on tight, but by conjuring a steering wheel and bending them to his will.

Yet at its beginning the wild ride was further complicated by attempts to put Ike in handcuffs by members of his own team. Now that he was president, his instinct was to ignore Joe McCarthy. But when the senator threatened to block his choice for ambassador to the Soviet Union, Charles Bohlen, largely because he had been a trusted interpreter for FDR at Yalta and Truman at Potsdam, but also stoking rumors of homosexuality, the president jumped to his defense. He liked Bohlen and with Stalin's death needed him in Moscow immediately. Nixon was dispatched to the Senate for damage control, and Ike told the press he had played golf with Bohlen and knew him to be a good family man, never mentioning McCarthy. It worked, and the next day the Senate confirmed the new ambassador 74 to 13. "I just won't get into a pissing contest with that skunk," Ike told brother Milton.

He preferred the silent approach of an anaconda as his prey's charges of Communist infiltration became ever more reckless and strident. Finally, McCarthy zeroed in on the Army, citing the promotion of a slightly pink service dentist from captain to major as evidence of the insidious nature of the Communist threat, and later calling his commanding general, a D-Day hero, "not fit to wear the uniform" and having "the brains of a five-year old child." The gauntlet was at Ike's feet.

"This guy McCarthy is going to get into trouble over this," he told Hagerty, then simply waited for him to talk his way to destruction as millions of Americans watched on TV. The Army's counsel, the wily Joseph Welch, helped the process considerably through a combination of gay-bashing McCarthy's young assistants Roy Cohn and David Schine, plus honest outrage when accused of having a staff attorney with Communist ties: "Senator, until this moment I never really gauged your cruelty or your recklessness." Ike was watching, and seeing it was time to apply the stranglehold, ordered first the military and then all members of the executive branch to stop testifying before McCarthy's committee. Barely lifting a finger that anyone could see, Eisenhower not only cut off Joseph McCarthy's red-baiting blood supply, effectively ending his political career, but also got away with an unprecedented expansion of executive privilege with barely anybody noticing.

Yet Ike still had to deal with the Bricker Amendment. The other manacle, it was waiting for him when he took office—a constitutional change that was the brainchild of Ohio senator John Bricker to curtail the president's power to conclude not just treaties, but all sorts of foreign agreements, and then subject them to continuous congressional scrutiny. The Bricker Amendment was popular with the public, and already had the sixty-four Senate cosponsors needed for passage—a done deal, it seemed. "Can't we find a way to avert a head-on collision over this thing?" Ike first suggested. But by April getting around the Bricker Amendment and looking the part of a loyal Republican was proving tougher than expected, and the president fumed to his cabinet: "I'm so sick of this I could scream. . . . We talk about the French not being able to govern themselves—and we sit here wrestling with the Bricker Amendment." At this point it looked as if the Republican right wing just about had their president pinned.

But even they underestimated Ike's willingness and ability to walk on the wild side, the forbidden other side among Democrats ready to make a deal. When it came to partisanship, Ike was not afraid to reveal his bipartisan side, and the Bricker Amendment was one of those instances. In the tradition of politics making for strange bedfellows, Ike soon found himself cuddling up to Democratic Majority Leader Lyndon Johnson, the master of the senatorial stall and convoluted schemes in general. For this occasion, LBJ came up with a plan complicated enough to enervate the Bricker Amendment over the course of a year, leaving it to go down to defeat in the Senate 42 to 50—not having even majority support—on February 26, 1954. Ike's cuffs were off.

IN PART, THE NEW MANAGER PULLED OFF HIS ESCAPES FROM MCCARTHY AND the Bricker Amendment because he sensed the political winds were shifting, that the rampage of the right wing was just about over. He was correct, and in November 1954 the Republicans lost both houses of Congress, leaving Ike a bipartisan path to govern. And though it was a narrow one, it left him actually more comfortable than he felt in the right wing's closet.

One example was George Marshall: the new climate allowed Ike the cloud cover to resume efforts to rehabilitate his reputation; besides, though tired, he was not without utility. As far as Ike could see, Marshall's fall from grace was strictly domestic, and his reputation overseas, particularly among the European allies, remained intact. Consequently, the Marshalls had been immediately designated fixtures at White House dinners for visiting heads of state and monarchs. And not even a month into the administration, when news arrived from the United Kingdom of George VI's death, Ike had grabbed the opportunity to appoint Marshall to head a delegation, including soon-to-be chief justice Earl Warren and Omar Bradley, to represent the United States at the June 1953 coronation of his successor, Elizabeth II.

The trip was a great success for George and Katherine, exactly the people the British wanted to see at the installation of their new queen, at least if Winston Churchill's behavior at the ceremony was any indicator. Once again prime minister, Churchill stopped as he walked toward Westminster's altar and drew even with the Marshalls: "He dignified me in the Abbey by turning out of the procession to shake hands with me after he had reached the dais." Montgomery and now viscount Alanbrooke followed suit, which must have appealed to Marshall's sardonic side, though he failed to mention it.

The year 1953 continued George's rehabilitation, but his health went in the opposite direction. In September, he and Katherine both caught severe colds on a short vacation to Pinehurst, North Carolina, and George's turned into bad flu. Ike found out when the Marshalls sent their regrets for the state dinner of the king and queen of Greece; he immediately sent a plane for Marshall and installed him in the presidential suite at Walter Reed. George's recovery was a slow one, and though it actually marked the beginning of the long slide down, he still had good reason to want to get out of the hospital. News arrived that he had just won the Nobel Peace Prize for the Marshall Plan, which would be awarded on December 10 in Oslo.

Katherine knew George was in no shape to make the trip, since she didn't feel well enough to go herself, but he was adamant. The doctors suggested a voyage via the southern route to Europe, and George booked passage on the new and soon-to-sink *Andrea Doria*, planning to write his acceptance speech while soaking up some therapeutic rays and sea air. In keeping with this hard-luck ship, the weather was cold and miserable, and after eight days Marshall landed in Naples without having written a line.

He then proceeded to Paris to stay with Ike's successor, Alfred Gruenther, at NATO headquarters. The morning after he arrived he wrote Katherine he "grit his teeth" and dictated a draft of his Nobel lecture. Meanwhile, Gruenther, sensing Marshall's lassitude, sent some help in the form of his special assistant, Colonel Andrew Goodpaster, who possessed both a PhD in international relations and a way with words. Instead of a draft, the colonel found Marshall had merely "jotted down a few thoughts on some notepaper." Having worked with him before, Goodpaster later wrote in the *New York Times*, "I had never seen the general so tired." It took three meetings, but the hired pen finally squeezed out enough words for the speech, and Marshall flew to Oslo the morning before the award ceremony.

As it turned out, nobody remembers what he said, since, as he was called forward to speak, three young Communist journalists in the balcony began screaming, "Murderer! Murderer!" and simultaneously bombarding the audience with leaflets accusing Marshall of war crimes. The US agricultural attaché yanked one back from the railing by his hair, while Marshall's pilot punched another, before the police cleared them out and the hall broke into thunderous applause. Not a great atmosphere to get a Peace Prize.

Poor George, bashed from the right and now from the left. Still, if you were a European, you might have had reason to question the choice. The Marshall Plan had been a brilliant success, and its recipients had every reason to be grateful to its namesake, who in turn deserved a substantial measure of the credit. But the strategic architecture that had emerged subsequent to the disaster of World War II had to give Europeans pause—they were caught between two nuclear-armed giants, on ground that was likely to be the first to burn should the edifice collapse into war. Marshall didn't design it, or preside over it, but if there could be a retrospective title "chief facilitator," he probably deserved that prize too. He had managed the delivery of the first nuclear weapon; pitched in to ensure the Cold War construction project featured NATO and was pointed at Russia; and for a long while, he was always available for congressional testimony when it came budget time to keep the funds flowing.

But those days were over now—George Marshall the public person had faded into the Virginia countryside, within a shell of privacy and in increasingly bad health. Since he never wrote a memoir, and through it all kept his Zen-like sense of detachment, it's hard to say what he actually thought of his career. But as far back as his days at VMI and his competition with his brother, George had been an achiever, and what he got out of life probably far exceeded what he ever expected. He was never a reflective man, more a reactive man, so his silence now made sense. So we'll leave him to enjoy the remainder of his life as an adopted Virginia gentleman, with a wife who suffered more from bullets than he ever did, in a nice house with columns out front.

Meanwhile, in another one just over thirty miles southeast, Dwight Eisenhower, the last active member of Team America's murderers' row, was just getting the hang of his role as manager in a bipartisan dugout. Though he presided over prosperous times and did manage to balance the budget, Ike would not be remembered for his domestic achievements. Nonetheless, they existed, and two in particular bore the stamp of the man in charge.

In 1919 Ike had been part of the Army's transcontinental convoy that averaged ten miles per hour, and many attribute what he later recalled as this "difficult, tiring and fun" journey as the genesis of the biggest public works project ever attempted: the Interstate Highway System. The timing was right. Auto ownership was exploding in the early fifties—a true engine of the economy—and Ike always liked cars. This was a program likely to make friends on both sides of the aisle—the Democrats with jobs for their voters in mind, and the Republicans with visions of a mountain of contracts for business—with Ike leading the bipartisan way, but always with his own agenda.

In the summer of 1954, Eisenhower brought in Lucius Clay to put together an advisory committee that would provide guidance. Clay stacked it with the heads of big construction companies, heavy equipment manufacturers, and the Teamsters union, and they in turn recommended spending $101 billion (nearly $1 trillion in 2022 dollars) over ten years. This was to be no second-rate system. Looking back on his war experience and "after seeing the autobahns of modern Germany and knowing the asset those highways were to the Germans, I decided, as President, to put an emphasis on this kind of road building." And his justification to Congress revealed another facet of Eisenhower's thinking—he wanted big, multilane highways to evacuate US cities in case of nuclear war, the military big picture never far from his mind.

So it was with Ike as reluctant champion of civil rights. Like most Americans of his time, Dwight Eisenhower was an unconscious racist, simply accepting an unjust system without giving it much thought. The Army during most of his career was strictly segregated, and as with Marshall and Patton, Ike became interested in sending Black troops to white units only when battle casualties made it operationally advisable. Personally, he seems to have always treated Black people well enough, but without much insight or empathy. And though Ike showed every sign of enjoying Sergeant Moaney's company, happily cooking and barbecuing with him, he plainly never questioned the underpants ritual every morning.

But it's also important to add that Eisenhower was not necessarily opposed to racial justice, and as a military man he was inclined to follow orders. His civil rights record as president was plainly driven by these factors. Harry Truman is remembered for having integrated the Army. But when Harlem congressman Adam Clayton Powell Jr. pointed out that two-thirds of the units still remained segregated, it was Ike who moved quickly to finish the job, Secretary of Defense Wilson announcing on October 24, 1954, that the last Army unit had been integrated.

Meanwhile, Eisenhower had appointed California governor Earl Warren as chief justice of the nation's highest judicial body, telling his brother he was a "liberal-conservative" representative of the "kind of political, economic, and social thinking that I believe we need on the Supreme Court." Ike knew a major civil rights case aimed at desegregating America's schools was looming on the court's docket, and in February 1954 invited Warren to the White House for dinner and advice: "These are not bad people. All they are concerned about is to see that their sweet little girls are not required to sit in school alongside some big overgrown Negroes." These are toxic words to contemporary ears, and if he thought they made an impression on the chief justice, he was wrong. On May 17, the Supreme Court ruled unanimously in *Brown v. Board of Education* that segregation in public schools was unconstitutional, setting in place the cornerstone of the modern social justice movement.

Ike hardly praised the decision, but he was still all about carrying out orders: "The Supreme Court has spoken and I am sworn to uphold the constitutional process in this country, and I will obey." Spoken like the soldier he was, and it was exactly what he would do.

Orval Faubus didn't have the slightest idea what Ike was really like. As governor of Arkansas he had no intention of desegregating Central High

School in Little Rock, and blithely defied a court order issued by a judge Eisenhower had appointed. Instead, he complained to Ike that the FBI was harassing him. The presidential response was a clear statement of escalation dominance: "The only assurance I can give is that the Federal Constitution will be upheld by me by every legal means under my command." Faubus's reaction was to send Arkansas National Guardsmen to surround Central High, where they prevented nine Black students from entering as they were being harassed by a white mob. Ike could have deputized the Arkansas Guard himself or sent in a massive police presence, but he told Attorney General Brownell he had something else in mind: "In my career I have learned that if you have to use force, use overwhelming force and save lives thereby." Eisenhower sent in the 101st Airborne, the stick in the Wehrmacht's spokes at Bastogne, to exert a calming influence on Little Rock. The violence stopped immediately, and the nine Black children took their places at desks in Central High. Georgia's senator Richard Russell Jr. complained that Ike's tactics reminded him of "the manual issued the officers of Hitler's storm troopers"—vile hyperbole, but still on the mark. Ike was a soldier, and this was a soldier's solution.

IN THE SUMMER OF 1961 EISENHOWER'S RECENTLY INSTALLED SUCCESSOR, John Kennedy, began to realize that a war with Russia was more of a possibility than he had ever suspected, and called in Dean Acheson to help him get a better grasp of the nuclear threshold and when and how it might be crossed. According to Kennedy's national security adviser, Acheson replied that "he believed the president should himself give that question the most careful and private consideration, . . . that he should reach his own clear conclusion in advance as to what he would do, and that he should tell no one at all what his conclusion was." Ike would have told JFK the same thing, and it has been the mantra of every president since: when it comes to nuclear deterrence, silence is golden.

As the Chinese poured into Korea, Harry Truman had tried to explain his position, only to contradict himself and scare everybody. Ike figured it was better to say nothing. This did not preclude nuclear threats, only specific, advance statements of how and when they might be applied. At the height of the Cold War this was the code of silence, the loneliest secret in the world.

Yet it's hard to square this calculating assessment, not just with Ike's "everybody's grandpa" schtick, but also with what was demonstrably a warm and loving side to his nature and an honest revulsion at what nuclear weapons represented. Then again, when son John was asked by an interviewer to consider the balance between the open and sunny Ike and the cold-blooded one, he initially stated it was about even, thought for a moment more, and said: "Make that 75 percent cold-blooded." Maybe it was parallel processing, but whatever it was, it left Ike holding his nuclear card close to his vest and leaving subsequent analysis of his strategic approach a matter of deduction. Nonetheless, a retrospective of Eisenhower's crisis behavior reveals a clear and consistent path, not always the right one, but close enough to get us through alive.

On his first morning as president he walked into the Oval Office to learn the Chinese had shot down a B-29. Ike let it pass—he would not allow that to shoot down the negotiations to end the fighting in Korea. Less immediately threatening was America's first encounter with the neocolonial quicksand of Vietnam, since it was the French who were caught and sinking, claiming to want only a helping hand.

In a notably shortsighted expedition against the tide of history, the French had attempted to reoccupy and re-create colonial Vietnam after World War II. Instead, they encountered a population bent on independence and an enervating guerrilla war. It ground on until 1953, when, in a notably delusional move, the French commander sought to generate a conventional battle by luring the Vietnamese into attacking a remote and strategically worthless position, Dien Bien Phu. Instead, the Vietnamese general Vo Nguyen Giap surrounded the French with fifty thousand troops covered by copious artillery in the encircling mountains.

When he was head of NATO Ike had thought the Gallic effort in Vietnam folly, and when as president he learned of the plan for Dien Bien Phu he told the French ambassador, "You cannot do this. The fate of troops invested in an isolated fortress is almost inevitable." Now in the spring of 1954 they were caught and wanted America's strong hand to pull them out.

Ike offered not a lifesaving grip, but a fig leaf—twenty-five World War II–vintage medium bombers and all the conventional ordnance they could drop on a doomed operation, with the French surrender coming on May 7,

1954. There is a story that during the deliberations over what to do, Eisenhower reacted to a Joint Chiefs of Staff plan to save the situation with a nuclear strike incredulously: "You boys must be crazy. We can't use those awful things against Asians for the second time in less than ten years. My God." Yet the source of the quote has been largely discredited, and it has been cogently argued elsewhere that Ike was otherwise scrupulously careful not to tell even his closest advisers whether he would or would not use nuclear weapons.

The specter rose again just months later in September, when the mainland Chinese appeared ready to invade a tiny offshore island, Quemoy, held by Chiang Kai-shek's forces. When the Joint Chiefs of Staff recommended defending the island with air strikes and tactical nuclear weapons if necessary, Eisenhower's reaction was instructive: "We are not talking about a brush-fire war. We're talking about going to the threshold of World War III. If we attack China, we're not going to impose limits on our military actions, as in Korea. [And] if we get into a general war, the logical enemy is Russia, not China, and we'll have to strike there." Inevitable escalation, in other words. That was enough to cause the chiefs to put the nukes back in their holsters, but the president was still confronted with the problem of not being seen as backing down in the face of the Chinese.

Ike called forth a basic skill—he bluffed his way through. First, he got Congress to pass a resolution to defend Formosa and maybe Quemoy and a similar island, Matsu (Mazu), at once satisfying the GOP's right wing and putting the Chinese on notice exactly where the red line in the water lay. To reinforce the point, Ike told an interviewer that we might use nuclear weapons "just exactly as you would use a bullet or anything else."

Then, as the situation played out in the spring of 1955, the president remained sufficiently ambiguous to keep not only the Chinese guessing, but pretty much everybody else. Before one news conference Hagerty warned his boss: "Some of the people in the State Department say that the Formosa Strait issue is so delicate that no matter what question you get on it, you shouldn't say anything at all."

"Don't worry Jim," Ike shot back. "If that question comes up I'll just confuse them." That he did, stringing everybody along until Zhou Enlai signaled publicly that China did not want war with the United States and the shelling of Quemoy and Matsu slowed, then ceased by mid-May.

It was a conclusion that epitomized Eisenhower's strategic approach, or at least its overt side. Nuclear weapons were part of the arsenal, but only the psychological one; otherwise they were so dangerous that any commitment of US forces could well lead to explosive and disastrous escalation. This extreme and almost paradoxical version of deterrence was a product of Eisenhower's past, a nightmare derived from the era of industrial warfare, when crises accelerated into all-out war with blinding speed. Squinting at the future from the shadow of two world wars, Ike would remain supremely cautious in any situation that could plausibly draw America's military into actual combat.

But he was not about to renounce the use of force; he just wanted to keep it covert. Eisenhower always had a penchant for the dark side, international intrigue and special operations, so much so that in North Africa George Marshall had to remind him to pay more attention to the conventional war. Now that he was managing Team America, there was nobody to restrain his weakness for "dirty deeds, done dirt cheap." In Ike's narrowed strategic vision, as long as nobody found out who was behind them, they were a safe, economical, and effective means of coercion. The problem was that inevitably American fingerprints would be revealed, and then there would be hell to pay.

But for a while, it worked great. Just days before Ike's inauguration, Beetle Smith, still at the CIA, called one of Allen Dulles's best spooks, Kermit Roosevelt, into his office. Every bit as aggressive and impulsive as his grandfather Theodore, this version devoted himself to the Middle East and the arts of subversion. His mission would be to topple Mohammad Mossadegh, Iran's left-leaning popularly elected prime minister who had recently nationalized the British-controlled Anglo-Iranian Oil Company, and replace him with Mohammad Reza Pahlavi, the once and future shah. "Pull up your socks and get going," Beetle told him. "You won't have any trouble in London. They'll jump at anything we propose. And I'm sure you can come up with something sensible enough for Foster to OK. Ike will agree." Since Smith was no man to speak out of school, it's likely Ike already had, despite

making some distracting subsequent noises that he really wanted to prop
Mossadegh up with a $100 million loan, and plainly not wanting to know
too many specifics.

They were of the comic opera variety, at one point featuring Roosevelt
being carted around Tehran rolled up in a Persian carpet, busily managing a
race to hire enough thugs to stage fake street demonstrations and eventually
attack Mossadegh's house, leading to his capture in a pair of pink pajamas,
the only thing close to being Red about him. At his moment of triumph, Roosevelt got a cable from Washington urging him to flee. He replied: "Yours
of 18 August received. Happy to report. . . . KGSAVOY [the shah] will be
returning to Tehran in triumph shortly. Love and kisses from all the team."

There ensued for Roosevelt the spook equivalent of a victory tour: a quiet
dinner with and toast from Winston Churchill; a detailed briefing for both
Dulles brothers, during which he noticed they "seemed almost alarmingly
enthusiastic"; and finally, on September 23, 1953, a secret medal presentation
by Ike himself. From a presidential perspective, it was an ideal outcome—
presto chango statecraft, without any of the muss and fuss. And perhaps
best of all, nothing to escalate. There would be a price to pay in the sullen
rage of the Iranian people, one that came due a quarter-century later with a
revolutionary invoice. But that was a long way off, and meanwhile there was
more low-hanging fruit to be picked.

United Fruit, actually. In Guatemala, near the end of World War II, the
country's nascent middle class had overthrown a long and brutal dictator-
ship, and instituted a democracy based largely on American New Deal principles. But it had always been handicapped by the vast power of the United
Fruit Company—*El Pulpo* ("the Octopus") to natives, controlling 42 percent
of the arable land and exempt from all taxes. Playing roughly the same role
as the Anglo-Iranian Oil Company in Iran, United Fruit looked to Washington for help when in 1953 and 1954 the Guatemalan government expropri-
ated more than 400,000 acres of its fallow land as part of an agricultural
reform program—or "Communism against the right of property," as the
company's president put it.

These were fighting words to the Eisenhower administration, but ones
whispered rather than shouted. The Dulles brothers asked Kermit Roosevelt
if he was interested in a Central American reprise at covert state flipping,

but he was already deeply involved in the underside of Egyptian revolutionary politics and begged off. It didn't matter. United Fruit's tentacles were hard not to trip over. The Dulles boys had managed its legal affairs while on Wall Street; the husband of Ike's personal secretary, Ann Whitman, was head of its public relations; Bobby Cutler had been United Fruit's banker back in Boston; Beetle Smith wanted a job with the company after he left government; and on it went, even to brother Milton, whom Ike sent to Latin America to report on conditions and returned convinced that the land reforms had pushed Guatemala into the Communist camp.

During early planning aimed at reversing this perceived condition, Allen Dulles told Ike the chances of staging a successful coup were "better than forty percent, but less than even," and Ike told him to keep working. Then on June 16, 1954, after the Guatemalans brought in a shipload of Eastern Bloc arms, the president gave his final approval: "I want all of you to be damn good and sure you succeed. . . . When you commit the flag, you commit it to win," which, of course, was exactly what they weren't doing.

Poised on such a mandate, it's not surprising that the Guatemalan counterrevolution had trouble getting started, much less gaining momentum. When the 150-man surrogate "national liberation force" came across the border from Honduras, pretty much nobody noticed. The army stayed in its barracks, and the general calm was interrupted only by occasional bombing runs by the guerrillas' three-plane air force, two of which were soon out of commission.

But once committed, Ike wasn't giving up: "My duty was clear to me. We would replace the airplanes," which he did, with two P-51 Mustangs borrowed from the Nicaraguan dictator Anastasio Somoza Debayle. However, it was the CIA's operations directorate, one of whom was Allen Dulles's brightest boy, Richard Bissell, that turned the tide by jamming the government's radio while, out of a station of their own in Honduras, the guerrillas broadcast a narrative featuring the "national liberation army" relentlessly forging ahead; it was entirely fictitious, but realistic enough for the Guatemalan president to drink himself into an ill-considered resignation. Not exactly a glorious outcome, but one good enough for Ike, who now wanted to be immediately briefed, which had not been the case with the Iran operation.

The president and his strategic approach were becoming increasingly intertwined with a brash and burgeoning intelligence component. Still, Ike was not entirely spellbound by Allen Dulles's hail-fellow facade, and he even set up a board of civilian advisers to keep an eye on the CIA—one headed by the veteran diplomat David Bruce and George Marshall's alter ego, Bob Lovett. They duly warned him in late 1956: "The intrigue is fascinating— considerable self-satisfaction, sometimes with applause, derives from suc- cesses," but covert actions were "responsible in great measure for stirring up the turmoil and raising the doubts about us that exist in many countries of the world today. Where will we be tomorrow?" Just a hop, skip, and a jump from the Bay of Pigs, where the habit of covert operations caught up with America and Ike's successor, JFK. But while Eisenhower would escape Allen Dulles's ultimate covert crackup in Cuba, he still fell prey to his addic- tive wares aimed in a direction far beyond those tropical shores.

TESTING AN ATOMIC BOMB IN 1949, THE SPECTACULAR PERFORMANCE OF THE MiG-15 in Korea, a thermonuclear device by 1953—the Soviet Union looked the part of a frighteningly capable technological competitor just as the Cold War's signature arms race took off, both literally and figuratively. In the spring of that year the mathematical side of the Air Force's new think tank, RAND, stood the nuclear deterrence concept of its now colleague Bernard Brodie on its head, arguing that in a surprise attack the Soviets, by dropping only fifty atomic bombs, would destroy not only America's cities, but also enough of its long-range bombers to draw into question the certainty of ef- fective retaliation. The report had a seismic impact, and its aftershocks were plainly felt in Washington and the White House. Yet to deliver those nukes the Soviets needed a flock of intercontinental bombers, of which there was no sign. But that didn't mean they didn't exist—perhaps we just couldn't see them. So when presented with the opportunity to peek over the hith- erto hermetically sealed Iron Curtain, Ike was ready to brave the potential consequences.

Actually, he was there at the inception, or close enough. In the summer of 1954, Eisenhower characteristically asked the president of MIT, James Kil- lian, to form the Technological Capabilities Panel to examine the issue of surprise attack. Among those gathered around the problem was Edwin Land, founder of Polaroid and unquestionably America's leading imagery technologist. To Land, the problem of finding bombers was reduced to a matter of putting a sufficiently powerful camera over the Russian landmass, which in turn required a high-flying plane to take it there. This proposition involved Kelly Johnson, the iconoclastic designer of the P-38 and now head of Lockheed's top secret Skunk Works, who suggested grafting an eighty-foot wing to the slender fuselage of an F-104 fighter, to create what amounted to a jet-powered glider with a projected range of nearly 5,000 miles and a ceiling of 70,000 (soon 80,000) feet. But the Air Force wasn't interested in developing the plane—Curtis LeMay, the founder and head of the increasingly powerful Strategic Air Command, called it a "pile of bullshit."

But not Ike. In late October, he met with Land and Killian in the Oval Office and informed them he was approving the new spy plane. But "he stipulated that it should be handled in an unconventional way so that it would not become entangled in the bureaucracy of the Defense Department," Killian remembered. In other words, a black program, the purview of the CIA.

Allen Dulles, "the Great White Case Officer," had little interest in the concept, maintaining later that Ike too had reservations. But Dulles, always a man to pursue his bureaucratic interests, nonetheless saluted smartly, in part because he had just the man to make it happen.

The most flamboyant of the pack of Yalies drawn to the CIA during its early years, Richard Bissell managed by sheer energy, intellectual arrogance, and cheek to become both a fixture of Georgetown's social set and also pivot man at the intersection of the nation's darkest secrets—an exotic creature only the Cold War could have invented. A bureaucrat who not only despised rules but had the wit to break them mostly without consequence, Bissell was perfect for his new role. Hovering over the project like the "Mad Stork" his colleagues nicknamed him, he had the plane, known subsequently as the U-2, in the air in just eighteen months, $3 million under budget. But for Bissell that was just the beginning: he wanted his creation flying over Russia, and to make it happen he was willing to cut the same kind of corners it took to bring it into being, a cascade of shortcuts that would eventually land on Ike's head.

Meanwhile, the president's need-to-know only grew. On July 15, 1955, the Soviets held their annual Aviation Day celebration in Moscow, during which the American military attaché observed multiple waves of new M-4

long-range jet bombers fly overhead. While actually the same wave circling
back repeatedly, it was to Air Force intelligence the beginning of the Bomber
Gap, one in which according to their estimates the Russians would soon
have six hundred to eight hundred M-4s matched against just the first few
Boeing B-52s coming off the assembly line.

Even in the face of the Air Force's Soviet bomber nightmare, flying over
sovereign airspace without permission was not only strictly forbidden by
international law, it was a daunting proposition in a standoff laced with nuclear weapons. But Ike thought he might be able to finesse the situation with
a diplomatic hedge.

Stalin had been dead for two years, and Soviet leadership seemed in
Washington to have stabilized into government by committee, with Nikolai
Bulganin, a man who bore a striking resemblance to Colonel Sanders, acting
as first among equals. It looked like a propitious time to get to know the
Russian team better, and in early May 1955 the United States, Britain, and
France sent simultaneous, identical pitches proposing a meeting of the four
heads of state "to remove sources of conflict between us." The Soviet Union
accepted a week later, and it was announced the summit would convene in
Geneva on July 18.

Ike brought the family, both John and Mamie, and they settled into a villa
on the shores of Lake Geneva, touring the city in the 1942 Cadillac he had
used during the war—someone's version of a surprise, but one that likely
brought back uncomfortable memories of Kay Summersby. Maybe it was an
omen. Over cocktails at the villa, the president sought to chat up Marshal
Zhukov on the political situation in Russia. Zhukov, whom Ike had gotten to
know and like on his postwar trip to Moscow, was subdued but to the point:
"Things are not as they seem."

The next day in the great hall of the Palace of Nations, Ike unveiled his
big surprise. Putting aside his glasses and speaking without notes, he pos-
ited that in a time of "new and terrible weapons," the greatest danger was
"surprise attack." To render that impossible, he suggested not only "immediate practical steps," such as exchanges of charts and locations of military
airfields, but an "Open Skies" treaty allowing both sides to conduct aerial
inspections of the other's defense installations. As if that wasn't dramatic
enough, the instant the American president stopped speaking a thunderous

bolt of lightning shook the hall and the lights went out. In the darkness Ike was ready with a one-liner: "Well I expected to make a hit but not that much of one," once again shaking the hall, this time with laughter.

The combination of Ike's curveball, the blackout, and praise for the proposal by the other delegations seems to have knocked Bulganin off message, leaving him to venture that "Open Skies" had merit, and that his government would give it close and sympathetic study. But Eisenhower soon found this not to be the last Russian word on the subject.

At a tea (really drinks) held immediately after the formal session, he was approached by a short, stout, thoroughly spun-up Nikita Khrushchev: "I don't agree with the chairman"—probably a dip-speak translation of what he really said. The version recounted by Andrew Goodpaster, now Ike's staff secretary, was likely more accurate, with Khrushchev wagging his finger at the president and repeating "Nyet, nyet, nyet. . . . You're simply trying to look in our bedrooms," though the interpreter may have cleaned that up also. Khrushchev was certainly crude and also easy to underestimate. The normally astute Bohlen had cabled back from Moscow that he "wasn't especially bright," while the British foreign secretary and soon-to-be prime minister Harold Macmillan wrote in his diary at Geneva, "Khrushchev is a mystery. How can this fat, vulgar man with his pig eyes and ceaseless flow of talk, really be the head, the head—the aspirant Tsar—of all those millions of people of this vast country?" Ike was a better judge of gravitas; as he later wrote, after that first meeting, "I saw clearly then for the first time, the identity of the real boss of the Soviet delegation." Yet like most everybody else, including many Russians, Ike underrated Khrushchev's toughness and persistence. In fact, he had just met his foil and nemesis.

Meanwhile, there was a new game taking shape, one that would only add to Dwight Eisenhower's nuclear albatross, and also his temptation to get a glimpse of the other side's hand.

Ever since World War I and Germany's eighty-mile Paris cannon, it had been apparent just how far ballistic projectiles could fly once free of the atmosphere. But only so much velocity could be had from a single blast. Much better was the sustained thrust of a rocket motor, which, because it carried not only propellant but oxidizer, could operate in a vacuum, making unlimited terrestrial range and even space travel a possibility. While Hitler's

Germans brought theory into partial practice with the two-hundred-mile V-2, the conquering Americans and Russians soon discovered much more ambitious models on their drawing boards, among them a ballistic missile of intercontinental range, along with one launched from a submarine. Both sides also inherited a share of the engineering talent behind these concepts, cadres soon utilized to explore the possibilities of military rocketry in a Cold War context.

Not only did the Russians grab most of the Germans, but it is now known that they quickly prioritized long-range rocketry as the strategic future. The M-4 gambit, which produced the Bomber Gap, was largely a ruse; the Soviets never invested heavily in long-range military aircraft. Instead, throughout the late forties and early fifties, they worked steadily off the V-2 model to improve the range, reliability, and payload of their own rockets. By the time of the Geneva Conference, the Soviet Union not only fielded a ballistic missile capable of reaching London; it also had developed a very capable team of Russian-only rocket scientists and engineers working on several projects that, when revealed, would prove spectacularly threatening. But in the meantime, it was all just a dark ball of suspicion to Dwight Eisenhower—inchoate but sufficient for him to up the ante when he got the chance.

That came on August 4, three days after he had returned from Geneva, when Ike and the rest of the National Security Council met with the Air Force's head of research and development, Trevor Gardner; General Bernard Schriever, the energizing force behind the service's ballistic missile program; and Princeton's John von Neumann, the legendary mathematician and inventor of game theory. The meeting was scheduled to last thirty minutes but instead stretched on for hours, for the three wise men had arrived to announce the birth of what would amount to the ultimate weapon.

Edward Teller's original H-bomb was a pure thermonuclear design, astonishing in its power but at over eighty tons impossible to deliver in anything but a boat. But von Neumann was convinced it could be made far smaller and lighter, essentially by adding the hydrogen isotopes deuterium and tritium to a pit in a compact fission initiator, small enough to fit in the nose cone of a rocket. Suddenly, not only was the development of a nuclear-armed intercontinental ballistic missile feasible, it was absolutely imperative.

"Because gentlemen," Gardner stated, "this technology is also known to the Soviets—and our intelligence tells us they are going full out to develop it. It means . . . that it is now possible to send a ballistic missile armed with a nuclear warhead from the continental United States to Soviet Russia—or vice versa—in roughly thirty minutes." Not averse to piling on, von Neumann added that since US radar could not pick up Soviet missiles until they reached the top of their ballistic parabola, the warning time would be cut to fifteen minutes—a span far too short to get bombers off the ground, something everyone in the room, including Ike, understood.

"This has been impressive, most impressive," the president said, concluding the meeting amid the ruins of his daily schedule. He ordered Admiral Radford, chairman of the Joint Chiefs of Staff, to stage war games with intercontinental-range missiles, and to "do it right away."

There was more to the story, further devils in the details to be learned. The ICBM's champion, "Benny" Schriever, was locked in an uneven bureaucratic struggle with Curtis LeMay, head of the Strategic Air Command and the Air Force's spiritual bomber baron. Having burned down much of Japan using such aircraft, LeMay looked to a future studded with bombers—specifically, the Mach 3 B-70 and even a nuclear-powered model. LeMay may have had a pilot-centric service largely on his side, but Schriever plainly had the better of the argument.

But at the time of the initial Eisenhower briefing the two ICBM programs he was pushing toward fruition, Atlas and Titan, were highly questionable as weapons. Atlas employed liquid oxygen, which had to be kept at –300 degrees Fahrenheit and therefore could not be fueled until shortly before launching. Titan's storable propellant and oxidizer were hypergolic—or in simpler terms, really explosive. Liquid-fueled rockets were also slow to accelerate in the earliest stages of launch, making them vulnerable to Soviet infiltrators with sniper rifles, according to some in Air Force intelligence—this being a paranoid period.

But waiting in the wings was the solution. Until this point solid-fuel rockets—combining propellant and oxidizer in a plasticized matrix—had been a nonstarter, because of their low energy output and cracks stemming from the casting process, which resulted in an uneven burn and associated guidance problems. But by mid-1955 both issues had been resolved: new

casting techniques had largely removed the cracks, while an amazingly simple design change banished the slows from solid propellant. Engineers just
drilled a star-shaped hole at the center of the matrix and extended it along
the inner axis, enabling ignition not just at the bottom, but along the entire
length of the charge, thereby not just dramatically increasing thrust, but
turning the contemplated missile, soon to be designated Minuteman, into a
veritable sprinter out of the blocks. Even more to the point, such a missile
was safe and fast-accelerating enough to carry and launch from a submarine.

How much of this was presented to the president subsequent to the August 4 National Security Council meeting goes unrecorded. But NSC Action No. 1433, signed by Ike on September 13, argues that most of it was. It
not only designated the ICBM "a research and development program of the
highest priority above all others" but also followed the recommendation of
James Killian that the Air Force and the Navy work together on solid-fuel
rocketry. By the time Eisenhower left office in 1961, both the Minuteman
ICBM and the Polaris A-1 intermediate-range ballistic missile were ready for
deployment in silos and on submarines respectively, constituting the inevitable one-two punch of mutually assured destruction—and for sixty years
nothing much has changed. It should be noted, however, that just eleven
days after signing the ICBM memo, Ike was profoundly distracted from nuclear strategizing. Lady Luck caught his attention with a heart attack.

HE WAS IN DENVER ON A GOLFING VACATION, STAYING AT THE DOUDS', WHEN,
on the night of September 24, 1955, Mamie looked in on him, thinking he was
having a nightmare. He wasn't, instead complaining about pains in his upper
abdomen. Familiar with his stomach maladies, Mamie fed Ike milk of magne-
sia and called his personal physician, Dr. Howard Snyder. At age seventy-four
and on the ragged edge of incompetence, Snyder arrived around two a.m. and
agreed with Mamie that it looked like indigestion. Two hours later, when the
president's blood pressure dropped, he ordered Mamie to climb into bed and
wrap herself around her husband to keep him from shaking. Ike finally fell
asleep around five, and at eight Snyder sent a message to the press that he was
suffering from "digestive upset." It was only when Eisenhower awoke at one
in the afternoon and an EKG was administered that it became obvious that
his problem was coronary thrombosis, and he was immediately hospitalized.

"It hurt like hell, Dick" is the only time he seems to have taken his vice president into his deepest confidence. "I never let Mamie know how much it hurt." Today it seems apparent that the president suffered a major heart attack, followed by the acute depression that often accompanies such events. Ike later maintained that, being aware of the confusion caused by not making Woodrow Wilson's stroke public, he ordered Hagerty to "tell the truth, the whole truth, don't try to conceal anything."

The wily press secretary responded with a fire hose of information, all of it supporting the exaggerated notion of a steady recovery. In fact, Eisenhower was hospitalized for a full seven weeks before being gingerly returned to the White House for a long weekend, then removed to Gettysburg for another protracted period of rest and recuperation. Yet the cover story was important and necessary, justified both by Ike's eventual recovery and also because the American people found Eisenhower's steady hand at the wheel in dangerous times deeply reassuring. At this point the public didn't simply like Ike, he helped them sleep at night, and the thought of losing him could easily be presumed to have the opposite effect. That was a particularly pregnant proposition, since 1956 was an election year.

Gang members Ellis Slater and George Allen, a Gettysburg neighbor, both thought Ike would call it quits after one term, as did brother Milton and son John. But not Mamie. She had been talking to Dr. Snyder, whose medical shortcomings were balanced by a deep understanding of his patient, and he had warned her that sudden inactivity for anyone as driven as her husband could be fatal. Most important, as Ike gathered strength, he began to think he could, should, and probably would run again. That decision became final on January 13 at a secret meeting of close advisers, one Ike packed with reelection supporters, and then public at a special press conference held on February 29, ensuring that 1956 would be an especially eventful year for our last man standing, however shakily. Perhaps, if he'd known just how eventful, he would have stayed in Gettysburg.

As a preview, Ike's spring was blighted by an annoying and ultimately unsuccessful effort to scrape the perceived barnacle, Richard Nixon, off the ticket. "I've watched Dick a long time, and he just hasn't grown," he told his speechwriter Emmet Hughes. "So I just haven't honestly been able to believe that his is presidential timber." But Eisenhower's strategy, the same cold-shoulder approach he had applied to McCarthy, failed to account for Nixon being a much better politician and nowhere near as reckless. Both Sherman Adams and Hagerty wanted Ike to take the direct approach and simply dump him. "He has his own way to make," the president told Hagerty, "but there is nothing to be gained by ditching him." At a press conference on April 25, a reporter reminded Ike that he had earlier said Nixon would have to chart his own course and wanted to know if he had done so. "Well, he hasn't reported back in the terms in which I used the expression . . . no."

Nixon saw his opening and the very next day reported to the Oval Office: "I would be honored to continue as Vice President under you. The only reason I waited this long to tell you was that I didn't want to . . . make you think I was trying to force my way onto the ticket." Checkmated, Ike called in Hagerty: "Dick has just told me, he'll stay on the ticket. Why don't you take him out right now and let him tell the reporters himself. And you can tell them I'm delighted with the news."

Ike's political heartburn was soon compounded by the real thing, a lot of it, shortly after midnight on June 8. Dr. Snyder arrived and initially saw little cause for alarm; but when the pain failed to subside he began to worry, calling it "chronic ileitis." At noon, the president was rushed to Walter Reed, where he was diagnosed with an intestinal blockage, potentially fatal if not removed, which it was with major surgery early the next morning, the doctors pronouncing the procedure a complete success.

Yet, just as with the heart attack, Ike's recovery was slow. After a month, he was still attached to a surgical drainage tube and effectively detached from the job. To Nixon, Ike "looked far worse than he had in 1955. The ileitis was not half as serious, but he suffered more pain over a longer period of time." The problem was that as manager he had fielded the members of his team as extensions of various aspects of his own persona. Internationally, things had been relatively quiet during his coronary recovery, but now they were heating up, and without Ike's steadying hand his underlings began going off in their own directions. Nowhere was this truer than in the Middle East, and in particular Egypt.

In 1952 Lieutenant Colonel Gamal Abdel Nasser and his fellow Free Officers had overthrown the corrupt and British-dominated regime of King Farouk. Soon enough, Kermit Roosevelt, up to his old tricks, was funneling millions to Nasser, some of which was used to build a minaret CIA wags dubbed "Roosevelt's erection." But Nasser was primarily a nationalist and not about to trade British for American domination—meaning he would not stay bought, at least not by just one side. Instead, he began striking deals with the Communist bloc, trading cotton for weapons. In the middle of May 1956 Nasser put his neutralist sentiments to the test by recognizing Red China, a move that absolutely infuriated John Foster Dulles. Ike was more tolerant, telling the press two weeks later, "If you are waging peace, you can't be too particular. . . . We were a young country once, and our whole policy for the first 150 years was, we were neutral." But then he got sick and fell out of the loop.

Perceived critical to Egypt's future was the Aswan High Dam project, both in its promise of massive water storage for irrigation and as a bonanza of hydroelectricity, and the United States, seeing it as both viable and politically advantageous, offered Nasser a loan of $270 million to help with the financing. But in June the Russians came up with a competing bid of more than $1 billion, one that raised the hackles of Dulles, the administration's designated evangelical anti-Communist and temporarily off his leash. On July 19, at the president's first National Security Council meeting since the operation—one billed as a twenty-minute pro forma affair—Dulles slipped Ike a statement withdrawing support for the dam project. Ike gave it a cursory glance and nodded, leaving his secretary of state with the happy duty of informing the Egyptian ambassador, just hours later, that the deal was off. A week passed and Nasser responded with a statement that Egypt was nationalizing the Suez Canal, and, as Ike later ruefully put it: "The fat was in the fire."

Now fully alert to the implications of the crisis, Ike reacted calmly. He asked Brownell for a legal reading, and when told "The entire length of the Canal lay within Egyptian territory," Ike informed Dulles that "Egypt is within its rights," and that would be subsequent US policy.

Yet this was not the case with Britain and France, co-owners and co-operators of the canal, who were not about to be evicted, though they were keeping that very quiet. On October 24, at Sèvres, near Paris, British and French delegates finalized a secret protocol with David Ben-Gurion by which Israeli troops would first invade Egypt's Sinai Peninsula, then head toward the canal, a move scripted to provoke an Anglo-French ultimatum issued to both sides to stop fighting, to be followed by a joint invasion of the Suez Canal Zone by the two European powers.

For a while the deception held. The obvious mobilization of Israeli troops was explained to the Dulles brothers as a move against Jordan. But when they invaded Sinai on October 30, there were no more excuses, and Ike let loose at his secretary of state: "Foster, you tell 'em, God-damn-it, that we're going to apply sanctions, that we're going to the United Nations, we're going to do everything there is so we can stop this thing!" But that was before he learned the Israeli thing was just the tip of an Allied iceberg of betrayal.

The Anglo-French invasion, that last spasm of European colonialism, went pretty much as planned, except that it did them absolutely no good, since Nasser responded by blocking the Suez Canal with the sunken hulls of thirty-two cement-laden ships, rendering it useless for the foreseeable future. Meanwhile, there was their NATO ally and ultimate protector Team America to answer to.

It would be an exaggeration to say that hell hath no fury like Eisenhower upon learning that Britain and France had made their play behind his back, but he did react with a realpolitik that might have made Metternich or even Bismarck blush. "Nothing justifies double-crossing us," he was heard to say, and then acted accordingly. Knowing the chaos in the canal would cut off Middle Eastern oil and that Britain would soon turn to Venezuela for supplies, which would require dollars—lots of them—the president instructed relevant officials to make sure nothing was forthcoming: not oil, not dollars. Then he had Treasury Secretary George Humphrey, in response to an urgent phone call from the deputy prime minister, offer the British a loan of $1.5 billion contingent on a cease-fire and withdrawal of the invasion force. While the noose tightened, he had the Gang in for a weekend of bridge, obviously feeling a lot better.

On Tuesday, November 6, Prime Minister Anthony Eden announced that Britain would accede to a cease-fire. Ike called him immediately: "Anthony I can't tell you how pleased we are that you found it possible to accept the cease fire." But when Eden got squirrelly about the withdrawal, the president came down hard: "If you don't get out of Port Said tomorrow, I'll cause a run on the pound and drive it down to zero." There was nothing to do but fold, followed by France, and Israel the next day. Eden was history, replaced by Harold Macmillan two months later, as was his French counterpart, Guy Mollet—the price paid for swimming against the riptide of history.

Ike would later write that the Suez crisis constituted "the most demanding three weeks of my entire presidency." There was a reason for this that had nothing to do with the Middle East. The Eastern Bloc was simultaneously boiling.

In Soviet terms at least, Nikita Khrushchev was something of a reformer, or seemed that way to the people of Poland and Hungary following the publication of his "secret speech"—delivered in February 1956 before the twentieth Communist Party congress—detailing the crimes of Stalin, and seeming to promise a looser hold on Eastern Europe. The response in Warsaw was riots in late June that replaced one Communist premier with the slightly less Communist Władysław Gomułka, an action that had reverberations in Budapest, where massive demonstrations led to the installation of a much more reform-minded Communist, Imre Nagy, on October 23. The Soviets sent troops to restore order in both countries and, when this only inflamed the situation, appeared to be withdrawing the forces; *Pravda*'s new line was that their future presence would come "only with the consent" of the host.

Allen Dulles was practically gaga, telling Ike, "This utterance is one of the most significant to come out of the Soviet Union since the end of World War II."

"Yes, if it is honest," the president replied.

It wasn't. On November 4, the Soviets sent 200,000 troops backed by 4,000 tanks into Budapest to crush the Hungarian Revolution. Vastly outnumbered and outgunned, the freedom fighters, remembering John Foster Dulles's promises of liberation over Voice of America, looked to the United States for support. Ike the strategist gave them the cold shoulder. To him this was exactly the kind of crisis that could easily lead to nuclear war; plus in conventional military terms Ike noted that Hungary was "as inaccessible as Tibet." When the CIA asked for approval of a covert airdrop of weapons to the Hungarians, the president said no, an indication of Allen Dulles's diminished standing. The brothers Dulles had not fared well in this autumn of crises, one that left Allen the last Dulles standing when Foster was hospitalized with what was initially thought to be a kidney stone, but proved to be colon cancer. The tumor was removed, but from this point his health spiraled downward. Ike, on the other hand, emerged energized, so much so that the third of the bowling balls he was juggling that fall, his own re-election, was carried off more or less effortlessly.

Nominated by acclamation, Ike made only seven campaign appearances, relying mostly on TV—either scripted takes at the White House or slick ads put together by Madison Avenue and Hollywood. His opponent was the same Adlai Stevenson he had walloped four years earlier. This time, as he stumped the land Stevenson made the need for a nuclear test ban his chief campaign cudgel, and the press began to suggest he was picking up momentum. Meanwhile, Eisenhower was putting in a virtuoso performance doing the job Stevenson wanted. He wasn't worried. But son John, who read the newspapers, was: "One day in early October I marched into Dad's Oval Office considerably agitated. 'You've got to get moving,' I said. 'You're going to fall behind.'

"[Father] sat back and roared with laughter. 'This fellow's licked and what's more he knows it! Let's go to the ball game.'" So instead of worrying, they flew to New York for the first game of the '56 World Series, Yankees versus Dodgers. They might as well have stayed for the whole of the famed Subway Series, the one that featured Don Larsen's perfect game. For when the election results rolled in the night of November 6, Ike swamped Stevenson 35 million to 26 million, the biggest presidential victory in twenty years: his opponent didn't even win his home state of Illinois.

It had been a horrific thirteen months for Ike, one studded with crises and punctuated by two major illnesses, but as time spit him out at the other end, at sixty-six, he emerged reinvigorated, ready for more. That was a good thing, since the job and the times would not get any easier.

IN THE MONTHS LEADING UP TO THE SUEZ CRISIS, ABOUT THE ONLY HINT THE United States had that its allies were up to something fishy came in early October, when an unusually large number of French Mystère fighter aircraft were spotted at Israeli air bases. This was because the U-2 was lurking unseen, thirteen miles above, its camera shooting photos in the highest detail.

When it became operational in early 1956, Richard Bissell wanted above all to fly his creation where it was designed to go, high over Russia. But there was no guarantee that the president, though he had authorized the plane's construction, would let it go. To bring home—literally—the spy plane's potential, Bissell had a U-2 flown over the Gettysburg farm at seventy thousand feet, and at a subsequent briefing let Ike marvel at his own cows feeding. "This is close to incredible," he said. But it was also dangerous.

On May 28, Bissell and Allen Dulles went to the White House seeking the president's authorization for U-2 flights over Russia. Ike was frustrated by the Bomber Gap and how little he knew whether it actually existed; until he found out more, he would have to continue spending hundreds of millions of dollars churning out B-52s, not to mention the supersonic bombers the Air Force had waiting in the wings. Meanwhile, Senate Democrats Stuart Symington and Henry "Scoop" Jackson were picking up the Bomber Gap beat. So the temptation to let the U-2 off the leash was certainly present.

The president quizzed the CIA duo on what would happen if the plane was somehow intercepted and the pilot captured. Their reply was that the U-2 was so flimsy it would disintegrate and the pilot would be killed; if not, he was equipped with a suicide pill. But such a scenario remained highly unlikely, they continued, since Soviet radar was not assessed capable of spotting an aircraft flying that high (the CIA's technical directorate, on the day of the briefing, assessed a high probability of detection, a judgment Bissell deliberately kept from the president). Still, he hesitated and sent them back to the CIA without an answer.

Then Ike's intestinal blockage intervened, and a month passed with only silence from the White House. In "Mad Stork" mode, Bissell finally importuned Andy Goodpaster with a forecast of unusually clear weather over Russia in early July, one sufficient to grant him some face time with bedridden Ike on July 2. Judging by his passivity at the National Security Council meeting on July 19, the president can be assumed to have been, if not a pushover, certainly down on resistance. Still he sent Bissell back to the CIA without an answer, only to reverse himself the next day. The die was cast, the bird would fly.

On July 4, Bissell celebrated his own Independence Day by sending a U-2 from an air base in West Germany deep inside Russia, followed the next day by a mission that flew directly over Leningrad and Moscow. "Oh my Lord. Do you think that was wise the first time?" Dulles wondered aloud to

Bissell, who replied, "Allen, the first is the safest." Yet this was the second, neither leveling with the other, which was characteristic. But what mattered to both was they had what Ike wanted, or would after five U-2 missions had flown in rapid succession by mid-July. Bissell brought to the White House giant blowups of multiple Soviet air bases displaying the merest scattering of M-4s—nothing like the multitude predicted by Air Force intelligence. The Bomber Gap was a myth and now Ike knew it. But there were also things he wasn't shown.

Bissell didn't bring over blowups of photos taken on the first flight clearly showing MiG-17Fs flaming out and falling away 15,000 feet below the U-2, because if he had, it would have also revealed that Soviet radar had not only detected the intruder but also pinpointed its location. This the president only learned on July 10, when the Soviets formally protested, but in secret channels to avoid appearing weak to the world. Not unexpectedly, after his reaction to Ike's "Open Skies" proposal, Nikita Khrushchev took the flights as a personal insult, and one he quietly became determined to avenge.

"I don't like a thing about it," was Ike's reaction to the Soviet protest and being blindsided by his own people. Quietly, he took Allen Dulles aside after the National Security Council meeting on July 19 and told him he felt deceived by the CIA and had "lost enthusiasm" for the program. He suspended further flights and stipulated that future missions be individually approved by him. Hardly a pat on the back, but far from a fatal blow. The U-2 had not just cleared the air on bombers; the obvious conclusion now was that the Soviets must be putting their energies and priorities into rocketry, and this too would soon demand answers. So Ike holstered the U-2, keeping it close to the vest.

After the Hungarian Revolution was put down, Soviet military deployments in the Eastern Bloc assumed a new importance, and Ike slowly loosened his grip on the U-2, authorizing flights over areas right up to the USSR's border. But by May 1957, fear of a new gap, a Missile Gap, loomed, mostly on the basis of Soviet boasting. Again, Ike hesitated to authorize missions deep into Russia on the solid ground that all the preceding ones had been detected, but Bissell had an answer in the CIA's stealthy new "dirty bird," a U-2 the Skunk Works had rigged with radar-absorbing features—or at least thought to be radar absorbing.

Deep penetration overflights (Operation Soft Touch) began in the summer of 1957 and reached a climax in late August. Each was personally approved by the president, who, often bent over a map with John and Goodpaster, sometimes modified the routes. Pilots located a space and ballistic missile launch facility near Tyuratam, a nuclear test facility at Semipalatinsk (Semey), and a defensive missile center at Saryshagan, photographing them all in great detail—a bonanza of information. When the USSR unexpectedly announced the successful test of an ICBM at full range, two days later a U-2 passed over the launch site, and later the impact point. While the massive trove of photos would take a year to fully analyze, the most important lesson was immediately apparent: the Soviet Union certainly had a missile and space program, but it was no juggernaut, certainly not compared with the American equivalent. These photos revealed no Missile Gap—in fact, quite the opposite—but Ike had to keep that secret, one that soon turned into a lodestone.

Meanwhile, it was glaringly apparent that Bissell's "dirty birds," their stealth features canceling only a narrow band of Soviet radar, had been immediately detected and targeted, witnessed by imagery of Soviet fighters, including the new twin-engine supersonic MiG-19, flaming out several miles below during most missions. The flights tailed off in September and stopped again in October; Ike had seen enough for a while, and would authorize only six more in the next thirty-two months. He had good reason to distrust Bissell, and perhaps more to the point, the U-2 had failed to inform him of Khrushchev's biggest surprise, this in spite of the preparations taking place right under its lens at Tyuratam. On October 4, 1957, the Soviet Union launched Sputnik, the world's first artificial satellite, and Ike would face the consequences, and do it with lips sealed.

IT WOULD BE INTERESTING TO KNOW IF DWIGHT EISENHOWER EVER SAW THE 1940 Disney classic *Fantasia*, because if he had, it's likely he would have remembered the sorcerer's apprentice, Mickey Mouse—having appropriated his master's magic cap, he unwittingly sets off uncontrollable forces that soon find him hopelessly awash in unintended consequences—and felt the mouse's mess roughly approximated his own. Sputnik, relentlessly circling the globe at 18,000 miles per hour, let loose a wave of Cold War dread that at times looked like it might swamp his presidency. "I can't understand why the American people have got to be so worked up over this thing. It's certainly not going to drop on their heads," he complained to Goodpaster twelve days after the launch.

But the bad news kept cascading down. On November 3, the Soviet Union launched Sputnik II, carrying a dog and a television camera broadcasting his live image, but more important, weighing over 1,100 pounds, six times as heavy as its predecessor and large enough to carry a nuclear weapon in ICBM mode. Just over a month later, the Navy's Vanguard rocket, attempting to launch the first US earth satellite—a three-pounder Khrushchev labeled "the grapefruit"—blew up on the pad and across America on the nightly news. In the Senate Lyndon Johnson wailed the Missile Gap blues—"How long, how long, oh God, how long will it take us to catch up?"—while politicians on both sides of the aisle took up the backbeat. The president's almost 80 percent approval rating nose-dived to below 50.

Yet the day after the Vanguard fiasco, Dr. Snyder—always better at reading Ike's mood than his actual health—recorded in his log that his patient "doesn't seem to be taking the matter too seriously this morning." He wasn't. He had another satellite waiting in the wings—a stand-in that would soon be launched successfully by a German-derived Jupiter-C—and also a much more important one up his sleeve.

At a facility in Menlo Park, California, Lockheed was working on a project that amounted to putting the U-2 in orbit—a twenty-foot-long satellite stuffed with super-high-resolution cameras, to be launched by a modified intermediate-range ballistic missile, an undertaking so secret that at this point it didn't even have a name, just the alphanumeric WS-117L. The program was extremely ambitious, involving dropping canisters of exposed film from orbit and catching them in midair. But the payoff, continuous, invulnerable coverage of the Soviet landmass, would be invaluable, and Eisenhower understood from the beginning that WS-117L, not U-2, was the future. He just had to wait for it, but as usual in total silence.

The enormous strain Ike was living under became starkly apparent on the morning of November 26, when his personal secretary Ann Whitman looked up as he entered the Oval Office, tilted to one side, then greeted her with a jumble of words that made no sense. Snyder this time correctly diagnosed that the president was "aphasic"—having trouble understanding and using speech—a likely symptom of a stroke, soon confirmed by a panel of physicians. Besides putting him at a loss for words, it mainly infuriated Ike; but the stroke was mild, and in a matter of days he wanted to get back to work. Finally, on Saturday, November 30, after reading a *New York Times* article that intimated Nixon would soon be taking over, he told Snyder, "Howard, there can't be two Presidents of the United States. I think I'll go down to the Cabinet meeting Monday morning."

That afternoon Whitman noted with relief that after a full day in the Oval Office, Ike left with his hat set at a jaunty angle and muttering about unnecessary weapons that were "nuts." He would suffer headaches for weeks, but nothing else. Ike was back, but so were his frustrations.

Eisenhower had come into office determined to save money on defense by a heavier reliance on nuclear weapons—that was what the New Look was all about. But by mid-1957 nukes seemed to be multiplying like the endlessly replicating water-carrying brooms Mickey Mouse set into motion with his master's magic cap, the arsenal having grown to over six thousand and metastasized to all manner of missions—short-range missiles, artillery shells, land mines, depth charges, torpedoes, and so forth. When Lewis Strauss, chairman of the Atomic Energy Commission and czar of nuclear weapons production, asked him for more tests and many more warheads dedicated to air defense, Ike responded: "You've been giving us a pretty darn fine arsenal of atomic weapons," and wondered why we needed more, telling him instead that reducing nuclear weapons "would be a fine thing." Mickey, at least, tried his mouse best to stop the brooms from proliferating, but not Ike. After this mild admonition, he didn't lift a finger to halt or even slow production, apparently never coming to grips with the prospect that a nuclear weapon for every purpose would bring the dreaded threshold closer and more likely to be breached. Instead, like the football coach he had once been, Ike punted.

And while the nukes multiplied, he had to answer to those who would seek shelter from them. Few things strategic frustrated Eisenhower more than civil defense. The whole proposition that a society could ride out a nuclear attack and emerge anything but totally wrecked struck him as not only absurd, but potentially dangerous, since a substantial civil defense program on one side could convince the adversary of intentions to win such an endeavor, an utterly destabilizing proposition.

Yet fear being nature's most powerful motivator, "Gimme Shelter" was a cry that could not be suppressed. Congress built a 120,000-square-foot bunker retreat under the Greenbrier, the West Virginia hotel that Ike and Mamie had trouble standing for just a couple of days. All over the country, schoolchildren were drilled to huddle under their desks if the big bang came. Popular magazines featured do-it-yourself fallout shelters. At a National Security Council session considering the issue, a notetaker wrote: "The President wondered how far we could go until we reached a state of complete futility."

Hoping to head off that condition, Ike chose a characteristically oblique approach—he created a panel of experts who could talk the problem to death and issue a report he could ignore. To chair the group, he took James Killian's recommendation and chose H. Rowan Gaither, the head of the Ford Foundation. This is where the scheme went off the tracks. Gaither was also chairman of the board of RAND, and he stocked his eponymous commission at the working level with its analysts. RAND at this point was chockablock with theoreticians happy to explore the possibilities of nuclear war—mathematicians and physicists such as Albert Wohlstetter and Herman Kahn cloaking near-paranoid predictions of Soviet perfidy in an impenetrable panoply of numbers—and even Bernard Brodie was now talking about nuclear war in terms of coitus interruptus: wage a little, stop, and see if the enemy surrenders. This dark perspective suffused the subsequent Gaither Committee report, penned by Paul Nitze, who authored 1950's NSC-68, which might well be called the founding document for America's version of the Cold War. To add to Ike's discomfort, the committee's report was delivered to him just a month after Sputnik had raised the hairs on the back of America's collective neck.

Ike listened quietly as the Gaither Committee unwound its argument: The Missile Gap was portrayed as gaping and demanding major and immediate investment, but even that would not eliminate the Soviet ICBM lead until 1960, creating a period of maximum danger, one requiring a massive program of shelter construction—the total bill for the committee's many recommendations being pegged at $44 billion over five years. Knowing secretly that this was balderdash, the president first sought to bring the presenters back from wonderland, reminding them that a billion dollars was a stack of tens that would reach the top of the Washington Monument. But rather than go into any detail as to why all the money would be wasted anyway, the president had John Foster Dulles, at his dismissive best, trash the findings in general terms. As they left, Ike thanked the commissioners for their six months of work and added: "You can't have this kind of war. There just aren't enough bulldozers to scrape the bodies off the streets."

On the other hand, there was Curtis LeMay, also very much a part of Ike's grand mechanism of nuclear deterrence. As part of the workup for the Gaither Committee report, its deputy director Robert Sprague had decided to put to the test the RAND mathematicians' proposition that the Strategic Air Command's bomber force was vulnerable to Soviet surprise attack. On September 16, 1957, he staged a strategic alert exercise, giving the bombers the six hours' warning time they could expect from forward radar, and not a single aircraft got off the ground under the deadline—mainly because their nuclear weapons were stored separately and had to be loaded, not to mention fuel and crews. When Sprague presented all of this to LeMay, he responded by claiming the Air Force had intelligence-gathering aircraft flying over Soviet territory twenty-four seven: "If I see that the Russians are amassing their planes for an attack, I'm going to knock the shit out of them before they take off the ground."

"But General LeMay," Sprague countered, "that's not national policy."

"I don't care," LeMay responded. "It's my policy. That's what I'm going to do."

Since his claim of continuous air coverage was a fabrication, this incident raises a serious question. It is possible that the Strategic Air Command was already keeping a certain number of bombers in the air at all times but not over the USSR, and LeMay simply didn't want to reveal this, or how few there were; but the alternative was that the general was contemplating strategic freelancing—preemption when he thought the time was ripe. It's a chilling thought, especially coming from LeMay, a man who had already incinerated large portions of Japan with unremitting enthusiasm.

Ike had a lifetime of service connections to draw on: if not this specific statement, he must have been aware of others LeMay, hardly the most discreet of souls, had scattered along the same lines. Yet he did nothing about him, apparently placing his faith instead on the perceived unbreakable nature of the chain of command, the iron imperative of officers to obey orders. It worked. LeMay was certainly credible as a threat instrument and never broke his leash, but the thought remains: Why did his master parade him with such a loose hand?

This also extended to the Strategic Air Command institutionally and the Air Force in general. Eisenhower had already placed his bet on strategic missiles, but during his years as president he continued to allow the Air Force to crank out six hundred B-52 heavy bombers, along with one hundred B-58 Hustlers (basically a supersonic fuel tank with a nuclear weapon attached). He did refuse to support the even faster B-70 along with LeMay's nuclear-powered model, but nevertheless backed the Air Force program to fill the wild blue yonder with a host of new supersonic jet fighters, plus the general-purpose aircraft such as tankers and surveillance planes needed to complete its profile as a credible instrument of strategic retribution. That it was, but considered in light of a coming wave of American ICBMs and ballistic missile submarines, probably an unnecessary one. Yet, in the face of the presumed Missile Gap, politically it represented the easiest way forward for Ike, so he put away his cost-cutting scissors and took it, heralding the emergence of the strategic triad of ICBMs, submarine-launched ballistic missiles, and bombers that persists to this day.

Ike was a strategist as true strategists have to be, small creatures seeking to somehow guide vast forces, compromising always in the cause of the central objective, which in his case was to avoid nuclear war. On the face of it the Eisenhower doctrine of massive retaliation left no room for anything less than that, causing critics, the most notable being Army chief of staff Maxwell Taylor, to call for something in between, the capacity for a so-called flexible response, using conventional forces only to achieve traditional military objectives. Since Eisenhower's central fear was rapid and uncontrolled escalation, his strategic instinct told him that a "flexible response" might stiffen quickly to become a bridge to Armageddon. More specifically, according to Goodpaster:

He thought General Taylor's position was dependent on an assumption that we are opposed by a people who think as we do with regard to the value of human life. But they do not, as shown in many incidents from the last war. We have no basis for thinking that they abhor destruction as we do. In the event they should decide to go to war, the pressure on them to use atomic weapons in a sudden blow would be extremely great.

On the other hand, Ike was at his core an Army officer, and his entire career had taught him to value "feet on the ground." So, yet again, he prevaricated, not only letting what amounted to flexible response into the policy planning process, but essentially adopting it as policy, along with the attached funding and weapons-acquisition implications that would ensure a large and continuously modernized array of ground forces. And since subsequent events would reveal the structure of nuclear deterrence to be far sturdier than Eisenhower feared, he and his successors would find plenty of uses for those conventional forces in a pockmarked but holocaust-free future.

MEANWHILE, AS THIS WAS THE COLD WAR, THE CRISES JUST KEPT COMING like waves in an angry sea. In the summer of 1958 the simmering Middle East threatened to go full boil, with Nasser, the force behind the newly formed United Arab Republic (UAR), calling for regicide in US-allied Jordan and Saudi Arabia, while a mid-July coup in Iraq overthrew and executed the Hashemite King Faisal and his family.

Surrounded geographically and politically, the pro-Western regime in Lebanon, claiming the United Arab Republic was about to pounce, pleaded for American intervention. Eisenhower surprised everybody by doing just that, sending the Sixth Fleet armed with assorted tactical nuclear weapons and three battalions of Marines, two airborne battle groups from Germany— a total of around fourteen thousand men—to storm ashore on the beaches of Beirut. But when General Taylor suggested sending troops into the mountains behind the city to secure them, Ike was adamant: "They stay on the beaches so they can get away just as fast as they can if anything goes wrong." This wasn't about securing bridgeheads; this was about sending a message to Nasser, and also to Congress, that he could flex when necessary. The troops were out by October 25, their mission accomplished.

But no sooner had that wave broken than Ike was hit by another, in the early fall of 1958. This time it came from a familiar direction, Quemoy and Matsu, where the precarious standoff in the Formosa Strait threatened to collapse after Chiang Kai-shek, going against warnings from Washington, deployed over a hundred thousand men on the two worthless and vulnerable offshore islands. The Chinese responded with continuous shelling, and Ike smelled a rat: "The Orientals can be very devious," he told one Pentagon functionary. "If we give Chiang our full support he would then call the tune."

Therefore, when John Foster Dulles and the Joint Chiefs suggested authorizing the Seventh Fleet to nuke the Chinese if worse came to worst, the president instead had his secretary of state reaffirm in early September the US intention to protect the offshore islands, but also include a scarcely disguised olive branch signaling negotiations. Zhou Enlai bit two days later with a positive response, and soon after the Joint Chiefs of Staff changed their tune, indicating the offshore islands were not necessary to the defense of Formosa, and that the latter's garrisons could safely be reduced. Chiang got the message and began withdrawing troops, a process aided by the mainland Chinese, who announced they would fire on Nationalist convoys only during odd days of the month. "I wondered if we were in a Gilbert and Sullivan war," Ike later maintained, but what mattered was that yet another crisis had broken and dissipated.

But only to be slammed by a bigger, more dangerous example of one that could well have left Dwight Eisenhower wondering if they were beginning to repeat themselves. "Berlin is the testicles of the West," Nikita Khrushchev would later maintain. "Every time I give them a yank, they holler." It was Ike's turn as the year turned toward 1959.

As usual, the Russian leader's bravado was meant to disguise his own weakness. In almost every respect—especially when compared with its Western alternative—Communist East Germany was a failure, one whose dysfunction could be measured by the ever-increasing number of its citizens voting against it with their feet simply by crossing over into West Berlin. Since the end of the war to 1958, the population of the East had already shrunk by 15 percent, and now the diaspora showed signs of turning into a stampede. On November 10, 1958, Khrushchev sought to corral it by delivering an ultimatum to the Western Allies that in six months the Soviets would allow East Germany to control its own borders, a move meant to force the French, British, and Americans to vacate Berlin, thereby sealing the escape route.

Ike responded with a resolute facade, telling advisers in early December: "In order to avoid beginning with the white chips and working up to the blue we should place them on notice that our whole stack is in play." The game face hadn't changed on January 22, when the National Security Council notetaker recorded the president saying the United States would "hit the Russians as hard as we could. [They] will have started the war, we will finish it. That is all the policy the President said he had."

Of course, it wasn't. Behind the adamant mask was still a man with poker never far from his thoughts. But at this point he was alone at the table, John Foster Dulles being eaten up by abdominal cancer and destined to die days before Khrushchev's deadline in late May. Ike would miss Dulles; he was more than just a loyal and hardworking functionary; in private his counsel was often wise, and in public he could be depended on to play bad cop to the president's better angel.

Now without him, Ike reverted to his "just confuse them" mode in public, especially when it came to the nuclear-use question posed by a *Washington Post* reporter: "I must say, to use that kind of nuclear war as a general thing looks to me a self-defeating thing for all of us. . . . I don't know what it would do to the world and particularly the Northern Hemisphere; and I don't think anyone else does. But I know it would be quite serious. Therefore, we have got to stand right ready and say, 'We will do what is necessary to protect ourselves, but we are never going to back off on our rights and responsibilities.'" This was a statement suitable for all sides of the issue, including those fond of black humor—a tour de force in muddled prose ambiguity and, most important, the opposite of an ultimatum. Ike was playing for time. He knew Khrushchev was overextended, a fat man way out on a skinny limb, and he needed to be given time to crawl back. The May 27 deadline came and went, with Khrushchev's deputy premier announcing it only related to beginning negotiations, not settling the issue of Berlin. Fine with Ike; he was glad to kick that very explosive can down the road as far as he could, far enough that it eventually landed in the lap of his successor, JFK.

Knowing Ike, we can assume it's a safe bet that by this time he wanted to get better acquainted with his redoubtable adversary. The opportunity would soon arise courtesy of Richard Nixon. During the course of a heated and nationally televised debate on July 25, one that took place in the model kitchen of an American exhibit of consumer products in Moscow, the vice president invited Khrushchev to visit the United States, an offer almost immediately accepted. Perhaps because it was Nixon, Eisenhower registered only chagrin, but he must have known the visit was the right move, not only providing time to size up the Soviet, but also the opportunity to expose Khrushchev to the sprawling landscape of power and prosperity that was midcentury America.

Whatever his mood, the president conceded the importance of the visit by flying to Europe—his first time aboard the brand-new presidential jet, a Boeing 707—to consult with the leadership of the key Allies: Harold Macmillan in London, Konrad Adenauer in Bonn, and in Paris, Charles de Gaulle, now back in power after a hiatus of twelve years. Of the wartime leaders, no one owed more to Eisenhower than de Gaulle. Ever since they had bonded before Overlord, all signs had pointed to these two soldiers destined to turn presidents sharing a true friendship.

Yet states have no friends, only interests, and on this trip de Gaulle made that clear: "You, Eisenhower, would go to nuclear war for Europe because you know what is at stake. But as the Soviet Union develops the capability to strike the cities of North America, one of your successor[s] [might not]. . . . When that comes, I or my successor must have in hand the nuclear means to turn what the Soviets may want to be a conventional war into a nuclear war." Prior to that, he added, no nuclear-armed American aircraft could be stationed or land on French soil without a host nation veto over their use. It was a preface to France's force de frappe and a cold lesson in the logic of nuclear deterrence among relative small fry—the capability may not make you safe, but the alternative is helplessness in the face of giants, even if one of those giants is on your side. For his part, Ike must have wondered if the proliferation would ever stop.

"But time was running out on Eisenhower's presidency," writes Evan Thomas, whose book *Ike's Bluff* is a penetrating guide to the tortured path of his subject's nuclear anabasis. "He needed to start taking some positive steps toward defusing the tensions of the Cold War and slowing the arms race that seemed to be heading for some terrible last spasm." Khrushchev's trip provided that opening.

He had insisted on flying from Moscow in the brand-new Tu-114, a plane derived from a Russian bomber, arriving nonstop in Washington on September 15, a strategic message in itself. Ike countered by helicoptering Khrushchev over a rush-hour traffic jam on the way to the White House— broad highways stuffed with American automobiles carrying commuters to expanses of newly constructed single-family dwellings—a vista the Russian would look down on repeatedly during his ten-day coast-to-coast very guided tour—one that also included Manhattan capitalists, a Hollywood mix-and-mingle with film stars, a stop at an Iowa corn farm, and a visit with Eleanor Roosevelt at Hyde Park.

But the main event was saved for the last two days: face-to-face meetings between the two leaders at the rural and newly renovated Camp David, Ike's version of a dacha. The environment seems to have relaxed the Russian. Their talks spanned the full range of issues, but were not productive—no new thing during the Cold War, but also not the point. This was an information-gathering session, each not only taking the measure of the other, but both mutually trying to come to grips with what gave every appearance of being the most dangerous aspect of their adversarial relationship—a runaway nuclear arms race.

"Tell me, Mr. Khrushchev, how do you decide on funds for the military?" Ike asked, but without waiting for an answer moved on. "My military leaders come to me and say, 'Mr. President, we need such and such a sum for such and such a program. If we don't get the funds we need, we'll fall behind the Soviet Union.' So I invariably give in."

"It's just the same," answered Khrushchev. "Some people from our military department come and say, 'Comrade Khrushchev, look at this! The Americans are developing such and such a system but it would cost such and such.' I tell them there's no money. So they say, 'If we don't get the money we need and if there's a war, then the enemy will have superiority over us.' So we talk some more, I mull over their requests and finally come to the conclusion that the military should be supported with whatever funds they need."

"That's what I thought," Ike ventured. "You know, we really should come to some sort of an agreement in order to stop this fruitless, really wasteful rivalry."

The dialogue constituted an early and accurate description of the action-reaction dynamic that drove the US-USSR weapons competition, and perhaps also the seed of arms control. But if so, it was an entirely premature one. Eventually limitations would be negotiated, but at levels that had already reached the absurd numbers apparently psychologically necessary to enforce the unyielding gridlock of nuclear deterrence. But it was too early for either Eisenhower or Khrushchev to grasp or even guess at the paradoxical notion of more being better; instead, they may well have thought they made some sort of breakthrough.

If so, it was one the president opened up a bit more by being Ike and knowing how to treat a guest. He took the Soviet leader on a little road trip up to Gettysburg, not to his own manicured Dorothy Draper–decorated farm, but to the much less impressive residence of his son John, to meet his wife and four young children. On the way back to Camp David, the Soviet leader seemed moved by the experience, the next day agreeing to abandon his Berlin ultimatum and also inviting Ike to Russia. Very suddenly substantial chunks of ice seemed to be melting off the Cold War's glacial face. Newspapers and the airwaves were filled with commentary on "the Spirit of Camp David," and it was announced that the wartime Big Four—the United States, the Soviet Union, the United Kingdom, and France—would convene again the following May at a summit in Paris. Unfortunately, what looked like the springtime of Eisenhower's dreams would be revealed as a nightmare.

AS HE HEADED INTO THE LAST LAP OF HIS PRESIDENCY, IKE WAS TIRED, FA-tigued almost beyond description; the only prolonged relaxation he'd had in over a decade had been the results of life-threatening illness. Compounding that, Eisenhower came to strategic maturity in an era strewn with crises that led to major conflict; hence the dangers he perceived during each one he encountered in the Cold War were very real to him, and the cumulative effect absolutely exhausting. Dulles was dead and Ike's domestic shield, Sherman Adams, had been driven from office over a scandal involving a vicuña coat, leaving him alone at the helm trying to steer a course through a gale that roared ever louder, "Missile Gap." He barely held on to the finish line.

Boosted by lurid press reporting of a gap projected only to loom larger with time, Democrats, led by Stuart Symington and Lyndon Johnson, had used it to drub the Republicans in the November 1958 midterm elections. Pivoting off the panic, the Joint Chiefs wanted a $50 billion defense budget for 1960, when Eisenhower knew $40 billion would cover just about every contingency except major war. Meanwhile, Air Force and CIA estimates of Soviet ICBM capabilities were little more than whispered versions of the missile envy portrayed in public lamentations. Finally, in early April 1960, hounded by Bissell and prodded by an intelligence community judgment that the Soviets were about to put their first ICBM sites on operational status, the president agreed to renewed U-2 missions aimed at locating them before they could be camouflaged.

It was a terrible decision, and more than just a matter of endangering "his reputation for honesty," which he perceived as his key asset on the eve of the Paris summit. Eisenhower was a techie and the Army already had a well-developed defensive missile program in the Nike series. Hence, even though Bissell failed to inform him of an Air Force warning of "a high probability of intercept" on future U-2 flights, he should have put two and two together and concluded that out-of-breath Soviet fighters would soon be replaced by guided missiles carrying their own oxidizer and capable of much higher altitudes—a prospect made obvious when three such rockets shot down a Taiwanese spy aircraft over China at 65,600 feet in early October 1959.

Meanwhile, Ike's alternative to the U-2 had morphed from WS-117L into a named, though still super-secret, program, "Corona," and hardware in the form of the KH-1 series of spy satellites, which were launched beginning in early 1959, resulting in a string of failures that included booster malfunctions, guidance problems, recovery overshoots, and even film breaking up in the cold vacuum of space. But the arc of Corona was plainly in a positive direction, and after thirteen tries, an intact film canister would be recovered from orbit by a transport plane on August 18, 1960. But that was four months too late for Ike. He could have waited. He should have waited. But he didn't, and would face instead the wrath of Khrushchev.

The first flight, on April 9, was almost immediately detected and once again intercepted by Soviet fighters that couldn't reach it. The U-2 returned safely and the Russians said nothing, but the incursion gave Khrushchev time to plot his revenge when the next one came.

Which it did, after Bissell, pressing his own and Ike's luck, got his approval on April 25 for "just one more" flight over Russia—"Operation Grand Slam" he code-named it, a 3,800-mile stratospheric trek across Russia in search of launchpads. On May 1, veteran U-2 pilot Francis Gary Powers took off from Peshawar, Pakistan, headed for Bodø, Norway. He made it less than halfway before he was shot down by an S-75 Dvina surface-to-air missile, and the trap snapped shut on Dwight D. Eisenhower.

Khrushchev was joyously enraged. On one hand, as his son Sergei later reported, this was "a betrayal by General Eisenhower, a man who had referred to him as a friend, a man with whom he only recently sat at the same table. . . . He would never forgive Eisenhower for the U-2." But on the other, his strategic opponent had just committed a grave blunder, and he had him exactly where he wanted. And having been a minion of Stalin and survived the experience, he was a virtuoso at retribution.

First, he waited until the U-2 was reported missing in Washington, and a prearranged press release was issued that a NASA weather research plane had crashed in Turkey. Then on May 5, during a long speech in the Supreme Soviet, he announced that an American spy plane had been shot down, blaming it on "Pentagon militarists," and providing no further details—a gambit that prompted a change in the American story to the effect that the weather plane was now believed to have gone off course and accidentally drifted into the Soviet Union. Khrushchev's response was devastating, announcing to the Supreme Soviet on May 7 that enough of the U-2's wreckage had survived to demonstrate this was no weather plane—"And we also have the pilot, who is quite alive and kicking! . . . The pilot's name is Francis Gary Powers. He is thirty years old and works for the CIA." He displayed the pilot's personal effects, including a suicide needle, and also blowups of Soviet air bases taken by the plane's cameras—an airtight case.

Official Washington continued to search for wiggle room, conceding that an intelligence aircraft had "probably" flown into the Soviet Union, but that "there was no authorization for any such flight," a trial balloon that convinced practically no one. Finally, perhaps remembering the note he wrote and stuck in his wallet prior to D-Day accepting personal responsibility for its failure, Ike decided to own up, or at least admit the plane had flown under broad presidential authority, not his specific instructions, hoping to give Khrushchev the cover to proceed with the summit in Paris. He might as well have stonewalled.

On May 15, he arrived at the French capital, only to be presented with a long letter from Khrushchev refusing to participate unless Eisenhower publicly apologized for the overflights and agreed to halt them, along with punishing those responsible. Then, when finally talked into attending the opening session, Khrushchev continued the diatribe for forty-five loud and angry minutes, at one point prompting Ike to pass a note to his new secretary of state, Christian Herter: "I think I'm going to take up smoking again." Khrushchev closed by repeating his demand for an Eisenhower apology, his pound of flesh for continued participation, also not forgetting to rescind his invitation for the president to visit Russia. Shortly after, the Soviet delegation walked out, never to return. Once they left, de Gaulle went over to Ike and took his arm: "I do not know what Khrushchev is going to do, nor what is going to happen, but whatever he does, or whatever happens, I want you to know that I am with you to the end." It was the only positive in a day that left the Cold War at one of its most frigid moments.

Shortly after he returned from Paris, Ike, alone in the Oval Office with his science adviser George Kistiakowsky, attempted to berate him and the scientific community for failing to inform him of the U-2's vulnerability.

> I responded that the scientists had consistently warned about the U-2 eventually being shot down and it was the management of the project that failed. . . . Cooling off, the President began to talk with much feeling about how he had concentrated his efforts the last few years on ending the cold war, how he felt that he was making big progress, and how the stupid U-2 mess had ruined all his efforts. He ended very sadly that he saw nothing worthwhile left for him to do now until the end of his presidency.

But there was definitely one more thing that grabbed his attention. Once the U-2 had at least cleared the air over the nonexistent Missile Gap, it caused Ike to consider more carefully the vast array of American weapons, many nuclear, that had grown in its shadow, plunging him into a sorcerer's apprentice moment.

When Eisenhower took office, the United States had around one thousand atomic bombs; by 1960 the arsenal had grown to twenty thousand, with a nuke for practically every tactical occasion, growing like the mushrooms their explosions resembled. On April Fools' Day 1960, Ike had listened with growing impatience to Defense Department officials arguing for more than doubling the Minuteman solid-fuel ICBM force from 150 to 400, finally interrupting sarcastically: "Why don't we go completely crazy and plan for a force of 10,000?"

On November 25, he had Kistiakowsky, not defense officials, brief him on the Single Integrated Operational Plan (SIOP), the new blueprint for massive nuclear retaliation, devised primarily by the Strategic Air Command and revised by RAND analysts. It called for all 3,200 strategic bombs and warheads—a total of nearly eight thousand megatons—to be delivered against targets in Eastern Europe, China, and, most of all, Russia in generous portions, Leningrad receiving nine nukes, Moscow twenty-three, and even Kaliningrad getting seven. "Overkill," the science adviser called it, "megatons to kill four or five times over somebody who is already dead."

After the briefing Ike told his naval aide that the SIOP "frighten[ed] the devil out of me"; he wanted to allow SAC "to have just one whack—not ten whacks" at each target. Later, Dr. Snyder heard Ike complaining about the SIOP, saying he was thinking about refusing to sign off on the document. But it had been Ike—always a man with a plan—who had proposed the idea in August as a means of rationalizing retaliation. Then, when the SIOP arrived, it was already three weeks after the presidential election, and, rationalizing himself, he decided it was too late to change it before the arrival of the next administration. So he put pen to paper, going with the tide of history on its way to making all-out war absurd. But he didn't know it.

The epic mobilizations of World Wars I and II being his context, Eisenhower had long worried that a huge peacetime defense establishment would lead to an Orwellian outcome—permanent and pervasive militarization of the United States and an end to democracy. Now, having succumbed repeatedly to this establishment's relentless logic and appetite, feeling himself and the country caught in its implacable gearset, Ike decided to issue a warning before it was too late. He worked on his jeremiad for a month and through twenty-one drafts, assisted by his brother Milton and a young political scientist from Johns Hopkins, using as his model Washington's Farewell Address, warning the nation against entangling alliances. Finally, on the evening of January 17, 1961, Ike delivered his own farewell from the Oval Office, and a somber one it was. "Our military organization today bears little relation to that known by any of my predecessors . . . [T]he United States had no armaments industry. American makers of plowshares could . . . make swords as well. . . . [Now] we have been compelled to create a permanent armaments industry of vast proportions. . . . This conjunction of an immense military establishment and a large arms industry is new to the American experience. . . ." Then, having brought his audience to the brink, Ike delivered his most memorable and widely quoted admonition: "In the councils of government, we must guard against the acquisition of unwarranted influence, whether sought or unsought, by the military-industrial complex. The potential for the disastrous rise of misplaced power exists and will persist."

There is a picture of Ike taken three days later at JFK's inauguration; he looks the part of an embittered old man, exhausted by the travails of the Cold War and not sure if his successor was up to the task. He shouldn't have been so glum. The arms race would gradually stabilize at the necessarily ridiculous levels, the military-industrial complex would retreat to being just another component of the national mix, and the Cold War would eventually end peacefully. Now ex-president Eisenhower, having gotten America safely through its most formative stage, should have been all smiles on the way back to Gettysburg in the Chrysler Imperial Mamie had given him for his sixty-fifth birthday. He wasn't. But either way, the public Ike was over, the last of our four superstars to leave the stadium.

CHAPTER 13

THE END

"Stay away from Walter Reed" would seem like good advice from a future biographer, since the remaining three of our key players died there. But it would be worthless, since the hospital was and remains excellent, death is inevitable, and time runs only forward. Still, as had been the case in their stellar and intertwined lives, all three ended their days there in a similar fashion, but in ways that also reflected the inner recesses of each.

George Marshall had continued to fade slowly through various health problems, but in early 1959 he suffered an incapacitating stroke in Pinehurst, North Carolina. It took until the spring for him to be safely transported to Walter Reed, where he declined from wheelchair to full-time bed rest to oxygen, catheter, and feeding tubes, in and out of consciousness. John Foster Dulles was also on the same floor, near the end of his life, and Ike brought the visiting Winston Churchill in for one final meeting with his faithful secretary of state. On the way out, they looked in on George Marshall, the man Churchill had honored at the coronation of his own queen. The general had not the slightest idea who either of them were. They both left somberly, Churchill wiping tears from his eyes.

Marshall hung on until the fall and died quietly in the early evening of October 16, 1959. He had left explicit instructions forbidding a funeral at the National Cathedral or lying in state in the Capitol Rotunda. He wanted no eulogy, and his interment was to be private. He left a short guest list that included Bob Lovett "if it is convenient" and Beetle Smith "if he is in town," but not Ike, though he left it up to Katherine to invite more guests as she thought appropriate.

On October 20, a brilliant autumn day, his coffin was brought to the Fort Myer chapel for a short service. Before the family entered, Harry Truman and his former military aide, along with Dwight Eisenhower with his own aide, settled into the same pew in icy silence. When the abbreviated Episcopal rites were over, a caisson took the coffin down the hill from Arlington National Cemetery's Tomb of the Unknown Soldier to a grave next to the ones Lily and her mother had been buried in decades earlier. Katherine would join them there in 1978. Fittingly, his legacy is preserved primarily at the George C. Marshall Research Library on the campus of the Virginia Military Institute.

FOR DOUGLAS MACARTHUR, EXILE WAS MORE ELBA THAN SAINT HELENA, NOT Washington and the White House but the Waldorf and New York City—both certainly bearable. He never meant to stay, yet settled in comfortably. First Army Headquarters in Lower Manhattan gave him a four-room office suite, and each morning he arrived to read the cable traffic. Later in the day he and Jean might be seen prowling the smart shops along Fifth Avenue, lunching at someplace elegant, followed, perhaps, by a night on Broadway or at a sporting event, seated in the owner's box, cheering on Gotham's teams with the intensity of a teen, especially if it was Jackie Robinson, his favorite and most admired athlete. On the other hand, nanny Ah Chue never went out, becoming a Waldorf recluse, while little Arthur quietly drifted into his own world of music and anonymity, choosing Columbia, not West Point, and eventually living under an assumed name, his father's persona too much for him.

But it attracted Ike's successor, JFK, who, like Doug, had had a dicey PT boat experience and doubtless had felt his presence looming over the Pacific War. So, in April 1961, he paid him a courtesy call at the Waldorf Towers and found himself sucked into a two-hour colloquy. Kennedy told aides that Mac was one of the most captivating conversationalists he had ever met, and, wanting more, an invitation to the White House for lunch soon followed.

The chat that ensued captured the president's entire afternoon. "You're lucky to have that mistake happen in Cuba," MacArthur told him, not hesitant to broach a sore subject—the abortive Bay of Pigs invasion that Kennedy had approved but a tired, distracted Ike had funded, and that Allen

Dulles and Richard Bissell had orchestrated, the last of their covert extravaganzas. He then pivoted to the East, no doubt drawing on his own experience, to warn Kennedy that anyone willing to commit ground troops to Asia "should have his head examined." Speaking with a passion that impressed Kennedy's aides Arthur Schlesinger Jr. and Theodore Sorenson, MacArthur apparently achieved the desired effect. Shortly after, Kennedy informed deputy national security adviser Walt Rostow that he would not risk sending combat troops to Indochina, and that the ten thousand Marines on alert in Okinawa could stand down.

A few weeks later MacArthur's old compadre and aide Carlos Romulo, now Philippine ambassador to the United States, approached Kennedy's press secretary to inform him that as part of his country's celebration of its fifteenth year of independence he wanted to invite Doug and Jean to join the official festivities. When the president heard, he broke into a broad smile, and put a Boeing 707 at MacArthur's disposal. Filipinos still loved Mac and were always happy to see him, but he returned home a sick, frail octogenarian plainly on the downward slope.

But being Doug, he said his goodbyes in a monumental fashion. On May 12, 1962, he arrived at West Point to inspect the Corps of Cadets. Leaning on Jean's arm as he made his way to the rostrum, MacArthur made it clear to the 2,200 cadets and others assembled on the Plain that this would be his last inspection. But then speaking without notes and even pacing at times, Doug came alive, giving yet another boffo farewell, concluding with:

> The shadows are lengthening for me. The twilight is here. My days of old have vanished, tone and tint; they have gone glimmering through the dreams of things that were. . . . I listen vainly, but with thirsty ear, for the witching melody of faint bugles blowing reveille, of far off drums beating the long roll. In my dreams I hear again the crash of guns, the rattle of musketry, the strange, mournful mutter of the battlefield. But in the evening of my memory, I always come back to West Point. Always there echoes and re-echoes in my ears—Duty, Honor, Country.
>
> Today marks my final roll call with you. But I want you to know that when I cross the river my last conscious thoughts will be of the Corps; and the Corps; and the Corps. I bid you farewell.

But he was still not finished. If the body was weakening, the mind remained strong, and in October he began to compose his memoirs. Sitting in the same chair, he filled hundreds of legal pads with virtually no cross-outs or erasures before he was finished the following August. The result of the ten-month marathon, *Reminiscences*, was pure MacArthur, and also an excellent summary of his life, if you weren't interested in his first marriage, which he left out entirely. But that didn't bother conservative publishing magnate Henry Luce, who bought the rights to be serialized in seven installments for *Life* magazine and then presented in book form, for a cool $900,000 (almost $8,000,000 in 2022). Reviews were predictably mostly anti-Mac, but the book sold well and probably earned Luce his money back.

It was MacArthur's last kudo. Lifelong good health had taught him to avoid doctors, and by this time his liver was shot. There was nothing left but the one-way ticket to Walter Reed in early March 1964. He lasted just over a month, dying of biliary cirrhosis on April 5.

No surprise, Doug left elaborate instructions for his funeral, which Kennedy's successor, Lyndon Johnson, only elaborated on, leading to a three-stage event. It began in Manhattan with a public viewing and televised procession wheeling the coffin to a special train, which then took it to Washington. There it lay in state in the Capitol Rotunda, with 150,000 people filing past, before it was taken to Pinky's hometown, Norfolk, Virginia, where the old city hall had been turned into the Douglas MacArthur Memorial. Here, in an elaborate sepulcher, he was laid to rest, joined finally by Jean, who continued living in the Waldorf Towers until she died at the age of 101.

LIKE THE OTHERS, IKE DECLINED TRUE TO FORM. WHEN HE REQUESTED THAT he be returned to his five-star rank—reassuming the identity of general rather than Mr. President, about the only tangible benefit being that he got to keep Sergeant Moaney as his valet—JFK was flabbergasted. But it was completely in character. Dwight Eisenhower was a soldier at the core, and this was the identity he intended to die with. It was largely in this role that he was consulted by his successors, particularly Lyndon Johnson, now the one caught in Vietnam's quicksand. Eisenhower thought he was too involved in the day-to-day running of the war, and advised him to "go for victory," suggestions Johnson was not about to follow.

Meanwhile, Ike's remaining nine years were generally happy ones. He presided over the farm, and Mamie the house at Gettysburg. He painted frequently, a hobby begun back at Columbia and pursued without much talent, almost purely for relaxation. There was still the Gang for golf and bridge, Scrabble with Mamie in her sunroom, and there were grandchildren to indulge. The Eisenhowers also began spending the winters at Eldorado Country Club in California's high chaparral, doing much the same things.

Being Ike, writing was an obvious recourse. But while *Crusade in Europe* had been dictated in a blazing three months, Eisenhower now struggled for three years with his White House memoirs, and the two-volume product was not nearly so crisp, bordering on turgid, actually, robbed of energy and coherence by security considerations. But later he bounced back and produced *At Ease: Stories I Tell to Friends*, a look back much more in the spirit of MacArthur's *Reminiscences*, and accordingly more readable.

Unfortunately, Ike's good times were punctuated by heart attacks, one in November 1965 he recovered well enough from to play golf again; but a coronary in April 1968 landed him in his last stop, Walter Reed, ward 8, VIP suite. Here he suffered a third the day after giving a televised speech to the GOP convention in Miami about to nominate Richard Nixon for president. It was from the same sickbed that he watched Nixon win the election and the White House, and then his daughter Julie marry Ike's grandson David, tying these two now presidential families together with bonds of matrimony. For Ike, there was no escape from Nixon.

And nothing more to do but wait for the end in the suite Mamie had decorated in soft pinks and greens. On March 27, he told his son he wanted to be taken off life support: "I've had enough, John. Tell them to let me go." He then added, "Now Mamie, don't forget that I have always loved you." The next morning he summoned John and grandson David and had them lower the shade and stand at attention while he uttered his last words: "I want to go; God take me," then lapsed into unconsciousness. He died just past noon, March 28, 1969.

The next day his body was placed in the Capitol Rotunda, in a standard-issue Army coffin, clad in his Ike jacket uniform, devoid of decorations, only his five stars. Among the mourners was Charles de Gaulle, as promised, with him to the end.

The Washington ceremonies concluded, his flag-draped casket was placed on a funeral train for Abilene, where two days later Dwight David Eisenhower was laid to rest on the property of his boyhood home in a simple funeral for just family and close friends, next to the grave of little Icky. A decade later Mamie joined them there. But before she did, when asked by grandson David if she had ever truly known her husband, Mamie replied: "I'm not sure anyone did."

THE END HAD FINALLY COME FOR OUR FOUR ALL-STARS, THE GREATEST GROUP of soldiers to ever serve together in the United States Army, maybe any army, any time.

Of the four, George Patton was probably the most gifted pure warrior, although Douglas MacArthur would undoubtedly dispute this claim. Unlike the others, Patton truly loved war, believed he had been fighting them since the time of Caesar, and likely died happily thinking he would be fighting them long into the future. In between, Patton spent his life preparing to fight, and was ready when America needed him, almost the perfect combat general to wage all-out warfare—his mask of command sufficiently awe-inspiring and magnetic enough to induce armies of hundreds of thousands to self-organize around him and become his terrible swift sword. But like some other great actors, his signature role left him typecast and the future held no parts for him. George Patton and the Cold War's gridlock were antithetical, and should he have somehow drifted into politics, a logical enough alliance with Joe McCarthy is not something America needed. Better he exited the stage when he did.

Our remaining three, though no less significant militarily if the entire scope of industrialized warfare is considered, were far more broadly endowed with skills that translated well in the postwar environment, which was really the beginning of a transformed environment.

More than any other triple-play combination that comes to mind, Douglas MacArthur, George Marshall, and Dwight Eisenhower helped guide us safely across the uncharted and treacherous terrain that would mark the basepaths during the early stages of the Cold War—not to mention playing very significant roles in the restoration and reintegration of Japan and Western Europe, today with the United States the heart of the developed democratic world. Together they led Team America through the last of its glory days, but escaped Vietnam, the disaster that was the beginning of team spirit's end, its final act being the peaceful termination of the Cold War, a historically unprecedented event, which essentially finalized the regime of mutually assured destruction that has rendered defunct all-out warfare.

At this moment of pandemic and deeply divided politics, there is likely to be at least some nostalgia for a time of unity, one in which the metaphor Team America fit the national mood. Don't be fooled. Americans joined the team because they had to—in part because they were drafted, or otherwise channeled into defense industries—but really because they understood the alternative was something like national destruction.

That was the price you paid for losing in this league, one dedicated to perpetual gamesmanship at the most destructive levels possible, backed by entire nations, pulled from their peaceful pursuits to provide the cornucopia of weapons and the human cannon fodder necessary to keep the team going, through extra innings if necessary, to achieve victory rendered almost meaningless in the face of all the carnage. And there seemed no way off this industrialized Möbius strip, marching in cadence through successive world wars until life became truly dystopian. But then an explosion beyond all expectations blew us free and we were saved.

Sort of. Our mostly peaceful world is held together, or, more aptly perhaps, held apart by a strategic nuclear structure, the failure of which in a worst-case scenario would be utterly catastrophic, literally the end of civilization, most of it accomplished inside of a week. For the sake of sanity and defense budgets, practically all the necessary implements remain largely hidden from our purview and consideration. The chances of any of them being used remains very low. But they really do exist—Minuteman ICBMs buried in silos, Trident equivalents lined up on submarines quietly cruising the world's oceans, B-52s and stealth bombers on runways, along with their Russian, Chinese, British, French, Israeli, Pakistani, Indian, and North Korean counterparts—all presumably kept in tip-top world-destroying condition. Given this situation, in the same spirit God told Adam and Eve not to eat the apple, a final admonition in closing: "Don't ever press the button."

ACKNOWLEDGMENTS

Writing a book in the middle of a pandemic has been a disorienting and mostly electronic experience—gathering a burgeoning harvest of books from Amazon and searching the internet seemingly for endless hours. Just to make sure I stayed home, across town the University of Virginia library was in the midst of a major renovation with the collection packed in a warehouse. Nonetheless, they managed to get me most of what I needed, for which I am profoundly grateful. Special thanks also to my pal Gordon Latter, who undertook the role of test reader, providing me with chapter-by-chapter comments promptly and with his usual brand of sarcasm and sagacity. A number of other friends and colleagues also read the manuscript in whole or part, including David Kalergis, Hy Rothstein, Michael Freeman, and Sterling Deal. I am grateful to all of you. Thanks also to the editorial staff at HarperCollins, who waged war relentlessly against my exaggerations, mistakes, misspellings, and grammatical folly. Also to Gene Thorp, who composed and provided the maps needed to orient the reader to the geography of mayhem the text describes. And above all, to my wife, Benjie, who not only sat through all of this with me, but provided daily comments as I trudged along.

NOTES

Introduction

xv To do so, generals: John Keegan, *The Mask of Command* (New York: Viking Penguin, 1987).

xix "Best clerk I ever had": Both quotations cited in William Manchester, *American Caesar: Douglas MacArthur, 1880–1964* (New York: Little, Brown, 1978), 166.

xx Douglas MacArthur, when he was chief of staff: Arthur Herman, *Douglas MacArthur: American Warrior* (New York: Random House, 2016), 250–51.

xxiii True enough, the country: Richard Overy, *Why the Allies Won* (New York: W. W. Norton, 1995), 190.

Chapter 1: Dawn

1 Until then, we'll focus: Marshall and MacArthur did meet briefly in 1909 at Fort Leavenworth, where the former was an instructor and the latter was in charge of a company of soldiers based there. But because their jobs were different, and "even then" they apparently "rubbed each other the wrong way," the two remained at arm's length, with only superficial contact. William Manchester, *American Caesar: Douglas MacArthur, 1880–1964* (New York: Little, Brown, 1978), 71.

1 "Douglas Macarthur was born": Douglas MacArthur, *Reminiscences* (New York: McGraw-Hill, 1964), 14.

2 "My first recollection": Manchester, *American Caesar*, 39–40.

2 "It was here I learned": MacArthur, *Reminiscences*, 15.

3 "beyond doubt the most distinguished": Ibid., 16.

3 "that whirlpool of glitter": Ibid.

4 "was doing Conic Sections": Both quotations from D. Clayton James, *The Years of MacArthur*, vol. 1, *1880–1941* (Boston: Houghton Mifflin, 1970), 29.

4 More impressive still: Manchester, *American Caesar*, 45.

4 Promoted to lieutenant colonel: Arthur Herman, *Douglas MacArthur: American Warrior* (New York: Random House, 2016), 68.

5 "worked together for a year": Ibid., 47.

5 "Every day I trudged there": MacArthur, *Reminiscences*, 18.

5 "Doug you will win": Ibid.

6 "almost all that can be reported": Forrest C. Pogue, *George C. Marshall*, vol. 1, *Education of a General, 1880–1939* (New York: Viking Press, 1963), 17.

7 "that it could be remembered by Marie": Ibid., 31.

7 This became apparent when his father: Debi and Irwin Unger with Stanley Hirshson, *George Marshall: A Biography* (New York: HarperCollins, 2014), 9.

7 "made fun of me a great deal": Cited in Pogue, *Education of a General*, 20.

7 but in general Marshall remained: Unger and Unger, *George Marshall: A Biography*, 9.

7 "we had to economize very bitterly": Cited in Pogue, *Education of a General*, 35.

8 "Flicker": David L. Roll, *George Marshall: Defender of the Republic* (New York: Caliper, 2019), 2.

8 "My father was so keen": Cited in Pogue, *Education of a General*, 5.

8 If this wasn't insulting enough: Unger and Unger, *George Marshall: A Biography*, 3.

9   "made more impression on me": Forrest C. Pogue, "George C. Marshall Interviews and Interview Notes, 1956–57," George C. Marshall Foundation, Lexington, VA, March 6, 1957, tape 4, 116.
9   This was no small thing": Ibid.
10   "remember thinking that I must be": Cited in Fred Ayer, *Before the Colors Fade* (London: Cassell, 1964), 18.
10   Soon enough the private caught on: Martin Blumenson, *Patton: The Man behind the Legend, 1885–1945* (New York: William Morrow, 1985), 16.
12   "While ages of gentility": Ibid., 43.
13   Jacob, David's father: Michael Korda, *Ike: An American Hero* (New York: HarperCollins, 2007), 62.
13   Neither Ida nor David: Jean Edward Smith, *Eisenhower: In War and Peace* (New York: Random House, 2013), 6.
14   Roughly two years passed: Dwight D. Eisenhower, *At Ease: Stories I Tell to Friends* (Garden City, NY: Doubleday, 1967), 72, 303.
14   Throughout their travails: Still another son, Paul, would die in infancy of diphtheria.
14   Energized by love: Eisenhower, *At Ease*, 72–3.
15   At the end of his days: Ibid., 83.
15   "a remarkably solid little library": Korda, *Ike*, 77.
16   "the battles of Marathon": Eisenhower, *At Ease*, 40.
16   Yet David doggedly: Ibid., 303.
17   While most instruction was based: Ibid., 100.
17   Yet his real academic strength: Ibid., 42.
17   "with one arm cut off": Ibid., 102.
18   But it was Ida who: Korda, *Ike*, 87.
19   "I'm just folks": Cited in John Gunther, *Eisenhower: The Man and the Symbol* (New York: Harper and Row, 1952).

Chapter 2: Long Gray Lines

21   Tall, diffident George began: Debi Unger and Irwin Unger with Stanley Hirshson, *George Marshall: A Biography* (New York: HarperCollins, 2014), 14–15.
21   Yet Marshall simply went: David L. Roll, *George Marshall: Defender of the Republic* (New York: Caliper, 2019), 68.
21   "I think I was more philosophical": Forrest C. Pogue, *George C. Marshall*, vol. 1, *Education of a General, 1880–1939* (New York: Viking Press, 1963), 44.
22   "What I learned at VMI": Ibid., 46.
23   "as a kind of romantic": See ibid., 55; Unger and Unger, *George Marshall: A Biography*, 16–17; Roll, *Defender of the Republic*, 70–71.
23   "I was very much in love": Forrest C. Pogue, "George C. Marshall Interviews and Interview Notes, 1956–57," George C. Marshall Foundation, Lexington, VA, March 6, 1957, tape 4, 91.
24   In order to cope: Pogue, *Education of a General*, 62–63.
24   And now he was opposed: Rose Page Wilson, *General Marshall Remembered* (Englewood Cliffs, NJ: Prentice-Hall, 1968), 175.
24   "Come on, George": As reported by Pogue, *Education of a General*, 68.
25   The place fit Georgie: Martin Blumenson, *Patton: The Man behind the Legend, 1885–1945* (New York: William Morrow, 1985), 43.
25   "I am treated almost": Ibid., 46.
26   "I HAVE NOMINATED YOUR SON": Cited in Carlo D'Este, *Patton: A Genius for War* (New York: HarperCollins, 1995), 67.
26   "Please thank the California Club": George S. Patton to George S. Patton II, March 18, 1904, quoted ibid., 68.
26   "nice fellows but very few": George S. Patton to Ruth Wilson Patton, June 21, 1904, quoted ibid., 72.
26   "I belong to a different class": George S. Patton to George S. Patton II, July 3, 1904, quoted ibid., 73.
27   "I am absolutely worthless": George S. Patton to George S. Patton II, October 5, 1904, quoted ibid.

28   "If I never amount to anything": George S. Patton to Beatrice Ayer, March 30, 1905, quoted ibid., 83.

28   Considering his penchant: quoted ibid., 95.

29   "Don't argue with a man": Blumenson, *Man behind the Legend*, 53–54.

29   He even clawed: D'Este, *A Genius for War*, 107.

29   If this wasn't enough to mark: Arthur Herman, *Douglas MacArthur, American Warrior* (New York: Random House, 2016), 38.

30   He was still jerking: D. Clayton James, *The Years of MacArthur*, vol. 1, *1880–1941* (Boston: Houghton Mifflin, 1970), 70.

30   "By your plucky work": Frazier Hunt, *The Untold Story of Douglas MacArthur* (New York: Devon-Adair, 1954), 25.

30   Suddenly it was a national scandal: Herman, *American Warrior*, 43.

30   "I would do anything in the way": Douglas MacArthur, *Reminiscences* (New York: McGraw-Hill, 1964), 26.

31   But he had done something more: William Manchester, *American Caesar: Douglas MacArthur, 1880–1964* (New York: Little, Brown, 1978), 52–53.

31   When all the numbers: MacArthur, *Reminiscences*, 54.

31   "the finest drill master": Ibid.

32   "Without doubt, the handsomest": Robert E. Wood, "An Upperclassman's View," *Assembly* 23, no. 1 (Spring 1964): 4.

32   "He had style": Manchester, *American Caesar*, 51.

32   "a slightly faraway": Herman, *American Warrior*, 48.

32   One rumor had him: Manchester, *American Caesar*, 60.

33   "Where else could you get": Dwight D. Eisenhower, *At Ease: Stories I Tell to Friends* (Garden City, NY: Doubleday, 1967), 4.

33   "From here on it would": Ibid.

33   "I was, in matters of discipline": Ibid., 10.

34   "I'm never going to crawl": Susan Eisenhower, *Mrs. Ike: Memories and Reflections on the Life of Mamie Eisenhower* (New York: Farrar, Straus and Giroux, 1996), 64.

34   His days in the woods: Michael Korda, *Ike: An American Hero* (New York: HarperCollins, 2007), 102.

34   Ike responded like Ike: Jean Edward Smith, *Eisenhower: In War and Peace* (New York: Random House, 2012), 23.

35   "I couldn't get up, so they took": Dwight D. Eisenhower, *In Review: Pictures I've Kept: A Concise Pictorial "Autobiography"* (Garden City, NY: Doubleday, 1969), 14.

35   "We saw in Eisenhower": Lieutenant Colonel Morton F. Smith, commandant of cadets, "1915 US Military Academy Efficiency Report, Dwight David Eisenhower," personnel file, Eisenhower Library, Abilene, KS.

35   And as befit such: Jean Edward Smith, *Eisenhower*, 25.

35   "I said that this was all right": Eisenhower, *At Ease*, 24.

36   "He brought the interview": Ibid.

36   "Mr. Eisenhower, if you will not": Ibid.

36   "From the first day": Ibid., 25–26.

Chapter 3: First Base

37   "The Philippines charmed": Douglas MacArthur, *Reminiscences* (New York: McGraw-Hill, 1964), 29.

38   "Begging thu Loo'tenant's": Cited ibid.

38   "First, I'd round": Cited in John Hersey, *Men on Bataan* (New York: Knopf, 1942), 76.

38   Meanwhile, using his father's contacts: Arthur Herman, *Douglas MacArthur: American Warrior* (New York: Random House, 2016), 59.

38   He was ordered to proceed: Ibid., 60.

39   "It was crystal clear": MacArthur, *Reminiscences*, 31–32.

39   "I heard the bullets": George Washington to John Washington, May 31, 1754, George Washington Papers; MacArthur, quoted in William Manchester, *American Caesar: Douglas MacArthur, 1880–1964* (New York: Little, Brown, 1978), 65.

40   "He [TR] was greatly": MacArthur, *Reminiscences*, 33.

40 "I am sorry to report": D. Clayton James, *The Years of MacArthur*, vol. 1, *1880–1941* (Boston: Houghton Mifflin, 1970), 95.

40 "wished to be undisturbed": Cited in Manchester, *American Caesar*, 69.

41 "Lieutenant MacArthur . . . did not": Cited in Herman, *American Warrior*, 71.

41 "I could not have been happier": MacArthur, *Reminiscences*, 34.

41 "A most excellent": Ibid., 35.

43 "How soon can you leave?": Ibid., 40.

43 "just what we needed": Ibid., 42.

44 "It is a mystery": James, *The Years of MacArthur*, 1:120.

44 "Captain MacArthur was not a member": Herman, *American Warrior*, 87.

45 "the sin of 1898": Cited in Martin Blumenson, *Patton: The Man behind the Legend, 1885–1945* (New York: William Morrow, 1985), 62.

45 "I only feel it": Ibid., 84.

45 "I would not sell": Ibid.

46 That's exactly what he did: Ibid., 61.

46 "Attack . . . push forward": Ibid., 69.

47 "nearer God than": Ibid., 70.

48 "Possibly suffering from nerves": Carlo D'Este, *Patton: A Genius for War* (New York: HarperCollins, 1995), 132–33.

48 "Once I came to": From Patton manuscript, "My Father As I Knew Him," quoted ibid., 134.

49 That would be George: Ibid., 140.

49 Actually, this had been: Robert L. O'Connell, *Of Arms and Men* (New York: Oxford University Press, 1989), 200.

49 But combatants on: Robert L. O'Connell, *Ride of the Second Horseman: The Birth and Death of War* (New York: Oxford University Press, 1995), 71–83.

50 He took and passed: Blumenson, *Man behind the Legend*, 78.

51 "I really think that Villa": D'Este, *A Genius for War*, 171.

52 "We have a bandit": George S. Patton Jr., *War as I Knew It* (Boston: Houghton Mifflin, 1947), xvii.

53 By the spring: Blumenson, *Man behind the Legend*, 726.

53 "the hardest service": Forrest C. Pogue, *George C. Marshall*, vol. 1, *Education of a General, 1880–1939* (New York: Viking Press, 1963), 86.

53 "the best one received": Marshall to General Scott Shipp, March 3, 1906, in Larry I. Bland and Sharon Ritenour Stevens, eds., *The Papers of George Catlett Marshall*, vol. 1, *"The Soldierly Spirit": December 1880–June 1939* (Baltimore: Johns Hopkins University Press, 1981), 34.

54 In an effort to get back: Debi Unger and Irwin Unger with Stanley Hirshson, *George Marshall: A Biography* (New York: HarperCollins, 2014), 17.

54 "I taught myself to study": Forrest C. Pogue, "George C. Marshall Interviews and Interview Notes, 1956–57," George C. Marshall Foundation, Lexington, VA, 14 April 1957, tape 5, 156.

55 Even better, he: Pogue, *Education of a General*, 102.

56 "there are not five": William Frye, *Marshall: Citizen Soldier* (Indianapolis: Bobbs-Merrill, 1947), 110.

56 "acute dilation": Unger and Unger, *George Marshall: A Biography*, 32.

56 When he emerged: Pogue, *Education of a General*, 124.

56 Early in the trip: Bland and Stevens, *Papers of George Catlett Marshall*, 1:81–84.

56 Faced with the deadest: Marshall to Bruce Magruder, ibid., August 7, 1939, 2:31–32.

58 "Colonel Moriarity . . . was happy": Dwight D. Eisenhower, *At Ease: Stories I Tell to Friends* (Garden City, NY: Doubleday, 1967), 119.

59 He indicated the conversation: Ibid., 122.

59 One Doud in: Ibid., 113.

59 "was just about the handsomest": Vivian Cadden, "Mamie and Ike Talk about Fifty Years of Marriage," *McCall's*, September 1966; Jean Edward Smith, *Eisenhower: In War and Peace* (New York: Random House, 2012), 31.

60 "If [Eisenhower] were so irresponsible": Eisenhower, *At Ease*, 117–18.

60 "a decision that brought": Ibid., 118.

60 "to perform every": Ibid.

61 "on a wartime footing": Michael Korda, *Ike: An American Hero* (New York: HarperCollins, 2007), 118.

61 "All right, you may": Eisenhower, *At Ease*, 122.

Chapter 4: Total War

63 Here on the western front: J. F. C. Fuller, *The Conduct of War: 1789–1961* (Boston: DaCapo Press, 1992), 160.

65 "MacArthur, in his *Reminiscences*": Douglas MacArthur, *Reminiscences* (New York: McGraw-Hill, 1964), 47.

65 "forces of the United States": Frederick Palmer, *Newton D. Baker: America at War*, 2 vols. (New York: Dodd, Mead, 1931) 1:170–71.

65 "I shall give you": Carlo D'Este, *Patton: A Genius for War* (New York: HarperCollins, 1995), 189.

66 He tried, but narrowly: George C. Marshall, *Memoirs of My Services in the World War, 1917–1918* (Boston: Houghton Mifflin, 1976), 3.

66 "They were such a dreadful-looking": David L. Roll, *George Marshall: Defender of the Republic* (New York: Caliper, 2019), 16.

66 Desperately needing help: Forrest C. Pogue, *George C. Marshall*, vol. 1, *Education of a General, 1880–1939* (New York: Viking Press, 1963), 143.

67 "We were sure that": Dwight D. Eisenhower, *At Ease: Stories I Tell to Friends* (Garden City, NY: Doubleday, 1967), 132–33.

67 "Our new Captain": Lieutenant Edward C. Thayer to Mother, January 1918, Correspondence Related to Leavenworth Field Training Mission, Presidential Papers, Eisenhower Presidential Library, Abilene, KS.

67 "Too much depended": Eisenhower, *At Ease*, 137.

68 "My mood was black": Ibid.

68 In actual fact: Arthur Herman, *Douglas MacArthur: American Warrior* (New York: Random House, 2016), 95.

68 "I agree with you": MacArthur, *Reminiscences*, 45.

69 "I disclosed my puzzle": Henry J. Reilly, *Americans All: The Rainbow at War; The Official History of the 42nd Rainbow Division in the World War* (Columbus, OH: F. C. Heer, 1936), 26.

69 "Fine, that will stretch": Ibid.

69 "No, the Infantry": MacArthur, *Reminiscences*, 45.

69 While forty-two states: Herman, *American Warrior*, 99.

70 And leading the mad: Ibid., 100.

70 Things apparently hadn't: Ibid., 101.

71 If the Americans basically: D'Este, *A Genius for War*, 189.

71 "a rat chewing": Ibid., 198.

71 "George is eager": John J. Pershing to George S. Patton II, August 17, 1917, quoted ibid., 198.

72 "the only American": George S. Patton to John J. Pershing, October 3, 1917, George S. Patton Papers, Library of Congress; Martin Blumenson, *Patton: The Man behind the Legend, 1885–1945* (New York: William Morrow, 1985), 96.

73 "if resistance is broken": George S. Patton, "Light Tanks," December 12, 1917, Box 9, Patton Papers, Library of Congress.

73 "I ran this show," D'Este, *A Genius for War*, 225.

74 "he was very severe": Pogue, *Education of a General*, 152.

75 "Yes, General, but": Ibid., 153.

75 "My action was": MacArthur, *Reminiscences*, 53.

76 "Flares soared and": Ibid., 54.

77 "went over the top": Ibid.

77 Also, his reputation: Herman, *American Warrior*, 112; William Manchester, *American Caesar: Douglas MacArthur, 1880–1964* (New York: Little, Brown, 1978), 89.

77 "the most brilliant young officer": Jules Archer, *Front-Line General: Douglas MacArthur, America's Most Controversial Hero* (New York: Messner, 1963), 53–54.

78 "For eighty two": MacArthur, *Reminiscences*, 56.

79 "a most brilliant officer": Herman, *American Warrior*, 116.

79 "This division is a disgrace": Manchester, *American Caesar*, 92.

80 In Marshall's eyes: Marshall, *Memoirs of My Services*, 90.

80 "The success of this phase": Ibid., 95.

80 "continuous bombardment by 210-mm": Ibid., 96.

80 "were not justified by the importance": Roll, *Defender of the Republic*, 33.

81 "It may have been the vision": MacArthur, *Reminiscences*, 58.

82 "The dead were so": Ibid., 60.

83   "Well, I'll be damned!": Ibid., 61.
83   The Germans were gone: Herman, *American Warrior*, 128; D. Clayton James, *The Years of MacArthur*, vol. 1, *1880–1941* (Boston: Houghton Mifflin, 1970), 191.
83   "it was officially learned": Manchester, *American Caesar*, 99.
84   "are devoted to him": James, *Years of MacArthur*, 1:195.
84   The Germans built: Blumenson, *Patton*, 108.
84   He had to organize: Roll, *Defender of the Republic*, 35–36.
84   "a fine piece of work": Marshall, *Memoirs of My Services*, 138–39.
85   "If your gun": D'Este, *A Genius for War*, 233.
86   "I joined him and the creeping": George S. Patton to George S. Patton II, September 20, 1918, quoted ibid., 235.
86   "Don't worry, Colonel": Quoted ibid., 235; Herman, *American Warrior*, 230.
87   As for the battle: Pogue, *Education of a General*, 174.
87   "Here was an unparalleled": MacArthur, *Reminiscences*, 63–64.
87   "Get out! And stay out!": Herman, *American Warrior*, 131.
88   "as though every gun": David McCullough, *Truman* (New York: Simon & Schuster, 1993), 129.
88   "Some put on gas": George S. Patton to Beatrice Patton, September 28, 1918, quoted in D'Este, *A Genius for War*, 256.
88   "I think I killed": D'Este, *A Genius for War*, 257.
88   "there were other faces": Ruth Ellen (Patton) Totten, "Ma: A Button Box Biography," quoted in, D'Este, *A Genius for War*, 257–58.
89   "came out just": George Patton to Beatrice Patton, September 28, 1918, D'Este, *A Genius for War*, 259.
89   "one of my tanks": George Patton to Nita Patton, October 26, 1918, quoted in D'Este, *A Genius for War*, 260.
89   "the Dr. says": George Patton to Beatrice Patton, October 24, 1918, quoted in D'Este, *A Genius for War*, 262.
90   "All right, General, we'll": MacArthur, *Reminiscences*, 66.
91   "It was God": William Ganoe, *MacArthur Close-Up: Much Then and Some Now* (New York: Vantage, 1962), 143–44.
91   "The two battalions": MacArthur, *Reminiscences*, 67.
91   For this the cost: Military Records, Meuse-Argonne Offensive, US National Archives website.
91   "General Pershing desires": Pogue, *Education of a General*, 186.
92   "Boundaries will not": Marshall, *Memoirs of My Services*, 189–90.
93   "The following evening": Eisenhower, *At Ease*, 149.
93   "Get me out of here": Ibid., 150.
94   "By God, from": Evan Thomas, *Ike's Bluff: President Eisenhower's Secret Battle to Save the World* (Boston: Back Bay Books, 2012), 5.

Chapter 5: Halftime

96   "How would you": Forrest C. Pogue, *George C. Marshall*, vol. 1, *Education of a General, 1880–1939* (New York: Viking Press, 1963), 196.
97   "the doctors pulled": Arthur Herman, *George Marshall: American Warrior* (New York: Random House, 2016), 153.
97   "One little urchin": Douglas MacArthur to Weller, May 13, 1919, in Douglas MacArthur, *Reminiscences* (New York: McGraw-Hill, 1964), 72.
97   "might injure the dance floor": Faubion Bowers, "The Late General MacArthur, Warts and All," *Esquire*, January 1967.
98   "You will not find our Boy": Mary Hardy MacArthur to John L. Pershing, in D. Clayton James, *The Years of MacArthur*, vol. 1, *1880–1941* (Boston: Houghton Mifflin, 1970), 171.
99   MacArthur later claimed: Edward Coffman, *The Hilt of the Sword: The Career of Peyton C. March* (Madison: University of Wisconsin Press, 1966), 186.
99   "I'm not an educator": Ibid.
101 "If he hadn't proposed": *New York Times*, February 10, 1922.
101 "all poppycock": Ibid.
102 "But for Rose": Rose Page Wilson, *General Marshall Remembered* (Upper Saddle River, NJ: Prentice-Hall, 1968), 12, 22–24.

102 He also learned about Congress: Debi Unger and Irwin Unger with Stanley Hirshson, *George Marshall: A Biography* (New York: HarperCollins, 2014), 49.

103 So hasty was it: Pogue, *Education of a General*, 224–25.

103 "Part of an audience": Dwight D. Eisenhower, *At Ease: Stories I Tell to Friends* (Garden City, NY: Doubleday, 1967), 165.

104 "The trip had been": Ibid., 166.

104 "a fellow named Patton": Ibid., 169.

105 "the flying end": Ibid., 171.

106 "anything incompatible with": Ibid., 195.

107 "The marriage was clearly": Irene and Lester David, *Ike and Mamie: The Story of a General and His Lady* (New York: Putnam, 1981), 90.

108 "Wouldn't you like": Eisenhower, *At Ease*, 185.

109 "After Johnny was": Virginia Conner, *What Father Forbad* (Philadelphia: Dorrance, 1951), 120–21.

109 "I was down to skin": Susan Eisenhower, *Mrs. Ike: Memories and Reflections on the Life of Mamie Eisenhower* (New York: Farrar, Straus and Giroux, 1996), 83.

109 "In the new war": Eisenhower, *At Ease*, 195.

110 "the same old tanks": Ibid., 198.

110 Having protested to: Ibid., 199.

110 "I was ready to fly": Ibid., 200.

110 "You may not know": Ibid.

111 But not frequent enough: Martin Blumenson, *Patton: The Man behind the Legend, 1885–1945* (New York: William Morrow, 1985), 123.

111 At a formal dinner: Ibid.

112 "Will you give me": Carlo D'Este, *Patton: A Genius for War* (New York: HarperCollins, 1995), 329.

113 So began what one: Herman, *American Warrior*, 178; MacArthur, *Reminiscences*, 84.

114 "Won't you be real": Mrs. Arthur MacArthur to John Pershing, n.d., Pershing Papers, Notebooks, Library of Congress;. quoted in William Manchester, *American Caesar: Douglas MacArthur, 1880–1964* (New York: Little, Brown, 1978), 134.

114 "excellent troops, completely": Herman, *American Warrior*, 183.

115 "I snaffled a": Marshall to John C. Hughes, January 2, 1925, quoted in Pogue, *Education of a General*, 234.

116 "How the Powers": Marshall to Pershing, December 29, 1926, Larry I. Bland and Sharon Ritenour Stevens, eds., *The Papers of George Catlett Marshall*, vol. 1, *"The Soldierly Spirit": December 1880–June 1939* (Baltimore: Johns Hopkins University Press, 1981), 294.

116 Meanwhile, Lily had: Lily Marshall to John C. Hughes, November 25, 1926, quoted in Pogue, *Education of a General*, 245.

117 "No one knows": Ibid., 246.

117 "It was as if": D'Este, *A Genius for War*, 336.

117 "Beatrice Ayer Patton": Martin Blumenson, *The Patton Papers: 1885–1940* (Boston: Houghton Mifflin, 1972), 2:6.

118 "too positive in his": D'Este, *A Genius for War*, 341.

118 "Invaluable in war": Blumenson, *Man behind the Legend*, 128.

118 "I have known him": Roger H. Nye, *The Patton Mind: The Professional Development of an Extraordinary Leader* (New York: Avery, 1993), 82.

119 "a model command post": Eisenhower, *At Ease*, 202.

119 "a battlefield guide": Michael Korda, *Ike: An American Hero* (New York: HarperCollins, 2007), 176.

120 Then seven-year-old: John S. D. Eisenhower, *Strictly Personal* (Garden City, NY: Doubleday, 1974), 304.

120 "I think they're interesting": Eisenhower, *At Ease*, 208.

120 "He thought General Pershing": Ibid., 209.

121 "Hell, he's not": John S. D. Eisenhower to Jean Edward Smith, March 10, 2008, quoted in Smith, *Eisenhower: In War and Peace* (New York: Random House, 2012), 91.

121 "one of the most distasteful": MacArthur, *Reminiscences*, 85.

122 To prove his point: Robert L. O'Connell, *Sacred Vessels: The Cult of the Battleship and the Rise of the US Navy* (New York: Oxford University Press, 1993), ch. 9.

122 "the direct result": Douglas Waller, *A Question of Loyalty: Gen. Billy Mitchell and the Court-Martial That Gripped the Nation* (New York: HarperCollins, 2004), 21–22.

122 "MacArthur looks like": Manchester, *American Caesar*, 136.
122 "his features as": Waller, *A Question of Loyalty*, 323–25.
123 "The General is": Herman, *American Warrior*, 191.
124 "The outlook was not": MacArthur, *Reminiscences*, 86.
124 "Athletes are among": Ibid.
124 "To portray adequately": Herman, *American Warrior*, 194.
124 "No assignment could": MacArthur, *Reminiscences*, 87.
124 But his third: Herman, *American Warrior*, 191.
124 "Meanwhile, War Plan": Smith, *In War and Peace*, 127.
126 "I therefore searched the Army": Herbert Hoover, *Memoirs*, vol. 2, *The Cabinet and the Presidency, 1920–1933* (New York: Macmillan, 1952), 339.
126 "He's one of my boys": *New York Times*, August 7, 1930.
126 His aim was: Bland and Stevens, "The Soldierly Spirit," 320.
128 "took over the place": Pogue, *Education of a General*, 273.
129 "a work of exceptional": D'Este, *A Genius for War*, 349.
129 "Why Francis, you": Ibid., 344.
131 "On any subject": Eisenhower, *At Ease*, 214.
133 "organized and promoted": Paul Johnson, *Modern Times: The World from the Twenties to the Eighties* (New York: Harper and Row, 1983), 24.
133 He also arranged: James and Jean Vivian, "The Bonus March of 1932: The Role of General George Van Horn Moseley," *Wisconsin Magazine of History* 51, no. 1 (Autumn 1967); J. Edgar Hoover to Colonel William H. Wilson, Military Intelligence Division, General Staff, July 11, 1932, quoted in Smith, *In War and Peace*, 108.
133 "The incipient revolution": James, *Years of MacArthur*, 1:399.
134 "By this time": Eisenhower, *At Ease*, 216.
134 Except for the fashion: Smith, *In War and Peace*, 111–12.
134 "Shame! Shame!": Gene Smith, *The Shattered Dream: Herbert Hoover and the Great Depression* (New York: William Morrow, 1970), 161.
134 "In neither instance": Eisenhower, *At Ease*, 217.
135 "The veterans . . . were": Ibid.
135 "That mob down": James, *Years of MacArthur*, 1:404.
135 "When Major Patton saw": Lucian K. Truscott, *Twilight of the US Cavalry: Life in the Old Army, 1917–1942* (Lawrence: University of Kansas Press, 1989), 129.
136 When queried who: Manchester, *American Caesar*, 152.

Chapter 6: Spring Training

139 "When we lost the next": Douglas MacArthur, *Reminiscences* (New York: McGraw-Hill, 1964), 101.
139 "You must not talk": Ibid.
139 "I told him he had": Ibid.
140 labeled "legalized murder": "Hits 'Legalized Murder' Rickenbacker's Comment on Army Fliers' Deaths," *New York Times*, February 18, 1934.
140 "When are those airmail killings": Arthur Herman, *Douglas MacArthur: American Warrior* (New York: Random House, 2016), 236–37.
140 "I knew nothing": Ibid., 237.
141 "Ike got so": Frazier Hunt, *The Untold Story of Douglas MacArthur* (New York: Manor Books, 1977), 171.
142 "he didn't want his mother": William Manchester, *American Caesar: Douglas MacArthur, 1880–1964* (New York: Little, Brown, 1978), 156.
143 "I have sent a letter": Press Conference, December 12, 1934, quoted in D. Clayton James, *The Years of MacArthur*, vol. 1, *1880–1941* (Boston: Houghton Mifflin, 1970), 446.
143 "This year definitely": Herman, *American Warrior*, 243.
144 "Douglas, if war": James, *Years of MacArthur*, 1:491–92.
144 "Although I was not": Dwight D. Eisenhower, *At Ease: Stories I Tell to Friends* (Garden City, NY: Doubleday, 1967), 219.
145 "We will lose no": Herman, *American Warrior*, 230–31.
145 "The greatest social": Forrest C. Pogue, *George C. Marshall*, vol. 1, *Education of a General, 1880–1939* (New York: Viking Press, 1963), 280.

145 "I think he regarded": Interview with Don Mace, November 14, 1958, quoted in Pogue, *Education of a General*, 308.

146 "General Pershing asks": Franklin D. Roosevelt to George Dern, May 24, 1935, quoted in Pogue, *Education of a General*, 295.

147 "did the major portion": Pogue, *Education of a General*, 323.

147 "He finally came": Ibid.

149 "I told him I wanted": Ibid., 330.

149 "I feel deeply honored": Ed Cray, *General of the Army: George C. Marshall, Soldier and Statesman* (New York: Cooper Square Press: 2000), 139.

149 "go and drown without": Carlo D'Este, *Patton: A Genius for War* (New York: HarperCollins, 1995), 358.

149 "We can learn": Ibid.

149 "Where the hell": Ibid.

150 "Whatever had happened": Ibid., 359.

150 "Your father needs": Ibid.

151 "arrest and intern": Martin Blumenson, *Patton: The Man behind the Legend* (New York: William Morrow, 1985), 136.

151 "Goddammit, Walter": Ibid.

152 "I guess they're not": Fred Ayer, *Before the Colors Fade: A Portrait of a Soldier, George S. Patton, Jr.* (Boston: Houghton Mifflin, 1964), 104.

153 "My, I certainly enjoyed": D'Este, *A Genius for War*, 364.

153 "He was one of Pancho": Ibid., 367.

153 "just consummated": George S. Patton to Beatrice Patton, July 27, 1939, Letters File, Patton Papers, US Military Academy.

154 "George, you mustn't": Pogue, "George C. Marshall Interviews," 510.

154 "Well, he was the epitome": Ibid., February 15, 1957.

155 "Her departure from": Eisenhower, *At Ease*, 224.

156 "I'm going to New York": Herman, *American Warrior*, 273.

156 "I didn't know you had it": Ibid., 282.

157 "Why in hell do you want a banana": Peter Lyon, *Eisenhower: Portrait of the Hero* (Boston: Little, Brown, 1974), 78.

158 "Every scrap of": Jean Edward Smith, *Eisenhower: In War and Peace* (New York: Random House, 2012), 137.

159 "I gather I have grounds": Susan Eisenhower, *Mrs. Ike: Memories and Reflections on the Life of Mamie Eisenhower* (New York: Farrar, Straus and Giroux, 1996), 143.

159 "terse but jocular,": Michael Korda, *Ike: An American Hero* (New York: HarperCollins, 2007), 218.

160 "We've been financially": Mamie Doud Eisenhower to the Douds, February 8, 1938, quoted in Smith, *In War and Peace*, 136.

160 "So I'd like to go back": Herman, *American Warrior*, 279.

162 "It is almost incomprehensible": Eisenhower, Philippine Diary, November 10, 1938, Eisenhower Presidential library, Abilene, KS.

162 "I feel like a boy": Dwight Eisenhower to Mark Clark, September 23, 1939, quoted in Smith, *In War and Peace*, 149.

162 "I got out clean": Ibid., 149.

163 "Defeat": Manchester, *American Caesar*, 168.

165 "I very pointedly": Pogue, "George C. Marshall Interviews," 302.

165 "Marshall would sell": Harry Woodring to J. A. Harris, June 23, 1954, inserted into *Congressional Record* by Senator Joseph McCarthy, August 2, 1954, quoted in David L. Roll, *George Marshall: Defender of the Republic* (New York: Caliper, 2019), 140.

166 "We do not have": Larry I. Bland and Sharon R. Stevens, eds., *The Papers of George Catlett Marshall*, vol. 2, *"We Cannot Delay," July 1, 1939–December 6, 1941* (Baltimore: Johns Hopkins University Press, 1986), 48.

166 "It is only through": Radio Address on Selective Service, September 16, 1940, Electronic Article, George Marshall Foundation, Lexington, VA.

167 "the morale of the hostile": George C. Marshall to Franklin D. Roosevelt, November 9, 1942, quoted in Forrest C. Pogue, *George C. Marshall*, vol. 3, *Organizer of Victory, 1943–1945* (New York: Viking Press, 1973), 12–13.

168   "It is the policy": Bland and Stevens, *"We Cannot Delay,"* 336–37.
170   He even designed: D'Este, *A Genius for War*, 387.
171   "Have you learned": Eisenhower, *At Ease*, 236.
171   "Of course, in": Ibid., 240.
171   Then his study partner: Ibid., 239.
172   "combat college for": Bland and Stevens, *"We Cannot Delay,"* 94; Smith, *In War and Peace*, 167.
173   "I'll get you for": Merle Miller, *Ike the Soldier* (New York: Putnam, 1987), 328–29.
174   "safe and unassailable": Sean Usher, *Letters of Note: An Eclectic Collection of Correspondence Deserving of a Wider Audience* (San Francisco: Chronicle Books, 2014), 151.
176   "First Bill Knudsen": Norman Beasley, *Knudsen: A Biography* (New York: McGraw-Hill, 1947), 228.
177   "Knudsen? I want": Ibid., 234.
177   "And still you": Arthur Herman, *Freedom's Forge: How American Business Produced Victory in World War II* (New York: Random House, 2012), 68.
177   "I don't expect any paycheck": Beasley, *Knudsen*, 237.
178   "Our greatest need": Herman, *Freedom's Forge*, 79.
178   "The first thing": Ibid., 80.
179   "He seemed to carry": Jesse H. Jones and Edward Angly, *Fifty Billion Dollars: My Thirteen Years with the RFC, 1932–1945* (New York: Macmillan, 1951), 271–75.
179   "I told Knudsen": Stimson, Diary, December 21, 1940, *The Politics of Integrity: The Diaries of Henry L. Stimson, 1931–1945* (NY: McGraw-Hill, 1976), 51.
180   With good reason: Herman, *Freedom's Forge*, 154.
180   "Alvan, I'm leaving": Beasley, *Knudsen*, 265.
180   "We want more": *Time*, January 19, 1942, 10.
180   "Look here Judge": Beasley, *Knudsen*, 341.
182   "We must fight": George S. Patton Jr., *War as I Knew It* (Boston: Houghton Mifflin, 1947), 366.
182   "credit for the original": Lt. E. P. Hogan, US Quartermaster's Corps, Herman, *Freedom's Forge*, 123.
183   As did 2,000: Figures derived from Mark Harrison, "Resource Mobilization for World War II: The USA, UK, USSR, and Germany, 1938–1945," *Economic History Review* 41, no. 2 (May 1988): 184, https://doi.org/10.2307/2596054.
183   "The Americans, as": Patton, *War as I Knew It*, 366.

Chapter 7: Devil Season

188   "Then I said to General Groves": David Lilienthal, *The Journals of David Lilienthal*, vol. 6, *Creativity and Conflict, 1964–1967* (New York: Harper and Row, 1976), 200.
188   "interested in the evolution": Vannevar Bush, *Pieces of the Action* (New York: William Morrow, 1970), 306.
188   "I would spend": Forrest C. Pogue, *George C. Marshall*, vol. 4, *Statesman, 1945–1959* (New York: Viking, 1987), 11.
189   Ultimately, the Manhattan: Frank A. Settle Jr., *General George C. Marshall and the Atomic Bomb* (Santa Barbara: Praeger, 2016), 82; Stephen I. Schwartz, *Atomic Audit: The Cost and Consequences of U.S. Nuclear Weapons since 1940* (Washington, DC: Brookings Institution Press, 1999), 58; Cameron Reed, "Kilowatts to Kilotons: Wartime Electricity Use at Oak Ridge," American Physical Society, forum on the History of Physics website.
190   "With its occupation": William Manchester, *American Caesar: Douglas MacArthur, 1880–1964* (New York: Little, Brown, 1978).
190   "Is that you Ike?": Dwight D. Eisenhower, *Crusade in Europe: A Personal Account of World War II* (Garden City: Doubleday, 1948), 14.
190   "Give me a few hours": Ibid., 18.
190   "General, it will be": Ibid., 21–22.
191   "I agree with you": Manchester, *American Caesar*, 242.
191   "Eisenhower . . . I must": Forrest C. Pogue, *George C. Marshall*, vol. 2, *Ordeal and Hope, 1939–1942* (New York: Viking Press, 1966), 239.
192   "might offer the best": Ibid., 247.
192   "I immediately discarded": Pogue, Ibid., 247–48.
192   "keep our flag": Manchester, *American Caesar*, 248.
192   "are both babies": David L. Roll, *George Marshall: Defender of the Republic* (New York: Caliper, 2019), 386.

192 "Bataan is made": Dwight D. Eisenhower, *The Eisenhower Diaries* (New York: Norton, 1981), 49.

193 "I make quite clear": Herman, *American Warrior*, 387–88.

194 "Bulkeley, I'm giving": Manchester, *American Caesar*, 263.

194 "I knew this train": Sid Huff with Joe Alex Morris, *My Fifteen Years with General MacArthur* (New York: Paperback Library, 1964), 72.

195 "fleeing general": Manchester, *American Caesar*, 275.

198 "Would General Sutherland": Charles Bohlen, *Witness to History, 1929–1960* (New York: W. W. Norton, 1973), 268–69.

199 "Bob take Buna": Robert Eichelberger, *Our Jungle Road to Tokyo* (New York: Gorget Books, 2017), 48; Manchester, *American Caesar*, 325.

200 "I really heard about": George C. Kenney, *The MacArthur I Know* (New York: Duell, Sloan and Pearce, 1951), 39; Manchester, *American Caesar*, 301; George C. Kenney, *Air War in the Pacific: The Journal of General George Kenney, Commander of the Fifth U.S. Air Force* (Lulu.com, 2018), Diary, July 28–29, 1942.

200 "I think we're going to get along all right": Manchester, *American Caesar*, 301.

200 "I want to find": George C. Kenney, *General Kenney Reports: A Personal History of the Pacific War* (Washington, DC: Office of Air Force History, 1987), 52–53.

200 "I don't care whether": Kenney, diary, August 3, 1942, quoted in Herman, *American Warrior*, 449.

204 "a new type of": MacArthur, *Reminiscences*, 166.

204 On the way to: Manchester, *American Caesar*, 280.

205 "Bombs came down": Herman, *American Warrior*, 466.

206 "The weather will": Kenney, diary, February 25, 1943.

206 "Are you calling": Thomas Griffith, *MacArthur's Airman: General George Kenney and the War in the Southwest Pacific* (Lawrence: University of Kansas Press, 1998), 98.

206 "The Japs are going": Kenney, diary, February 28, 1943.

207 Some of the Allied: Edward J. Drea, *MacArthur's ULTRA: Codebreaking and the War against Japan, 1942–45* (Lawrence: University of Kansas Press, 1991), 71.

207 "Air Power has": Kenney, *General Kenney Reports*, 295–96.

210 "not only wants": Manchester, *American Caesar*, 360.

Chapter 8: The Main Event

211 "I am convinced": Foreign Relations of the United States, Conference at Washington, 1941–42, 93.

212 "What the devil": Forrest C. Pogue, *George C. Marshall*, vol. 2, *Ordeal and Hope, 1939–1942* (New York: Viking Press, 1966), 280.

212 "I told him": Ibid.

213 "the most even tempered": Mark A. Stoler, *George C. Marshall, Soldier-Statesman of the American Century* (New York: Twayne Publications, 1989), 117.

214 "finished war materiel": J. R. M. Butler and M. A. Gwyer, *Grand Strategy* (London: Her Majesty's Stationery Office, 1964), 397.

214 "I got Pearl Harbor": Dwight D. Eisenhower, diary, February 16, 1942.

214 "as his [Marshall's]": Dwight D. Eisenhower, *Crusade in Europe: A Personal Account of World War II* (Garden City, NY: Doubleday, 1948), 31.

215 He did make the invasion: Ibid., 46.

215 "must not be permitted": Steven E. Ambrose, *The Supreme Commander: The War Years of Dwight D. Eisenhower* (Garden City, NY: Doubleday, 1970), 33

215 "This is it. I approve": Eisenhower, *Crusade in Europe*, 47.

216 "the United States should": Eisenhower to chief of staff, March 25, 1942, OPD 381, Bolero sec. 1., US National Archives.

216 "a good general": Alex Danchev and Daniel Todman, *War Diaries, 1938–1945: Field Marshal Lord Alanbrooke* (Berkeley: University of California Press, 2001), 248–49.

217 "The men who are going": Dwight D. Eisenhower, *At Ease: Stories I Tell to Friends* (Garden City, NY: Doubleday, 1967), 248.

217 "Take your case": Ibid., 249.

217 "General . . . I don't": Ibid.

217 "A tiny smile quirked": Ibid., 249.

218 "Stop it": Mark W. Clark, *Calculated Risk* (New York: Harper, 1950), 19.

218   "That son of": Kay Summersby Morgan, *Past Forgetting: My Love Affair with Dwight D. Eisenhower* (New York: Simon & Schuster, 1976), 28.

219   "Specifically, they seemed": Eisenhower, *Crusade in Europe*, 49.

219   "on a slow boat": Kay Summersby Morgan, *Eisenhower Was My Boss* (New York: Prentice-Hall, 1948), 16.

219   "I certainly do": Eisenhower, *Crusade in Europe*, 56.

219   "I'm going to command": Carlo D'Este, *Eisenhower: A Soldier's Life* (New York: Holt Paperbacks, 2015), 307.

219   "I particularly appreciate": Eisenhower to Patton, July 20, 1942, quoted in Jean Edward Smith, *Eisenhower: In War and Peace* (New York: Random House, 2012), 200.

220   "My sergeant has": Michael J. McKeogh and Richard Lockridge, *Sgt. Mickey and General Ike* (New York: G. P. Putnam's Sons, 1946), 29.

220   "I don't think my": Morgan, *Past Forgetting*, 43.

220   "Some of these": Ibid., 81.

221   Sex, it seems: Ibid., 194.

222   Meanwhile, like Kay: Ibid., 92.

223   "Give it 'em back": Robert L. O'Connell, *Of Arms and Men: A History of War, Weapons, and Aggression* (New York: Oxford University Press, 1989), 284.

224   Extrapolating to Germany: Ronald H. Bailey, *The Air War in Europe* (London: Time-Life UK, 1979), 54.

224   It all added up: Hans Sperling, "Die Luftkriegsverluste während des zweiten Weltkriegs in Deutschland," *Wirtschaft und Statistik*, October 1956, journal published by Statistisches Bundesamt Deutschland (German Statistical Office).

226   "certain to lead": Winston Churchill, *The Second World War*, vol. 4, *The Hinge of Fate* (Boston: Houghton Mifflin, 1950), 381–82.

226   "gone off the rails": Ibid.

226   "Can We afford": Ibid.

226   "Tobruk has surrendered": Ibid., 343.

226   "one of the heaviest": Ibid.

227   "Order him back": Carlo D'Este, *Patton: A Genius for War* (New York: HarperCollins, 1995), 415.

227   "That's the way": Ladislas Farago, *Patton: Ordeal and Triumph* (New York: Ivan Obolensky, 1963), 174–75.

227   "If the United States": Marshall and King to President, July 10, 1942, quoted in Pogue, *Ordeal and Hope*, 341.

227   "Exactly what Germany": Roll, *Defender of the Republic*, 236.

227   "Roosevelt C-in-C": Ibid., 235.

229   "If some little fart": George S. Patton Jr. to George S. Patton III, July 13, 1942, Blumenson, *Patton*, 144.

229   "convinced they could": Farago, *Ordeal and Triumph*, 96.

229   "Patton is indispensable": D'Este, *A Genius for War*, 421.

229   "Come in Skipper": Eric Larrabee, *Commander in Chief: Roosevelt, His Lieutenants, and Their War* (New York: Touchstone Books, 1988), 486.

230   "Sir, all I want": Farago, *Ordeal and Triumph*, 191–92.

230   "I can always pick": George Patton, diary, October 21, 1942, quoted in D'Este, *A Genius for War*, 424.

230   "When I think": D'Este, *A Genius for War*, 426.

232   "This is the first": Eisenhower, *Crusade in Europe*, 97.

232   "Giraud will be": Ibid., 101.

235   "we have now settled": Martin Blumenson, *Patton: The Man behind the Legend, 1885–1945* (New York: William Morrow, 1985), 172.

236   "Not a pretty sight": Michael Korda, *Ike: An American Hero* (New York: HarperCollins, 2007), 330.

237   "You are doing an excellent": Marshall to Eisenhower, December 22, 1942, quoted in Pogue, *Ordeal and Hope*, 423.

238   "moving spaghetti": Blumenson, *Man behind the Legend*, 175.

240   "Ike seems jittery": Robert E. Sherwood, *Roosevelt and Hopkins: An Intimate History* (New York: Harper, 1948), 676.

240   "Maybe as early": Elliot Roosevelt, *As He Saw It* (New York: Duell, Sloan and Pearce, 1946), 79.

240 "the President told General Marshall": Sherwood, *Roosevelt and Hopkins*, 689.

240 "He and I talked until": Martin Blumenson, ed., *The Patton Papers*, vol. 2, 1940–1945 (Boston: Houghton Mifflin: 1974), 154–55.

241 "I always had": Morgan, *Past Forgetting*, 120.

241 "Fine . . . Agreed": Ibid., 126.

242 "Well, maybe you don't": Eisenhower, *At Ease*, 261.

242 "as if to halt": Omar N. Bradley, *A Soldier's Story* (New York: Rand McNally, 1952), 27.

243 Then, on February 22: Margaret Bourke-White, "Women in Lifeboats: Torpedoed on an African-Bound Troopship," *Life*, February 22, 1942.

245 "Each time a soldier": Bradley, *A Soldier's Story*, 44–45.

245 "They knew that there": Brig. Gen. Paul M. Robinett, *Armor Command: The Personal Story of a Commander in the 13th Armored Regiment of the CCB, 1st Armored Division, and of the Armored School during World War II* (n.p.: Literary Licensing, 2012), 198.

246 "unfortunate results as to": Larry I. Bland and Sharon Ritenour Stevens, eds., *The Papers of George Catlett Marshall*, vol. 3, *"The Right Man for the Job," December 7, 1941–May 31, 1943* (Baltimore: Johns Hopkins University Press, 1991), 643–44.

247 "I should say he": Nigel Hamilton, *Master of the Battlefield: Monty's War Years, 1942–1944* (New York: McGraw-Hill, 1983), 213.

247 "He is so proud": Eisenhower to Marshall, April 5, 1943, quoted in Korda, *Ike*, 372.

247 "Like frogs about": Plato, *Phaedo*, 109b.

249 "Let's go": Smith, *In War and Peace*, 279.

250 "not aggressive enough": D'Este, *A Genius for War*, 507.

251 "We were told that General Patton": Ibid., 510.

251 "This is a horse": Blumenson, *Man behind the Legend*, 202.

252 "I could smell dead": George S. Patton Jr., *War as I Knew It* (Boston: Houghton Mifflin, 1947), 58.

253 "slapped his face": Report of Col. F. Y. Leaver, MC, CO, Fifteenth Evacuation Hospital, August 4, 1943, quoted in D'Este, *A Genius for War*, 533.

253 "Your nerves": Ibid., 534.

253 "I meant what I": Ibid.

253 "They were flatly": Korda, *Ike*, 408.

254 "chagrin and grief": Blumenson, *Man behind the Legend*, 212–23.

254 "complete interchange of": US Department of State, Foreign Relations of the United States, Conferences at Washington and Quebec, 1943 (Washington, DC: US Government Printing Office, 1970), 1117–19.

256 "To transfer him": John J. Pershing to Franklin D. Roosevelt, September 16, 1943, quoted in Katherine Tupper Marshall, *Together: Annals of an Army Wife* (n.p.: Franklin Classics, 2018), 156–57.

257 "He has qualities": Eisenhower to Marshall, August 24, 1943, quoted in Smith, *In War and Peace*, 284.

257 "Headquarters should be": D'Este, *A Soldier's Life*, 453.

Chapter 9: The Overlord

259 "We British will": Dwight D. Eisenhower, *Crusade in Europe* (Garden City, NY: Doubleday, 1948), 197.

259 "That's a bargain, Sir": Elliot Roosevelt, *As He Saw It* (New York: Duel, Sloan and Pearce, 1946), 136–37.

260 "I've heard quite": Kay Summersby Morgan, *Eisenhower Was My Boss* (New York: Prentice-Hall, 1948), 89.

260 "Won't you come": Ibid., 94.

260 "Roosevelt enjoyed himself": Kay Summersby Morgan, *Past Forgetting: My Love Affair with Dwight D. Eisenhower* (New York: Simon & Schuster, 1976), 152.

261 "Well, who knows?": Ibid., 174.

261 "Ike you and I": Robert E. Sherwood, *Roosevelt and Hopkins: An Intimate History* (New York: Harper, 1948), 770.

261 "But it's dangerous": Eisenhower, *Crusade in Europe*, 197.

262 "It got hotter": Forrest C. Pogue, *George C. Marshall*, vol. 3, *Organizer of Victory, 1943–1945* (New York: Viking Press, 1973), 307.

262 "I think Winston": Jean Edward Smith, *Eisenhower: In War and Peace* (New York: Random House, 2012), 312.
262 "The probabilities are": Joseph Stilwell to Currie, August 1, 1942, quoted in Forrest C. Pogue, *George C. Marshall*, vol. 2, *Ordeal and Hope, 1939–1942* (New York: Viking Press, 1966), 366.
263 "basis for all": *Foreign Relations of US, Conferences at Cairo and Tehran, 1943*, First Plenary Meeting, 494–95.
263 "Nothing [will] come": Ibid., Second Plenary Meeting, 535.
264 "the war would": Ibid., 469.
264 "in some concern": Sherwood, *Roosevelt and Hopkins*, 803.
264 "Well, I didn't feel": Ibid.
264 "Well, Ike, you": Eisenhower, *Crusade in Europe*, 207.
264 "wanted to show MacArthur": William Manchester, *American Caesar: Douglas MacArthur, 1880–1964* (New York: Little, Brown, 1978), 352.
265 "my staff": Ibid.
265 "from that time": Douglas MacArthur, *Reminiscences* (New York: McGraw-Hill, 1964), 184.
266 "Every time I": Morgan, *Past Forgetting*, 176.
266 They chatted until ten p.m.: Dwight D. Eisenhower, *At Ease: Stories I Tell to Friends* (Garden City, NY: Doubleday, 1967), 268.
266 "please tell Colonel": Eisenhower to Bradley, January 15, 1944, quoted in Smith, *In War and Peace*, 331.
266 "could belong to": Morgan, *Eisenhower Was My Boss*, 123.
268 "General Patton is not": Carlo D'Este, *Patton: A Genius for War* (New York: HarperCollins, 1995), 585.
268 "since it is": Ibid., 586.
268 "Patton is the only": Omar N. Bradley, *A Soldier's Story* (New York: Holt, 1951), 231.
268 "When I gave him": Eisenhower, *At Ease*, 270–71.
269 "where his staff": Morgan, *Eisenhower Was My Boss*, 128–29.
269 "Staged a meeting of": Ibid., 136.
270 Calling them both: Michael Korda, *Ike: An American Hero* (New York: HarperCollins, 2007), 457.
270 memorandum for the record: The memo is dated March 22, 1944, quoted in Smith, *In War and Peace*, 818, n64.
271 "history will never": Arthur Bryant, *Triumph in the West: A History of the War Years Based on the Diaries of Field-Marshal Lord Alanbrooke, Chief of the Imperial General Staff* (Garden City, NY: Doubleday, 1959), 134.
271 "Dragooned": Pogue, *Organizer of Victory*, 413.
272 "You were originally": Charlies de Gaulle, *War Memoirs*, 2 vols. (New York: Simon & Schuster, 1955), 2:241.
272 That exchange turned: Harry Butcher, *My Three Years with Eisenhower: The Personal Diary of Captain Harry C. Butcher, USNR, Naval Aide to General Eisenhower, 1942 to 1945* (New York: Simon & Schuster, 1946), 473; De Gaulle, *War Memoirs*, 2:241.
273 "the southern portion": Eisenhower, *Crusade in Europe*, 238.
273 "What do you think I should do?": De Gaulle, *War Memoirs*, 2:254.
274 "Go!": Forrest C. Pogue, *The Supreme Command* (Washington, DC: Office of the Chief of Military History, Department of the Army, 1954), 170.
274 "I never realized before": Walter Bedell Smith, *Eisenhower's Six Great Decisions: Europe, 1944–1945* (New York: Longmans, Green, 1956), 53–54.
274 "Okay, we'll go": John S. D. Eisenhower, *Allies: Pearl Harbor to D-Day* (Garden City, NY: Doubleday, 1982), 469.
274 "Our landings in": Stephen E. Ambrose, *Eisenhower: Soldier and President* (New York: Simon & Schuster, 1990), 140.
275 "It's very hard": Morgan, *Past Forgetting*, 215.
275 "I have as": Eisenhower to Marshall, June 6, 1944, quoted in *Papers of George Marshall: War Years*, III, 1914–15.
276 "Ike's visit had": Omar N. Bradley and Clay Blair, *A General's Life: An Autobiography* (New York: Simon & Schuster, 1983), 257.
277 "heartening to the troops": Eisenhower, *Crusade in Europe*, 254.
277 "Ike was tired": Morgan, *Past Forgetting*, 219.

277 "Ike was not a particularly": Ibid., 221.

277 "What was going through": Ibid.

279 "There can be no": Smith, *In War and Peace*, 359.

280 "actually allowed it": Eisenhower, *Crusade in Europe*, 269.

282 "I'm almost sorry": D'Este, *A Genius for War*, 668.

284 "for about two": George S. Patton, *War as I Knew It* (Boston: Houghton-Mifflin, 1947), 108.

284 "If I had": Ibid., 119.

284 "Touring France with": Ibid., 91.

284 "Forty-eight hours": Eisenhower, *Crusade in Europe*, 278–79.

286 "I pissed in": Blumenson, *Patton Papers*, 2:521.

287 "Tell the Fuhrer": Larry Collins and Dominique Lapierre, *Is Paris Burning?* (New York: Simon & Schuster, 1965), 222.

287 "What the hell": Ibid., 194.

287 "I did this": Carlo D'Este, *Eisenhower: A Soldier's Life* (New York: Holt Paperbacks, 2015), 576.

288 "unique kind of": Korda, *Ike*, 497.

288 "We planned, following": Eisenhower, *Crusade in Europe*, 278.

290 "Give us gasoline": Alden Hatch, *General George Patton: Old Blood and Guts* (New York: Sterling Publishing, 1996), 184.

291 "Georgie composed an": D'Este, *A Genius for War*, 653.

291 "grip and operational direction": Larry I. Bland and Sharon Ritenour Stevens, eds. *The Papers of George Catlett Marshall*, vol. 4, *"Aggressive and Determined Leadership," June 1, 1943–December 31, 1944* (Baltimore: Johns Hopkins University Press, 1996), 624.

291 "came pretty near": Ibid.

292 "The combination of": Eisenhower, *Crusade in Europe*, 453.

292 When Eisenhower rejected: Andrew Rawson, ed., *Eyes Only: The Top Secret Correspondence Between Marshall and Eisenhower, 1943–1945* (Stroud, UK: Spellmount, 2012), 606.

292 "Steady, Monty": D'Este, *A Genius for War*, 650.

292 "will be off their": Ibid., 657.

294 Meanwhile, calling nuclear: Albert Speer, *Inside the Third Reich: Memoirs* (London: Avon, 1971), 225.

295 "It's all so terrible": Dwight D. Eisenhower to Mamie Eisenhower, November 12, 1944, quoted in Dwight D. Eisenhower, *Letters to Mamie*, ed. John S. D. Eisenhower (Garden City, NY: Doubleday, 1978), 219–20.

295 "The old thrill": MacArthur, *Reminiscences*, 234.

296 "a combination of thug": Korda, *Ike*, 534.

299 On the other hand, Ibid., 373.

299 "When it is considered": Patton, *War as I Knew It*, 190.

299 "Chaplain, I want": Ibid., 184–55, n1.

299 "The present situation": Martin Blumenson, *Patton: The Man behind the Legend, 1885–1945* (New York: William Morrow, 1985), 247.

299 "On December 22": D'Este, *A Genius for War*, 679–80; Blumenson, *Man behind the Legend*, 247.

300 "Funny thing George": D'Este, *A Genius for War*, 681.

300 "sort things out": Nigel Hamilton, *Monty: Final Years of the Field Marshal, 1944–1976* (New York: McGraw-Hill, 1986), 218.

300 "Brad, I—not you": Smith, *In War and Peace*, 412; Korda, *Ike*, 540.

301 "Any man who": D'Este, *A Genius for War*, 684.

301 "One sentinel, reinforced": Patton, *War as I Knew It*, 207.

301 "Stalingrad No. 2": Basil H. Liddell Hart, *The Other Side of the Hill: The Classic Account of Germany's Generals, Rise and Fall, with Their Own Account of Military Events, 1939–45* (London: Pan Books, 1978), 464.

302 "But it had one": Paul Fussell, *Doing Battle: The Making of a Skeptic* (Boston: Little, Brown, 1996), 27.

302 "I spent long hours": Stephen E. Ambrose, *Citizen Soldiers: The U.S. Army from the Normandy Beaches to the Bulge to the Surrender of Germany, June 7, 1944–May 7, 1945* (New York: Simon & Schuster, 1997), 285.

303 "The new shell": Logan Nye, "The Revolutionary Fuse That Won World War II," *Mighty History*, https://www.wearethemighty.com/mighty-history/proximity-fuse-world-war-2/.

303 "On the night of December 25": Patton, *War as I Knew It*, 205.

303   One biographer notes: George Patton to Beatrice Patton, January 22 and 24, 1945, quoted in D'Este, *A Genius for War*, 700.

303   "sublime moment": Blumenson, *Patton Papers*, 2:599.

303   "If those Hun": D'Este, *A Genius for War*, 683.

303   "Don't you love": Pogue, *Organizer of Victory*, 121.

304   "I shall merely": Marshall to Eisenhower, December, 22, 1944, quoted ibid., 486.

304   "Roosevelt didn't send": Ibid.

304   "My feeling is this": Marshall to Eisenhower, December 30, 1944, quoted ibid., 487.

304   Guingand: Korda, *Ike*, 546.

304   "a most interesting little battle," Alistair Horne, *Monty: The Lonely Leader, 1944–1945* (New York: HarperCollins, 1994), 284.

304   "This incident caused": Ibid.

304   "a tired little": Ibid.

305   "There can be no": Eisenhower to Brooke, February 16, 1945, quoted in Smith, *In War and Peace*, 425.

305   "to express his full": D'Este, *A Genius for War*, 679–80; Blumenson, *Man behind the Legend*, 247.

305   "Thank God Ike": Eisenhower, *Crusade in Europe*, 372.

306   "As we watched": Edward R. Stettinius, *Roosevelt and the Russians* (Westport, CT: Praeger, 1970), 33–35.

306   "We'll deal with": Ibid.

307   "I must tell you in my": Eisenhower, *Crusade in Europe*, 371.

308   "I fairly shouted": Ibid., 379–80.

308   "Brad, don't tell anyone": D'Este, *A Genius for War*, 712.

308   "Time for a short": Ibid.

308   "the true organizer": Pogue, *Organizer of Victory*, xi.

309   "He would eat lunch": Morgan, *Eisenhower Was My Boss*, 245.

310   "I will be out of": Blumenson, *Man behind the Legend*, 268.

311   "make every effort to perfect": Eisenhower to Stalin, March 28, 1945, quoted in Smith, *In War and Peace*, 426.

311   "The battle for Germany": Memorandum, US Army Chief of Staff, March 30, 1945, quoted ibid., 428.

311   "I regret even more": Franklin D. Roosevelt to Winston Churchill, April 4, 1945, quoted ibid., 429.

312   "the first horror": Patton, *War as I Knew It*, 292–93.

312   "It later developed": Ibid.

314   "ordered the Regular Army": Alice Kamps, "The 1932 Bonus Army: Black and White Americans Unite in March on Washington," *Pieces of History* (blog), US National Archives, July 15, 2020, https://prologue.blogs.archives.gov/2020/07/15/the-1932-bonus-army-black-and-white-americans-unite-in-march-on-washington/.

314   "Don't see how": D'Este, *A Genius for War*, 755.

314   "and told me that": Harry S. Truman, *Memoirs*, vol. 1, *Year of Decisions* (Garden City, NY: Doubleday, 1955), 10–11.

314   "even he told me": Ibid.

315   However tough his rhetoric: History.com Editors, "This Day in History, April 23, 1945: President Truman Confronts Soviet Foreign Minister Vyacheslav Molotov," History, November 13, 2009, updated April 21, 2021, https://www.history.com/this-day-in-history/truman-confronts-molotov.

315   "I was anxious to get": Ibid., 245.

315   "Stimson seemed at least": Ibid., 87.

316   "You will, officially": Eisenhower, *Crusade in Europe*, 426.

316   "General Ike's face": Morgan, *Eisenhower Was My Boss*, 243.

316   "It was a somber": Ibid., 2.

317   "the mark of death": Manchester, *American Caesar*, 368.

317   "Well, Douglas, where": Ibid.

317   "Mindanao, Mr. President": Ibid.

318   "Mr. President, my losses": MacArthur, *Reminiscences*, 198.

318   "We talked of everything": Ibid., 198–99.

320 "She's ten miles": Manchester, *American Caesar*, 402–3.

321 "People of the Philippines": Arthur Herman, *MacArthur: American Warrior* (New York: Random House, 2016), 538–39.

321 "The Sixteenth Division": Manchester, *American Caesar*, 287.

324 "Tommy I love you": D. Clayton James, *The Years of MacArthur*, vol. 2, *1941–1945* (Boston: Houghton-Mifflin, 1975), 607.

324 "What you have": MacArthur, *Reminiscences*, 238.

324 "muster all strength": Manchester, *American Caesar*, 403.

324 "The hour of total": MacArthur, *Reminiscences*, 234.

325 "Running to Manila": Herman, *American Warrior*, 568.

325 "Get to Manila!": Manchester, *American Caesar*, 410.

325 "Yamashita's inability": Ibid., 411.

327 "My country has kept": MacArthur, *Reminiscences*, 252.

327 "Filipino guerillas": Manchester, *American Caesar*, 428–29.

330 "It was as though": Curtis LeMay, *Superfortress: The Story of the B-29 and American Air Power* (New York: McGraw-Hill, 1988), 123.

330 "offered the only": "Minutes of Meeting Held at the White House on Monday, 18 June 1945 at 1530," quoted in Bland and Stevens, "*Aggressive and Determined Leadership*," 233.

331 "You might tell Douglas": Frazier Hunt, *The Untold Story of Douglas MacArthur* (New York: Devin-Adair, 1954), 375.

331 "'One school of thought'": MacArthur, *Reminiscences*, 360.

331 After the meeting with Truman: Frank A. Settle Jr., *General George Marshall and the Atomic Bomb* (Santa Barbara: Praeger, 2016), 93–94.

332 "Is there any reason": Leslie R. Groves, *Now It Can Be Told: The Story of the Manhattan Project* (New York: Harper, 1962), 266–67.

332 "Don't ask *me*": David L. Roll, *George Marshall: Defender of the Republic* (New York: Caliper, 2019), 364; Forrest C. Pogue, *George C. Marshall*, vol. 4, *Statesman, 1945–1959* (New York: Viking, 1987), 550, n30.

332 Later, when Stimson's chief: McCloy to Hadsel, April 8, 1985, quoted in Settle, *Marshall and the Atomic Bomb*, 116.

333 "I voiced to him": Dwight D. Eisenhower, *The White House Years*, vol. 1, *Mandate for Change, 1953–1956* (Garden City, NY: Doubleday, 1963), 312–13.

333 "Mr. President I don't": Bradley and Blair, *A General's Life*, 444–45; Eisenhower, *Crusade in Europe*, 444.

333 "All he said was": Truman, *Year of Decisions*, 416.

334 "tell Kurchatov": Georgy K. Zhukov, *The Memoirs of Marshal Zhukov* (London: Jonathan Cape, 1971), 675.

334 "Even if we went": Harry S. Truman, *Memoirs*, vol. 2, *Years of Trial and Hope* (Garden City, NY: Doubleday, 1956), 420–21; Stimson, diary, July 24, 1945, quoted in Pogue, *Statesman*, 20–21.

334 "Gentlemen, on the first": Maxwell D. Taylor, *The Uncertain Trumpet* (New York: Harper & Brothers, 1960), 3.

334 "So far as the atomic": Charles M. Province, "General George S. Patton, Jr.," speech presented to the British Patton Society, http://www.pattonhq.com/textfiles/myspeech.html.

335 "It will seem superhuman": Roger Olaf Egeberg, *The General: MacArthur and the Man He Called "Doc"* (New York: Hippocrene Books, 1984), 192.

335 "In my opinion there": Commander in Chief Pacific Command, transmission C-3187, quoted in Settle, *Marshall and the Atomic Bomb*, 128.

336 "enduring the unendurable": "Text of Hirohito's Radio Rescript," *New York Times*, August 15, 1945, 3.

337 "Take them off": Manchester, *American Caesar*, 444–45; see also Herman, *American Warrior*, 620, for a slightly different wording.

337 "It seems to me": Pogue, *Statesman*, 23–24.

338 "I simply could not": Jonathan M. Wainwright, *General Wainwright's Story: The Account of Four Years of Humiliating Defeat, Surrender, and Captivity*, ed. Robert Considine (Garden City, NY: Doubleday, 1946), 279–80.

338 "No, I'll be all right": Herman, *American Warrior*, 629.

338 "There were a million": Quoted in MacArthur, *Reminiscences*, 274.

339   "Sutherland, show him": George C. Kenney, *General Kenney Reports: A Personal History of the Pacific War* (Washington, DC: Office of Air Force History, 1987), 577.

339   "Let us pray": Herman, *American Warrior*, 632–33.

Chapter 10: Aftermath

341   "Everything that I have": Fred Kaplan, *The Wizards of Armageddon* (New York: Simon & Schuster, 1983), 10.

341   "Regardless of American decisions": Arnold Wolfers, Percy Corbett, and William Fox, *The Absolute Weapon: Atomic Power and the World Order*, ed. Bernard Brodie, preliminary draft, Yale Institute of International Studies, New Haven, CT, February 15, 1946, 17, 19, 25, 36, 41, 51. This report was also published in book form by Harcourt Brace, with a publication date of January 1, 1946.

343   "The shortest road": John Adams to John Taylor, January 13, 1815, Founders Online, US National Archives, https://founders.archives.gov/.

343   "Give me bread or": D. Clayton James, *The Years of MacArthur*, vol. 3, *1945–1965* (Boston: Houghton Mifflin, 1970), 156.

344   "He looked frightened": Arthur Herman, *Douglas MacArthur: American Warrior* (New York: Random House, 2016), 650.

345   "I come to you": Douglas MacArthur, *Reminiscences* (New York: McGraw-Hill, 1964), 288.

345   "He was an Emperor": Ibid.

345   Mac read him: James, *Years of MacArthur*, 3:114.

345   "It turned out to be nothing": MacArthur, *Reminiscences*, 300.

346   "budget to be patterned": SWNCC 228, quoted in James, *Years of MacArthur*, 3:124–25.

346   "No Japanese Army": Ibid.

346   "my desire that": John W. Dower, *Embracing Defeat: Japan in the Wake of World War II* (New York: W. W. Norton, 2000), 384–85.

346   "Then I should say": MacArthur, *Reminiscences*, 305.

347   "the delicate and difficult": William Manchester, *American Caesar: Douglas MacArthur, 1880–1964* (New York: Little, Brown, 1978), 483.

347   "I'm going to do something": James, *Years of MacArthur*, 3:18–19.

347   "Most farmers in Japan": Manchester, *American Caesar*, 508.

348   "I don't think that since": Ibid., 507.

349   "Why, it's the General!": Ibid., 518.

349   "You must be hungry": Ibid.

350   "Now, because of Russia's": Robert Harvey, *American Shogun: A Tale of Two Cultures* (New York: Harry N. Abrams, 2006), 386.

350   "we parted with a common": Herman, *American Warrior*, 690.

350   "a high degree of instability": George F. Kennan, *Memoirs, 1925–1950* (New York: Pantheon, 1983), 388–39.

351   "got those letters from": Merle Miller, *Plain Speaking: An Oral Biography of Harry S. Truman* (New York: Berkeley, 1974), 339–40.

354   "any type of combat": Carlo D'Este, *Patton: A Genius for War* (New York: HarperCollins, 1995), 713.

354   "That man died": Ibid., 741.

354   "I'll bet you goddam": Ibid., 749.

355   "There's a German village": Ibid., 762–63.

355   "We are going to have to fight": Ladislas Farago, *The Last Days of Patton* (New York: McGraw-Hill, 1981), 207.

355   "Shut up, Georgie": Ibid.

355   "The Nazi thing": Martin Blumenson, *Patton: The Man behind the Legend, 1885–1945* (New York: William Morrow, 1985), 287.

355   "If you are spending": Farago, *Last Days of Patton*, 216–17.

356   "I took the train": George Patton, diary, September 29, 1945, quoted in D'Este, *A Genius for War*, 772.

356   "Look at all those": D'Este, *A Genius for War*, 785.

356   "This is a helluva": Fred Ayer, *Before the Colors Fade* (London: Cassell, 1964), 263; D'Este, *A Genius for War*, 785.

357   "I hope he makes a better": Charles B. Odom, *General George S. Patton and Eisenhower* (New Orleans: Word Picture Productions, 1985), 79–80.

358 "There is no place": Blumenson, *Man behind the Legend*, 307.

358 "May the Great Worm": D'Este, *A Genius for War*, 806.

358 "I will be with Uncle Georgie": Ibid.

358 "As I gnawed": Dwight D. Eisenhower, *At Ease: Stories I Tell to Friends* (Garden City, NY: Doubleday, 1967), 301.

359 "Take care of yourself": Dwight Eisenhower to Kay Summersby, November 22, 1945, quoted in Jean Edward Smith, *Eisenhower: In War and Peace* (New York: Random House, 2012), 44.

359 "George Patton would have": Ibid.

360 "I soon got": Kay Summersby Morgan, *Past Forgetting: My Love Affair with Dwight D. Eisenhower* (New York: Simon & Schuster, 1976), 276.

360 "I'm astounded and appalled": Dwight D. Eisenhower, *Eisenhower Diaries*, ed. Robert H. Ferrell (New York: W. W. Norton, 1980), 136 (December 15, 1945).

360 But it put him in: Eisenhower, *At Ease*, 316.

360 "Since then, I've": Ibid., 318.

361 "Let's be realistic": House Committee on Military Affairs, Hearings on HR 515, 77–8, 79th Congress, 1st Session, 1945, quoted in Smith, *In War and Peace*, 458.

361 Then, in late March 1946: Document searched under The Absolute Weapon, Preliminary Draft for Restricted Distribution New Haven, Connecticut, February 15, 1946, https://www.osti.gov/opennet/servlets/purl/16380564 produced a copy labeled PROPERTY OF GENERAL D.D. EISENHOWER. Although it is possible someone else could have marked the document, it seems more likely that subordinates would be wary of doing something that might be seen as defacing it.

361 "renunciation of the use": Frank A. Settle Jr., *General George Marshall and the Atomic Bomb* (Santa Barbara: Praeger, 2016), 152.

361 Then, in his own report: Ibid.

362 "He came and told the soldiers": Manchester, *American Caesar*, 530.

362 "the greatest military man": Forrest C. Pogue, *George C. Marshall*, vol. 4, *Statesman, 1945–1959* (New York: Viking, 1987), 1.

364 "Henceforth it would seem": Dwight D. Eisenhower, *Crusade in Europe* (Garden City, NY: Doubleday, 1948), 456.

366 "just as disappointed": Clifford Roberts, interview, September 12, 1968, Columbia Oral History Project.

366 "I always stand ready": Eisenhower to Truman, November 18, 1948, quoted in Smith, *In War and Peace*, 480.

367 "next six months": Eisenhower, *Eisenhower Diaries*, 158 (March 19, 1949).

367 "his stomach was bloated": Smith, *In War and Peace*, 487.

367 "General Howard Snyder": Eisenhower, *At Ease*, 354.

367 "I gave myself an order": Evan Thomas, *Ike's Bluff: President Eisenhower's Secret Battle to Save the World* (Boston: Back Bay Books, 2013), 342.

367 "I feel stronger every day": Eisenhower to Johnson, April 20, 1949, Letters, 1948–50, Eisenhower Presidential Library.

368 "Without any preparation": Harry S. Truman, *Memoirs*, vol. 2, *Years of Trial and Hope* (Garden City, NY: Doubleday, 1956), 66.

368 "Mentally, he had aged": MacArthur, *Reminiscences*, 320.

368 "I am going to accomplish": Albert C. Wedemeyer, *Wedemeyer Reports!* (New York: Henry Holt, 1958), 363.

369 "His reputation was so great": Ibid., 370.

369 He had been warned: David L. Roll, *George Marshall: Defender of the Republic* (New York: Caliper, 2019), 384.

369 Zhou even concluded: Ibid.

370 "prospects are favorable": Ibid., 390.

370 "I can only trust": Pogue, *Statesman*, 95.

370 "remained a mystery": Daniel Kurtz-Phelan, *The China Mission: George Marshall's Unfinished War, 1945–1947* (New York: W. W. Norton, 2018), 138.

370 Nonetheless, he did: Roll, *Defender of the Republic*, 392.

370 "the United States and Chiang Kai-shek": Michael M. Sheng, *Battling Western Imperialism: Mao, Stalin, and the United States* (Princeton, NJ: Princeton University Press, 1997), 126.

370 "We have not the authority": Pogue, *Statesman*, 100.

370   "The Americans tend to be naïve": Ed Cray, *General of the Army: George C. Marshall, Soldier and Statesman* (New York: Cooper Square Press, 2000), 566.

371   "Great goodness, Eisenhower": Roll, *Defender of the Republic*, 398.

371   "But this gives me": Alfred Chandler, ed., Papers of Dwight David Eisenhower, vol. 7 (Baltimore: Johns Hopkins University Press, 1978), 1085.

372   "You have broken agreements": Marshall Interviews, George Catlett Marshall Library, 607.

372   "Congratulations, Mr. Secretary Marshall": Roger B. Jeans, ed., *The Marshall Mission to China, 1945–1947: The Letters and Diaries of Colonel John Hart Caughey* (Lanham, MD: Rowman and Littlefield, 2011), 194.

372   "the military is taking over": Roll, *Defender of the Republic*, 413.

372   "I am assuming that the office": Marshall quoted in the *Washington Star*, January 22, 1947; Pogue, *Statesman*, 145.

372   "The more I see and talk": David McCullough, *Truman* (New York: Simon & Schuster, 1992), 535.

373   "It was a striking and communicated force": Dean Acheson, *Present at the Creation: My Years in the State Department* (New York: W. W. Norton, 1987), 140–41.

373   "He said to me": Ibid., 213.

374   A Russian expert: "George Kennan and Containment," Office of the Historian, US Department of State, https://history.state.gov/departmenthistory/short-history/kennan.

375   "to advise them of": Truman, *Years of Trial and Hope*, 103.

375   While all agree: Roll, *Defender of the Republic*, 424, quoted from the three-page prepared statement Marshall read.

375   Acheson thought his boss: Acheson, *Present at the Creation*, 219.

375   Prior to delivery: Charles E. Bohlen, *Witness to History, 1929–1969* (New York: Norton, 1973), 261.

375   "as rigid as the Washington Monument": James Reston, "Marshall Held Too Aloof at Conference in Moscow," *New York Times*, April 30, 1947.

376   "We thought they could": Pogue, *Statesman*, 196.

376   "It was the Moscow": Ibid.

376   "The Policy Planning": US Department of State, *Foreign Relations of the United States (FRUS)— The British Commonwealth: Europe (1947)* (Washington, DC: US Government Printing Office, 1947), 225.

376   Bohlen did the drafting: Pogue, *Statesman*, 209.

376   "a crowd of 15,000": Roll, *Defender of the Republic*, 441.

377   "soldier and statesman": Pogue, *Statesman*, 212.

377   "The whole world's": Ibid., 213–14.

378   "I've decided to give": Clark Clifford and Richard Holbrooke, *Counsel to the President: A Memoir* (New York: Random House, 1991), 144.

378   "We couldn't have gotten": Marshall interview with Harry Price and Ray Foulke, February 18, 1953, George C. Marshall Research Library.

378   "Left to their own": Pogue, *Statesman*, 240.

379   "Oh Lord I traveled": Marshall Interviews, 557, 556, George C. Marshall Research Library.

379   "God bless democracy!": George Marshall to Spencer L. Carter, June 14, 1948, quoted in *The Words of George C. Marshall* (Lexington, VA: George C. Marshall Foundation, 2014), 84.

379   "a spiritual federation": Pogue, *Statesman*, 285.

380   "a final militarization": John Lewis Gaddis, *George F. Kennan: An American Life* (New York: Penguin Press, 2011), 333.

380   "seemed to indicate general approval": Roll, *Defender of the Republic*, 536.

380   "I felt the Soviet leaders": Memorandum of Conversation 20 November, 1948, FRUS, 1948, III, 270, quoted in Pogue, *Statesman*, 334.

Chapter 11: Fire on Ice

383   "General, we have just": Douglas MacArthur, *Reminiscences* (New York: McGraw-Hill, 1964), 327.

383   "I had an uncanny feeling": Ibid.

383   Doug tried to: Ibid.

384   "I wouldn't put my foot": Faubion Bowers, "The Late General MacArthur, Warts and All," in *MacArthur and the American Century*, ed. William M. Leary (Lincoln: University of Nebraska Press, 2001), 255.

384 In a January 1950: "Excerpts from Acheson's Speech to the National Press Club," January 12, 1950, web.viu.ca/davies/H102/Acheson.speech1950.htm.

385 "I don't believe that war": C. L. Sulzberger, *A Long Row of Candles: Memories and Diaries* (New York: Macmillan,1969), 561.

385 The conversation was short: Forrest C. Pogue, *George C. Marshall*, vol. 4, *Statesman, 1945–1959* (New York: Viking Press, 1973), 420.

386 "They are charging me with the downfall": Ibid., 422.

386 "I can't, and he's of the *great*": Harry S. Truman, *Off the Record: The Private Papers of Harry S. Truman*, ed. Robert H. Ferrell (New York: Harper and Row, 1980), 189.

386 "General Marshall is not only": *Congressional Record*, US Senate, September 15, 1950, quoted in Pogue, *Statesman*, 427.

386 "Jenner? Jenner?": Pogue, *Statesman*, 428.

386 "The South Korean forces": MacArthur, *Reminiscences*, 334.

387 "Let's think of a couple": Arthur Herman, *Douglas MacArthur: American Warrior* (New York: Random House, 2016), 728.

387 "I'd land 'em here": Ibid.

388 "The only alternative": D. Clayton James, *The Years of MacArthur*, vol. 3, *1945–1965* (Boston: Houghton Mifflin, 1970), 470.

389 "Accept my personal": All quotes from MacArthur, *Reminiscences*, 356–57n.

389 "Thanks George for": Ed Cray, *General of the Army: George C. Marshall, Soldier and Statesman* (New York: Cooper Square Press, 2000), 695.

392 And when one member: Ronald Powaski, *March to Armageddon: The United States and the Nuclear Arms Race, 1939 to the Present* (New York: Oxford University Press, 1987), 56.

392 Later, remembering the meeting: David E. Lilienthal, *The Journals of David E. Lilienthal*, vol. 2, *The Atomic Energy Years, 1945–1950* (New York: Harper and Row, 1964), 633.

393 "make plans for the occupation": William Manchester, *American Caesar: Douglas MacArthur, 1880–1964* (New York: Little, Brown, 1978), 583.

393 Then, on September 27: Debi Unger and Irwin Unger with Stanley Hirshson, *George Marshall: A Biography* (New York: HarperCollins, 2014), 461.

393 When the general wanted: Manchester, *American Caesar*, 584.

393 "We desire you": Ibid.

393 "a unified, independent": Ibid., 585.

393 A word of warning: Ibid., 586.

393 "A bald attempt": Harry S. Truman, *Memoirs*, vol. 2, *Years of Trial and Hope* (Garden City, NY: Doubleday, 1956), 362.

394 "I am sure the President": MacArthur, *Reminiscences*, 360.

394 "If he'd been a lieutenant": Merle Miller, *Plain Speaking: An Oral Biography of Harry S. Truman* (New York: Berkley Books, 1973), 315–16.

394 "unsinkable carrier-tender": Manchester, *American Caesar*, 568.

394 "Only 50,000 or 60,000": Courtney Whitney, *MacArthur: His Rendezvous with History* (New York: Knopf, 1956), 392.

394 "It just wouldn't work": Manchester, *American Caesar*, 592; Herman, *American Warrior*, 760.

394 "He was the most persuasive": Manchester, *American Caesar*, 592.

396 "drive forward with": Sidney L. Mayer, *MacArthur in Japan* (New York: Ballantine Books, 1973), 103–4.

397 "I decided to reconnoiter": Herman, *American Warrior*, 774.

398 And so Mac remained: J. Lawton Collins, *War in Peacetime: The History and Lessons of Korea* (Boston: Houghton Mifflin, 1969), 141–42.

398 "It was Inchon": Herman, *American Warrior*, 782.

399 "The Eighth Army": Matthew Ridgway, *The Korean War: How We Met the Challenge* (Garden City, NY: Doubleday, 1967), 83.

400 "includes every weapon": Pogue, *Statesman*, 464.

400 A month later George: MacArthur to the Joint Chiefs of Staff, January 10, 1951, Foreign Relations of the US, 1951, VII, 56–57.

401 "I've never heard anything": Greg Herkin: *The Atomic Bomb and the Cold War, 1945–1950* (Princeton, NJ: Princeton University Press, 1981), 332.

401 "the wrong war": Military Situation in the Far East, hearings, 82nd Congress, 1st session, part 2, 732 (1951).

401 "exercise extreme caution": Manchester, *American Caesar*, 615.

401 "accomplishment of our": Ibid., 633.

401 The administration chose: Dean Acheson, *Present at the Creation: My Years in the State Department* (New York: W. W. Norton, 1987), 518.

402 "ready at any time to confer": *Foreign Relations of the United States*, 1951, Korea and China, vol. VII, part 1, pp. 265–66.

402 "I couldn't send a message": Margaret Truman, *Harry S. Truman* (New York: William Morrow, 1973), 513.

402 "It created a very serious": Senate Hearings: Testimony before the Armed Forces and Foreign Relations Committees, 82nd Congress, 1951, 483–86, quoted in Manchester, *American Caesar*, 637.

402 "on the side of sin": Lovett, Memorandum of Conversation, March 24, 1951, quoted in Roll, *Defender of the Republic*, 581.

402 "There is no substitute": *Foreign Relations of the United States*, 1951, Korea and China, 1951, vol. VII, part 1, p. 299.

402 "This look[s] like the last": Harry S. Truman, diary, April 6, 1951, quoted in Roll, *Defender of the Republic*, 582.

402 Lovett claimed Marshall: Ibid., 581.

402 "for consultations and reaching": Acheson, *Present at the Creation*, 521.

403 "Brave, brilliant, and majestic": Manchester, *American Caesar*, 617.

404 "You will turn over your commands": Ibid., 642.

404 "Jeannie, we're going home": James, *Years of MacArthur*, 3:600.

404 "George Marshall pulled": Manchester, *American Caesar*, 646.

404 "Nobody takes the place": Ibid., 653.

404 "When you put on a uniform": Ibid., 648.

405 "Once war is forced upon us": Ibid., 659.

406 "The world has turned over": Vorin E. Whan, ed., *A Soldier Speaks: Public Papers and Speeches of General of the Army Douglas MacArthur* (New York: Praeger, 1965), 251–52.

406 "MacArthur had hoped to breach": Manchester, *American Caesar*, 672.

407 "It was so dangerous it was silly": Colonel James T. Quirk to wife, June 12, 1951, quoted in Pogue, *Statesman*, 488.

408 "to diminish the United States": *Congressional Record*, 82nd Congress, 1st Session, part 5, pp. 6556–603.

408 "If I have to explain at this point": Unger and Unger, *George Marshall: A Biography*, 477.

408 "At eleven o'clock I cease": Pogue, *Statesman*, 491.

Chapter 12: Ike

411 "The seeker is never so popular": David Halberstam, *The Fifties* (New York: Villard Books, 1993), 209.

411 "Ike was no fool": Jean Edward Smith, *Lucius D. Clay: An American Life* (New York: Henry Holt, 1990), 591–92.

412 The New Hampshire primary: William I. Hitchcock, *The Age of Eisenhower: America and the World in the 1950s* (New York: Simon & Schuster, 2018), 63.

412 "Under no circumstances will I": *New York Times*, January 8, 1952; Jean Edward Smith, *Eisenhower: In War and Peace* (New York: Random House, 2012), 510.

413 "Mary Martin, star": Charles J. Kelly, *Tex McCrary: Wars, Women, Politics: An Adventurous Life across the American Century* (Lanham, MD: Hamilton Books, 2009), 135.

413 "Tell General Clay": Interview with Jacqueline Cochran, Eisenhower Library, quoted in Smith, *In War and Peace*, 512.

413 "I thought it was a lot": Smith, *Lucius D. Clay*, 593.

414 "a rose among cabbages": Evan Thomas, *Ike's Bluff: President Eisenhower's Secret Battle to Save the World* (Boston: Back Bay Books, 2013), 106.

415 Still, Ike procrastinated: William Manchester, *American Caesar: Douglas MacArthur, 1880–1964* (New York: Little, Brown, 1978), 685.

415 "Absolutely dismal . . .": Hitchcock, *Age of Eisenhower*, 67.

416 "I hope you never use": Richard Norton. Smith, *Thomas E. Dewey and His Times* (New York: Simon & Schuster, 1982), 586.

416 "Eisenhower was certainly the calmest": Smith, *Lucius D. Clay*, 599.

417   "He'd never seen anything": Smith, *In War and Peace*, 519.

418   "I felt because of the vigorous": Marshall to Eisenhower, July 12, 1952, quoted in Smith, *In War and Peace*, 523–24.

418   "How dare anyone say": Peter Lyon, *Eisenhower: Portrait of the Hero* (Boston: Little, Brown, 1974), 448–49.

418   Appropriately enough, the candidate: Hitchcock, *Age of Eisenhower*, 76.

419   Not naming names: Smith, *In War and Peace*, 530.

419   "I felt dirty": Emmet John Hughes, *The Ordeal of Power: A Political Memoir of the Eisenhower Years* (New York: Atheneum, 1963), 41.

419   "a sobering lesson": Ibid., 42.

419   "Eisenhower was forced": Rose Page Wilson, *General Marshall Remembered* (Englewood Cliffs, NJ: Prentice-Hall, 1968), 371.

420   "Don't attack President Eisenhower": Forrest C. Pogue, *George C. Marshall*, vol. 4, *Statesman, 1945–1959* (New York: Viking, 1987), 497.

420   "I am extremely sorry that you": Truman to Eisenhower, August 13, 1952; Eisenhower to Truman, August 14, 1952; Truman to Eisenhower, August 16, 1952; all quoted in Smith, *In War and Peace*, 528–29.

421   "Of what avail": Lyon, *Portrait of the Hero*, 456.

421   "at the close": Smith, *In War and Peace*, 537.

421   "Just tell them": Roger Morris, *Richard Milhous Nixon: The Rise of an American Politician* (New York: Henry Holt, 1990), 827.

421   "Wire and write": "Nixon Leaves Fate to G.O.P. Chiefs; Eisenhower Calls Him to a Talk; Stevenson Maps Inflation Curbs," *New York Times*, September, 24, 1952.

422   "That job requires a personal": *New York Times*, October 25, 1952.

422   "For all practical purposes": Sherman Adams, *Firsthand Report: The Story of the Eisenhower Administration* (New York: Harper Brothers, 1961), 43–44.

422   "pure showbiz": Omar N. Bradley and Clay Blair, *A General's Life: An Autobiography* (New York: Simon & Schuster, 1983), 656.

422   "Dad could have seen": Thomas, *Ike's Bluff*, 11.

423   "Clark and Van Fleet": James E. Schnabel and Robert J. Watson, *History of the Joint Chiefs of Staff*, vol. 3, *The Korean War* (Wilmington, DE: Glazer, 1979), 932–34.

423   "I know just how": Smith, *In War and Peace*, 559.

424   "I am looking forward": Douglas MacArthur, *Reminiscences* (New York: McGraw-Hill, 1964), 409.

425   "Your present plan": Ibid., 412.

425   As they got up: Manchester, *American Caesar*, 689.

426   "He'll sit here": Michael Korda, *Ike: An American Hero* (New York: HarperCollins, 2007), 665.

427   "Dull, duller, Dulles": Smith, *In War and Peace*, 552.

427   "his speech was slow": Harold Macmillan, *Riding the Storm: 1956–1959* (New York: Harper & Row, 1971), 321.

427   "Dulles's grandfather": Thomas, *Ike's Bluff*, 51.

427   "he was a hail": Evan Thomas, *The Very Best Men: The Daring Early Years of the CIA* (New York: Simon & Schuster, 2006), 73.

427   "Dulles, goddamnit": Thomas, *Ike's Bluff*, 92.

428   "Ike always had": David J. Rothkopf, *Running the World: The Inside Story of the National Security Council and the Architects of America's Power* (New York: Public Affairs, 2006), 73.

428   "Therefore let him": "Si vis pacem, para bellum," Publius Flavius Vegetius Renatus, *De Re Militari*, Book III.

428   "for us to build enough": Andrew P. N. Erdman, "War No Longer Has Any Logic Whatever," in *Cold War Statesmen Confront the Bomb: Nuclear Diplomacy since 1945*, ed. John Lewis Gaddis et al. (Oxford: Oxford University Press, 1999), 96.

429   This radical alternative: Hitchcock, *Age of Eisenhower*, 109.

430   "Oh, goddamnit, we forgot": Herbert S. Parmet, *Eisenhower and the American Crusades* (New York: Macmillan, 1972), 176.

431   "What we found was the result": Emmet Hughes, diary, March 16, 1953, quoted in Thomas, *Ike's Bluff*, 59.

431   "Let's not make our mistakes": Dennis E. Showalter, *Forging the Shield: Eisenhower and National Security for the 21st Century* (Chicago: Imprint Publications, 2005), 3.

432   "I have but one career": J. B. West and Mary Lynn Kotz, *Upstairs at the White House: My Life with the First Ladies* (New York: Coward, McCann, and Geoghegan, 1973), 140–41.

432   "pat Ike on his": Susan Eisenhower, *Mrs. Ike: Memories and Reflections on the Life and Times of Mamie Eisenhower* (New York: Farrar, Straus and Giroux, 1996), 276.

432   "I must have this place": Lester David, *Ike and Mamie The Story of the General and His Lady* (New York: Putnam, 1983), 216.

434   "I just won't get": Milton S. Eisenhower, *The President Is Calling* (Garden City, NY: Doubleday, 1974), 318.

434   "not fit to wear": Thomas, *Ike's Bluff*, 133.

434   "This guy McCarthy": James C. Hagerty, *The Diary of James C. Hagerty: Eisenhower in Mid-Course, 1954–1955*, ed. Robert H. Ferrell (Bloomington: Indiana University Press, 1983), 20.

434   "Senator, until this moment": Smith, *In War and Peace*, 594.

435   "Can't we find a way": Hughes, *Ordeal of Power*, 143.

435   "I'm so sick": Cabinet minutes, April 3, 1953, quoted in Smith, *In War and Peace*, 597.

436   "He dignified me": Pogue, *Statesman*, 503.

437   "I had never seen the general": Andrew Goodpaster, "George Marshall's World and Ours," *New York Times*, December 11, 1953.

438   "difficult, tiring and fun": Dwight D. Eisenhower, *At Ease: Stories I Tell to Friends* (Garden City, NY: Doubleday, 1967), 166.

438   "after seeing the autobahns": Ibid., 166.

439   "kind of political": Ed Cray, *Chief Justice: A Biography of Earl Warren* (New York: Simon & Schuster, 1997), 261–62.

439   "These are not bad people": Michael O'Donnell, "When Eisenhower and Warren Squared Off over Civil Rights," *Atlantic*, March 9, 2020, https://www.theatlantic.com/magazine/archive/2018/04/commander-v-chief/554045/.

439   "The Supreme Court has spoken": Tim Hrenchir, "Biographers Debate Eisenhower's Effect on Desegregation," *Topeka Capital-Journal*, May 5, 2019.

440   "The only assurance": Thomas, *Ike's Bluff*, 674.

440   "In my career I have learned": Herbert Brownell and John P. Burke, *Advising Ike: The Memoirs of Attorney General Herbert Brownell* (Lawrence: University of Kansas Press, 1993), 211.

440   "the manual issued": Richard Russell to Eisenhower, September 27, 1957, quoted in Smith, *Eisenhower*, 727. Recently there has been a tendency to interpret Eisenhower as more on the side of expanding Black civil rights than hitherto thought. For this point of view, see David A. Nichols, *A Matter of Justice: Eisenhower and the Beginning of the Civil Rights Revolution* (New York: Simon & Schuster, 1990).

440   "he believed the president": McGeorge Bundy, *Danger and Survival* (New York: Vintage, 1990), 375.

441   "Make that 75 percent": Interview of John Eisenhower by Evan Thomas, quoted in Thomas, *Ike's Bluff*, 13.

441   "You cannot do this": Dwight D. Eisenhower, *The White House Years*, vol. 1, *Mandate for Change, 1953–1956* (Garden City, NY: Doubleday, 1963), 339, 351.

442   "We are not talking": Ibid., 464.

442   "just exactly as you": Thomas, *Ike's Bluff*, 158.

442   "Don't worry Jim": Eisenhower, *Mandate for Change*, 477–78.

443   "You won't have any": Kermit Roosevelt, *Countercoup: The Struggle for Control of Iran* (New York: McGraw-Hill, 1979), 115–16.

444   "Yours of 18 August": Thomas, *Ike's Bluff*, 96.

444   "Communism against": Smith, *In War and Peace*, 629.

445   "better than forty percent": Halberstam, *The Fifties*, 376.

445   "I want all of you": Smith, *In War and Peace*, 631.

445   "My duty was clear": Eisenhower, *Mandate for Change*, 425–26.

446   "responsible in great measure": Report quoted in Thomas, *Ike's Bluff*, 237.

447   But the Air Force: Dino A. Brugioni, *Eyes in the Sky: Eisenhower, the CIA, and Cold War Aerial Espionage* (Annapolis, MD: Naval Institute Press, 2010), 97.

447   "he stipulated that": Thomas, *Ike's Bluff*, 150; Hitchcock, *Age of Eisenhower*, 173.

448   "to remove sources": Smith, *In War and Peace*, 664–65.

448 "Things are not as": John S. D. Eisenhower, *Strictly Personal* (Garden City, NY: Doubleday, 1974), 175–76.

448 To render that impossible: Thomas, *Ike's Bluff*, 175.

449 "Well I expected": Eisenhower, *Mandate for Change*, 612.

449 "I don't agree with the chairman": Ibid., 621.

449 "Nyet, nyet": Andrew Goodpaster, "Cold War Overflights: A View from the White House," in *Early Cold War Overflights, 1950–1956: Symposium Proceedings*, vol. 1, ed. R. Cargill Hall and Clayton D. Laurie (Washington, DC: Office of the Historian, National Reconnaissance Office, 2003), 41.

449 The normally astute Bohlen: Charles Bohlen, *Witness to History: 1929–1969* (New York: W. W. Norton, 1973), 369–70.

449 "Khrushchev is a mystery": Michael R. Beschloss, *Mayday: Eisenhower, Khrushchev, and the U-2 Affair* (New York: Harper and Row, 1988), 104.

449 "I saw clearly then": Eisenhower, *Mandate for Change*, 621.

449 In fact, he had just met: I owe much of this interpretation to Evan Thomas and his book *Ike's Bluff*.

451 "Because gentlemen": Memorandum of Discussion, National Security Council meeting August 4, 1955, quoted in Thomas, *Ike's Bluff*, 184–85.

451 "do it right away": Ibid.

452 "a research and development": Ibid., 186.

452 Ike finally fell asleep: Ibid., 188.

453 "It hurt like hell": Ibid.

453 "tell the truth": Eisenhower, *Mandate for Change*, 638.

453 "So I just haven't": Hughes, *Ordeal of Power*, 173.

454 "but there is nothing": Ann Whitman, diary, March 13, 1956, quoted in Smith, *In War and Peace*, 684–85.

454 "Well, he hasn't reported": Smith, *In War and Peace*, 685.

454 "I would be honored": Richard Nixon, *RN: The Memoirs of Richard Nixon* (New York: Grosset and Dunlap, 1978), 172.

454 "Dick has just told": Ibid., 172–73.

454 "looked far worse": Richard Nixon, *Six Crises* (Garden City, NY: Doubleday, 1962), 168.

455 "If you are waging": Press conference, June 6, 1956, quoted in Smith, *In War and Peace*, 692–93.

455 "The fat was": Dwight D. Eisenhower, *The White House Years*, vol. 2, *Waging Peace, 1956–1961* (Garden City, NY: Doubleday, 1965), 34.

455 "The entire length of the Canal": Whitman, diary, July 28, 1956, quoted in Smith, *In War and Peace*, 694–95.

456 "Foster, you tell": Thomas, *Ike's Bluff*, 220.

456 "Nothing justifies double-crossing": Conference with the president, October 29, 1956, quoted in Smith, *In War and Peace*, 697.

456 "If you don't get out of Port": Ibid., 704.

457 "the most demanding three": Eisenhower, *Waging Peace*, 58.

457 "Yes, if it is honest": Ibid., 78–79.

457 "as inaccessible as Tibet": Ibid., 95.

458 "[Father] sat back": Eisenhower, *Strictly Personal*, 189.

459 "This is close to incredible": Thomas, *Ike's Bluff*, 212.

460 "Allen, the first": Ibid., 214.

460 "I don't like a thing": "U-2 Operations in the Soviet Bloc and Middle East, 1956–1968," George Washington University, 110, https://nsarchive2.gwu.edu/NSAEBB/NSAEBB434/docs/U2%20-%20Chapter%203.pdf.

460 "Quietly, he took Allen": Ibid.

460 He suspended further flights: Hitchcock, *Age of Eisenhower*, 385.

462 "I can't understand": Andrew Goodpaster, Memorandum of Conversation, October 16, 1956, Anne Whitman File, Eisenhower Library, quoted in Thomas, *Ike's Bluff*, 254.

462 "How long, how long, oh God": William Manchester, *The Glory and the Dream: A Narrative History of America, 1932–1972* (New York: Bantam, 1974), 813.

462 "doesn't seem to be taking": Howard Snyder, diary, December 7, 1957, quoted in Thomas, *Ike's Bluff*, 268.

463 "Howard, there can't": Howard Snyder, diary, November 30, 1957, quoted in Thomas, *Ike's Bluff*, 265.

463 "You've been giving us": Notes of Cabinet Meeting, June 3, 1957, Eisenhower Library, quoted in Thomas, *Ike's Bluff*, 290.

464 "The President wondered": Memorandum of Discussion, National Security Council meeting, April 8, 1955, Eisenhower Library.

465 "You can't have this kind": Gregg Herken, *Counsels of War* (New York: Oxford University Press), 116.

465 "If I see that the Russians": Fred Kaplan, *The Wizards of Armageddon* (New York: Simon & Schuster, 1983), 134.

466 "He thought General Taylor's": *Foreign Relations of the United States*, 1955–1957, vol. 19, 302; quoted in Thomas, *Ike's Bluff*, 205.

467 "They stay on the beaches": Kenneth W. Thompson, *Portraits of American Presidents: The Eisenhower Presidency* (Lanham, MD: University Press of America, 1984), 48.

468 "The Orientals can be": Memorandum of Conference, August 25, 1958, quoted in Smith, *In War and Peace*, 742.

468 "I wondered if we": Eisenhower, *Waging Peace*, 304.

468 "Berlin is the testicles": Beschloss, *Mayday*, 172.

469 "In order to avoid": *Foreign Relations of the United States*, 1958–1960, vol. 8, 172–77, quoted in Thomas, *Ike's Bluff*, 321.

469 "hit the Russians as hard": Ibid., 8: 178; quoted in *Ike's Bluff*, 178.

469 "I must say, to use": Campbell Craig, *Destroying the Village: Eisenhower and Thermonuclear War* (New York: Columbia University Press, 1998), 99.

470 "You, Eisenhower, would go": Vernon A. Walters, *Silent Mission* (Garden City, NY: Doubleday, 1978), 491.

470 "He needed to start taking": Thomas, *Ike's Bluff*, 332.

471 "That's what I thought": Nikita S. Khrushchev, *Khrushchev Remembers: The Last Testament*, ed. Strobe Talbott (Boston: Little, Brown, 1974), 412.

473 "his reputation for honesty": Andrew Goodpaster, Memorandum of Conversation, February 8, 1960, Eisenhower Library, quoted in Thomas, *Ike's Bluff*, 369.

474 "a betrayal by General Eisenhower": William Taubman, *Khrushchev: The Man and His Era* (New York: W. W. Norton, 2004), 455.

474 "And we also have the pilot": Beschloss, *Mayday*, 59; *Current Digest of the Soviet Press*, June 3–7, 1960, quoted in Smith, *Eisenhower*, 753.

474 "there was no authorization": Thomas, *Ike's Bluff*, 375.

475 "I think I'm going to take up": Beschloss, *Mayday*, 344.

475 "I do not know what Khrushchev": Walters, *Silent Mission*, 346.

475 "I responded that the scientists": George Kistiakowsky, *A Scientist at the White House: The Private Diary of President Eisenhower's Special Assistant for Science and Technology* (Cambridge, MA: Harvard University Press, 1976), 375; Hitchcock, *Age of Eisenhower*, 474.

476 "Why don't we go": Kistiakowsky, 293.

476 "megatons to kill": David Alan Rosenberg, "The Origins of Overkill: Nuclear Weapons and American Strategy, 1947–1960," *International Security* 7, no. 4 (Spring 1983): 8, https://doi.org/10.2307/2626731.

476 "frighten[ed] the devil": Ibid.

477 "In the councils": Dwight David Eisenhower, "Farewell Radio and Television Address, January 17, 1961," *The Public Papers of the President of the United States, Dwight D. Eisenhower . . . 195 1961, 8 vols. (Washington, D.C.: National Archives and Records Service, 1952–61), 1035–40.

Chapter 13: The End

478 He left a short guest: Forrest C. Pogue, *George C. Marshall*, vol. 4, *Statesman, 1945–1959* (New York: Viking, 1987), 512.

479 "You're lucky to have": Arthur Herman, *Douglas MacArthur: American Warrior* (New York: Random House, 2016), 829.

480 "should have his head examined": William Manchester, *American Caesar: Douglas MacArthur, 1880–1964* (New York: Little Brown, 1978), 696.

480 "The shadows are lengthening": Douglas MacArthur, *Reminiscences* (New York: McGraw-Hill, 1964), 426.

481 Eisenhower thought he: Michael Korda, *Ike: An American Hero* (New York: HarperCollins, 2007), 722.

482 "I've had enough, John": Evan Thomas, *Ike's Bluff: President Eisenhower's Secret Battle to Save the World* (Boston: Back Bay Books, 2013), 412.

482 "I want to go; God take me": Ibid.

483 "I'm not sure anyone did": David and Julie Nixon Eisenhower, *Going Home to Glory* (New York: Simon & Schuster, 2010), 65.

INDEX

Index

ABOUT THE AUTHOR

ROBERT L. O'CONNELL received a PhD in history at the University of Virginia and spent three decades as a senior analyst at the National Ground Intelligence Center, followed by fourteen years as a visiting professor at the Naval Postgraduate School. He is the author of numerous books, two of which, *The Ghosts of Cannae: Hannibal and the Darkest Hour of the Roman Republic* and *Fierce Patriot: The Tangled Lives of William Tecumseh Sherman*, were national bestsellers, the latter winning the 2015 William H. Seward Award for Excellence in Civil War Biography. He lives with his wife, Benjie, in Charlottesville, Virginia.